Superconductivity and Magnetism in Skutterudites

T0256323

Superconductivity and Magnetism in Skutterudites

Ctirad Uher

CRC Press
Taylor & Francis Group
Boca Raton London New York

CRC Press is an imprint of the
Taylor & Francis Group, an **informa** business

First edition published 2022
by CRC Press
6000 Broken Sound Parkway NW, Suite 300, Boca Raton, FL 33487-2742

and by CRC Press
2 Park Square, Milton Park, Abingdon, Oxon, OX14 4RN

© 2022 Ctirad Uher

CRC Press is an imprint of Taylor & Francis Group, LLC

Library of Congress Cataloging-in-Publication Data
Names: Uher, Ctirad, author.
Title: Superconductivity and magnetism in skutterudites / Ctirad Uher.
Description: First edition. | Boca Raton : CRC Press, 2022. | Includes
bibliographical references and index.
Identifiers: LCCN 2021044472 | ISBN 9781032116884 (hardback) | ISBN
9781032127200 (paperback) | ISBN 9781003225898 (ebook)
Subjects: LCSH: Skutterudite. | Materials--Thermal properties. |
Thermoelectric materials.
Classification: LCC TK2950 .U343 2022 | DDC 621.31/243--dc23/eng/20211108
LC record available at https://lccn.loc.gov/2021044472

ISBN: 978-1-032-11688-4 (hbk)
ISBN: 978-1-032-12720-0 (pbk)
ISBN: 978-1-003-22589-8 (ebk)

DOI: 10.1201/9781003225898

Typeset in Times
by SPi Technologies India Pvt Ltd (Straive)

To my sons Peter and Andrew

Content

Preface

Skutterudites have been the subject of keen scientific studies for the past 40 years or so. The interest was precipitated by the discovery of Prof. Wolfgang Jeitschko and his colleagues (1977), revealing that an open structure of skutterudites typified by two large voids in the unit cell can be filled with foreign species (ions), giving rise to what are called filled skutterudites. Throughout the years, it has been shown that the filling dramatically modifies the physical properties of the original binary form of skutterudites and gives rise to a plethora of fascinating electronic, magnetic, and superconducting characteristics.

The presence of a weakly bonded filler in the oversized void of the crystalline lattice of skutterudites very strongly degrades the ability of the structure to propagate acoustic phonons and, coupled with the outstanding electronic properties (electrical conductivity and the Seebeck coefficient), it resulted in exceptional thermoelectric performance of filled skutterudites that rivals the properties of the best thermoelectric materials. This is particularly so in the temperature range of 500 K–800 K, the regime where industrial processes generate plentiful waste heat that can be harvested and converted to useful electrical energy. The topic of thermoelectric energy conversion using skutterudites has been extensively covered in my book Thermoelectric Skutterudites (2021).

In the present monograph, the emphasis is on exciting developments in superconducting and magnetic properties of skutterudites that followed the initial unexpected detection of a superconducting state in $LaFe_4P_{12}$ with a transition temperature of 4.1 K by Meisner (1981). Since then, the progress has been rapid, and superconductivity has been observed in many other filled skutterudites, including arsenide- and antimonide-based structures, and later in various forms of skutterudites having the $[Pt_4Ge_{12}]$ framework. The superconducting state has often been characterized by a rather high transition temperature that in the case of $La_xRh_4P_{12}$ reached 17 K, Shirotani et al. (2005). Moreover, two skutterudite compounds, $LaFe_4P_{12}$ and YFe_4P_{12}, displayed a highly unusual increase in the transition temperature with applied pressure, DeLong and Meisner (1985) and Cheng et al. (2013), respectively. The climax of the studies was the discovery of a highly exotic and unconventional nature of the superconducting state in $PrOs_4Sb_{12}$, Maple et al. (2002) and Bauer et al. (2002), and subsequently in the Pr-filled $[Pt_4Ge_{12}]$ structures, Gumeniuk et al. (2008).

Extensive studies of the superconducting state in $PrOs_4Sb_{12}$, particularly when aided by an external magnetic field, revealed a fascinating crossover from the superconducting state to the high field-induced ordered phase taking place below 1.5 K and fields in the range between about 4.5T and 14T. The phase is also often referred to as the field-induced ordered phase, Aoki et al. (2002), Ho et al. (2002), and Kohgi et al. (2003). This momentous discovery stimulated even more intense efforts that led to a realization that $PrOs_4Sb_{12}$ is a unique chiral heavy fermion superconductor. Although the subsequently discovered superconductivity in the Pr-filled $[Pt_4Ge_{12}]$ framework shared similarities with $PrOs_4Sb_{12}$, namely, as the pair-breaking and time-reversal symmetry are concerned, the latter framework differed in one important aspect – the mass of the charge carriers was only mildly enhanced, lacking the status of heavy fermions typical in pnicogen-based skutterudites.

Filling the skutterudite voids also brings into play a variety of magnetic states, depending on the nature of the filler species and how the fillers interact with the neighboring ions of the framework, Sales (2003), Leithe-Jasper et al. (2004), and Yoshizawa et al. (2007). Understanding the magnetic state of skutterudites is not only of intrinsic interest, but it also reveals the nature of bonding and aids in illuminating the transport properties. One of the key issues is the crystalline electric field (CEF) potential of 12 surrounding pnicogen (or germanium in the case of APt_4Ge_{12} skutterudites) ions exerted on a filler located in the structural void of skutterudites. Although skutterudites are cubic structures, the void is a slightly distorted icosahedron, and this gives rise to non-spherical charge distribution around each filler. If the filler is a rare-earth element, the CEF splits the f-electron multiplet and results in what is referred to as a CEF energy scheme. Particularly relevant is the energy separation between the ground state and the first excited state, which governs the magnetic and transport properties. How various fillers respond to the CEF (in other words, what kind of the CEF energy scheme develops) dictates the nature of magnetic interactions in the structure. Because the same filler in the voids of different skutterudites (altered frameworks) may have a very different CEF energy scheme, there are a myriad of unique magnetic phases observed in skutterudites.

Chapter 1 of the monograph starts with a brief review of the key features of the atomic structure and electronic bands of skutterudites. An extensive discussion of superconducting properties of the known superconducting skutterudites is presented in Chapter 2. Magnetic properties of skutterudites are covered in Chapter 3. The latter two chapters also include the basic aspects relevant to superconductivity and magnetism, respectively.

The target audiences of the book are researchers and graduate students working in the field of skutterudites who wish to have a broad understanding of the transport, superconducting, and magnetic properties of this exciting class of materials.

It is my distinct pleasure to acknowledge the work of my graduate students and postdocs, who worked with me during the past 25 years on the physical properties of skutterudites. I have also much enjoyed and benefitted from collaborations with numerous scientists, among them, Prof. Mercouri Kanatzidis, Prof. Pierre Ferdinand Poudeau, Prof. Xinfeng Tang, and Prof. Lidong Chen. I also want to acknowledge funding support for over 40 years extended to me by the US Federal Agencies and notably the US Department of Energy.

REFERENCES

Aoki, Y., T. Namiki, S. Ohsaki, S. R. Saha, H. Sugawara, and H. Sato, *J. Phys. Soc. Jpn.* **71**, 2098 (2002).

Bauer, E., N. A. Frederick, P.-C. Ho, V. S. Zapf, and M. B. Maple, *Phys. Rev. B* **65**, 100506(R) (2002).

Cheng, J.-G., J.-S. Zhou, K. Matsubayashi, P. P. Kong, Y. Kubo, Y. Kawamura, C. Sekine, C. Q. Jin, J. B. Goodenough, and Y. Uwatoko, *Phys. Rev. B* **88**, 024514 (2013).

DeLong, L. E. and G. P. Meisner, *Solid State Commun.* **53**, 119 (1985).

Gumeniuk, R., W. Schnelle, H. Rosner, M. Nicklas, A. Leithe-Jasper, and Y. Grin, *Phys. Rev. Lett.* **100**, 017002 (2008).

Ho, P.-C., V. S. Zapf, E. D. Bauer, N. A. Frederick, M. B. Maple, G. Giester, P. Rogl, S. T. Berger, C. H. Paul, and E. Bauer, *Int. J. Mod. Phys. B* **16**, 3008 (2002).

Jeitschko, W. and D. J. Braun, *Acta Crystallog.* **B33**, 3401 (1977).

Kohgi, M., K. Iwasa, M. Nakajima, N. Metoki, S. Araki, N. Bernhoeft, J. M. Mignot, A. Gukasov, H. Sato, Y. Aoki, and H. Sugawara, *J. Phys. Soc. Jpn.* **72**, 1002 (2003).

Leithe-Jasper, A., W. Schnelle, H. Rosner, M. Baenitz, A. Rabis, A. A. Gippius, E. N. Morozova, H. Borrmann, U. Burkhardt, R. Ramlau, U. Schwarz, J. A. Mydosh, Y. Grin, V. Ksenofontov, and S. Reiman, *Phys. Rev. B* **70**, 214418 (2004).

Maple, M. B., P.-C. Ho, V. S. Zapf, N. A. Frederick, E. D. Bauer, W. M. Yuhasz, F. M. Woodward, and J. W. Lynn, Proc. Int. Conf. Strongly Correlated Electrons with Orbital Degree of Freedom (ORBITAL 2001), *J. Phys. Soc. Jpn.* **71**, Suppl, 23 (2002).

Meisner, G. P., *Physica B&C* **108**, 763 (1981).

Sales, B. C., in *Handbook on the Physics and Chemistry of Rare Earths*, eds. K. A. Gschneidner, J.-C. Bunzli, and V. K. Pecharsky, Elsevier, Amsterdam, Vol. **33**, p.1 (2003).

Shirotani, I, S. Sato, C. Sekine, K. Takeda, I. Inagawa, and T. Yagi, *J. Phys.: Condens. Matter* **17**, 7353 (2005).

Uher, C., in *Thermoelectric Skutterudites*, CRC Press, Taylor & Francis Group, Boca Raton, FL, 2021.

Yoshizawa, M., Y. Nakanishi, T. Fujino, P. Sun, C. Sekine, and I. Shirotani, *J. Magn. Magn. Mater.* **310**, 1786 (2007).

About the Author

Ctirad Uher is a C. Wilbur Peters Professor of Physics at the University of Michigan in Ann Arbor.

He earned his BCs in physics with a University Medal from the University of New South Wales in Sydney, Australia. He carried out his graduate studies at the same institution under Professor H. J. Goldsmid on the topic of "Thermomagnetic effects in bismuth and its dilute alloys" and received his PhD in 1975. Subsequently, Professor Uher was awarded the prestigious Queen Elizabeth II Research Fellowship, which he spent at Commonwealth Scientific and Industrial Research Organization (CSIRO), National Measurement Laboratory (NML), in Sydney. He then accepted a postdoctoral position at Michigan State University, where he worked with Profs. W. P. Pratt, P. A. Schroeder, and J. Bass on transport properties at ultra-low temperatures.

Professor Uher started his academic career in 1980 as an assistant professor of Physics at the University of Michigan. He progressed through the ranks and became a full professor in 1989. The same year, the University of New South Wales awarded him the title of DSc for his work on the transport properties of semimetals. At the University of Michigan, he served as an associate chair of the Department of Physics and subsequently as an associate dean for research at the College of Literature, Sciences and Arts. In 1994, he was appointed as chair of physics, the post he held for the next ten years.

Professor Uher's 46 years of research is described in more than 530 refereed publications in the areas of transport properties of solids, superconductivity, diluted magnetic semiconductors, and thermoelectricity, and his h-index currently stands at 88 (Web of Science). He has written a number of authoritative review articles and has presented his research at numerous national and international conferences as an invited and plenary speaker. In 1996, he was elected fellow of the American Physical Society. Professor Uher was honored with the title of Doctor Honoris Causa by the University of Pardubice in the Czech Republic in 2002, and in 2010 he was awarded a named professorship at the University of Michigan. He received the prestigious China Friendship Award in 2011. Professor Uher supervised 17 PhD thesis projects and mentored numerous postdoctoral researchers, many of whom are leading scientists in academia and research institutions all over the world.

Professor Uher served on the Board of Directors of the International Thermoelectric Society. In 2004–2005, he was elected vice-president of the International Thermoelectric Society, and during 2006–2008, he served as its president.

1 Brief Review of the Structure and Electronic Bands in Skutterudites

1.1 INTRODUCTION

Skutterudites have a long history in the annals of science. The word skutterudite was first used by Wilhelm Karl von Haidinger (1845) to describe minerals of composition (Co, Ni, and Fe)As$_3$ that were mined in a small town called Skuterud in Norway. However, this class of minerals was known well before the time of von Haidinger under various names spanning from cobaltum eineraceum, the name given to it in 1529 by Georgius Agricola, fondly known as the father of mineralogy, to a simple Arsenikkobalt, the name coined by Gustav Rose (1852). Other synonyms occasionally encountered in the literature are modumite, smaltite, kieftite (referring specifically to CoSb$_3$), and chloanthite (for Ni-rich forms of the structure).

As accessory minerals of hydrothermal origin, they are found in many regions of the world usually accompanied by other Ni-Co minerals. The current description given by the International Mineralogical Association distinguishes four main groups of the mineral: skutterudites (CoAs$_{3-x}$), nickelskutterudites (NiAs$_{3-x}$), ferroskutterudites ((Fe-Co)As$_3$), and kieftite (CoSb$_3$). As recently pointed out by Schumer et al. (2017), because neither the mineral nor synthetic form of NiAs$_3$ exists, the nickelskutterudite group should properly be classified as (Ni, Co, and Fe)As$_3$.

One of the first synthetic forms of skutterudites was prepared by Jolibois (1910), who made single crystals from Sn flux; variants of this technique (Sb instead of Sn) have been used even today. Considering that the crystal structure and the phase diagram were not known at that time, Jolibois had to be a quite skilled researcher.

Appropriate structural characterization of synthetic skutterudites was performed by Ivar Oftedal (1928), who identified skutterudites to possess a body-centered cubic lattice belonging to the space group $Im\bar{3}$.

The scientific interest in skutterudites has been driven primarily by two phenomena: the fascinating and exotic superconducting properties of filled skutterudites and the prospect that by filling the structural void of the skutterudite lattice, the thermal conductivity can be dramatically reduced, making such filled skutterudites outstanding thermoelectric materials for power generation. The latter aspect, the development and properties of skutterudites as viable thermoelectric materials, has been extensively described in my recent book titled *Thermoelectric Skutterudites*, published by Taylor & Francis, Uher (2021). In the present monograph, the focus is on a plethora of fascinating magnetic properties and on the truly exotic nature of superconductivity observed in several filled skutterudites. However, to aid in an understanding of magnetic and superconducting properties of skutterudites, it is necessary to consider structural aspects and electronic bands of skutterudites. Thus, here the essential features of the crystalline lattice of skutterudites and the structure of their electronic bands are included. A detailed account of the above two aspects in given in the aforementioned monograph *Thermoelectric Skutterudites*.

In general, skutterudites can be divided into three main categories: binary skutterudites, ternary skutterudites, and filled skutterudites, and they are briefly reviewed in turn.

DOI: 10.1201/9781003225898-1

1.2 BINARY SKUTTERUDITES

As the term binary implies, the structure consists of two elements, and its general formula is MX_3, where M stands for an element of the column-9 transition metals Co, Rh, or Ir, and X represents a pnicogen (also pnictogen or pnictide) element P, As, or Sb. Thus, there are nine possible combinations constituting structurally stable bulk binary skutterudites. Occasionally, one finds reports of a stable form of NiP_3, Jolibois (1910), Biltz and Heimbrecht (1938), and Rundqvist and Larsson (1959). Because it is a thin film, it is possible to synthesize also metastable $FeSb_3$ (Daniel et al. 2015).

The arrangement of atoms in the crystalline lattice is depicted in Figure 1.1, where the left-hand side panel shows the unit cell with the transition metal atoms M occupying 8c (¼, ¼, ¼) sites (Wyckoff notation) that are octahedrally coordinated by pnicogen atoms at 24 g (0, y, z) sites, where y and z are the positional parameters that specify the exact location of a pnicogen atom. Typical of the skutterudite structure is the tilt of the MX_6 octahedrons, which gives rise to two large structural voids in the unit cell at position 2a (0, 0, 0) and leads to the formation of near-square pnicogen rings. The panel on the right-hand side of Figure 1.1 shows the unit cell shifted by one-quarter distance along the body diagonal. In this view, the transition metal atoms M form a simple cubic sublattice, and the near-square planar pnicogen rings are clearly depicted. There are six such pnicogen rings in the unit cell. Two of the eight small cubes do not contain the pnicogen rings and are locations of the two structural voids mentioned above. The chemical formula describing the unit cell of binary skutterudites is $\square_2 M_8[X_4]_6 = 2(\square M_4[X_4]_3) = 2(\square M_4 X_{12})$, where the empty square $\square$ stands for a structural void. It is customary to take just one-half of the unit cell, i.e., $\square M_4 X_{12}$, the structure that has a valence electron count (VEC) of 72. The complete specification of binary skutterudites is given by the lattice parameter a accompanied by the positional parameters y and z.

The original structural analysis of skutterudites by Oftedal (1928) assumed a square planar configuration of the pnicogen rings, resulting in the so-called Oftedal relation

$$2(y+z) = 1. \tag{1.1}$$

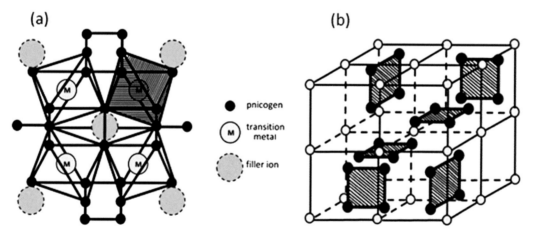

FIGURE 1.1 (a) Atomic arrangement in the unit cell of a binary skutterudite centered over a void at (0, 0, 0). The picture highlights the octahedral coordination of transition metals by pnicogen atoms and the tilt of the MX_6 octahedrons leading to the formation of two structural voids and four-membered pnicogen rings. (b) The unit cell of a binary skutterudite shifted by one-quarter distance along the body diagonal depicting the four-membered pnicogen rings that occupy six of the eight small cubes. The front upper left cube and the back bottom right cube do not contain pnicogen rings and are sites of two structural voids, often called cages.

More precise structural measurements performed by Kjekshus and Rakke (1974) revealed that the sides of the pnicogen rings are not equal, and all binary skutterudites possess rectangular rather than square planar ring configurations.

Binary skutterudites (except for NiP_3 and $FeSb_3$) are diamagnetic semiconductors. This follows from the mostly covalent nature of bonding and the fact that there are no unpaired electrons in the structure. Pnicogen atoms with their ns^2np^3 valence electron configuration use two electrons to form bonds with their two nearest neighbors on the pnicogen ring and three electrons to bond with two nearest transition metal atoms. Of the nine valence electrons of a transition metal atom of column 9, three electrons are used to establish bonds with the six neighboring pnicogen atoms, giving rise to octahedral d^2sp^3 hybrid orbitals, and the remaining six non-bonding electrons adopt a maximum spin-pairing d^6 configuration with zero spin. Structural parameters of binary skutterudites, including the void radius, are given in the monograph by Uher (2021).

While no bulk binary skutterudites form with the transition elements of columns 8 and 10 of the periodic table, (except for the already noted NiP_3), it is possible to partly replace Co, Rh, and Ir with their immediate transition metal neighbors Ni and Fe, Ru and Pd, and Os and Pt, respectively. Of course, such partial replacements alter the VEC and lead to a metallic behavior and paramagnetism. The solid solution limits in antimony skutterudites were established by Dudkin and Abrikosov (1957) and in arsenide skutterudites by Pleass and Heyding (1962). Roseboom (1962) pointed out that it is possible to make a coupled replacement of two Co atoms by one Fe and one Ni according to the equation

$$2Co^{3+}\left(d^6\right) \rightleftarrows Fe^{2+}\left(d^6\right) + Ni^{4+}\left(d^6\right). \tag{1.2}$$

Structures with symmetrically replaced Co preserve the total number of electrons, and all transition metal ions are in the low-spin d^6 configuration. This was, indeed, confirmed for arsenides by magnetic measurements performed by Nickel (1969), and Jeitschko et al. (2000) observed the same for phosphides.

Solid solubility limits for various families of skutterudites have been extensively explored. Lutz and Kliche (1981) established that phosphide and arsenide skutterudites form a complete series of solid solutions obeying Vegard's law. Unfortunately, solid solutions $MAs_{3-x}Sb_x$ between arsenides and antimonides are more restricted, and a miscibility gap exists for $0.4 < x < 2.8$. Solid solutions on the transition metal site were explored by Borshchevsky et al. (1995) with the emphasis on $Co_xIr_{1-x}Sb_3$, where a miscibility gap was observed in the range of $0.2 < x < 0.65$. Slack and Tsoukala (1994) were able to prepare $Rh_{0.5}Ir_{0.5}Sb_3$, but the full range of solubility was not explored. In $Rh_xCo_{1-x}Sb_3$, Wojciechowski (2007) observed a linear dependence of the cell parameter for all contents x, suggesting that solid solutions are unrestricted.

Binary skutterudites contain intrinsic defects that play an important role in the transport properties. The presence of intrinsic defects should not be surprising given that binary skutterudites are composed of rather high-melting-point transition metals and low-melting-point pnicogen species that have a high vapor pressure. The loss of pnicogen atoms during the synthesis is thus highly probable. However, energetically it is not favorable to create a pnicogen vacancy because this means severing a strong bond holding together the X_4 ring. Rather, it is Co interstitials and interstitial pairs that have the lowest formation energy and are therefore the dominant defects. Detailed accounts of defect formation, the temperature evolution of defects, and their strong effect on the charge carrier transport were extensively evaluated by Park and Kim (2010) and by Li et al. (2016).

Realistic band structure calculations must reflect the key structural aspects of skutterudites, namely, the existence of the near-square pnicogen rings, the octahedral coordination of transition metals by pnicogen atoms, and the fact that pure binary skutterudites are diamagnetic semiconductors. Although there were early attempts to elucidate the structure of electronic bands in binary skutterudites, Jung et al. (1990), real advances had to wait until the algorithms for first-principles

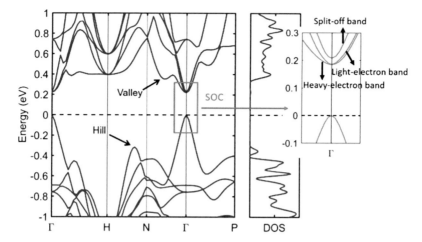

FIGURE 1.2 The band structure and the density-of-states of CoSb$_3$ around the Fermi energy. Of interest is a "valley" band some 0.13 eV above the triply degenerate conduction band at Γ. The inset on the right shows the conduction band calculated under the influence of split-orbit coupling. Reproduced from C. Z. Hu et al., *Physical Review B* 95, 165204 (2017). With permission from the American Physical Society.

calculations became robust enough to capture the complexity of the atomic structure of skutterudites, including their rather open environment. In the first such study, Singh and Pickett (1994) have shown that binary skutterudites possess a considerable indirect pseudogap around the Fermi level, which is crossed by a single linearly dispersing band that touches (nearly so in CoSb$_3$) the conduction band at Γ. With the improvements of functionals used in density functional theory (DFT) calculations, subsequent numerous treatments sharpened the position of the bands in various binary skutterudites and revealed their detailed features. The reader might find useful articles by Llunell et al. (1996), Sofo and Mahan (1999), Fornari and Singh (1999), Lefebvre-Devos et al. (2001), Takegahara and Harima (2003a), Kurmaev et al. (2004), Chaput et al. (2005), Wei et al. (2009), Smith et al. (2011), Pardo et al. (2012), Hammerschmidt et al. (2013), Nieroda et al. (2013), Khan et al. (2015), Tang et al. (2015), Hu et al. (2017), Kolezynski and Szczypka (2017), Yang et al. (2019), and Isaacs and Wolverton (2019).

Of the plethora of band structure calculations, two studies are of particular interest. Tang et al. (2015) drew attention to the importance of multi-valley bands in binary skutterudites, where the Γ-point conduction band in CoSb$_3$ is triply degenerate and one of the bands develops a higher-lying minimum along the Γ-N direction. The minimum is strongly temperature-dependent, moves down in energy as the temperature increases, and converges with the Γ-point bands at about 700 K. Subsequent band structure calculations of CoSb$_3$ by Hu et al. (2017) are depicted in Figure 1.2. Another interesting feature of the skutterudite band structure was pointed out by Smith et al. (2011) and expanded by Pardo et al. (2012). They considered the formation of the band gap as the structure evolves from a highly metallic perovskite to a semiconducting skutterudite and have shown that CoSb$_3$ is very near a conventional-to-topological transition. With a small tetragonal strain and spin-orbit coupling, CoSb$_3$ was predicted to become a topological insulator phase.

1.3   TERNARY SKUTTERUDITES

Ternary skutterudites are synthetically modified binary skutterudites where isoelectronic substitutions are made either on the cation site M by a pair of elements from columns 8 and 10 or on the pnicogen site with a pair of elements from columns 14 and 16. An example of the former is Fe$_{0.5}$Ni$_{0.5}$Co$_3$, Kjekshus and Rakke (1974), and that of the latter is CoGe$_{1.5}$Se$_{1.5}$, Korenstein et al.

(1977). The primary interest in ternary skutterudites was driven by a possibility of significantly lowering the lattice thermal conductivity, hoping that it would make the structure more suitable for thermoelectric applications. Unfortunately, a concomitant strong degradation of the electrical conductivity was a major impediment, and no ternary skutterudites have emerged as truly promising thermoelectric materials. While some structural modifications (distortion of the ring structure, deviation of the dihedral angle from the original 90°, and even lowering of the symmetry) take place because of substitutions, particularly when they are conducted on the pnicogen site where they distort the four-membered pnicogen rings, Lyons et al. (1978), Lutz and Kliche (1981), Partik et al. (1996), Fleurial et al. (1997), Vaqueiro et al. (2006), Volja et al. (2012), and Zevalkink et al. (2015), the isoelectronic nature of substitutions leaves the VEC unchanged and the undoped ternary skutterudites offer no special interest for magnetic studies as they remain diamagnetic materials. Moreover, detailed structural measurements performed by Vaqueiro et al. (2008) revealed that many ternary skutterudites formed by a substitution on the pnicogen site, including $IrGe_{1.5}S_{1.5}$ and $RhSn_{1.5}Te_{1.5}$, often described as single phase structures, in reality contain some 10% of IrGe and $RhTe_2$, respectively.

Changes in the electronic band structure of ternary skutterudites were first investigated by Bertini and Cenedese (2007) in $CoSn_{1.5}Te_{1.5}$ and indicated some 22% reduction in the band gap compared to $CoSb_3$. Moreover, the authors noted a significant charge transfer, which suggested a stronger ionicity of the bonds. Subsequent detailed DFT calculations by Volja et al. (2012) on several ternary skutterudites revealed significantly larger rather than smaller band gaps due to the more ionic bonding. The single band emerging from the valence band manifold toward the Γ point conduction band was preserved, but its linear dispersion was weaker. The study by Zevalkink et al. (2015) agreed substantially with the findings of Volja et al. A comparison of band structures of alkali earth-filled ternary skutterudites with the unfilled forms of the structure was made using DFT calculations by Bang et al. (2016). The study was aimed at improving the thermoelectric properties, but the outcome was uncertain because the exact degree of degradation of the thermal conductivity could not be determined based on the analysis of phonon dispersion only.

1.4 FILLED SKUTTERUDITES

As noted in Section 1.1, the tilt of the MX_6 octahedrons creates large icosahedral voids at the 2a sites in the skutterudite structure. The voids are large enough to accommodate a large variety of ions of alkali, alkaline, and rare earth species as well as elements, such as Li, Tl, Sn, In, and Ga. Filling of the voids was demonstrated first by Jeitschko and Braun (1977) with phosphide-based skutterudites and soon after replicated with antimonides and arsenides, Braun and Jeitschko (1980a and 1980b). Skutterudites with the 2a sites occupied are referred to as filled skutterudites.

If not for two rather unrelated findings, skutterudites would likely be an interesting open structure compound but mostly inconsequential as far as prominent physical properties and practical potential are concerned. The first momentous event was the discovery of superconductivity in $LaFe_4P_{12}$, Meisner (1981), with a rather high transition temperature of 4.1 K in a structure that contained Fe, the element that typically breaks the Cooper pairs. The discovery opened the door to intensive worldwide efforts to explore other filled skutterudites, which culminated with revealing $PrOs_4Sb_{12}$ as a truly exotic heavy fermion superconductor, Maple et al. (2002) and Bauer et al. (2002). The present monograph is dedicated to the superconducting and magnetic properties of filled skutterudites. The other important milestone was the idea of Slack and Tsoukala (1994) and Slack (1995) that filling the voids of the skutterudite structure might make skutterudites an example of the phonon-glass electron-crystal material that maintains the outstanding electronic properties of binary skutterudites, while its thermal conductivity is drastically degraded by vibrations of loosely bonded filler species. Thermal conductivity measurements of filled skutterudites confirmed the exceptionally low values of the thermal conductivity, Morelli and Meisner (1995), Fleurial et al. (1996), Sales et al. (1996), Tritt et al. (1996), and Nolas et al. (1996), and since then, filled skutterudites have become one of the most intensely studied thermoelectric materials intended for power generation applications. The properties and the development of skutterudites as promising thermoelectric materials are described in a book by Uher (2021).

The presence of a filler alters the structural parameters, Chakoumakos and Sales (2006), typified by an increased lattice constant a and an unequal rise in the positional parameters y and z that makes the pnicogen ring structure more square as the filling increases. Filled skutterudites can be categorized into several groups, depending on the framework they are made of.

1.4.1 Filled Skutterudites with the $[M_4X_{12}]$ Framework

The $[M_4X_{12}]$ framework is isoelectronic with that of the binary skutterudites, and skutterudites of the form $R_yM_4X_{12}$ can be filled only partially, i.e., $y < 1$, with electropositive fillers R. As the electrons of the filler R are stripped off and donated to the structure, filled skutterudites $R_yM_4X_{12}$ are invariably n-type conductors. The filling fraction y is a strong function of the valence state of the filler with trivalent rare earth ions having a particularly low filling fraction. Moreover, due to the lanthanide contraction, the smaller and heavier rare earth fillers beyond Sm are too small to be trapped in the large skutterudite voids. Divalent Eu or intermittent valence Yb and all alkaline earth fillers have a significantly higher filling limit up to 45%. Alkali ions, particularly Na and K, can fill over 60% of voids in $CoSb_3$. Theoretical criteria for filling were developed by Chen (2002), and a particularly useful form was presented by Shi et al. (2005). Making use of high pressure during the synthesis, it is possible to fill the voids with even smaller and heavier rare earth fillers, including Lu, Sekine et al. (1998 and 2000) and Shirotani et al. (2003), and enhance the filling fraction of many fillers, Takizawa et al. (2002) and Tanaka et al. (2007).

Reports on the band structure of filled skutterudites that have an electrically neutral $[M_4X_{12}]$ framework are few. The first one was by Akai et al. (2002), who considered Yb-filled $CoSb_3$, but the calculations were made for an unrealistic case of 100% Yb filling. A large family of new filled skutterudites based on the column-9 transition metals (Co, Rh, and Ir) was revealed by Luo et al. (2015) who strictly adhered to the Zintl concept in their band structure calculations and confirmed their findings by synthesizing the structures. In general, the filled $[M_4X_{12}]$ framework does not support superconductivity. However, when the filler content is dramatically enhanced by applying high pressure during the synthesis, as shown by Shirotani et al. (2005) in the case of $La_{0.6}Rh_4P_{12}$ and by Imai et al. (2007) with $La_{0.8}Rh_4P_{12}$, the transition temperature attains the record-high values among all skutterudites of 17 K and 14.9 K, respectively. A similar approach, this time with a very high content of Ba fillers in the Ir-based phosphide and arsenide skutterudites (supported by DFT calculations of the density-of-states), led to a superconducting transition temperature of 5.6 K for $Ba_{0.89}Ir_4P_{12}$ and 4.8 K for $Ba_{0.85}Ir_4As_{12}$, Qi et al. (2017).

1.4.2 Filled Skutterudites with the $[T_4X_{12}]^{4-}$ Framework

When one speaks of filled skutterudites, it is generally understood that the structure contains the $[T_4X_{12}]^{4-}$ framework rather than the neutral $[M_4X_{12}]$ framework. The difference is in the transition metal that is now a Fe-like column-8 element (Fe, Ru, and Os), which makes the framework deficient of four electrons. The role of the filler ion R is to supply the missing electrons, saturate the bond, and electrically neutralize the structure. This requires a tetravalent filler ion R^{4+}, which would bring the VEC to 72. In reality, this is plausible only with Th and U. All other fillers, including Ce, assumed originally by Jeitschko and Braun (1977) to be tetravalent, cannot supply four electrons. The VEC of the fully filled skutterudites will then be less than 72, and the structure will be a paramagnetic metal, unless magnetic interactions set in at lower temperatures or the structure becomes a superconductor. It is possible to bring the structure back into its semiconducting regime (and the VEC of 72) by charge compensation. This means replacing some pnicogen atoms in the

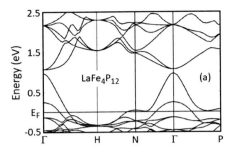
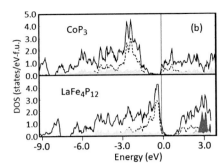

FIGURE 1.3 (a) Band structure of $LaFe_4P_{12}$ near the Fermi level. (b) Density-of-states of $LaFe_4P_{12}$ compared to the density-of-states of CoP_3. The solid lines are the total density-of-states, the dashed lines indicate the contribution of Co-d states, the light gray regions designate the phosphorus p states, and the dark gray region represents the resonant La f-component. Adapted from M. Fornari and D. J. Singh, *Physical Review B* 59, 9722 (1999). With permission from the American Physical Society.

24g positions with elements from column 14 (e.g., Ge or Sn) or replacing a fraction of the Fe-like element at 8c sites by a Co-like transition metal. Making such substitutions in the spirit of the Zintl concept greatly enhances the family of skutterudites and gives rise to an abundance of fascinating physical properties.

In 1981, Meisner discovered superconductivity in $LaFe_4P_{12}$. This surprising observation was soon followed by the discovery of superconductivity in several other phosphide skutterudites. It is not surprising that then phosphide skutterudites became the primary target of band structure studies, hoping to shed light on the mechanism of superconductivity. The first among many studies were DFT calculations performed by Nordström and Singh (1996). The calculations revealed a greater complexity of the electronic bands due to the presence of a filler and indicated heavy electron masses associated with the flat conduction bands and a tendency of the pnicogen rings becoming more square upon filling. As an example of the electronic band structure of filled skutterudites, Figure 1.3 depicts the electronic bands in $LaFe_4P_{12}$ near the Fermi level as well as the density-of-states of $LaFe_4P_{12}$ compared to the density-of-states of binary CoP_3.

Band structure calculations have been performed for many different filled skutterudites, and to aid interested readers, the targeted filled skutterudite is indicated in the following references: Singh and Mazin (1997), Singh (2002), Takegahara and Harima (2002) [$LaFe_4Sb_{12}$]; Harima (1998) [$LaFe_4P_{12}$]; Harima and Takegahara (2003a) [$LaFe_4X_{12}$, X = P, As, and Sb]; Harima and Takegahara (2003b) [$PrRu_4P_{12}$, $PrFe_4P_{12}$, and $LaFe_4P_{12}$]; Saha et al. (2002), Harima (2008) [$LaRu_4P_{12}$]; Harima and Takegahara (2002a), Harima et al. (2002) [perfect nesting in $PrRu_4P_{12}$]; Harima and Takegahara (2002b) [$LaOs_4Sb_{12}$]; Akai et al. (2002), Takegahara and Harima (2002) [$YbCo_4Sb_{12}$ and $YbFe_4Sb_{12}$]; H. Sugawara et al. (2002), Harima and Takegahara (2005) [$PrOs_4Sb_{12}$]; Takegahara and Harima (2003b) [$ThFe_4P_{12}$]; Takegahara and Harima (2008) [$SmOs_4Sb_{12}$]; Yan et al. (2012) [$CeOs_4As_{12}$ and $CeOs_4Sb_{12}$]; Nieroda et al. (2013) [$Ag_xCo_4Sb_{12}$]; Ram et al. (2014) [$LaRu_4X_{12}$, X = P, As, and Sb]; Xing et al. (2015) [$LaFe_4X_{12}$ and $NaFe_4X_{12}$, X = P, As, and Sb]; Luo et al. (2015) [$LaT_4Sb_9Sn_3$]; Shankar et al. (2017) [$EuRu_4As_{12}$]; Hu et al. (2017) [$La_xCo_4Sb_{12}$]; Qi et al. (2017) [$Ba_xIr_4As_{12}$]; and Tütüncü et al. (2017) [$LaRu_4P_{12}$ and $LaRu_4As_{12}$].

1.4.3 FILLED SKUTTERUDITES WITH THE [PT_4GE_{12}] FRAMEWORK

Skutterudites with an entirely new polyanionic framework based on [Pt_4Ge_{12}] were reported by Bauer et al. (2007) and Gumeniuk et al. (2008a). Although the voids in this structure are somewhat smaller than those in $CoSb_3$, all other structural aspects are essentially identical to those of the

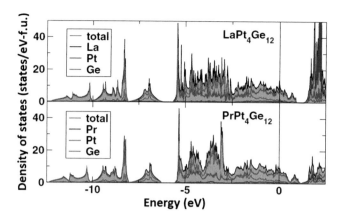

FIGURE 1.4 Total and atom-resolved electronic density-of-states for $LaPt_4Ge_{12}$ and $PrPt_4Ge_{12}$. Modified from R. Gumeniuk et al., *Physical Review Letters* 100, 017002 (2008). With permission from the American Physical Society.

pnicogen-based skutterudites. The filler species here transfer the charge to the polyanion and in the process stabilize the structure. Although not contemplated for use as thermoelectric materials (Pt and Ge are far too expensive), the filled forms of the $[Pt_4Ge_{12}]$ framework display equally exciting magnetic and superconducting properties as do their pnicogen-based structural relatives.

The first reports of the existence and properties of skutterudites with the $[Pt_4Ge_{12}]$ framework were accompanied by DFT calculations of the electronic bands and contributions of the respective constituting elements to the density-of-states, Bauer et al. (2007), Gumeniuk et al. (2008a), Grytsiv et al. (2008), and Tran et al. (2009a, 2009b). The most notable feature distinguishing $[Pt_4Ge_{12}]$-based skutteruddites from pnicogen-based skutterudites is the overwhelming dominance of Ge p states in the density-of-states near the Fermi energy. This suggests that the dominant role in the physical properties, such as superconductivity, is played by the $[Pt_4Ge_{12}]$ framework and the fillers have only a minor effect. The results also indicated a substantial electron transfer from the fillers to Pt that helps to stabilize the structure. An example of the computed density-of-states of skutterudites with the $[Pt_4Ge_{12}]$ framework is shown in Figure 1.4 for $LaPt_4Ge_{12}$ and $PrPt_4Ge_{12}$.

Band structure calculations devoted to skutterudites with the $[Pt_4Ge_{12}]$-based framework and filled with a specific filler can be found in the following publications: Bauer et al. (2007, 2008a), Grytsiv et al. (2008), Khan et al. (2008) $[BaPt_4Ge_{12}]$; Bauer et al. (2007), Gumeniuk et al. (2008b), Rosner et al. (2009), Khan et al. (2008) $[SrPt_4Ge_{12}]$; Gumeniuk et al. (2008a), Humer et al. (2013), Tŭtŭncŭ et al. (2017) $[LaPt_4Ge_{12}]$; Nicklas et al. (2012) $[CePt_4Ge_{12}]$; Gumeniuk et al. (2010) $[SmPt_4Ge_{12}]$; Gumeniuk et al. (2008a) $[PrPt_4Ge_{12}]$; Bauer et al. (2008b), Tran et al. (2009a, 2009b), Galvan (2009) $[ThPt_4Ge_{12}]$; and Bauer et al. (2008b) $[UPt_4Ge_{12}]$.

1.4.4 FILLED SKUTTERUDITES WITH ELECTRONEGATIVE FILLERS

The stability of all forms of filled skutterudites discussed above relied on the transfer of electrons to the framework. It thus came as a great surprise when Fukuoka and Yamanaka (2010) reported that using a high-pressure synthesis they filled $RhSb_3$ with iodine and the filler attained the valence state 1−, i.e., iodine acted as an acceptor. The discovery of an anion-filled skutterudite led to more studies. Li et al. (2014) documented that iodine can also be trapped as an electronegative filler in $CoSb_3$, and Zhang et al. (2015) showed the same for the compensated $Fe_4Co_{4-x}Sb_{12}$ framework. Reports of other halides as electronegative fillers were soon followed by reports of Ortiz et al. (2016) filling bromine into $CoSb_3$ and Duan et al. (2016) inserting chlorine into the same framework. By replacing a fraction of Sb with Te, Wang et al. (2018) were able to insert sulfur as an electronegative filler into $CoSb_3$, and Wan et al. (2018) and Li et al. (2019) achieved the same feat by substituting a small amount of Pd for Co and Ni for Co, respectively. Without doping electrons at the site of Sb or partly replacing Co with transition metals of column 10, the filling fraction limit of sulfur in $CoSb_3$ is only about 5%, Ghosh et al. (2020). Theoretical analysis of filling with sulfur and doping with Ni, Pd, and Pt at the site of Co was performed by Tu et al. (2019). Se is a particularly interesting electronegative filler because it plays a dual role, it fills the voids of $CoSb_3$ and also substitutes at the sites of Sb where it charge compensates the presence of Se in the voids. Additional electrons can be introduced by replacing a fraction of Co with, e.g., Pd, Bao et al. (2019).

With this brief summary of structural features and electronic bands of skutterudites, we now turn to the superconducting and magnetic properties of skutterudite compounds.

REFERENCES

Akai, K., K. Koga, K. Oshiro, and M. Matsuura, *Proc. 20th Int. Conf. on Thermoelectrics*, IEEE Catalog Number 01TH8589, Piscataway, NJ, p. 93 (2002).

Bang, S., D.-H. Wee, A. Li, M. Fornari, and B. Kozinsky, *J. Appl. Phys.* **119**, 205102 (2016).

Bao, X., Z. H. Wu, and H. Q. Xie, *Mater. Res. Express* **6**, 025511 (2019).

Bauer, E., N. A. Frederick, P.-C. Ho, V. S. Zapf, and M. B. Maple, *Phys. Rev. B* **65**, 100506(R) (2002).

Bauer, E., A. Grytsiv, X.-Q. Chen, N. Melnychenko-Koblyuk, G. Hilscher, H. Kaldarar, H. Michor, E. Royanian, G. Giester, M. Rotter, R. Podloucky, and P. Rogl, *Phys. Rev. Lett.* **99**, 217001 (2007).

Bauer, E., A. Grytsiv, X.-Q. Chen, N. Melnychenko-Koblyuk, G. Hilscher, H. Kaldarar, H. Michor, E. Royanian, M. Rotter, R. Podloucky, and P. Rogl, *Adv. Mater.* **20**, 1325 (2008a).

Bauer, E., X.-Q. Chen, P. Rogl, G. Hilscher, H. Michor, E. Royanian, R. Podloucky, G. Giester, O. Sologub, and A. P. Goncalves, *Phys. Rev. B* **78**, 064516 (2008b).

Bertini, L. and S. Cenedese, *Phys. Stat. Solidi (RRL)* **1**, 244 (2007).

Biltz, W. and M. Heimbrecht, *Z. Anorg. Allg. Chem.* **237**, 132 (1938).

Borshchevsky, A., J.-P. Fleurial, E. Allevato, and T. Caillat, *Proc. 13th Int. Conf. on Thermoelectrics*, American Institute of Physics, Kansas City, p. 3 (1995).

Braun, D. J. and W. Jeitschko, *J. Less-Common Met.* **72**, 147 (1980a).

Braun, D. J. and W. Jeitschko, *J. Solid State Chem.* **32**, 357 (1980b).

Chakoumakos, B. C. and B. C. Sales, *J. Alloys and Compd.* **407**, 87 (2006).

Chaput, L., P. Pecheur, J. Tobola, and H. Scherrer, *Phys. Rev. B* **72**, 085126 (2005).

Chen, L. D., *Proc. 21st Int. Conf. on Thermoelectrics*, IEEE Catalog Number 02TH8657, Piscataway, NJ, p. 42 (2002).

Daniel, M. V., L. Hammerschmidt, C. Schmidt, F. Timmermann, J. Franke, N. Jöhrmann, M. Hietschold, D. C. Johnson, B. Paulus, and M. Albrecht, *Phys. Rev. B* **91**, 085410 (2015).

Duan, B., Jiong Yang, J. R. Salvador, Y. He, B. Zhao, S. Wang, P. Wei, F. S. Ohuchi, W. Q. Zhang, R. P. Hermann, O. Gourdon, S. X. Mao, Y. W. Cheng, C. M. Wang, J. Liu, P. C. Zhai, X. F. Tang, Q. J. Zhang, and Jihui Yang, *Energy & Environ. Sci.* **9**, 2090 (2016).

Dudkin, L. D. and N. K. Abrikosov, *Zh. Neorg. Khim.* **2**, 212 (1957).

Fleurial, J.-P., A. Borshchevsky, T. Caillat, D. T. Morelli, and G. P. Meisner, *Proc. 15th Int. Conf. on Thermoelectrics*, IEEE Catalog Number 96TH8169, Piscataway, NJ, p. 91 (1996).

Fleurial, J.-P., T. Caillat, and A. Borshchevsky, *Proc. 16th Int. Conf. on Thermoelectrics*, IEEE Catalog Number 97TH8291, Piscataway, NJ, p. 1 (1997).

Fornari, M. and D. J. Singh, *Phys. Rev. B* **59**, 7922 (1999).

Fukuoka, H. and S. Yamanaka, *Chem. Mater.* **22**, 47 (2010).

Galvan, G. H., *J. Supercond. Nov. Magn.* **22**, 367 (2009).

Ghosh, S., K. K. Raut, A. Ramaksishnan, K.-H. Chen, S.-J. Hong, and R. C. Mallik, *AIP Conf. Proc.* **2265**, 030606 (2020).

Grytsiv, A., X.-Q. Chen, N. Melnychenko-Koblyuk, P. Rogl, E. Bauer, G. Hilscher, H. Kaldarar, H. Michor, E. Royanian, R. Podloucky, M. Rotter, and G. Giester, *J. Phys. Soc. Jpn.* **77**, 121 (2008).

Gumeniuk, R., W. Schnelle, H. Rosner, M. Nicklas, A. Leithe-Jasper, and Y. Grin, *Phys. Rev. Lett.* **100**, 017002 (2008a).

Gumeniuk, R., H. Rosner, W. Schnelle, M. Nicklas, A. Leithe-Jasper, and Y. Grin, *Phys. Rev. B* **78**, 052504 (2008b).

Gumeniuk, R., M. Schöneich, A. Leithe-Jasper, W. Schnelle, M. Nicklas, H. Rosner, A. Ormeci, U. Burkhardt, M. Schmidt, U. Schwarz, M. Ruck, and Y. Grin, *New J. Phys.* **12**, 103035 (2010).

Haidinger, Wilhelm Karl, in *Handbuch der Bestimmenden Mineralogie*, Vienna (1845).

Hammerschmidt, L., S. Schlecht, and B. Paulus, *Phys. Stat. Solidi A* **210**, 131 (2013).

Harima, H., *J. Mag. Magn. Mater.* **177–181**, 321 (1998).

Harima, H., *J. Phys. Soc. Jpn.* **77**, 114 (2008).

Harima, H. and K. Takegahara, *Physica B* **312–313**, 843 (2002a).

Harima, H. and K. Takegahara, *Physica C* **388–389**, 555 (2002b).

Harima, H., K. Takegahara, K. Ueda, and S. H. Curnoe, *Acta Phys. Pol. B* **34**, 1189 (2002).

Harima, H. and K. Takegahara, *Physica B* **328**, 26 (2003a).

Harima, H. and K. Takegahara, *J. Phys.: Condens. Matter* **15**, S2081 (2003b).

Harima, H. and K. Takegahara, *Physica B* **359–361**, 920 (2005).

Hu, C. Z., X. Y. Zeng, Y. F. Liu, M. H. Zhou, H. J. Zhao, T. M. Tritt, J. He, J. Jakowski, P. R. C. Kent, J. S. Huang, and B. G. Sumpter, *Phys. Rev. B* **95**, 165204 (2017).

Humer, S., E. Royanian, H. Michor, E. Bauer, A. Grytsiv, M.-X. Chen, R. Podloucky, and P. Rogl, in *New Materials for Thermoelectric Applications: Theory and Experiment*, NATO Science for Peace and Security Series B: Physics and Biophysics, Springer Science, Dordrecht, p. 115 (2013).

Imai, M., M. Akaishi, and I. Shirotani, *Supercond. Sci. Technol.* **20**, 832 (2007).

Isaacs, E. B. and C. Wolverton, *Chem. Mater.* **31**, 6154 (2019).

Jeitschko, W. and D. J. Braun, *Acta Crystallogr. B* **33**, 3401 (1977).

Jeitschko, W., A. J. Foecker, D. Paschke, M. V. Dewalsky, E. B. H. Evers, B. Künnen, A. Lang, G. Kotzyba, U. C. Rodewald, and M. H. Möller, *Z. Anorg. Allg. Chem.* **626**, 1112 (2000).

Jolibois, P., *C. R. Acad. Sci.* **150**, 106 (1910).

Jung, D. W., M.-H. Whangbo, and S. Alvarez, *Inorg. Chem.* **29**, 2252 (1990).

Khan, B., H. A. Rahnamaye Aliabad, Saifullah, S. Jalali-Asadabadi, I. Khan, and I. Ahmad, *J. Alloys Compd.* **647**, 364 (2015).

Khan, R. T., E. Bauer, X.-Q. Chen, R. Podloucky, and P. Rogl, *J. Phys. Soc. Jpn.* **77**, 350 (2008).

Kjekshus, A. and T. Rakke, *Acta Chem. Scand., Ser. A* **28**, 99 (1974).

Kolezynski, A. and W. Szczypka, *J. Alloys Compd.* **691**, 299 (2017).

Korenstein, R., S. Soled, A. Wold, and G. Collin, *Inorg. Chem.* **16**, 2344 (1977).

Kurmaev, E. Z., A. Moewes, I. R. Shtein, L. D. Finkelstein, A. L. Ivanovski, and H. Anno, *J. Phys.: Condens. Matter* **16**, 979 (2004).

Lefebvre-Devos, I., M. Lassalle, X. Wallart, J. Olivier-Fourcade, L. Monconduit, and J. Jumas, *Phys. Rev. B* **63**, 125110 (2001).

Li, G.-D., S. Bajaj, U. Aydemir, S.-Q. Hao, H. Xiao, W. A. Goddard III, P. C. Zhai, Q.-J. Zhang, and G. J. Snyder, *Chem. Mater.* **28**, 2172 (2016).

Li, J. L., B. Duan, H. J. Yang, H. T. Wang, G. D. Li, J. Yang, G. Chen, and P. C. Zhai, *J. Mater. Chem. C* **7**, 8079 (2019).

Li, X.-D., B. Xu, L. Zhang, F.-F. Duan, X.-L. Yan, J.-Q. Yang, and Y.-J. Tian, *J. Alloys Compd.* **615**, 177 (2014).

Llunell, M., P. Alemany, S. Alvarez, V. P. Zhukov, and A. Vernes, *Phys. Rev. B* **53**, 10605 (1996).

Luo, H. X., J. W. Krizan, L. Muechler, N. Haldolaarachchige, T. Klimczuk, W. W. Xie, M. K. Fuccillo, C. Felser, and R. J. Cava, *Nat. Commun.* **6**, 6489 (2015).

Lutz, H. D. and G. Kliche, *J. Solid State Chem.* **40**, 64 (1981).

Lyons, A., R. P. Gruska, C. Case, S. N. Subbarao, and A. Wold, *Mater. Res. Bull.* **13**, 125 (1978).

Maple, M. B., P.-C. Ho, V. S. Zapf, N. A. Frederick, E. Bauer, W. M. Yuhasz, F. M. Woodward, and J. W. Lynn, *Proc. Int. Conf. on Strongly Correlated Electrons with Orbital Degrees of Freedom* (ORBITAL 2001), *J. Phys. Soc. Jpn.* **71**, 23 (2002).

Meisner, G. P., *Physica B* **108**, 763 (1981).

Morelli, D. T. and G. P. Meisner, *J. Appl. Phys.* **77**, 3777 (1995).

Nickel, E. H., *Chem. Geol.* **5**, 233 (1969).

Nicklas, M., S. Kirchner, R. Borth, R. Gumeniuk, W. Schnelle, H. Rosner, H. Borrmann, A. Leithe-Jasper, Y. Grin, and F. Steglich, *Phys. Rev. Lett.* **109**, 236405 (2012).

Nieroda, P., K. Kutorasinski, J. Tobola, and K. Wojciechowski, *J. Electron. Mater.* **43**, 1681 (2013).

Nolas, G. S., G. A. Slack, D. T. Morelli, T. M. Tritt, and A. C. Ehrlich, *J. Appl. Phys.* **79**, 4002 (1996).

Nordström, L. and D. J. Singh, *Phys. Rev. B* **53**, 1103 (1996).

Oftedal, I., *Z. Kristallogr.* **A66**, 517 (1928).

Ortiz, B. R., C. M. Crawford, R. W. McKinney, P. A. Parilla, and E. S. Toberer, *J. Mater. Chem. A* **4**, 8444 (2016).

Pardo, V., J. C. Smith, and W. E. Pickett, *Phys. Rev. B* **85**, 214531 (2012).

Park, C.-H. and Y.-S. Kim, *Phys. Rev. B* **81**, 085206 (2010).

Partik, M., C. Kringe, and H. D. Lutz, *Z. Kristallogr.* **211**, 304 (1996).

Pleass, C. M. and R. D. Heyding, *Canad. J. Chem.* **40**, 590 (1962).

Qi, Y. P., H. C. Lei, J. G. Guo, W. J. Shi, B. H. Yan, C. Felser, and H. Hosono, *J. Am. Chem. Soc.* **139**, 8106 (2017).

Ram, S., V. Kanchana, and M. C. Valsakumar, *J. Appl. Phys.* **115**, 093903 (2014).

Rose, G., in *Das Krystallo-Chemishe Mineralsystem*, Verlagen Wilhelm Engelmann, Leipzig, (1852).

Roseboom, E. H. Jr., *Am. Mineral.* **47**, 310 (1962).

Rosner, H. J., D. Gerger, D. Regesch, W. Schnelle, R. Gumeniuk, A. Leithe-Jasper, H. Fujiwara, T. Hauptricht, T. C. Koethe, H.-H. Hsieh, H.-J. Lin, C. T. Che, A. Ormeci, Y. Grin, and L. H. Tjeng, *Phys. Rev. B* **80**, 075114 (2009).

Rundqvist, S. and E. Larsson, *Acta Chem. Scand.* **13**, 551 (1959).

Saha, S. R., H. Sugawara, R. Sakai, Y. Aoki, H. Sato, Y. Inada, H. Shishido, R. Settai, Y. Onuki, and H. Harima, *Physica B* **328**, 68 (2002).

Sales, B. C., D. Mandrus, and R. K. Williams, *Science* **272**, 1325 (1996).

Schumer, B. N., M. B. Andrade, S. H. Evans, and R. T. Downs, *Am. Mineral.* **102**, 205 (2017).

Sekine, C., H. Saito, T. Uchiumi, A. Sakai, and I. Shirotani, *Solid State Commun.* **106**, 441 (1998).

Sekine, C., T. Uchiumi, I. Shirotani, K. Matsuhira, T. Sakakibara, T. Goto, and T. Yagi, *Phys. Rev. B* **62**, 11581 (2000).

Shankar, A., D. P. Rai, Sandeep, M. P. Ghimire, and R. K. Thapa, *Indian J. Phys.* **91**, 17 (2017).

Shi, X., W. Q. Zhang, L. D. Chen, and J. Yang, *Phys. Rev. Lett.* **95**, 185503 (2005).

Shirotani, I., Y. Shimaya, K. Kihou, C. Sekine, and T. Yagi, *J. Solid State Chem.* **174**, 32 (2003).

Shirotani, I., S. Sato, C. Sekine, K. Takeda, I. Inagawa, and T. Yagi, *J. Phys.: Condens. Matter* **17**, 7353 (2005).

Singh, D. J. and W. E. Pickett, *Phys. Rev. B* **50**, 11235 (1994).

Singh, D. J. and I. I. Mazin, *Phys. Rev. B* **56**, R1650 (1997).

Singh, D. J., *Mat. Res. Soc. Symp. Proc.* **691**, 15 (2002).

Slack, G. A., in *CRC Handbook of Thermoelectrics*, ed. D. M. Rowe, CRC Press, Boca Raton, FL, pp. 407–440 (1995).

Slack, G. A. and V. G. Tsoukala, *J. Appl. Phys.* **76**, 1665 (1994).

Sofo, J. O. and G. D. Mahan, *Mat. Res. Soc. Symp. Proc.* **545**, 315 (1999).

Smith, J. C., S. Banerjee, V. Pardo, and W. E. Pickett, *Phys. Rev. Lett.* **106**, 056401 (2011).

Sugawara, H., S. Osaki, S. R. Saha, Y. Aoki, H. Sato, Y. Inada, H. Shishido, R. Settai, Y. Onuki, H. Harima, and K. Oikawa, *Phys. Rev. B* **66**, 220504 (2002).

Takegahara, K. and H. Harima, *J. Phys. Soc. Jpn.* **71**, 240 (2002).

Takegahara, K. and H. Harima, *Physica B* **328**, 74 (2003a).

Takegahara, K. and H. Harima, *Physica B* **329–333**, 464 (2003b).

Takegahara, K. and H. Harima, *J. Phys. Soc. Jpn.* **77**, 193 (2008).

Takizawa, H., K. Okazaki, K. Uhedu, T. Endo, and G. S. Nolas, *Mater. Res. Soc. Symp. Proc.* **691**, 37 (2002).

Tanaka, K., Y. Kawahito, Y. Yonezawa, D. Kikuchi, H. Aoki, K. Kuwahara, M. Ichihara, H. Sugawara, Y. Aoki, and H. Sato, *J. Phys. Soc. Jpn.* **76**, 103704 (2007).

Tang, Y. L., Z. M. Gibbs, L. A. Agapito, G. Li, H.-S. Kim, M. B. Nardelli, S. Curtarolo, and G. J. Snyder, *Nat. Mater.* **14**, 1223 (2015).

Tran, V. H., D. Kaczorowski, W. Miller, and A. Jezirski, *Phys. Rev. B* **79**, 054520 (2009a).

Tran, V. H., B. Nowak, A. Jezirski, and D. Kaczorowski, *Phys. Rev. B* **79**, 144510 (2009b).

Tritt, T. M., G. S. Nolas, G. A. Slack, A. C. Ehrlich, D. J. Gillespie, and G. L. Cohn, *J. Appl. Phys.* **79**, 8412 (1996).

Tu, Z. K., X. Sun, X. Li, R. X. Li, L. L. Xi, and J. Yang, *AIP Adv.* **9**, 045325 (2019).

Tütüncü, H. M., E. Karaca, and G. P. Srivastava, *Phys. Rev. B* **95**, 214514 (2017).

Uher, C., in *Thermoelectric Skutterudites*, CRC Press, Taylor & Francis, Boca Raton, FL (2021).

Vaqueiro, P., G. G. Sobany, A. V. Powell, and K. S. Knight, *J. Solid State Chem.* **179**, 2047 (2006).

Vaqueiro, P., G. G. Sobany, and M. Stindl, *J. Solid State Chem.* **181**, 768 (2008).

Volja, D., B. Kozinski, A. Li, D. Wee, N. Marzari, and M. Fornari, *Phys. Rev. B* **85**, 245211 (2012).

Wan, S., P. F. Qiu, X. G. Huang, Q. F. Song, S. Q. Bai, X. Shi, and L. D. Chen, *ACS Appl. Mater. Interfaces* **10**, 625 (2018).

Wang, H. T., B. Duan, G. H. Bai, J. L. Li, Y. Yu, H. J. Yang, G. Chen, and P. C. Zhai, *J. Electron. Mater.* **47**, 3061 (2018).

Wei, W., Z. Y. Wang, L. L. Wang, H. J. Liu, R. Xiong, J. Shi, H. Li, and X. F. Tang, *J. Phys. D: Appl. Phys.* **42**, 1 (2009).

Wojciechowski, K. T., *J. Alloys Compd.* **439**, 18 (2007).

Xing, G. Z., X. F. Fan, W. T. Zheng, Y. M. Ma, H. L. Shi, and D. J. Singh, *Sci. Rep.* **5**, 10782 (2015).

Yan, B., L. Müchler, X.-L. Qi, S.-C. Zhang, and C. Felser, *Phys. Rev. B* **85**, 165125 (2012).

Yang, X. X., Z. H. Dai, Y. C. Zhao, W. C. Niu, J. Y. Liu, and S. Meng, *Phys. Chem. Chem. Phys.* **21**, 851 (2019).

Zevalkink, A., K. Star, U. Aydemir, G. J. Snyder, J.-P. Fleurial, S. Bux, T. Vo, and P. von Allmen, *J. Appl. Phys.* **118**, 035107 (2015).

Zhang, L., B. Xu, X.-D. Li, F.-F. Duan, X.-L. Yan, and Y.-J. Tian, *Mater. Lett.* **139**, 249 (2015).

2 Superconducting Skutterudites

2.1 INTRODUCTION

Superconductivity is surely one of the most mysterious discoveries ever made. It must have been a great shock to Kamerlingh Onnes (1911) in his laboratory at the University of Leiden in Holland to see that his mercury sample apparently lost all its resistance when cooled down to near 4.5 K. It took several months and many experiments with other metallic samples before he was convinced that what he was observing is a genuine effect and not a glitch in the experimental setup. For a number of years, Kamerlingh Onnes was "the king of the world" with no competitors being able to liquefy helium and perform experiments below 10 K. For his low-temperature work and liquefaction of helium, he received the Nobel Prize in 1913. Perhaps not as dramatic but equally important was the discovery of the exclusion of magnetic flux from the interior of a metal some 20 years later by Meissner and Ochsenfeld (1933) as the metal underwent a transition to the superconducting state. The two effects, complete loss of resistance and flux exclusion, are macroscopic revelations of the mysterious phenomenon of superconductivity.

From our time in the middle school, we all know that electrons are negatively charged particles and, as such, repel each other by the Coulomb force. The essence of superconductivity is the formation of the so-called Cooper pairs whereby two electrons, under the right circumstances, bind together over a short distance called the coherence length ξ (a few tens to a few hundreds of nm, the latter value applicable to pure elemental superconductors), overcoming their Coulomb repulsion. In a simplistic but substantially correct picture of how such attractive interaction between two electrons arises, one can imagine an electron moving through a lattice of positively charged ions. The lattice gets polarized by the passage of the electron; i.e., the ions are drawn closer to the negative electron. However, because the ionic motion is retarded, meaning that the ions move much slower than electrons because of their nearly 2000 times heavier mass, the electron will be some 100–1000 nm ahead of the maximally perturbed ionic lattice. The locally greater density of positive ions thus attracts the second electron, effectively coupling the two electrons as they form a Cooper pair. The strength of this coupling is expressed by the electron–phonon coupling constant λ_{e-p}. In the Bardeen–Cooper–Schrieffer (BCS) theory, the Cooper pairs involve two electrons with opposite momenta $+\vec{k}$ and $-\vec{k}$ and opposite spins $\uparrow$ and $\downarrow$ (spin singlet). The requirement of opposite momenta implies that the orbital angular momentum $\vec{L}$ of the Cooper pair is zero, and because the paired electrons have opposite spins, so the spin angular momentum $\vec{S}$ is equal to zero. Consequently, the total angular momentum $\vec{J} = \vec{L} + \vec{S}$ is also zero. All Cooper pairs thus have the same zero total momentum, form a condensate of boson particles (particles with integer spin), and can be described by a macroscopic wavefunction Ψ. Superconductivity of this kind is referred to as the conventional *s*-wave superconductivity. The key finding of BCS theory is that Cooper pairs here have lower energy than two single electrons and are energetically separated by the superconducting gap 2Δ of the order of $10^{-4} - 10^{-3}$ eV. Consequently, superconductivity requires a low-temperature environment where the Cooper pairs are not destroyed by thermal energy (vibrations) of the crystal lattice. Apart from the coherence length, the properties of a superconductor are further characterized by the London penetration depth, λ, which indicates the distance in the interior of a superconductor over which the magnetic field drops to 1/e of the value it has at the surface. Typical values of the penetration length are in the range of 50 to 500 nm. The ratio of the penetration length and the coherence

DOI: 10.1201/9781003225898-2

length is called the Ginzburg–Landau (GL) parameter $\kappa = \lambda/\xi$, and it characterizes another peculiar aspect of superconductivity, namely, that certain superconductors remain superconducting even though pierced by an array of parallel narrow tubes (vortices) of the normal conducting state. The vortices first appear at the so-called lower critical magnetic field H_{c1} and grow in density as the magnetic field increases, and only at the upper critical magnetic field, H_{c2} is the superconductivity destroyed as the normal state takes over the entire volume of the sample. The magnitude of the GL parameter κ is used to classify superconductors into two categories. Superconductors for which $\kappa < 1/\sqrt{2}$ are called type-I superconductors and, except for niobium, vanadium, and technetium, are all elemental superconductors. In contrast, superconductors for which $\kappa > 1/\sqrt{2}$ are type-II superconductors and include all other superconducting structures, among them all superconducting skutterudites. The essential distinction between the two classes of superconductors is sketched in the phase diagrams in Figure 2.1.

The fundamental requirement for the s-wave superconductivity is a particular symmetry of the system. Specifically, the structure must possess inversion symmetry (imagine a mirror placed in the center of the structure) and time-reversal symmetry (TRS). The latter, a fancy sounding name, expresses the fact that the physical phenomena related to the structure should proceed in the same fashion when the arrow of time is reversed, $t \rightarrow -t$. If either of the two or both above symmetries are broken, the system belongs to the realm of unconventional superconductivity, characterized by the formation of nodes and lines of nodes on the Fermi surface where the superconducting gap function Δ_k becomes zero. Electrons at and close to such locations in k-space possess highly unusual properties, such as large effective masses (heavy fermions). Superconductors with broken TRS are often referred to as chiral superconductors. The excitement regarding the properties of chiral superconductors arises from a possibility that their electrons, given the right circumstances, can become their own antiparticles, referred to as Majorana fermions, the concept borrowed from high energy particle physics, Majorana (1937). While most superconductors and for that matter, most superconducting skutterudites are conventional s-wave superconductors, we shall see that there are some, particularly $PrOs_4Sb_{12}$, that show exotic behavior, including perhaps the best

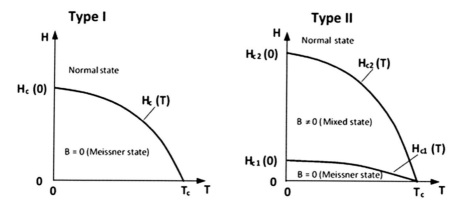

FIGURE 2.1 Schematic illustration of type-I and type-II superconductors. The superconducting state of a type-I superconductor is characterized by complete exclusion of flux (Meissner state). Its superconductivity is destroyed by magnetic field at the critical field $H_c(T)$. The highest magnetic field the type-I superconductor can survive is $H_c(0)$, the value at $T = 0$ K. In type-II superconductors, the Meissner state is destroyed; i.e., the vortices start to penetrate the superconductor, at a lower critical field $H_{c1}(T)$. However, superconductivity survives and is destroyed only at a higher critical field $H_{c2}(T)$. The highest magnetic field the type-II superconductor can withstand is $H_{c2}(0)$, the value at $T = 0$ K. The field range between H_{c1} and H_{c2} is called the mixed state of a superconductor.

prospect for harboring the Majorana fermions, Kozii et al. (2016). This has generated tremendous interest and fascination, provided theorists with innumerable opportunities to construct theoretical models to describe the novel physical phenomena, and offered experimentalists the challenging tasks of verifying them.

2.2 USEFUL RELATIONS FOR SUPERCONDUCTING PARAMETERS

In the analysis of experimental data, one aims to extract the key parameters characterizing the superconducting state. To achieve this aim, I summarize here several important relations I will refer to in the subsequent sections. Before I do so, a comment is in order regarding the use and dimension of a magnetic field, an important physical parameter when describing a superconducting state. A magnetic field H is a vector measured in units of Amperes per meter and represents the applied field, such as generated by a coil. Its value is unchanged whether or not a magnetic substance is placed inside the coil. In contrast, a magnetic induction B, also called a magnetic flux density, is a vector measured in units of Tesla and depends on the spatial distribution of all circulating electrical currents, both those in the coil and those in any magnetic medium placed inside the coil. In free space (empty coil), B and H are identical, except for their respective units, and $B = \mu_0 H$, where μ_0 is the permeability of free space, $4\pi \times 10^{-7}$ kgms^{-2}A^{-2} (Henry per meter). Magnetization M, defined as the magnetic moment m per volume of sample V, has the same dimension as H, i.e., Am^{-1}. In the presence of a magnetic material with magnetization M inside the coil, the magnetic induction becomes $B = \mu_0(H + M)$. It is a common practice to write and quote the lower and the upper superconducting critical fields, H_{c1} respectively H_{c2}, in units of Tesla (which, in fact, is $\mu_0 H$) rather than in strictly correct Amperes per meter. Magnetic susceptibility $\chi = M/H$ (a dimensionless quantity) relates the sensitivity of a material to magnetization M in the presence of the applied magnetic field H. The Meissner state (complete flux exclusion) has the susceptibility $\chi = -1$ in SI units and $\chi = -1/4\pi$ in cgs units. One often encounters a term "emu", which is an abbreviation for electromagnetic unit. Although it is not a unit in the conventional sense, the frequently quoted volume magnetization in cgs units, emu per cm^3, is equivalent to 10^3 Am^{-1} in SI units. More details of various susceptibilities and their conversion factors can be found in Chapter 3, describing magnetism in skutterudites.

The electron–phonon coupling constant λ_{e-p} that underpins the BCS theory is given by the product of the density of electronic states at the Fermi level $N(E_F)$ and the net-attractive potential between electrons at the Fermi surface V_0. This potential has an attractive electron–phonon part measured by λ_{e-p} and a repulsive screened Coulomb part μ_c^*. Superconductors are classified as weak-coupling ($\lambda_{e-p} \ll 1$), intermediate-coupling ($\lambda_{e-p} \sim 1$), and strong-coupling ($\lambda_{e-p} \gg 1$) superconductors. The coupling constant λ_{e-p} responsible for the attractive part of the Cooper pair bonding can be obtained from the McMillan (1968) formula relating it to the transition temperature T_c as

$$T_c = \frac{\theta_D}{1.45} exp\left[\frac{-1.04\left(1+\lambda_{e-p}\right)}{\lambda_{e-p} - \mu_c^*\left(1+0.62\lambda_{e-p}\right)}\right], \tag{2.1}$$

where θ_D is the Debye temperature, and the repulsive screened Coulomb part μ_c^* is usually taken as equal to 0.13.

One of the important and readily measured parameters is the upper critical field $H_{c2}(T)$ at which the superconducting state is destroyed and the sample reverts to the normal state. The maximum value of $H_{c2}(T)$ is attained at absolute zero temperature and is designated as $H_{c2}(0)$. Because achieving a temperature near 0 K is quite challenging, there is a useful relation, referred to as the WHH formula, standing for Werthamer–Helfand–Hohenberg (1966), that relates $H_{c2}(0)$ in units of Tesla to the initial slope of the T-dependent upper critical field in units of TK^{-1} taken at the transition temperature T_c,

$$H_{c2}^{orb}(0) = 0.693 T_c \left(-\frac{dH_{c2}(T)}{dT} \right)_{T_c} . \tag{2.2}$$

The upper critical field in Equation 2.2 expresses the resistance of the Cooper pairs against their destruction by the orbital motion of electrons. While the superscript "orb" is included in Equation 2.2 to indicate that the upper critical field here refers to the orbital pairbreaking mechanism, it will be dropped in subsequent discussions. One also often encounters the upper critical field that describes the ability of the Cooper pairs to withstand a spin flip of one of its electrons, referred to as the Pauli paramagnetic-limited upper critical field or, equivalently, as the Clogston–Chandrasekhar limit. The latter critical field $H_{c2}^P(0)$ is higher and in units of Tesla is given by

$$H_{c2}^P(0) = 1.84 \left(TK^{-1} \right) T_c . \tag{2.3}$$

The experimentally extrapolated upper critical fields of superconducting skutterudites lie much closer to the orbital estimate of Equation 2.2 than to the Pauli limited value given by Equation 2.3.

Once $H_{c2}(0)$ is known, the coherence length ξ_0 (in meters) can be obtained from, e.g., Tinkham (1983), as

$$\xi_0 = \left(\frac{\Phi_0}{2\pi H_{c2}(0)} \right)^{1/2} , \tag{2.4}$$

where $\Phi_0 = h/2e \approx 2.07 \times 10^{-15} \; Tm^2$ is the flux quantum.

The Fermi velocity of electrons v_F is related to the coherence length ξ_0 through

$$v_F = \frac{\xi_0 k_B T_c}{0.18\hbar} . \tag{2.5}$$

Assuming a simple spherical Fermi surface, the Fermi wave-vector k_F is given by

$$k_F = \left(\frac{3\pi^2 Z}{\Omega} \right)^{1/3} , \tag{2.6}$$

where Z is the number of electrons in the unit cell (in the case of superconducting skutterudites filled with Pr this means that two cages per unit cell each with three electrons from the Pr^{3+} ion for a total of $Z = six$ electrons) and Ω is the volume of the unit cell taken as the lattice constant to the power of three. Using Equations 2.5 and 2.6, the effective mass $m*$ follows from

$$m^* = \frac{\hbar k_F}{v_F} . \tag{2.7}$$

The electronic specific heat coefficient (Sommerfeld coefficient) γ_n is then obtained from

$$\gamma_n = \pi^2 \left(\frac{Z}{\Omega} \right) \frac{k_B^2}{\hbar k_F^2} \, m^*. \tag{2.8}$$

At zero temperature, the penetration depth is given by Gross et al. (1986) in the form

$$\lambda(0) = \frac{\left[\Phi_0 H_{c2}(0) \right]^{1/2}}{\sqrt{24} \, \delta_{sc} T_c \gamma_n^{1/2}}, \tag{2.9}$$

where $\delta_{sc} \equiv \Delta(0)/k_B T_c$, with $\Delta(0)$ being the superconducting band gap at zero temperature. In the BCS theory, the band gap at $T = 0$ K is given by

$$\Delta(0) = 1.76 \, k_B T_c, \tag{2.10}$$

and its temperature dependence close to T_c is of the form

$$\Delta(T) = 3.06 \, k_B T_c \left(1 - \frac{T}{T_c} \right)^{1/2}. \tag{2.11}$$

In zero magnetic field, the superconducting transition is a second-order phase transition, and, as such, the entropy $S(T)$ at T_c does not change. However, its derivative, related to the specific heat $C(T)$ = $TdS(T)/dT$, experiences a jump at the critical temperature designated as $\Delta C \equiv C_s - \gamma_n T$, where the term $\gamma_n T$ is the electronic specific heat in the normal state. The coefficient of the electronic specific heat γ_n is also directly related to the density of electronic states at the Fermi level,

$$\gamma_n = \frac{\pi^2}{3} D(E_F) k_B^2. \tag{2.12}$$

At the critical temperature T_c, the BCS theory predicts a jump in the specific heat of

$$\Delta C(T_c) = 4.68 \, D(E_F) k_B^2 T_c. \tag{2.13}$$

From Equations 2.12 and 2.13, the normalized jump in the specific heat at T_c is then a parameter-free expression in the BCS theory of magnitude

$$\frac{\Delta C(T_c)}{\gamma_n T_c} = 1.43. \tag{2.14}$$

In the BCS theory, the uniform energy gap $E_g = 2\Delta$ implies that the specific heat decreases exponentially with the decreasing temperature and is proportional to

$$C_s(T) \propto e^{-1.76 \frac{T_c}{T}}. \tag{2.15}$$

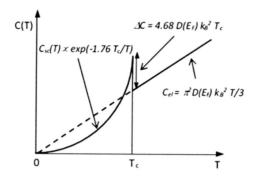

FIGURE 2.2   A sketch of the temperature dependence of the specific heat of a BCS superconductor.

A sketch of the temperature-dependent specific heat within the BCS theory is given in Figure 2.2. In contrast, the presence of nodes in the gap function of unconventional superconductors would result in a power law dependence of the specific heat below T_c.

Apart from the lattice specific heat that is a cubic function of temperature and rapidly diminishes as temperature decreases, the low-temperature specific heat measurements are often complicated by two additional Schottky-type contributions, i.e., specific heat associated with the crystalline electric field (CEF)-split levels of the rare-earth ions in the skutterudite cage, C_{CEF}, and, at very low temperatures, a nuclear contribution arising from the strong intra-site hyperfine coupling between the nucleus and conduction electrons, $C_{nucl}(T)$. Because we are interested in the specific heat at low temperatures, only the lowest two energy levels of the CEF-split multiplet are of relevance, namely, the ground state and the first excited state. The specific heat of such a two-level system was given by Gopal (1966) and can be expressed as

$$C_{CEF}\left(T\right) = R\left(\frac{\Delta_{0,1}}{T}\right)^2 \frac{g_0}{g_1} \frac{\exp\left(\dfrac{\Delta_{0,1}}{T}\right)}{\left[1 + \dfrac{g_0}{g_1}\exp\left(\dfrac{\Delta_{0,1}}{T}\right)\right]^2}. \tag{2.16}$$

Here, R is the universal gas constant, $\Delta_{0,1}$ is the energy difference (in units of temperature) between the ground state and the first excited state, and g_0 and g_1 are the respective degeneracies of the two states. Regarding the nuclear contribution, it can be written, following Aoki et al. (2002), as

$$C_{nucl}\left(T\right) = \frac{A_n}{T^2}, \tag{2.17}$$

where A_n is given by

$$A_n = R\left(\frac{A_{hf}\, M_{ion}}{g_J}\right)^2 \frac{I(I+1)}{3}, \tag{2.18}$$

with A_{hf} the magnetic dipole hyperfine coupling constant, M_{ion} the site-averaged magnitude of the filler ion's magnetic moment, g_J the Landé g-factor, and I the nuclear spin of the ion.

The overall low-temperature specific heat above T_c is then given by

$$C(T) = \gamma_n T + \beta T^3 + C_{CEF}(T), \tag{2.19}$$

where I have omitted the nuclear term as it is typically important only below 0.5 K. In the superconducting state, assuming an ordinary BCS state, the specific heat can be written as

$$C_{BCS}(T) = \gamma_s T + B exp\left(-\frac{\Delta(0)}{k_B T}\right) + \beta T^3 + C_{CEF}(T) + \frac{A_n}{T^2}, \tag{2.20}$$

with γ_s standing for the electronic specific heat coefficient in the superconducting state. For an unconventional superconductor, the exponential term is replaced by a power law dependence of the form DT^n, and the specific heat becomes

$$C_{uncony}(T) = \gamma_s T + DT^n + \beta T^3 + C_{CEF}(T) + \frac{A_n}{T^2}. \tag{2.21}$$

Usually, because the transition temperatures of heavy fermion superconductors are low, one can entirely neglect the phonon term βT^3, estimate, and subtract the CEF contribution and bring Equations 2.20 and 2.21 into a manageable form for fitting. Particularly useful fits are done when Equations 2.19–2.21 are divided by T, as this isolates the electronic specific heat coefficient.

2.3   LA-FILLED SUPERCONDUCTING SKUTTERUDITES

The early interest in filled skutterudites was primarily driven by spectacular observations of superconductivity with rather high transition temperatures in compounds that did not contain any Pb or Nb. Moreover, Meisner (1981) detected superconductivity initially in $LaFe_4P_{12}$ (T_c = 4.1 K), one of the rare situations (at that time) where a compound containing Fe was found to be a superconductor. Under the usual circumstance, Fe atoms carry a magnetic moment large enough to break the Cooper pairs. Shortly thereafter, Meisner (1982) observed superconducting transitions also in the sister elements $LaRu_4P_{12}$ (T_c = 7.2 K) and $LaOs_4P_{12}$ (T_c = 1.8 K), where Ru and Os are clearly non-magnetic. To document that Fe in $LaFe_4P_{12}$ cannot have any substantial magnetic moment, Shenoy et al. (1982) measured the ^{57}Fe Mössbauer spectrum and found that, indeed, the magnetic moment on a single Fe ion cannot be larger than 0.01 μ_B, where μ_B stands for the Bohr magneton. It is believed that the iron magnetic moment is quenched by the strong octahedral crystal field acting at the site of Fe. However, as discussed in section 1.1, the octahedrons are somewhat distorted, and this is reflected in an electric quadrupole interaction at the iron site that is temperature-dependent, specifically increasing with decreasing temperature. The implied "softness" of the crystalline structure of $LaFe_4P_{12}$ at low temperatures was tested by DeLong and Meisner (1985) *via* measurements of the pressure dependence of the superconducting transition temperature. In fact, they looked closely into the trend of all three superconducting filled phosphide skutterudites known at that time, LaT_4P_{12} with T = Fe, Ru, and Os. The results are depicted in Figure 2.3, where a shift in the transition temperature ΔT_c is plotted as a function of applied pressure up to 1.8 GPa. The respective pressure derivatives are $+7.2 \times 10^{-1}$ K(GPa)$^{-1}$ for $LaFe_4P_{12}$, -1.6×10^{-1} K(GPa)$^{-1}$ for $LaRu_4P_{12}$, and -9.5×10^{-2} K(GPa)$^{-1}$ for $LaOs_4P_{12}$. It is obvious that La in the $[Fe_4P_{12}]$ cage behaves very differently compared to when it is situated in the cages of $[Ru_4P_{12}]$ or $[Os_4P_{12}]$. In particular, its positive pressure derivative of the transition temperature, i.e., the increasing transition temperature with increasing applied pressure, is highly unusual, and its origin will be discussed shortly. More recent studies of the pressure derivative of the transition temperature of $LaFe_4P_{12}$ were extended

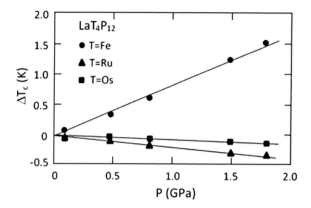

FIGURE 2.3   Pressure-induced shift in the superconducting transition temperature as a function of applied pressure for LaT_4P_{12}, T = Fe, Ru, and Os. Lines are guides to the eye. Redrawn from L. E. DeLong and G. P. Meisner, *Solid State Communications* **53**, 119 (1985). With permission from Elsevier.

to pressures up to 8 GPa by Kawamura et al. (2013), who confirmed the positive slope of dT_c/dP. However, the data also indicated that the T_c reaches its maximum value of about 8.2 K near 7 GPa and thereafter gradually decreases. What causes the rather atypical increase in the superconducting transition temperature of $LaFe_4P_{12}$ with applied pressure was rationalized by density functional theory (DFT) calculations of the electronic band structure by Nakazima et al. (2014). They computed angular momentum components of the density of states (DOS) of the highest occupied conduction bands in $LaFe_4P_{12}$ and $LaRu_4P_{12}$ and worked out the magnitude of the so-called Hopfield parameter η, defined as a product of the DOS at the Fermi level and the average squared electronic matrix element. Making calculations at ambient pressure as well as at 8 GPa, they were able to compare the effect of pressure for the two respective La-filled phosphide skutterudites. The difference between the two was dramatic. While the pressure tended to decrease the DOS and the Hopfield parameter of $LaRu_4P_{12}$, the same pressure increased the DOS and strongly enhanced the Hopfield parameter of $LaFe_4P_{12}$. Consequently, in those rare cases when the pressure increases the DOS at the Fermi level, one may observe an increase in the transition temperature as the pressure increases. I shall return to this issue when discussing properties of more recently discovered superconductivity in YT_4P_{12} (T = Fe, Ru, and Os) skutterudites.

With these interesting studies on phosphide-based skutterudites, the time was ripe for more explorations of superconductivity in filled skutterudites. The initial major effort was spearheaded by a group of Japanese scientists from the Muroran Institute of Technology, who expanded the realm of superconducting skutterudites to several arsenide and antimonide structures. Their work much benefitted from the use of high-pressure high-temperature (HPHT) synthesis. In short, while the reaction of transition metals with powders of pnicogens is rather sluggish at ambient pressures, it becomes quite vigorous when pressures of several GPa generated in special apparatus, such as a wedge-type cubic-anvil, are applied at high temperatures. Using the HPHT synthesis technique, Shirotani et al. (1996) confirmed the onset of superconductivity in $LaRu_4P_{12}$ at 7 K seen previously by Meisner (1982) and observed the superconducting transition at the same 7 K also in partly Os-substituted, Ru-based structures $LaRu_3OsP_{12}$ and $LaRu_2Os_2P_{12}$. It is rather surprising that a 50% replacement of Ru by Os did not affect the onset of superconductivity, while in $LaOs_4P_{12}$, where all Ru is replaced with Os, the superconductivity is suppressed down to 1.8 K. In the follow-up study on arsenides, Shirotani et al. (1997) observed superconductivity in $LaRu_4As_{12}$ with an impressive transition temperature of 10.3 K, more than 3 degrees higher than the transition seen in the sister phosphide compound. They also found the superconducting transition in the first antimonide skutterudite $LaRu_4Sb_{12}$ around 2.8 K and reported the superconductivity of $PrRu_4As_{12}$ setting in at 2.4 K.

The latter finding was particularly interesting for the fact that the superconductivity was observed in a skutterudite filled with a magnetic rare-earth element. The role of Pr ions deserves more discussion, especially in the context of the highly unconventional superconductivity of $PrOs_4Sb_{12}$, so an entire section 2.4 is dedicated to the superconducting properties of Pr-filled skutterudites. In a subsequent study, Uchiumi et al. (1999) inquired the effect of a small amount of Ce substituted for La in the framework of $[Ru_4P_{12}]$, i.e., the structures $La_{1-x}Ce_xRu_4P_{12}$ with $x = 0$, 0.05, 0.1, and 0.2. Here, the presence of a single f-electron of Ce^{3+} showed its destructive influence on superconductivity. With just $x = 0.1$, the onset of the superconducting transition was suppressed down to 5 K, and the transition was completed only below 2 K. A 20% replacement of La with Ce led to no detection of any significant diamagnetism, which must be interpreted as a lack of superconductivity, down to at least 0.5 K, the lowest temperature covered.

All early studies of La-filled skutterudites were performed for either polycrystalline samples often having significantly less than the theoretical density and containing impurities or small single crystals prepared by a flux growth method. The preparation of arsenide skutterudite single crystals is more challenging because of the toxicity of As and the fact that, at ambient pressure, As evaporates before it melts, thus hampering the flux growth. The problem was solved by Henkie et al. (2008), who developed the technique of mineralization in a molten Cd:As flux. Single crystals of $LaRu_4As_{12}$ prepared by this technique had a marginally higher $T_c = 10.45$ K than the previously studied polycrystalline samples. The high quality of the crystals was also attested by their residual resistance ratio (RRR) of 198 and an exceptionally narrow width of the superconducting transition $\Delta T_c = 0.03$ K. The crystals were used in the study of Bochenek et al. (2012) aimed at a closer look into the superconducting state of one of the highest T_c filled skutterudites and more specifically into the characteristics of its gap function. A very wide temperature range, over which the superconducting properties could be traced, offered a possibility to better ascertain the functional forms of various superconducting parameters. A large jump in the specific heat at T_c, $\Delta C_c/\gamma_n T_c = 2.26$, indicating a strong-coupling superconductor, was confirmed by the value of $\Delta(0)/k_B T_c \approx 2.18$, considerably larger than the 1.76 expected for a BCS weak-coupling superconductor. A wide superconducting range also helped in an easy determination of the residual electronic specific heat as small as 0.1 mJmol^{-1}K^{-2}, documenting that at 1 K all but 0.17% of the conduction electrons are condensed into Cooper pairs. However, the expected single superconducting gap did not provide a good fit to the C_e/T temperature dependence, while a fit with two gap functions of the BCS-type, $\Delta_1(0)/k_B T_c = 2.61$ and $\Delta_2(0)/k_B T_c = 1.85$ with the respective weights of 40% and 60%, resulted in an excellent fit. This suggested that $LaRu_4As_{12}$ is a multiband superconductor (MSBC), a feature we will see in a few other superconducting skutterudites in the sections to follow.

A parameter that contains useful information about the superconducting state is the magnetic field-dependent specific heat coefficient $\gamma(H)$ associated with quasiparticles that may be present in the superconducting condensate. For s-wave superconductors at fields $H < H_{c1}$, i.e., in the Meissner state, and at very low temperatures, essentially all quasiparticles have condensed into Cooper pairs and $\gamma(0) \rightarrow 0$. However, once the field exceeds H_{c1} and a superconductor enters the mixed state, vortices start to penetrate, and their density increases linearly with the magnetic field. Because the vortices have basically normal electrons in their cores, the number of quasiparticles also increases linearly with field. At magnetic fields approaching H_{c2}, the vortices start to overlap and, at $H = H_{c2}$, all Cooper pairs are destroyed. The electronic specific heat (the Sommerfeld coefficient) is then field independent ($\gamma(H_{c2}) = \gamma_n$). In contrast, for superconductors with nodes in the gap function, quasiparticles (essentially normal electrons) exist even at $T = 0$ K, making $\gamma \neq 0$. Moreover, the field dependence of $\gamma(H)$ is altered from the linear behavior to a distinctly sublinear dependence. For instance, for nodal d-wave pairing superconductivity, $\gamma(H)$ is expected to vary as $H^{1/2}$, according to Nakai et al. (2004). An example of the magnetic field-dependent electronic specific heat coefficient of $LaRu_4As_{12}$ plotted as γ/γ_n vs. $H/H_{c2}(0)$ is shown in Figure 2.4. The solid line through the data is

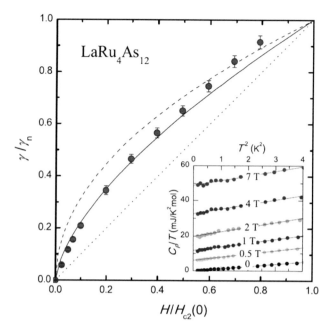

FIGURE 2.4 Normalized magnetic field-dependent electronic specific heat coefficient for a strong-coupling LaRu$_4$As$_{12}$ single crystal (filled red circles). The solid line through the data is a fit of the form $H^{0.65}$. The dotted and dashed lines are predictions of the field dependence of $\gamma(H)$ for isotropic s-wave and nodal d-wave superconductors, respectively. The inset shows the specific heat of LaRu$_4$As$_{12}$ in selected magnetic fields plotted as C_p/T vs. T^2. For each applied field, $\gamma(H) \equiv \lim_{T\to 0} C_p(T, H)/T$ was determined from a linear extrapolation of the experimental data obtained in the temperature range of 0.39 K–2 K. Reproduced from L. Bochenek et al., *Physical Review B* **86**, 060511(R) (2012). With permission from the American Physical Society.

a fit corresponding to the field dependence $\gamma(H) \sim H^{0.65}$. For comparison, the dotted line predicts the linear behavior for an isotropic s-wave superconductor, and the dashed line stands for the $H^{1/2}$ dependence expected for a nodal d-wave superconductor. Unfortunately, although one can fit a particular sublinear dependence, the differentiation between various superconducting gap functions and the multi-band forms of superconductivity is usually not sharp enough to make a definitive gap identification based on the field dependence of γ. However, given the negligibly small residual electronic specific heat, the field dependence of $\gamma(H)$ for LaRu$_4$As$_{12}$ points to either a highly anisotropic superconducting band or a multi-band superconductivity, the latter being consistent with the very good two-band fit of the temperature-dependent $C_e/\gamma_n T_c$ mentioned above. Further insight came from the temperature-dependent behavior of the upper critical field $H_{c2}(T)$. Again, an advantage of the high T_c offered three decades over which the functional form of $H_{c2}(T)$ could be conveniently followed in resistance measurements that extended down to a few tens of millikelvin using a dilution refrigerator. The data are displayed in Figure 2.5, where the experimental points (solid red circles) were taken at 50% values of the normal state resistance, and the transition temperatures from the specific heat measurements in magnetic field (open squares) were determined by equal entropy construction, examples of which are the three plots in the upper right corner of the figure. While the initial slope $-(dH_{c2}(T)/dT)|_{T_c} = 0.66$ TK^{-1} (the dashed line), the slope below 9 K is notably larger, giving the data a distinct positive curvature, typical of unconventional superconductors and often described in terms of two-band superconducting models. Resistance measurements of the upper critical field also established that the superconducting state possesses only a small anisotropy of ~

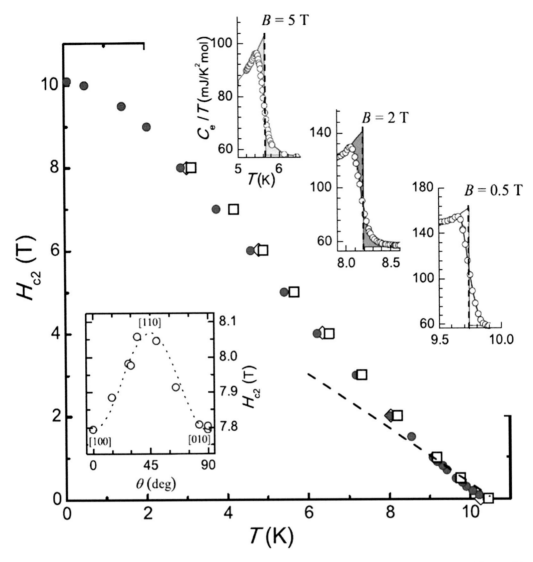

FIGURE 2.5 The upper critical field $H_{c2}(T)$ as a function of temperature for LaRu$_4$As$_{12}$ obtained from mid-point resistivity transitions (solid red circles) and equal entropy constructions (three panels above the main data) of transition temperatures in the specific heat (open squares). Open diamonds (mostly overlapped by open squares) are H_{c2} data from specific heat measurements on polycrystalline samples from Namiki et al. (2008). The inset shows the angular dependence of the upper critical field at $T = 3$ K. Reproduced from L. Bochenek et al., *Physical Review B* **86**, 060511(R) (2012). With permission from the American Physical Society.

4%, as documented by the data in the inset of Figure 2.5. Consequently, anisotropy is not an over-riding feature of the superconducting state in LaRu$_4$As$_{12}$, and attention was steered more toward the multi-band nature of superconductivity in this skutterudite. Indeed, there are other superconducting skutterudites where the multi-band character of the superconducting state is highly suspected, and one of them is LaOs$_4$Sb$_{12}$. Although the early studies by Shirotani et al. (2000) and the supercon-ducting properties of LaOs$_4$Sb$_{12}$ obtained from measurements of Pr$_{1-x}$La$_x$Os$_4$Sb$_{12}$, where LaOs$_4$Sb$_{12}$ is an end member of the solid solution series for $x = 1$, were consistently pointing at a conventional *s*-wave BCS-like superconductor with $T_c = 3.2$ K, more recent measurements for high-quality single crystals of $LaOs_4Sb_{12}$, extending to sub-Kelvin temperatures, suggest that the superconducting state in this skutterudite is more complex than once thought. Measurements of the penetration depth down to 35 mK by Tee et al. (2012) using a tunnel-diode-oscillator technique indicated an exponential behavior in the deviation of the penetration depth $\Delta\lambda = \lambda(T) - \lambda(0)$, consistent with the asymptotic approach expected for an isotropic s-wave superconductor,

$$\Delta(\lambda) = \lambda(0)\left(\frac{\pi\Delta(0)}{2k_BT}\right)^{1/2}\exp\left(-\frac{\Delta(0)}{k_BT}\right), \tag{2.22}$$

and the fit yielded $\Delta(0) = 1.34\ k_BT_c$, significantly smaller than the weak-coupling BCS value of $1.76\ k_BT_c$. From the penetration depth data, the temperature dependence of the superconducting condensate $\rho_s(T)$ follows from $\rho_s(T) = \lambda^2(0)/\lambda^2(T)$, and its T-dependence displayed a long tail (suppression of the condensate) near T_c, see Figure 2.6. While a much smaller value of $\Delta(0)$ could be due to either an anisotropic gap or multiple gaps, the suppression of the condensate near T_c has nothing to do with anisotropy but can be due to the presence of a smaller superconducting gap. Fitting the data to a two-band superconducting model with both gaps taken as having an s-wave symmetry, the fit returned the gap values of $\Delta_1(0) = 1.69\ k_BT_c$ and $\Delta_2(0) = 1.30\ k_BT_c$ with a weight ratio of 3.13 in favor of the smaller gap. Perhaps not surprisingly, the value of the second gap of $1.30\ k_BT_c$ matched very well with the value $\Delta(0) = 1.34\ k_BT_c$ from the low-temperature data of $\Delta\lambda(T)$ above, because at low temperatures it is the smaller gap that controls quasiparticle excitations across the gap. The larger gap of $1.69\ k_BT_c$, on the other hand, is very close to the BCS value of $\Delta(0) = 1.76\ k_BT_c$. The two-band model with the above gap magnitudes and the weight fitted the temperature dependence of the condensate very well, and the inset in Figure 2.6 shows the temperature dependence of each

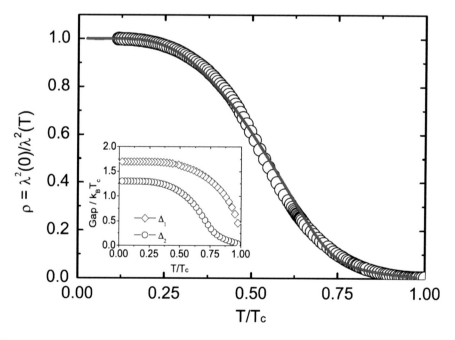

FIGURE 2.6 Experimental data (open circle) and theoretical fit (solid curve) for the normalized superfluid density of $LaOs_4Sb_{12}$. The inset shows the temperature dependence of the two energy gaps from the theoretical fit. Reproduced from X. Y. Tee et al., *Physical Review B* **86**, 064518 (2012). With permission from the American Physical Society.

gap separately. From the inset, it also follows that the suppression of the condensate density near T_c is caused by the vanishing of the smaller gap Δ_2 at higher temperatures due to its interband coupling with the larger gap Δ_1. Subsequent detailed studies of the specific heat in zero and non-zero magnetic fields on single crystals of $LaOs_4Sb_{12}$ by Juraszek et al. (2016) generated results that were also consistent with a two-gap superconducting model. In this case, the best fitted values of the gaps were $\Delta_1(0)/k_BT_c = 1.87$ and $\Delta_2(0)/k_BT_c = 1.23$ with a 40:60 split between the respective weights. Moreover, the upper critical field $H_{c2}(T)$ has displayed a similar positive curvature near T_c as in the case of $LaRu_4As_{12}$. Consequently, both $LaRu_4As_{12}$ and $LaOs_4Sb_{12}$ show very similar behavior that has all the telltale signs of a multi-band superconducting state.

Over a period of the past 30 years or so, a plethora of new superconducting filled skutterudites have been discovered, and they are listed in Table 2.1 with their respective superconducting temperatures and lattice constants. Where available, the table also provides values of the pressure derivative of the transition temperature and the initial slope of the dependence of the upper critical magnetic field H_{c2} on temperature.

A couple of comments are in order regarding the entries in Table 2.1. First, the table displays some quite high superconducting transition temperatures attained with filled skutterudites, particularly so for $LaRu_4As_{12}$ where the T_c is above 10 K. The reader may notice even a couple of much higher values of $T_c = 14.9$ K and $T_c = 17$ K reported for $La_{0.8}Rh_4P_{12}$ and $La_{0.6}Rh_4P_{12}$, respectively, by the same research team in papers published two years apart. The structure, in this case, is not the fully filled RT_4P_{12} skutterudite with T = Fe but, rather, a partially filled $R_yM_4P_{12}$ skutterudite with M = Rh, where the filling fraction of La was dramatically enhanced beyond the usual 20% filling limit because the samples were synthesized at high pressures of 9.4 GPa and 4 GPa, respectively. Although the higher synthesis pressure apparently squeezed in a larger fraction of fillers, the T_c of $La_{0.8}Rh_4P_{12}$ is actually lower by a couple of degrees compared to $La_{0.6}Rh_4P_{12}$ synthesized at less than half the pressure and having significantly less La in the voids. There might be an optimum La filling that maximizes the superconducting transition temperature, but what this magic filling fraction is awaits future experiments. Imai et al. (2007a) measured magnetization, electrical resistivity, and specific heat of a sample of $La_{0.8}Rh_4P_{12}$ and estimated the upper critical fields $H_{c2}(0) \approx 16.7$ T, the coherence length $\xi(0) \approx 44.1$ Å, the electronic specific heat coefficient $\gamma_n \approx 25.1$ mJmol^{-1}K^{-2}, and the electron–phonon coupling constant $\lambda_{e-p} \sim 0.7$, indicating a superconductor with medium strength of coupling.

The second comment concerns the sign of the pressure derivative of the transition temperature. The table shows two entries, $LaFe_4P_{12}$ and YFe_4P_{12}, where the pressure derivative is positive, i.e., the transition temperature of these two filled skutterudites increases with the increasing pressure. This is highly unusual as BCS theory usually assumes a negative pressure derivative of the superconducting transition temperature because of the stiffening of the crystalline lattice subjected to external pressure. Indeed, this is the case of all other entries in Table 2.1 for which the pressure derivative was measured. Prior to having available detailed band structure calculations for $LaFe_4P_{12}$ and YFe_4P_{12}, DeLong and Meisner (1985) offered a simple intuitive explanation why the pressure derivative of $LaFe_4P_{12}$ (and this presumably would apply to the later discovered YFe_4P_{12} also) might be positive. They assumed that there are two distinct and competing mechanisms present: i) as the crystal lattice shrinks under the influence of external pressure, the phonon modes stiffen and both the Debye temperature and the average of the squared phonon frequencies $\langle\omega^2\rangle$ increase. While in the weak-coupling limit of the BCS theory, a larger Debye temperature implies a larger T_c, the enhancement in $\langle\omega^2\rangle$ carries more weight and, as such, it strongly reduces the electron–phonon coupling constant, the essential component in the BCS theory. In turn, the transition temperature falls with the increasing pressure. ii) It is well known that pure La metal, Gardner and Smith (1965), and many of the La-based intermetallic compounds, Smith and Luo (1967), show an enhancement of T_c with applied pressure. Consequently, how $LaFe_4P_{12}$ and, presumably also YFe_4P_{12}, will respond to the external pressure depends on whether the pressure "affects" La and Y directly or whether the two kinds of fillers are substantially shielded from the pressure by the framework of $[Fe_4P_{12}]$. In the former case,

TABLE 2.1

Lattice Parameter a, Superconducting Transition Temperature T_c, the Pressure Derivative $(dT_c/dP)_{P=0}$ in Units of K(GPa)$^{-1}$, the Temperature Derivative of the Critical Field $-(dH_{c2}/dT)_{T_c}$ in Units of TK^{-1}, and the Synthesis Method Used in the Preparation of Filled Skutterudites. HPHT Stands for the High-Pressure High-Temperature Synthesis

Compound	a(Å)	T_c (K)	Synthesis	$(dT_c/dP)_{P=0}$	$-(dH_{c2}/dT)_{T_c}$	Reference
LaFe$_4$P$_{12}$	7.8316	4.1	Sn flux	$+7.2 \times 10^{-1}$		Meisner (1981)
LaRu$_4$P$_{12}$	8.0561	7.2	Sn flux	-1.6×10^{-1}		Meisner (1981)
LaRu$_4$P$_{12}$	8.0605	7.2	HPHT	–	–	Uchiumi et al. (1999)
LaRu$_4$As$_{12}$	8.5081	10.3	HPHT	-4.0×10^{-1}	0.08	Shirotani et al. (2000)
LaRu$_4$As$_{12}$	–	10.45	Cd/As flux	–	0.66	Bochenek et al. (2012)
LaRu$_4$Sb$_{12}$	9.2781	2.8	HPHT	–	0.21	Uchiumi et al. (1999)
LaRu$_4$Sb$_{12}$	9.2740	3.58	Sb flux	–	0.12	Takeda and Ishikawa (2000)
LaOs$_4$P$_{12}$	8.0844	1.8	Sn flux	-9.5×10^{-2}	–	DeLong and Meisner (1985)
LaOs$_4$As$_{12}$	8.5437	3.2	HPHT	-4×10^{-1}	–	Shirotani et al. (2000)
LaOs$_4$Sb$_{12}$	9.3029	0.74	Sb flux	–	0.095	Aoki et al. (2005)
La$_{0.6}$Rh$_4$P$_{12}$	8.0581	17	HPHT	–	–	Shirotani et al. (2005b)
La$_{0.8}$Rh$_4$P$_{12}$	8.0785	14.9	HPHT	-5×10^{-1}	1.2	Imai et al. (2007a)
CaOs$_4$P$_{12}$	8.0840	2.5	HPHT	$\sim -5 \times 10^{-1}$	0.88	Kawamura et al. (2018)
BaOs$_4$P$_{12}$	8.1240	1.8	HPHT	–	–	Deminami et al. (2017)
YFe$_4$P$_{12}$	7.7891	~ 7	HPHT	–	–	Shirotani et al. (2003a)
YFe$_4$P$_{12}$	7.7913	5.6	HPHT	$+1.0$	–	Cheng et al. (2013)
YRu$_4$P$_{12}$	8.0298	8.5	HPHT	–	–	Shirotani et al. (2005a)
YOs$_4$P$_{12}$	8.0615	~3	HPHT	–	–	Kihou et al. (2004)
PrRu$_4$P$_{12}$	8.0516	~2[a]	Sn flux	–	–	Miyake et al. (2004)
PrRu$_4$As$_{12}$	8.4963	2.4	HPHT	–	0.43[c]	Shirotani et al. (1997)
PrRu$_4$Sb$_{12}$	9.2648	1.1	HPHT	-2.1×10^{-1}	0.24[b]	Arii et al. (2008)
PrOs$_4$Sb$_{12}$	9.3017	1.85	Sb flux	-1.5×10^{-1}	1.9	Bauer et al. (2002)
Ba$_{0.89}$Ir$_4$P$_{12}$	8.1071	5.6	HPHT	–	–	Qi et al. (2017)
Ba$_{0.85}$Ir$_4$As$_{12}$	8.5605	4.8	HPHT	–	–	Qi et al. (2017)
SrPt$_4$Ge$_{12}$	8.6601	5.10	Arc melt.	–	0.31	Bauer et al. (2007)
BaPt$_4$Ge$_{12}$	8.6928	5.35	Arc melt.	–	0.53	Bauer et al. (2007)
LaPt$_4$Ge$_{12}$	8.6235	8.27	Arc melt.	–	0.19	Gumeniuk et al. (2008a)
PrPt$_4$Ge$_{12}$	8.6111	7.91	Arc melt.	-1.9×10^{-1d}	0.23	Gumeniuk et al. (2008a)
ThPt$_4$Ge$_{12}$	8.5924	4.62	Arc melt.	–	0.085	Kaczorowski (2008)

[a] The transition temperature measured under an applied pressure of 12 GPa.

[b] From Chia et al. (2004)

[c] From Sayles et al. (2010)

[d] From Foroozani et al. (2013)

the direct effect of pressure on La dominates, and the T_c should increase with pressure. In the latter case, a universal decrease of T_c should result as the pressure primarily compresses the framework rather than modifies the atomic structure of La. However, unlike the case of many La intermetallics where the transition temperature is, indeed, governed by the La–La distance (around 3.5 Å), in skutterudites the La–La separation is huge, standing at some 6.8 Å, and the Y–Y ions are comparably separated. This casts a serious doubt about La or Y playing the crucial role in the superconductivity of $LaFe_4P_{12}$ and YFe_4P_{12}. As reliable DFT-computed electronic band structures became available, it was clear that the key ingredient for superconductivity in skutterudites rests with the covalently bonded pnicogen rings that dominate the electronic DOS at the Fermi level. Detailed experimental studies by Cheng et al. (2013), looking closely into the pressure derivative of the transition temperature of YFe_4P_{12} by extending the range of external pressures to 8 GPa, with support from their DFT computations of the band structure, have confirmed that the positive pressure derivative arises from a dramatic enhancement of the DOS at the Fermi level $D(E_F)$ with applied pressure, at least for pressures up to 8 GPa. In comparison, similar band structure calculations for YRu_4P_{12} predicted negative dTc/dP. However, experimental verification of the sign of the pressure derivative for YRu_4P_{12} has not yet been obtained.

Table 2.1 also lists two filled skutterudites of the form $Ba_yM_4X_{12}$, where M = Ir and X = P and As with $y = 0.89$ and 0.85, respectively. The structures were prepared under a high pressure of 5 GPa resulting in the voids filled well above the filling limit of Ba inherent to ambient pressure synthesis conditions. Apparently, as the content of Ba increases in the voids of the $[Ir_4P_{12}]$ and $[Ir_4As_{12}]$ frameworks, the ground state of the skutterudite undergoes a dramatic transformation from a degenerate semiconductor to a substantially insulating structure followed by a non-metal-to-metal transition near $y = 0.5$ and, eventually, a superconducting state develops above $y = 0.67$. A sketch of the phase diagram for $Ba_yIr_4As_{12}$ is presented in Figure 2.7. The superconducting transitions are, respectively, 5.6 K for $Ba_{0.89}Ir_4P_{12}$ and 4.8 K for $Ba_{0.85}Ir_4As_{12}$. It is tantalizing to think what the eventual transition temperature might have been had an even larger synthesis pressure been used that would have presumably resulted in a 100% filling of the structural voids.

Finally, I wish to point out two most recent entries in Table 2.1, Ca- and Ba-filled superconductors with the $[Os_4P_{12}]$ framework. Again, both structures were synthesized under a high pressure of 3.5 GPa–4 GPa at temperatures around 1050°C and show superconducting transitions at 2.5 K and 1.8 K, respectively, Kawamura et al. (2018) and Deminami et al. (2017). Both are classified as weakly coupled, type-II superconductors of the BCS character with an upper critical field of 2.2 T

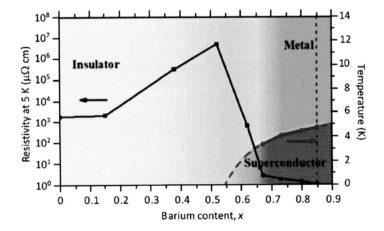

FIGURE 2.7 Phase diagram depicting the regimes of conductivity as a function of Ba content in the voids of the $[Ir_4As_{12}]$ skutterudite framework. Reproduced from Y. Qi et al., *Journal of the American Chemical Society* **139**, 8106 (2017). With permission from the American Chemical Society.

and less than 1 T, respectively. The more closely studied $CaOs_4P_{12}$ has the coherence length $\xi = 12$ nm and the penetration depth $\lambda = 400$ nm. The electron–phonon coupling constant λ_{ep} estimated by the McMillan formula, Equation 2.1 is, about 0.47.

2.4   SUPERCONDUCTIVITY OF $PrOs_4Sb_{12}$

We now return to the case of Pr-filled skutterudites and, specifically, to superconducting $PrOs_4Sb_{12}$, discovered by Maple et al. (2002) and Bauer et al. (2002). This interesting superconductor with a transition temperature of $T_c \approx 1.85$ K involves heavy fermions with effective masses equivalent to a mass of at least 50 free electrons. As such, it is the first Pr-based heavy fermion superconductor and represents an important new addition to the family of heavy fermion superconductors that, until 2002, comprised Ce- and U-based compounds only. Intensive research efforts over the past 30 years identified the hybridization of the conduction electrons with the f-electron orbitals of Ce and U ions in these two highly electron correlated systems as the essential ingredient that gives the itinerant particles their heavy effective mass. Technically, this is described in terms of the Kondo screening in the lattice of magnetic ions, e.g., Affleck (2010). The unique status of $PrOs_4Sb_{12}$ and its highly unusual transport and superconducting properties have generated tremendous scientific interest worldwide that continues to fascinate the community till this day. As the studies progressed, more and more intriguing findings have emerged, documenting that this skutterudite structure is a truly unique unconventional superconductor that does not conform to the theories developed for Ce- and U-based heavy electron superconductors.

2.4.1   Crystalline Electric Field in $PrOs_4Sb_{12}$

First of all, like the majority of other rare earths, Pr enters the voids of the skutterudite framework as a trivalent ion; i.e., it has two f-electrons in its valence shell, which are generally viewed as more localized and less hybridized with conduction electrons than in the case of Ce f-electrons on account of the larger binding energy. The Pr^{3+} state has been independently verified by susceptibility measurements, Bauer et al. (2002), that yielded the effective magnetic moment $\mu_{eff} = 2.97$ μ_B. Although not a perfect agreement with a free Pr^{3+} ion value of 3.58 μ_B, the trivalent state is the closest to the experimental value.

Skutterudites are classified as cubic structures and so is $PrOs_4Sb_{12}$. However, the local point symmetry of the Pr^{3+} ion sitting in the center of an icosahedral cage of Sb ions is the tetrahedral T_h rather than the cubic O_h and is missing an element of the fourfold rotation axis, Takegahara et al. (2001). Taking into account both spin-orbit coupling and Hund's-rule coupling, the Pr^{3+} ion has the total angular momentum $J = 4$. In free space, this implies a $2J + 1 = 9$-fold degeneracy. The local CEF lifts the degeneracy and results in a non-magnetic singlet designated as Γ_1, a non-magnetic doublet Γ_3, and two magnetic triplet states Γ_4 and Γ_5, the designation based on the cubic O_h symmetry group that is often used in the literature. In terms of the proper tetrahedral point group symmetry T_h, the corresponding states are labeled as Γ_1, Γ_{23}, $\Gamma_4^{(1)}$, and $\Gamma_4^{(2)}$. The crystal field levels in both the cubic and tetrahedral symmetries have the same degeneracy and comparable energy splitting. This is especially true for the all-important energy spacing of the ground state and the first excited state, with values of about 10 K and 8.2 K, respectively, on a temperature scale. Consequently, one can relate the two symmetry classes by $\Gamma_1(O_h) \rightarrow \Gamma_1(T_h)$; $\Gamma_3(O_h) \rightarrow \Gamma_{23}(T_h)$; $\Gamma_4(O_h) \rightarrow \Gamma_4^{(1)}(T_h)$; and $\Gamma_5(O_h) \rightarrow \Gamma_4^{(2)}(T_h)$. Indeed, treating the Pr^{3+} ion as having the locally cubic symmetry O_h is fine in most of the situations except in the case of assigning the order parameter to the magnetic field-induced phase of $PrOs_4Sb_{12}$, discussed shortly, where the use of the tetrahedral T_h symmetry is absolutely necessary. Throughout the text, I am using the tetrahedral T_h point group to describe the CEF-split states of the Pr^{3+} ion, but will also make a reference to the O_h cubic point group.

To glean the role of f-electrons in the superconductivity, the question of vital importance is which state has the lowest energy and forms the ground state of the system. From low-temperature susceptibility, and more specifically the diamagnetic nature of the $PrOs_4Sb_{12}$ crystal as it cools through its superconducting transition temperature, Bauer et al. (2002) established that the ground state is non-magnetic. This represented an entirely different situation than the one prevailing in essentially all unconventional superconductors, where magnetic interactions were believed to play the pivotal role in the formation of Cooper pairs. The non-magnetic ground state of the Pr^{3+} ion could be satisfied by both the Γ_1 singlet and the Γ_{23} doublet states. In the early studies, Maple et al. (2002), Vollmer et al. (2003), Maple et al. (2003), Miyake et al. (2003), Kotegawa et al. (2003), and many others, the Γ_{23} doublet (Γ_3 in the O_h notation) was favored on account of a possibility that it might give rise to quadrupolar Kondo fluctuations as the origin of heavy fermions, in a similar fashion to that in some heavy fermion systems, Cox and Zawadowski (1998). However, with reports of the crossover to a field-induced ordered state, Aoki et al. (2003b), Tayama et al. (2003), Kohgi et al. (2003), and Rotundu et al. (2004), the Γ_1 singlet started to gain preference as a better fitting option. The issue was settled by the inelastic neutron scattering studies of Goremychkin et al. (2004) and Kuwahara et al. (2004) that probed the temperature dependence of the crystal field transitions in $PrOs_4Sb_{12}$, and by the studies of Goto et al. (2005), who measured the elastic constants of $PrOs_4Sb_{12}$ in magnetic fields and compared the field-dependent quadrupolar susceptibility, $\chi_Q(O_2^2)$, with models based on $\Gamma_1 - \Gamma_4^{(2)}$ and $\Gamma_{23} - \Gamma_4^{(2)}$ energy level schemes. In all cases, the Γ_1 singlet was unequivocally identified as the ground state. The first excited state, a magnetic triplet $\Gamma_4^{(2)}$ (magnetic triplet Γ_5 in the O_h representation) sitting about 8.2 K above the ground state, was never in doubt. Higher energy levels lie more than 100 K above and, as such, are of no consequence for low-temperature phenomena.

2.4.2 Specific Heat of $PrOs_4Sb_{12}$

A plethora of experimental techniques have been applied to studies of $PrOs_4Sb_{12}$ to shed light on its superconducting properties, apart from its ground state. I shall focus here mainly on the pivotal studies that greatly advanced our understanding of the superconducting state of this fascinating filled skutterudite. I start with the specific heat, the measurement instrumental in revealing many of the unique properties of $PrOs_4Sb_{12}$.

As previously mentioned, the first report pointing out the heavy fermion nature of conduction electrons in $PrOs_4Sb_{12}$ came from measurements by Maple et al. (2002) and Bauer et al. (2002), where equal entropy construction (entropy is conserved just above and below T_c) was used to estimate the magnitude of the specific heat anomaly that indicated large values of $\Delta C/T_c$ of 500 $mJmol^{-1}K^{-2}$, implying enhanced effective masses of quasiparticles $m^* \sim 50\ m_e$. Moreover, the ratio $\Delta C_{sc}/\gamma T_c \approx 3$, where γ is the coefficient of the electronic specific heat (Sommerfeld coefficient), implied a strongly coupled superconducting state compared to $\Delta C_{sc}/\gamma T_c = 1.43$ predicted for the weak-coupling BCS superconducting state. Specific heat measurements of $PrOs_4Sb_{12}$ were subsequently performed several times, Vollmer et al. (2003), Aoki et al. (2003a), Méasson et al. (2004), Seyfarth et al. (2006), Méasson et al. (2008), Briarty et al. (2009), and Andraka and Pocsy (2012), and all showed a similar trend. An example of the specific heat, plotted as C/T vs. T for a range of $PrOs_4Sb_{12}$ crystals, all having the superconducting transition temperature $T_c \sim 1.85$ K, is shown in Figure 2.8.

A reader used to analyzing specific heat data of typical solids with their linear (electronic) and cubic (lattice) terms may find the temperature dependence of the specific heat of $PdOs_4Sb_{12}$ rather strange. It is given by a huge contribution of the CEF-excited two-level system (Schottky anomaly) that starts to dominate the energy landscape below 8 K, reaches a peak value around 2 K, and exponentially decreases thereafter as the thermal energy is no longer comparable to the energy

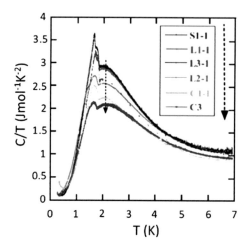

FIGURE 2.8 Specific heat plotted as *C/T vs. T* near $T_c = 1.85$ K for several $PrOs_4Sb_{12}$ crystals grown by the Sb flux method at three different laboratories (C standing for Canfield, S for Sugawara, and L for Lapertot). Samples C1-1 and C3 show a broad single transition while the rest of the samples display two transitions. Modified from M.-A. Méasson et al., *Physical Review B* **77**, 134517 (2008). With permission from the American Physical Society.

spacing between the ground state Γ_1 and the first excited state $\Gamma_4^{(2)}$. Although easily distinguished, the proximity of the superconducting transition to the peak value of the Schottky anomaly makes the signature mark of superconductivity somewhat less pronounced than in an ordinary *s*-wave super-conductor. The CEF-excited $\Gamma_4^{(2)}$ level also dominates the profile of the magnetic susceptibility and, as we will see later, is also reflected in the *T*-dependence of the spin-lattice relaxation rate in the normal state of $PrOs_4Sb_{12}$.

One of the surprising and long debated features in the specific heat measurements was an observation by Maple et al. (2002) and Vollmer et al. (2003) of two closely lying superconduct-ing transitions, $T_{c1} = 1.85$ K and $T_{c2} = 1.75$ K, rather than just one. The two-transition structure at virtually the same temperature was seen also in magnetic susceptibility by Vollmer et al. (2003), Méasson et al. (2004), and Cichorek et al. (2005), in dc magnetization measurements by Tayama et al. (2003), and thermal expansion measurements by Oeschler et al. (2004), to mention a range of experimental techniques. The distinction between the genuine two-transition temperature case, the single sharp superconducting transition, and a single broad superconducting transition is sketched in Figure 2.9. If substantiated in stoichiometric single crystals, the two-transition feature might be yet another footprint of the unconventional nature of superconductivity in $PrOs_4Sb_{12}$ as it would signal the presence of two distinct superconducting order parameters. Indeed, this feature was taken into account in many theoretical descriptions of superconductivity. However, it should be kept in mind that the real intrinsic double transition in a stoichiometric superconductor is a very rare event and, so far, has been confirmed only in UPt_3, Fisher et al. (1989), all other reported cases being the result of inhomogeneous structures. Because specific heat measurements are perhaps the best and clearest indication of the nature of superconducting transitions, a concerted effort has been made over a number of years to sort out this issue, Vollmer et al. (2003), Aoki et al. (2003a), Méasson et al. (2004), Seyfarth et al. (2006), Méasson et al. (2008), Briarty et al. (2009), and Andraka and Pocsy (2012).

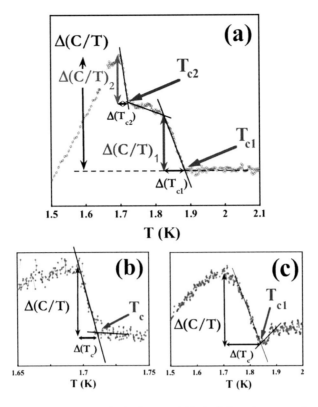

FIGURE 2.9 Illustration of (a) double, (b) single sharp, and (c) single broad superconducting transitions in the specific heat of $PrOs_4Sb_{12}$ single crystals. Specific heat jumps $\Delta(C/T)$ for respective transitions are illustrated together with the width of the transition ΔT. Composed from figures in M.-A. Méasson et al., *Physical Review B* **77**, 134517 (2008). With permission from the American Physical Society.

More insight into the origin of double transitions in $PrOs_4Sb_{12}$ was provided by studies of the specific heat under high pressure, to see if the two superconducting transitions respond the same way or differently to the applied pressure. The latter case would be strong evidence for the intrinsic nature of the two transitions. I should point out that such measurements are quite tedious as one must rely on ac techniques with a small sample quasi-hydrostatically compressed by pressure generated by opposed diamonds of a small diamond anvil cell and the pressure transmitted to the sample *via* a gasket filled with a suitable pressure-transmitting medium. The sample is typically heated by optical means, and the diamond cell is attached on a cold tip of a cryostat to cover the low temperatures of interest. Méasson et al. (2007) used such a technique to monitor relative changes in the two transition temperatures T_{c1}-T_{c2} as a function of pressure and the respective pressure derivatives dT_{c1}/dP and dT_{c2}/dP. The data are shown in Figure 2.10. The two pressure derivatives differ by about 20%, having magnitudes of 0.31 K(GPa)$^{-1}$ for dT_{c2}/dP and 0.23 K(GPa)$^{-1}$ for dT_{c1}/dP. This, together with a clearly rising T_{c1}-T_{c2} at pressures below 1 GPa, might imply a different behavior of the two superconducting transitions and, hence, their intrinsic origin. However, above 1 GPa, the difference in T_{c1}-T_{c2} settles at a constant value of about 200 mK (the width of each transition being about 35 mK), which the authors interpreted as an unlikely trend for the intrinsic origin of the double transitions. Moreover, they observed at least three high-quality crystals with just a

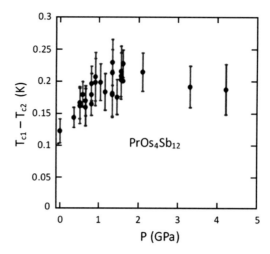

FIGURE 2.10 Pressure dependence of the temperature difference T_{c1}-T_{c2} between the two superconducting transition in $PrOs_4Sb_{12}$. The error bars are the average spacing of the two transitions. At ambient pressure, T_{c1} = 1.85 K and T_{c2} = 1.73 K, making T_{c1}-T_{c2} equal to 120 mK, with a transition width of about 35 mK. With increasing pressure, T_{c1}-T_{c2} clearly increases, reaching 200 mK at 1 GPa, before more or less maintaining this value at high pressures. Reproduced from M.-A. Méasson et al., *Journal of Magnetism and Magnetic Materials* **310**, 626 (2007). With permission from Elsevier.

single sharp superconducting transition at T_c that matched the value of T_{c2}. The two superconducting transitions therefore appeared to be an artifact of sample inhomogeneity. To press the point further, trying to come up with a feasible extrinsic origin of the two observed superconducting transitions, they invoked a plausible scenario of incomplete filling of cages by Pr. The presence of vacant void sites would create a negative chemical pressure around them, and because T_c decreases strongly with pressure, near such vacant voids the transition temperature may be somewhat higher than that in the bulk of the crystal. If one interprets the higher transition temperature as T_{c1}, and the transition temperature of the bulk as T_{c2}, one has a scenario for the extrinsic origin of the two transition temperatures.

Moreover, because the bulk modulus is expected to be smaller when all voids are filled with Pr, the slope dT_{c2}/dP for the fully filled part of the sample should be larger in magnitude than the slope dT_{c1}/dP corresponding to small regions of the sample with empty voids as, indeed, they observed. Of course, what is missing in the hypothesis is solid evidence for the actual density change due to the empty voids in $PrOs_4Sb_{12}$. However, some credence to the void-induced double T_c transition came from measurements of the specific heat, electrical resistivity, and susceptibility by Tanaka et al. (2009) on $PrOs_4Sb_{12}$ single crystals grown by the Sb flux technique under a high pressure of about 4 GPa. Such crystals have a higher Pr occupancy of voids of 96% compared to the usual 93% void filling in Sb flux grown crystals under ambient pressure. The higher Pr content in the structure leads to a marked decrease in T_c by as much as 0.2 K, a considerably narrower transition width ΔT_c by a factor of 3, significantly decreased effective mass, and a strong enhancement of the crystal field energy splitting Δ_{CF} between the ground state and the first excited state from the usual 8 K to about 13–14 K. The much sharper single transition in the electrical resistivity of these high pressure-prepared $PrOs_4Sb_{12}$ crystals is depicted in Figure 2.11 where, for comparison, a typical transition for a crystal synthesized under ambient pressure is also shown.

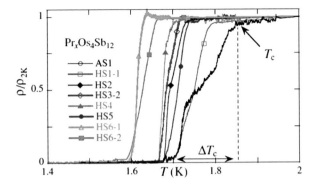

FIGURE 2.11   Superconducting transitions in the electrical resistivity of $PrOs_4Sb_{12}$ single crystals. The sample labeled AS is a crystal grown from Sb flux under ambient pressure and shows a broad, clearly defined double transition. Samples designated HS are single crystals grown from Sb flux under a pressure of 4 GPa. They have a distinctly higher Pr filling fraction that is reflected in lower temperature onsets of the superconducting transition, much narrower transition widths, and no signature of the presence of two transitions. Reproduced from K. Tanaka et al., *Journal of the Physical Society of Japan* **78**, 063701 (2009). With permission from the Physical Society of Japan.

To me, the most convincing experimental evidence for a single superconducting transition temperature of $PrOs_4Sb_{12}$ is, together with the above work on high pressure-synthesized samples, measurements of Seyfarth et al. (2006), Briarty et al. (2009), and Andraka and Pocsy (2012), all of whom searched for what happens to the specific heat or susceptibility at the transition temperature when a cluster of single crystals is progressively reduced in size to platelets of varying thickness by cutting or abrasion and, ultimately, when the crystal is powdered. In all such cases, the original two-transition nature of superconductivity of a crystal gradually gives way to a single superconducting transition as the thickness of the crystal is reduced to platelet sizes of about 50 μm or the powders are diminished to a typical size of 10–20 μm. Such studies also dispelled several scenarios attempting to explain the puzzling double-transition nature. Because the chemical content of the crystal and the powder made from it is identical, the double-transition is highly unlikely to arise from the presence of any secondary phases or impurities, as such would inevitably leave their mark on both the crystal and the powder. Although the increased void filling of Pr in high-pressure-synthesized $PrOs_4Sb_{12}$ crystals apparently sharpens the superconducting transition to the extent that only one transition is observed, this too cannot be the primary cause as the crystal and the powder made from it surely maintain the same Pr content. Thus, until someone puts a finger on and identifies the cause of the two-transition nature of superconductivity seen in the vast majority of studies and convincingly explains its origin, the controversy will likely not go away.

2.4.3   Nuclear Quadrupole Resonance in $PrOs_4Sb_{12}$

Another experimental technique that proved very important in revealing the microscopic nature of superconductivity in Ce- and U-based heavy fermion systems, and therefore was also applied to $PrOs_4Sb_{12}$, is the nuclear quadrupole resonance (NQR). The technique probes how the conduction electrons are distributed around a nucleus and extracts the relaxation time T_1 that characterizes the nuclear spin-lattice interaction. As the experiment can readily cover a wide range of temperatures, it yields information on the behavior of T_1 in both the normal and superconducting domains. Several such studies have been performed, Kotegawa et al. (2003), Nakai et al. (2005), Yogi et al. (2006), and Tou et al. (2011), and they, too, support the findings that the superconducting state in $PrOs_4Sb_{12}$ is unique. The experiments are particularly revealing when a comparison is made with an ordinary *s*-wave superconductor, such as isostructural $LaOs_4Sb_{12}$ with $T_c = 0.75$ K and no *f*-electrons, utilized in the study of Kotegawa et al. (2003). The difference in the behavior of the relaxation times can then be clearly assigned to the influence of the Pr^{3+} ion. An example is shown in Figure 2.12, where the relaxation rate $1/T_1$ obtained from the NQR spectrum of ^{123}Sb (nucleus with the nuclear spin $I = 7/2$ and about 43% natural abundance) is plotted as a function of temperature for both $PrOs_4Sb_{12}$ and $LaOs_4Sb_{12}$. There are four notable features in the data: i) while in the normal state of $LaOs_4Sb_{12}$, the usual Korringa relation T_1T = const. applies, indicating a conventional metallic system, the relaxation rate $1/T_1$ in the normal state of $PrOs_4Sb_{12}$ significantly exceeds the Korringa value. ii) The transition to the superconducting state in $LaOs_4Sb_{12}$ is marked by a distinct coherence peak (the so-called Hebel-Slichter peak observed in all *s*-wave superconductors) just below T_c, followed by exponentially diminishing relaxation frequency as the gap of magnitude $\Delta/k_BT \sim$ 1.6 opens, the value of the gap indicating ordinary weak-coupling BCS *s*-wave superconductivity. In contrast, although the relaxation frequency of $PrOs_4Sb_{12}$ falls over several orders of magnitude as the temperature drops below the $T_c = 1.85$ K, there are no hint of a coherence peak anywhere in the plot and no T^3 dependence of $1/T_1$ that is a characteristic feature of most Ce- and U-based heavy fermion superconductors. iii) The Arrhenius plot of the ratio $(T_1T)_{qp}/(T_1T)_{qp,n}$ vs. T_c/T, see the inset in Figure 2.12, where the subscript *qp* stands for quasiparticles and the subscript *qp,n* for quasiparticles in the normal state, indicates a large gap of size $2\Delta/k_BT \sim 5.4$ (strong coupling) that opens up in $PrOs_4Sb_{12}$ already at about $T_g = 2.3$ K, well above the transition at 1.85 K, as if Cooper pairs pre-formed prior to condensing. iv) A notable trend toward a constant relaxation rate sets in at temperatures below 0.6 K. This feature is of intrinsic origin as it is not suppressed by the magnetic field but, on the contrary, gets somewhat enhanced (not shown here). The $1/T_1 \approx$ const. behavior is likely related to a strong enhancement in $H_{c1}(T)$ observed in $PrOs_4Sb_{12}$ by Cichorek et al. (2005) below 0.6 K and perhaps also to a dip in the penetration depth noted by Chia et al. (2003) at the same temperature. The highly unusual features in the NQR data clearly point to unconventional strong-coupling superconductivity in $PrOs_4Sb_{12}$.

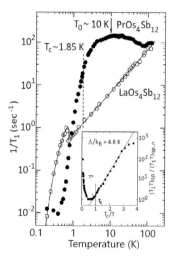

FIGURE 2.12 Temperature dependence of the relaxation rate $1/T_1$ obtained from ^{123}Sb NQR in $PrOs_4Sb_{12}$ (solid circles) and $LaOs_4Sb_{12}$ (open circles). The inset shows the Arrhenius plot of T_1T *vs.* T_c/T, indicating an exponential temperature dependence of T_1T with a gap value of $\Delta/k_BT_c = 2.7$ setting in already at $T_g = 2.3$ K, well above the superconducting transition temperature of $PrOs_4Sb_{12}$ $T_c = 1.85$ K. Adapted from H. Kotegawa et al., *Physical Review Letters* **90**, 027001 (2003). With permission from the American Physical Society.

An enormous amount of data was collected in the early years of the new millennium on the corre-lated aspects of filled skutterudites and on $PrOs_4Sb_{12}$, in particular. However, some key experimental input was still missing in order to make a reliable assessment of the nature of the superconducting state in this structure. It was clear that, in the case of $PrOs_4Sb_{12}$, one has on hand a heavy fermion system that is superconducting, but it does so in a different way from the well-researched cases of Ce- and U-based heavy fermion superconductors, for which exotic theories describing the nature of pairing, the order parameter, and the nodal structure of the superconducting band gap had been worked out over the last two decades of the twentieth century. With the development of sensitive measuring techniques, the ability to perform studies at sub-kelvin temperatures under several GPa of pressure, having available high-quality single crystals, and benefitting from improved DFT calcula-tions, the time was right to shed light on the pivotal issue of pairing, mass renormalization, and the symmetry of the superconducting order parameter.

2.4.4   HIGH FIELD-ORDERED PHASE (HFOP) IN $PrOs_4Sb_{12}$

One of the most impacting studies that provided an important clue toward identifying the mecha-nism that imparts heavy mass on electrons and enables them to stick together to form Cooper pairs was the discovery of the field-induced ordered phase in $PrOs_4Sb_{12}$. In the literature, this state is referred to as either FIOP or HFOP, the latter standing for the high field-ordered phase. A hint of a possible phase transition from the superconducting phase to a new phase induced at high magnetic fields was first noted by Maple et al. (2002) as a rapid decrease in the electrical resistivity below 1 K at a magnetic field of 6 T. It was further highlighted by a pronounced peak in $d\rho(T)/dT$ below 1 K at fields above 4.5 T observed by Ho et al. (2002). A detailed assessment of this surprising phase transition was made by Aoki et al. (2002) based on their low-temperature specific heat measure-ments on an oriented single crystal in the presence of intense magnetic field. As shown in Figure 2.8, at the superconducting transition at T_c = 1.85 K, the specific heat of $PrOs_4Sb_{12}$ acquires a distinct peak. Modest magnetic fields tend to rapidly suppress the peak, as well as dramatically diminish the Schottky anomaly, see Figure 2.13. At fields above 2 T, the peak is no longer noticeable anywhere on the plot of the specific heat because the superconducting state has been destroyed. Remarkably, as the field increased above about 4.5 T, a small kink appeared near 0.7 K and shifted close to 1 K, growing in strength as the magnetic field was raised toward the highest available field of 8 T. The large entropy released at the HFOP boundary, implied from the large jump in the specific heat, sug-gested that the field-induced phase is ordered, and its origin is the CEF of the Pr^{3+} ions. Carrying out specific heat studies in magnetic field applied parallel to the [001], [110], and [111] directions revealed the existence of the HFOP phase in all three magnetic field orientations with a modest anisotropy. Measurements were later extended to fields as high as 32 T by Rotundu et al. (2004) and were supplemented by magnetization and resistivity measurements by Tayama et al. (2003) and Sugawara et al. (2005), respectively. Based on these studies, a phase diagram was constructed depicting the applied magnetic field as a function of temperature, as shown in Figure 2.14. The exis-tence of the HFOP was further substantiated by a variety of other studies, Maple et al. (2003), Tenya et al. (2003), Vollmer et al. (2003), and Oeschler et al. (2004).

   The direct evidence for a long-range order in the HFOP of $PrOs_4Sb_{12}$ was provided by the neu-tron diffraction experiment of Kohgi et al. (2003), where scans in the reciprocal lattice plane (hk0) revealed peaks at superlattice positions with the wave vectors $\mathbf{q}$ = [100] and [010]. The scatter-ing pattern of the (210) superlattice reflection at 0.5 K at 8 T field and at zero field is shown in Figure 2.15(a). The field dependence of the peak intensity of the (210) reflection at 0.25 K in Figure 2.15(b) shows that the reflections start to develop when the field is near 5 T, consistent with the phase boundary of the HFOP in Figure 2.14. The temperature dependence of the peak intensity of the (210) reflection at 6 T and 8 T is displayed in Figure 2.15(c). The intensity of the reflec-tion falls to the background level at about 0.8 K and 1 K, respectively, again in good accordance with the corresponding boundaries of the HFOP. From the fact that the integrated intensities of all

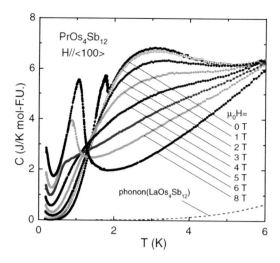

FIGURE 2.13 Specific heat vs. temperature in magnetic fields oriented along the [100] direction of the $PrOs_4Sb_{12}$ crystal. The superconducting transition is seen only in zero and 1 T fields. The magnetic field of 2 T destroys the superconducting state, and no kinks are observed at fields of 2 to 4 T. At 5 T, a small kink develops near 0.7 K, grows in magnitude, and shifts toward 1 K upon increasing the field to 8 T, the maximum field available in the experiment. The reader may note that the Schottky anomaly is dramatically suppressed compared to its contribution at zero and low fields. This makes the specific heat jump (a well-developed λ anomaly) associated with the transition to the field-induced ordered state so pronounced. The lattice specific heat of $LaOs_4Sb_{12}$, believed to be the same as that of the isostructural $PrOs_4Sb_{12}$, is shown by a dashed curve for comparison. Reproduced from Y. Aoki et al., *Journal of the Physical Society of Japan* **71**, 2098 (2002). With permission from the Physical Society of Japan.

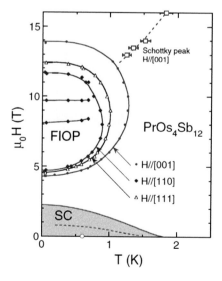

FIGURE 2.14 Magnetic field vs. temperature phase diagram for $PrOs_4Sb_{12}$ constructed based on the specific heat and magnetization measurements under magnetic field for three field orientations, showing a modest anisotropy of the field-induced phase designated here as FIOP, also known as HFOP. The diminishing Schottky anomaly for *H* parallel to the [001] crystal orientation is shown by a blue dashed curve. Reproduced from Y. Aoki et al., *Journal of the Physical Society of Japan* **76**, 051006 (2007). With permission from the Physical Society of Japan.

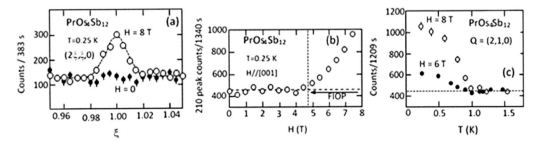

FIGURE 2.15 (a) Scattering pattern for the (210) superlattice reflection at 0.25 K in a magnetic field of 8 T (open circles) and in zero field (solid circles). (b) Magnetic field dependence of the peak intensity of the (210) reflection at 0.25 K. The range of the FIOP is delineated. (c) Temperature dependence of the peak intensity of the (210) reflection at 6 T (solid circles) and 8 T (open circles). Composed from the data of M. Kohgi et al., *Journal of the Physical Society of Japan* **72**, 1002 (2003). With permission from the Physical Society of Japan.

observed superlattice reflections (not shown here) followed the $sin^2\alpha$ dependence, where α is the angle between the scattering vector and the direction of the [010] axis, it became obvious that the reflections are due to magnetic Bragg scattering that has the origin in antiferromagnetic moments of Pr^{3+} ions parallel to the [010] direction with an estimated magnitude of $\mu_{AF} = 0.025 \pm 0.006\ \mu_B$ per Pr ion at 0.25 K and 8 T. It should be pointed out that the magnitude of the antiferromagnetic moment is quite small and it cannot arise from antiferromagnetic interactions among Pr^{3+} ions. Rather, it is the magnetic field $H \geq 4.5$ T applied parallel to the [001] direction that induces an ordered antiferromagnetic phase with the moments oriented along the [010] direction, the y-direction of the Pr^{3+} ion. This comes about as a result of Zeeman splitting of the magnetic $\Gamma_4^{(2)}$ triplet state (Γ_5 state in the O_h point group notation) where one of the split branches intersects the Γ_1 singlet near 8 T, see a sketch in Figure 2.16. The resulting pseudodoublet supports a variety of quadrupole moments, of which the tetrahedral T_h point group symmetry selects the antiferro-quadrupolar (AFQ) ordering of the O_{yz}-type moments stabilized at magnetic fields above about 4.5 T. Consequently, the HFOP is often referred to also as the AFQ phase. The key in identifying the order of the HFOP is the use of the tetrahedral T_h rather than the cubic O_h point group symmetry. Subsequent inelastic neutron studies by Kuwahara et al. (2005) explored the low-lying $\Gamma_1 - \Gamma_4^{(2)}$ excitations and found a softening of the excitations at a wavevector $\mathbf{Q} = (100)$, the same as the modulation vector of the field-induced AFQ ordering. The authors interpreted the behavior as direct evidence that the excitations are derived from non-magnetic, rather than magnetic, interactions between 4f-electrons. From the

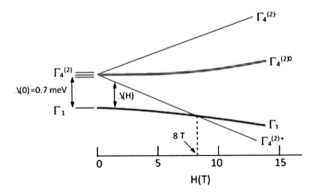

FIGURE 2.16 The two lowest energy levels of the Pr^{3+} ion. Zeeman splitting in magnetic field causes a crossover between the $\Gamma_4^{(2)+}$ level and the ground state Γ_1 at 8 T. Adapted from A. McCollam et al., *Physical Review B* **88**, 075102 (2013). With permission from the American Physical Society.

narrowing of the linewidth of the excitations below T_c, they also inferred a link to the heavy fermion superconductivity. For a theoretical justification of the emergence of the AFQ ordered phase within the singlet-triplet CEF level scheme that makes use of symmetry groups and utilizes pseudospin representations, the interested reader is referred to a paper by Shiina and Aoki (2004).

2.4.5 MAGNETO-THERMAL CONDUCTIVITY OF PrOs$_4$Sb$_{12}$

Studies of heat transport in superconductors, including high-T_c ceramics, over the past three decades have revealed the trend one may expect the thermal conductivity to follow when a superconductor is in the mixed state, i.e., the applied magnetic field is within $H_{c1} < H < H_{c2}$, Figure 2.17. S-wave superconductors (the vast majority of superconductors we know) are fully gapped and, as such, are typified by an exponential rise in the thermal conductivity in the applied magnetic field. In contrast, unconventional superconductors (d-wave superconductors among them) have nodes and perhaps even line nodes on the Fermi surface and should respond readily to even a small magnetic field. The behavior reflects the nature of quasiparticles in the system. In s-wave superconductors, the only quasiparticles that exist well below T_c are those tied to the vortex core in the mixed state. At low fields, the vortices are far apart and the quasiparticles are bound to them. As the magnetic field increases, the density of vortices increases and, near H_{c2}, the vortex cores start to overlap, rapidly enhancing the population of quasiparticles that now become free to carry heat. In stark contrast, the nodal structure of unconventional superconductors ensures the presence of quasiparticles near the nodes and line nodes in the gap at all temperatures below T_c, and they are easily disturbed by the applied magnetic field, often resulting in the $H^{1/2}$ field dependence of the thermal conductivity. Although this is a somewhat simplified viewpoint, especially regarding the thermal conductivity in high-T_c superconductors where, at much higher temperatures, the scattering effects modify the behavior, for low-transition-temperature skutterudites the picture emphasizing the effect of magnetic field on the quasiparticle DOS is basically correct.

An interesting study concerning measurements of the thermal conductivity in a magnetic field rotated with respect to the crystallographic axes of a single crystalline PrOs$_4$Sb$_{12}$ sample was performed by Izawa et al. (2003). The premise of the thermal conductivity studies was twofold: i) the fact that the thermal conductivity, unlike the electrical resistivity, does not vanish below T_c, and ii) the Cooper pairs do not carry entropy and, hence, do not participate in the transport of heat. Assuming an environment of very low temperatures and thus negligible conduction *via* lattice vibrations, the remaining contribution to thermal conduction is quasiparticle excitations. Their heat content and response to the applied magnetic field offered an opportunity to learn about the origin of the quasiparticles, information that should allow for the identification of the symmetry class of superconductivity.

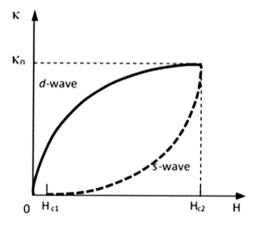

FIGURE 2.17   Schematic dependence of the thermal conductivity in the superconducting domain of s-wave and d-wave superconductors as a function of applied magnetic field.

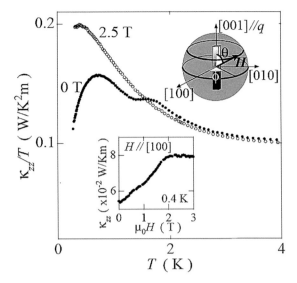

FIGURE 2.18 Temperature dependence of the thermal conductivity plotted in the form of κ_{zz}/T vs. T in zero magnetic field and in a field of 2.5 T, above the H_{c2}, applied perpendicular to the long axis (c-axis) of the sample. The top panel shows the experimental set up, and the lower inset depicts the magnetic field dependence at 0.4 K with the field [100] direction, perpendicular to the sample. Adapted from K. Izawa et al., *Physical Review Letters* **90**, 117001 (2003). With permission from the American Physical Society.

In Figure 2.18 is shown the thermal conductivity of $PrOs_4Sb_{12}$ measured by Izawa et al. (2003) along the c-axis (κ_{zz}) in zero magnetic field and in a field of 2.5 T, i.e., above the upper critical field H_{c2} of 2.45 T. The plot is in the form of κ_{zz}/T vs. T, a convenient way of extracting the residual electronic thermal conductivity $\kappa_{o,el}$ by extrapolating the expected linear dependence of conductivity to $T = 0$ from the temperature dependence of the electronic contribution and the cubic temperature

$$\frac{\kappa}{T} = \frac{\kappa_{0,el}}{T} + \beta T^2, \tag{2.23}$$

where β represents the phonon contribution. The onset of superconductivity is seen below 1.8 K as a shoulder on the curve. In a field of 2.5 T, the superconductivity is suppressed, and no distinct mark is observed on the curve. The field dependence of the thermal conductivity measured at 0.4 K, the inset shown in Figure 2.18, shows a substantially linear trend. Although not following the $H^{1/2}$ dependence illustrated in Figure 2.17, it is certainly not an exponentially rising thermal conductivity expected for an *s*-wave superconductor and, as such, suggests the presence of itinerant gapless quasiparticle excitations that carry heat perpendicular to the vortex lines. In other words, the inset reflects the unconventional nature of $PrOs_4Sb_{12}$. The low temperature of 0.4 K was not low enough to extrapolate convincingly the residual electronic contribution using Equation 2.23. Instead, the authors measured the thermal conductivity in a rotating magnetic field (the field rotated either in a plane perpendicular to the sample axis of an oriented crystal or in a conical fashion with the field and the sample axis at an angle less than 90 degrees), to explore the symmetry of the nodal points. One relies here on the Doppler energy shift of the quasiparticle spectrum in the circulating supercurrent, which alters the energy of the quasiparticle from $E_k = (\varepsilon_k^2 + \Delta_k^2)^{1/2}$ to $E_k + m\mathbf{v}_F \cdot \mathbf{v}_s$, where Δ_k is the order parameter (superconducting gap), $\mathbf{v}_F$ is the Fermi velocity, and $\mathbf{v}_s$ is the velocity of the circulating (shielding) current around vortices. The Doppler shift is most relevant where the local energy gap Δ_k becomes relatively smaller, such as near nodes and lines of nodes where $\Delta_k = 0$. Because the magnitude of the Doppler term depends critically on the angle between the node direction and the

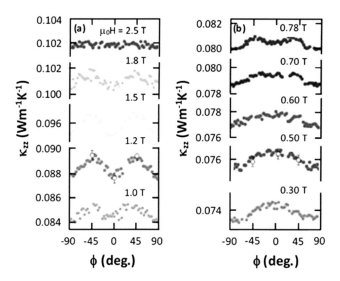

FIGURE 2.19 (a) and (b) Angular variation of the thermal conductivity κ_{zz} in magnetic fields rotated in the plane perpendicular to the sample axis (c-axis). The data are taken at 0.52 K in fields above and below H_{c2}. Reproduced from K. Izawa et al., *Physical Review Letters* **90**, 117001 (2003). With permission from the American Physical Society.

direction of the applied magnetic field, oscillations in the thermal conductivity are observed with a period corresponding to the angular position of the nodes as the magnetic field is rotated. The data are illustrated in Figure 2.19 for the case of the magnetic field (both above and below H_{c2}) rotated in the plane perpendicular to the c-axis.

While the thermal conductivity is essentially angle independent in fields higher than H_{c2}, in the superconducting state, it develops distinct oscillations. A clear fourfold variation is observed right below H_{c2} and persists down to fields of about 0.8 T, at which point it surprisingly switches to a twofold symmetry that grows stronger as the field decreases. This suggests that there is a phase boundary deep in the superconducting state between the fourfold and twofold symmetries. After evaluating all the data, Izawa et al. (2003) came to a conclusion that the observed symmetries originate from the gap nodes and therefore provide direct evidence for the change of the gap symmetry. From the orientation of the magnetic field where the minima in κ_{zz} occur, it follows that the nodes are located along the [100] and [010] directions in the high-field phase and only along the [010] direction in the low-field phase. Measuring the thermal conductivity also in a conically rotated field for several angles θ, (see the top panel in Figure 2.18 for the geometry of the setup), they were able to distinguish between point nodes and line nodes, and constructed the phase diagram shown in Figure 2.20. The magnetic field H^* delineates two phases: the A-phase where the gap function has fourfold symmetry and the B-phase where the gap function has twofold symmetry. Except for UPt$_3$, Machida et al. (1999), PrOs$_4$Sb$_{12}$ would be the only superconductor displaying multiple phases with different gap symmetries. The result drew much interest, and shortly thereafter, Thalmeier et al. (2004) proposed a model for the respective anisotropies of the two gap functions. Invoking hybrid gap functions $\Delta(\mathbf{k}) = \Delta f(\mathbf{k})$, with the form factor for phases A and B as

$$A : f\left(k\right) = 1 - k_x^4 - k_y^4 \quad B : f\left(k\right) = 1 - k_y^4, \tag{2.24}$$

they described the A-phase gap as the tetragonal $s + g$-wave type, and the B-phase gap as having only twofold symmetry. The computed angular variation of the thermal conductivity for the two phases is expressed through $I_A(\theta,\phi)$ and $I_B(\theta,\phi)$, respectively, in Figure 2.21.

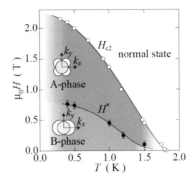

FIGURE 2.20 The phase diagram of the superconducting gap symmetry of $PrOs_4Sb_{12}$ obtained from measurements of the angular dependence of the thermal conductivity. Open circles represent the upper critical magnetic field H_{c2}. The magnetic field H^* designated by solid circles delineates the phase boundary between the regime of fourfold gap symmetry and twofold gap symmetry. Reproduced from K. Izawa et al., *Physical Review Letters* **90**, 117001 (2003). With permission from the American Physical Society.

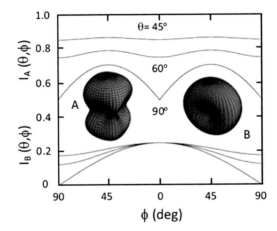

FIGURE 2.21 Calculated angular variation of the magneto-thermal conductivity $\kappa_{zz}(\theta,\phi)$ of $PrOs_4Sb_{12}$ showing fourfold (A) and twofold (B) oscillations in ϕ due to different node structures of the gap function. The polar plot shows nodes along two (A) or only one (B) cube axis. Here, θ and ϕ are polar and azimuthal angles, respectively, of the magnetic field H with respect to [001]. Reprinted from P. Thalmeier et al., *Physica C* **408–410**, 177 (2004). With permission from Elsevier.

On the experimental side, the results were followed by low-temperature thermal conductivity measurements by Seyfarth et al. (2005), Seyfarth et al. (2006), and Hill et al. (2008). Unfortunately, the reports have not resulted in a unified viewpoint but, rather, conflicted with each other. While the experiments extended to temperatures as low as a few tens of millikelvin and covered a magnetic field range from zero to several Tesla, none of them detected any sign of a phase transition in the mixed state of $PrOs_4Sb_{12}$ that would indicate separation into phases A and B in the magnetic field H^* supposed to form a boundary region between them. Moreover, the reports of Seyfarth et al. (2005, 2006), the more recent one describing thermal conductivity measurements on a high-quality crystal showing just a single superconducting transition, concluded that there is no residual thermal conductivity as T → 0. In other words, the superconducting gap is fully open and there are no nodes present. In their opinion, $PrOs_4Sb_{12}$ is a MBSC possessing two fully developed superconducting gaps, a larger gap Δ_l and a smaller gap Δ_s, that share the same transition temperature T_c. The analysis of the temperature and field dependence of the thermal conductivity indicated that the gap size ratio at $T = 0$ is about Δ_l/Δ_s ~3. On the other hand, Hill et al. (2008) performed comparative measurements of the thermal conductivity on $PrOs_4Sb_{12}$ and on $PrRu_4Sb_{12}$, an isostructural but non-heavy fermion s-wave superconducting skutterudite, using an identical experimental setup and being in contact with the samples in the same way. The experiments extended down to about 50 mK and provided an opportunity to extract the residual electronic contribution to the thermal conductivity, if any present, using Equation 2.23. By carefully extrapolating the data below 0.15 K to absolute zero, the residual electronic thermal conductivity divided by T of 0.46 ± 0.07 mWcm^{-1}K^{-2} was obtained for $PrOs_4Sb_{12}$, while $PrRu_4Sb_{12}$ possessed an order of magnitude smaller value of 0.058 ± 0.007 mWcm^{-1}K^{-2}. With the empirically extracted parameters for the power law to fit the phonon contribution, the latter value was further reduced down to 0.008 ± 0.017 mWcm^{-1}K^{-2}, the error bar clearly indicating a non-existent residual electronic contribution in $PrRu_4Sb_{12}$. The significant residual electronic conductivity in $PrOs_4Sb_{12}$ provided incontrovertible evidence for the presence of nodes on the Fermi surface of the superconducting order parameter of this skutterudite, especially given the fact that a sister skutterudite $PrRu_4Sb_{12}$ lacked $\kappa_{0,el}$, as it should have, given its s-wave superconducting nature. Moreover, by extrapolating the thermal conductivity measured at different magnetic fields to zero temperature and normalizing such values to the normal state thermal conductivity (value at H_{c2}), Hill et al. (2008) obtained the plot shown in Figure 2.22, where the magnetic field is also normalized to the upper critical field H_{c2}, taken as 2.5 T for $PrOs_4Sb_{12}$ and 0.2 T for $PrRu_4Sb_{12}$. In the case of $PrOs_4Sb_{12}$, the initial rise at low fields is dramatic and certainly in discord with the exponentially increasing residual electronic contribution expected for the simplest case of a fully gapped superconductor. At higher fields, the data reached a plateau followed by a superlinear dependence until the normal state was reached at H_{c2}. Such behavior is typical in MSBCs, and, in this aspect, Hill et al. (2008) and Seyfarth et al. (2006) have agreed. To explain the presence of the significant residual electronic thermal conductivity and the multi-band superconducting character, Hill et al. (2008) proposed a two-band structure for $PrOs_4Sb_{12}$, one band with nodes and the other one a fully gapped band with the gap maximum somewhat larger than the gap maximum of the nodal band. While the nodal character of the first band clearly contradicts the findings of Seyfarth et al. (2006), the scenario can explain most of the other features of the data, including, with some stretch, results of the angular-dependent thermal conductivity study of Izawa et al. (2003) discussed above. All one would need to do is to assign a degree of anisotropy to the gap structure of the fully gapped band,

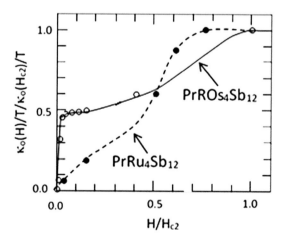

FIGURE 2.22 Magnetic field dependence of the extrapolated $T = 0$ K thermal conductivity normalized to the thermal conductivity at H_{c2}, and plotted as a function of reduced field H/H_{c2} for $PrOs_4Sb_{12}$ (open circles) and $PrRu_4Sb_{12}$ (solid circles). Adapted from the data of R. W. Hill et al., *Physical Review Letters* **101**, 237005 (2008). With permission from the American Physical Society.

such that its minimum is orthogonal to the node structure of the first band. In that case, the angular-dependent thermal conductivity measurements at not too low temperatures and fields should pick up the fourfold symmetry due to the gap nodes of the first band and the anisotropy of the fully gapped second band. At low fields and temperatures, only the anisotropy due to the nodal band would survive as quasiparticle excitations in the fully gapped band would be exponentially suppressed. Such a superconducting skutterudite would have not only two gaps but also two distinct gap symmetries. However, a direct challenge to the results of the angular-dependent thermal conductivity, and specifically to a crossover from the fourfold to the twofold symmetry that was interpreted as the existence of A and B phases of the superconducting condensate, came from angular-dependent studies of the specific heat by Custers et al. (2006) and Sakakibara et al. (2008). As indicated by the data in Figure 2.23, the fourfold symmetry of oscillations with the magnetic field rotated in the (001) plane at $T = 0.32$ K is maintained at all magnetic fields within the mixed state of $PrOs_4Sb_{12}$, and there is no suggestion of a crossover to the twofold symmetry oscillating regime that should have been observed in fields between 0.6 T and 1 T. The authors also confirmed that the same oscillating pattern is obtained when the magnetic field is rotated in the (010) plane, i.e., there is no preferred orientation in the superconducting state. Moreover, as indicated by the right-hand panel in Figure 2.23, the oscillations persist well into the normal state (well above H_{c2}), but their amplitude acquires an interesting trend. As the field approaches 3 T, the amplitude of oscillations diminishes, and at 3 T, the oscillations cease to exist. They reappear, but with a shifted phase (inverted) between 3 and 4 T and, as the AFQ phase boundary is crossed near 4.5 T, they switch back and attain very large amplitudes. The absolute amplitude of the oscillations of the specific heat of $PrOs_4Sb_{12}$ as a function of magnetic field at 0.32 K is depicted in Figure 2.24. Solid black circles indicate the measured amplitude, and open circles denote the electronic part of the oscillation amplitude corrected for the nuclear specific heat indicated in the figure by a dashed curve. The amplitude in the field-induced AFQ phase is very large, and in Figure 2.24 it is reduced by a factor of 30.

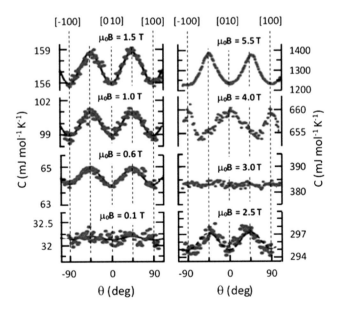

FIGURE 2.23 Angular dependence of the specific heat of $PrOs_4Sb_{12}$ measured at 0.32 K in a magnetic field rotated in the (001) plane. The left panel covers the mixed state domain while the right panel shows oscillations in the normal state at fields above H_{c2}. The pattern at 5.5 T shows oscillations obtained in the AFQ phase; please note their very large amplitude. Reprinted from T. Sakakibara et al., *Physica B* **403**, 990 (2008). With permission from Elsevier.

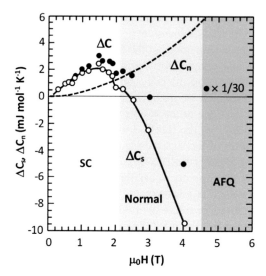

FIGURE 2.24 Absolute amplitude of the angular oscillations of the specific heat of $PrOs_4Sb_{12}$, $\Delta C = C_{[110]} - C_{[100]}$, as a function of magnetic field at 0.32 K. Closed circles denote the measured amplitude, and the dashed line is the estimated nuclear contribution ΔC_n. The electronic part of the oscillation amplitude $\Delta C_e = \Delta C - \Delta C_n$ is shown by open circles. The data point at 5.5 T in the AFQ phase is reduced by a factor of 30. The solid line is a guide to the eye. Reproduced from T. Sakakibara et al., *Physica B* **403**, 990 (2008). With permission from Elsevier.

2.4.6 BREAKDOWN OF TRS IN $PrOs_4Sb_{12}$

The pivotal question to ask when one suspects unconventional superconductivity is whether the TRS has been broken. One of the consequences of broken TRS is the non-zero magnetic moment of Cooper pairs that may align locally, giving rise to a tiny internal magnetic field. A technique that can probe and measure such minute magnetic fields is zero-field muon spin resonance (μSR), which was applied to small Sb flux-grown single crystals of $PrOs_4Sb_{12}$ by Aoki et al. (2003b, 2006). The spontaneous internal magnetic field in the superconducting state of $PrOs_4Sb_{12}$ provided unambiguous evidence of the broken TRS in the superconducting state of this skutterudite. Observing no such internal fields, i.e., no breaking of TRS, in $LaOs_4Sb_{12}$, a conventional superconductor with $T_c = 0.74$ K and no f-electrons, strongly suggested that Pr^{3+} ions are the origin of the dynamically fluctuating fields in $PrOs_4Sb_{12}$. Additional strong evidence for the broken TRS in the superconducting state of $PrOs_4Sb_{12}$ has come from recent studies of the polar Kerr angle by Levenson-Falk et al. (2018). The polar Kerr effect measures the rotation that a linearly polarized beam of light acquires as it gets reflected under normal incidence by a magnetized sample. More specifically, the Kerr angle θ_K measures the difference in the index of refraction between the right- and left-polarized circular beams of light that constitute the linearly polarized light. It should be noted that the Kerr angle in chiral superconductors is very small, merely a fraction of a micro-radian, and its measurement requires special, highly sensitive interferometers. For experimental details concerning measurements of the Kerr angle, the reader is referred to a highly readable article by Kapitulnik (2015). The relevance of the polar Kerr angle studies in the field of superconductivity rests in the fact that detection of any non-zero Kerr angle below T_c implies the breakdown of the TRS, Cho and Kivelson (2016). Polar Kerr angle measurements were essential in showing that a small subset of heavy fermion superconductors, most notably Sr_2RuO_4, Xia et al. (2006), breaks TRS, and such structures are referred to as chiral superconductors. As mentioned in section 2.1, the prospect that chiral superconductors may harbor Majorana fermions, particles that are their own antiparticles, has generated tremendous

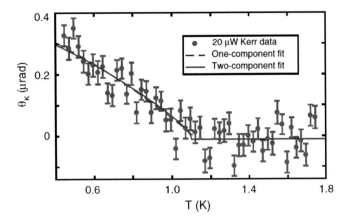

FIGURE 2.25 Averaged Kerr angle measured on a single crystal of $PrOs_4Sb_{12}$ with 20 µW incident power of the laser beam. The temperature where the Kerr angle starts to deviate from zero, here 1.167 K, is lower than the temperature of 1.672 K at which the susceptibility under irradiation undergoes the transition. This is to be expected in the presence of optical heating where the Kerr measurement focuses on the volume heated by the optical beam while the susceptibility measures the entire sample. A finite-element model used by the authors yields a "hot spot" that is about 0.5 K higher than the temperature of the bulk crystal. Error bars are statistical error (one-σ). A fit assuming two transition temperatures is indicated by a solid blue line. A fit done with a single transition temperature is shown by a dashed green line but is hardly visible as the two fits are essentially identical. Reproduced from E. M. Levenson-Falk et al., *Physical Review Letters* **120**, 187004 (2018). With permission from the American Physical Society.

excitement in the field of solid-state physics. Once Levenson-Falk et al. (2018) had observed a small Kerr angle in the superconducting domain of $PrOs_4Sb_{12}$, Figure 2.25, the skutterudite started to be viewed as the most prospective material in which the exotic Majorana particle might be observed, a status that has elevated filled skutterudites to the forefront of condensed matter research, rivaling their importance as highly efficient thermoelectric materials.

How actually $PrOs_4Sb_{12}$ crystals enter the time-reversed symmetry-broken phase as a function of temperature was explored in a recent study by Setty et al. (2017), who, for this purpose, re-derived the GL-like theory. In order to probe the pairing symmetry of the time-reversed symmetry-broken phase, they carefully monitored Raman responses in both A and B phases of $PrOs_4Sb_{12}$ under various light polarizations. The authors demonstrated that such a technique can access the different irreducible representations within the T_h point group. Depending on whether nodes on the Fermi surface were present or absent, they observed an enhancement or suppression, respectively, of the spectral weight in the Raman signal.

Having the broken TRS, it is obvious that the spin-singlet s-wave superconductivity cannot describe the superconducting state of $PrOs_4Sb_{12}$. But then, what is the pairing symmetry of the Cooper pairs? Odd parity pairing, such as the spin-triplet state, where the two electrons have spins oriented in the same direction, has been shown to exist in the d-electron Sr_2RuO_4 structure, Ishida et al. (1998), and the 5f-electron UPt_3 and UNi_2Al_3 compounds, Tou et al. (1996) and Ishida et al. (2002), but never before in a 4f-electron system.

One of the experimental techniques that can shed light on the pairing symmetry is the measurement of the Knight shift. The technique measures the change in the strength of the magnetic hyperfine field generated at the nuclear site by the presence of conduction electrons with spins oriented by the applied magnetic field (spin susceptibility). The change in the hyperfine field (compared to a situation where no polarized electrons are present) is reflected in a shift of the resonant frequency the nucleus generates. This rather sharp probe of the nuclear environment offers an opportunity to monitor spin susceptibility and its temperature dependence in both superconducting and normal states of the system. Typically, the Knight shift measurements would be performed as part of nuclear magnetic resonance (NMR) studies, and, in the case of $PrOs_4Sb_{12}$, one would use either the ^{121}Sb or ^{123}Sb nucleus. Such studies were, indeed, carried out by Tou et al. (2011), but with the limited accuracy of their NMR measurements, the pairing symmetry could not be determined. The problem rests with the large electric field gradients at the nucleus, which are also strongly temperature-dependent, that hinder precise determination of the resonant frequency. It is more advantageous to use μSR instead to determine the Knight shift generated by the polarized electrons (in the normal state) or quasiparticle excitations (in the superconducting state). Such an approach was chosen by Higemoto et al. (2007), who carried out the experiment over a wide temperature range, covering both the normal and the superconducting regimes of $PrOs_4Sb_{12}$. The value of the Knight shift ΔK_s is related to the spin susceptibility χ^{sp} via the equation

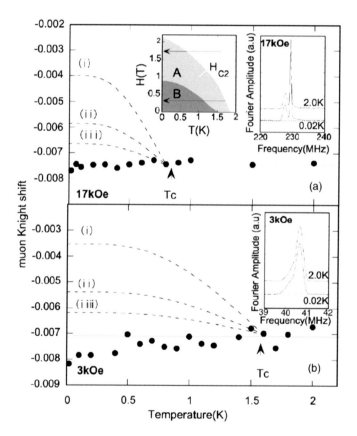

FIGURE 2.26 Temperature dependence of the muon Knight shift of $PrOs_4Sb_{12}$ at (a) 17 kOe and (b) 3 kOe. Note a temperature-independent Knight shift as the sample is cooled through the superconducting transition temperature at 1.85 K. In contrast, the dashed lines refer to the expected temperature dependence for s-wave pairing expressed through the Yosida function with the values of the parameter $g_j^2 J_{eff}^2$ corresponding to (i) the $4f^2$ configuration ($g_j^2 J_{eff}^2 = 12.8$), (ii) the $\Gamma_1 + \Gamma_5$ pseudoquartet state ($g_j^2 J_{eff}^2 = 6.0$), and (iii) a free electron ($g_j^2 J_{eff}^2 = 3.0$). The left-hand side inset in (a) shows the phase diagram labeled the same way as in Figure 2.20. The insets in (a) and (b) depict fast Fourier transfer spectra with the large peak originating predominantly from the silver sample holder. Reproduced from W. Higemoto et al., *Physical Review B* **75**, 020510(R) (2007). With permission from the American Physical Society.

$$\Delta K_s = A_{hf} \chi^{qp} = A_{hf} \frac{\gamma g_J^2 \mu_B^2 J_{eff}^2}{\pi^2 k_B^2} \times R \tag{2.25}$$

where A_{hf} is the hyperfine coupling constant, γ is the electronic specific heat coefficient, μ_B is the Bohr magneton, $g_J^2 J_{eff}^2$ is the effective spin, and R is the Wilson ratio. The Knight shift extracted from the measurements turned out to be essentially temperature independent as the single crystalline $PrOs_4Sb_{12}$ sample was cooled through the transition temperature. This was a very different temperature behavior from what one would expect for an s-wave state, in which the Knight shift has a distinct temperature dependence given by the Yosida function, Yosida (1958). The results are presented in Figure 2.26, together with the predicted behavior within the Yosida formalism for three different values of the effective spin, reflecting different scenarios of the quasiparticles' origin. The inset in Figure 2.26 shows the phase diagram virtually identical to that in Figure 2.20, but it is not clear whether it was independently generated or merely adapted from the work of Izawa et al. (2003).

2.4.7 FERMI SURFACE OF PrOs₄Sb₁₂

The electronic properties of any conducting solid depend critically on the size and shape of the Fermi surface. Perhaps the most powerful probe of the Fermi surface is de Haas-van Alphen (dHvA) measurements that can also determine the cyclotron effective mass of the carriers.

dHvA measurements of $PrOs_4Sb_{12}$ were reported by Sugawara et al. (2002) and by Aoki et al. (2005). With a single crystal sample cooled down to 30 mK, dHvA oscillations became detectable in a magnetic field above about 2.2 T. Three distinct branches emerged, and two of them, labeled as α and β in Figure 2.27(a), could be traced in the magnetic field covering the full angular range of the cubic symmetry. In other words, the two branches correspond to a closed Fermi surface. The third branch, labeled γ, had only a limited angular range around the [100] direction, suggesting that it was a part of the multiply connected Fermi surface. From the temperature dependence of the dHvA amplitude, the authors estimated the cyclotron effective mass and obtained the value 7.6 m_e. Although not as high as the estimate from the specific heat ($\sim 50\ m_e$), it was still substantial enough to document the heavy fermion nature of the electrons. The band structure of $PrOs_4Sb_{12}$ calculated by Sugawara et al. (2002) using DFT with the LDA + U method, shown in Figure 2.27(b), indicated three major crossings of the Fermi level. Two distinct valence bands, the 48th and the 49th, cross the Fermi level and are centered at the Γ point, and the 49th band also crosses around the N point.

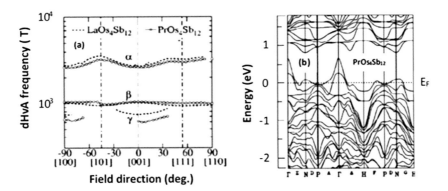

FIGURE 2.27 (a) Angular dependence of the dHvA frequencies for $PrOs_4Sb_{12}$ (open circles) and $LaOs_4Sb_{12}$ (dashed line). The α- and β-branches reflect a closed Fermi surface while the γ-branch indicates a part of the multiply connected Fermi surface. (b) DFT-computed band structure of $PrOs_4Sb_{12}$ using the local density approximation (LDA) + U method. Adapted from H. Sugawara et al., *Physical Review B* **66**, 220504(R) (2002). With permission from the American Physical Society.

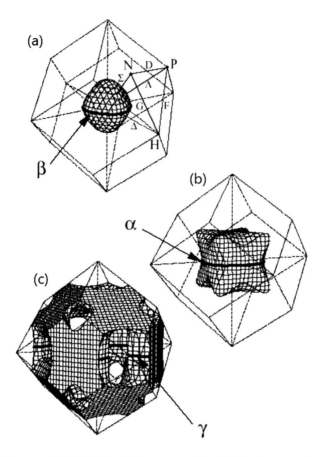

FIGURE 2.28 (a) The near spherical β-Fermi sheet centered at the Γ point and originating from the 48th valence band. (b) A rounded cubic α-sheet centered at the Γ point and originating from the 49th valence band. (c) Multiply connected γ-sheet centered at the N point and originating from the 49th valence band. Reproduced from H. Sugawara et al., *Physical Review B* **66**, 220504(R) (2002). With permission from the American Physical Society.

They form, respectively, the β-branch, the α-branch, and the γ-branch, each shown in Figure 2.27(a). The Fermi surface sheets obtained from the measurements are depicted in Figure 2.28.

Harima and Takegahara (2003), using the same LDA + U method, computed the Fermi surface of $LaOs_4Sb_{12}$ and similarly obtained two closed hole Fermi surfaces around the Γ point and a multiply connected sheet, as in the calculations of the Fermi surface of $PrOs_4Sb_{12}$ by Sugawara et al. (2002). The difference between the two skutterudites was the value of the cyclotron mass, which in the case of $LaOs_4Sb_{12}$ was enhanced only by a factor of 3 compared to a factor of 8 in the case of $PrOs_4Sb_{12}$. In their subsequent LDA + U calculations of the Fermi surface of $PrOs_4Sb_{12}$, Harima and Takegahara (2005) showed that the $4f$ electrons remain localized, and the angular dependence of the experimental dHvA oscillations could be replicated with the parameter U = 5.44 eV (0.4 Ry). However, the $4f$ electrons acquired an itinerant character when the Coulomb repulsion was small at U = 0.68 eV (0.05 Ry).

An interesting application of the dHvA oscillations was implemented by McCollam et al. (2013). The high sensitivity of their apparatus operating at temperatures down to 30 mK, and the use of the temperature-dependent phase shift of the dHvA oscillations (rather than the frequency itself), allowed detection of tiny changes in the volume of the Fermi surface in response to changes in the population of the crystal field levels Γ_1 and $\Gamma_4^{(2)}$ within the AFQ phase. Specifically, they found that the β-sheet of the Fermi surface shrinks as the occupancy of $\Gamma_4^{(2)}$ levels grows relative to that of Γ_1. Moreover, the technique was sensitive enough to probe the as yet unappreciated influence of the hyperfine coupling on the order parameter within the AFQ phase when operating below 300 mK. Furthermore, the measurement monitored the content of the $\Gamma_4^{(2)}$ level in six hyperfine states as a function of magnetic field and temperature. The technique might be an important tool to shed light on the coupling of nuclear states to electronic degrees of freedom in all strongly electron correlated systems.

2.4.8 POINT-CONTACT SPECTROSCOPY

One of the few sharp microscopic probes sensitive to the amplitude and phase of the order parameter that should directly address the presence or absence of nodes in the gap function is point-contact spectroscopy. The experiments are typically performed with metal/superconductor junctions formed by pressing fine normal metal tips against the surface of a superconductor and collecting the differential conductance dI/dV spectra of the junction. In ordinary isotropic s-wave superconductors, the point-contact conduction spectra reflect the bulk DOS. In high transparency junctions, the major contribution to the conductance comes from Andreev reflections, a process in which the normal current is converted to supercurrent at the junction. Because no low-energy single particle can navigate through the energy gap of a superconductor, the incoming electron (hole) from the metal side of the junction forms a Cooper pair with an electron (hole) of opposite spin at the superconductor side of the interface and, to preserve the particle count, a hole (electron) is retro-reflected into the normal metal (see a sketch in Figure 2.29). Andreev reflection is a unique probe of the order parameter because the resonance in the current at zero voltage bias at temperatures below T_c, the so-called zero bias conduction peak (ZBCP), is a feature observed exclusively in unconventional superconductors. In other words, the presence of the ZBCP is an unequivocal confirmation of the existence of nodes in the gap function.

The ZBCP is strongly temperature- and magnetic field-dependent and decreases in magnitude as the temperature approaches T_c. Moreover, by subtracting the normal state conductance from the spectrum and numerically integrating over a range of voltages near zero bias, one obtains the excess spectral area that, too, is strongly dependent on the temperature and magnetic field. Although Asano et al. (2003) drew attention to point-contact spectroscopy as something to be looked at in $PrOs_4Sb_{12}$, and even worked out the expected spectra for a few specific cases of the order parameter, the first measurements were not performed until some five years later by Turel et al. (2008). In Figures 2.30(a) and 2.30(b) are shown the temperature dependence of the conduction spectrum of a Pt-Ir/$PrOs_4Sb_{12}$ point-contact junction and the magnetic field dependence of the conductance, respectively. In the insets of the two figures are shown the temperature and field dependence of the excess

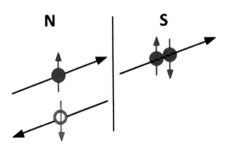

FIGURE 2.29 A sketch of the Andreev scattering. An electron (solid blue) approaching the junction from the metal side with energy less than the superconducting energy gap forms a Cooper pair just inside the superconducting side with an electron of opposite spin, and a hole (open red circle) with opposite spin is retro-reflected into the metal.

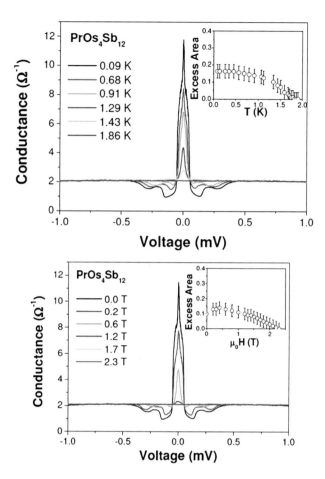

FIGURE 2.30 (a) Temperature dependence of the conductance spectrum for a Pt-Ir/PrOs$_4$Sb$_{12}$ point-contact junction. Note a pronounced zero-bias conductance peak, the height of which rapidly decreases as T_c is approached. (b) Magnetic field dependence of the conductance spectrum for the same junction. The insets in both figures present the (a) temperature and (b) field dependence of the excess spectral area. Both the zero-bias conductance peak and the excess spectral area are indisputable evidence for the unconventional nature of superconductivity in PrOs$_4$Sb$_{12}$. Reproduced from C. S. Turel et al., *Journal of the Physical Society of Japan* **77**, Suppl. A, 21 (2008). With permission from the Physical Society of Japan.

spectral area. The presence of two distinct effects, namely, the zero-bias conductance peak and the excess spectral area, was ascribed by Turel et al. (2008) to the existence of multiple order parameters with different symmetries. It is surprising to me that more point-contact measurements have not been performed on PrOs$_4$Sb$_{12}$ given their potential to couple to the order parameter.

2.4.9 SUPERCONDUCTING SOLID SOLUTIONS OF PR(OS$_{1-x}$RU$_x$)$_4$SB$_{12}$

Making solid solutions of praseodymium-containing skutterudites and exploring how substitutions on the site of Pr, or on the site of the transition metal, alter the superconducting properties turned out to be very illuminating. Frederick et al. (2004, 2005, 2007) studied changes in the superconducting state and the effective mass of electrons in Pr(Os$_{1-x}$Ru$_x$)$_4$Sb$_{12}$ as Os was gradually replaced with Ru. Pure PrRu$_4$Sb$_{12}$ is a superconductor with the transition temperature T_c = 1.1 K, not too different from the T_c = 1.85 K of PrOs$_4$Sb$_{12}$. Yet, its Sommerfeld coefficient γ is an order of magnitude

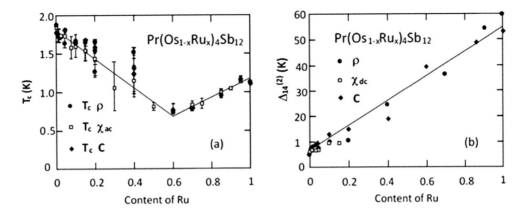

FIGURE 2.31 (a) Superconducting transition temperature of $Pr(Os_{1-x}Ru_x)_4Sb_{12}$ as a function of the Ru content x measured by electrical resistivity ρ, ac susceptibility χ_{ac}, and specific heat C. Some concentrations have more than one data point as different batches of crystals were tested. Note a distinct minimum at a Ru concentration of 60%. (b) Energy splitting $\Delta_{14}^{(2)}$ between the ground state Γ_1 and the first excited state $\Gamma_4^{(2)}$ as a function of Ru concentration x measured by electrical resistivity ρ, dc susceptibility χ_{dc}, and specific heat C. Adapted from the data of N. A. Frederick et al., *Physical Review B* **69**, 024523 (2004). With permission from the American Physical Society.

smaller, disqualifying $PrRu_4Sb_{12}$ from being classified as a heavy fermion system. The dependence of the superconducting transition temperature T_c on the content of Ru in the solid solution based on measurements of the electrical resistivity, ac susceptibility, and the specific heat is shown in Figure 2.31(a). Superconductivity is observed across the full range of compositions regardless of the experimental technique, but its trend is interesting; the T_c attains its minimum value at a Ru concentration of 60%. There could be several reasons why T_c is not a monotonic function of the composition. One of them is a possible existence of two-band superconductivity discussed in the context of thermal conductivity measurements by Seyfarth et al. (2005) and Hill et al. (2008). The other reason is a likely altered balance in the competition between the aspherical Coulomb scattering that promotes superconductivity and the exchange scattering that tends to suppress it, the result of significant changes in the CEF structure, see Figure 2.31(b). As the content of Ru increases, the energy separation between the Γ_1 ground state and the first excited state $\Gamma_4^{(2)}$, designated here as $\Delta_{14}^{(2)}$, increases linearly. The increasing $\Delta_{14}^{(2)}$ weakens both the aspherical Coulomb scattering and the exchange scattering but may initially suppress the former more, resulting in the decreased T_c. When $x > 0.6$, the large $\Delta_{14}^{(2)}$ may start to weaken the pair-breaking magnetic exchange interaction. While these are speculations that require further study, there is no dispute that the CEF energy levels are dramatically changed as Ru is introduced into the structure, documenting the great sensitivity of the Pr^{3+} ions in the local environment. It should also be mentioned that while the upper critical field $H_{c2}(0)$ decreases from about 2.3 T in $PrOs_4Sb_{12}$ down to about 0.25 T in $PrRu_4Sb_{12}$, the high field-induced AFQ phase is totally destroyed by the presence of merely 10% of Ru, Ho et al. (2008a).

Studies of the magnetic penetration depth λ in single crystals of $Pr(Os_{1-x}Ru_x)_4Sb_{12}$ down to 0.1 K by Chia et al. (2005) confirmed the crossover from the unconventional superconducting behavior in $PrOs_4Sb_{12}$, characterized by the presence of point nodes, to a BCS-like fully gapped superconductivity in $PrRu_4Sb_{12}$, which in these measurements took place near $x \approx 0.2$. Solid solutions with the concentration of Ru larger than 0.2 (structures with $x = 0.4$, 0.6, and 0.8) displayed a clear exponential behavior in the penetration depth and in the extracted density of the superconducting condensate $\rho_s(T)$ at all temperatures below T_c. In contrast, samples with $x \leq 0.2$ exhibited T^2 power law dependence at low temperatures. However, this quadratic temperature variation of $\rho_s(T)$ changed into an exponential one above a certain temperature, designated as T_{c3}, which depended on x. The data are

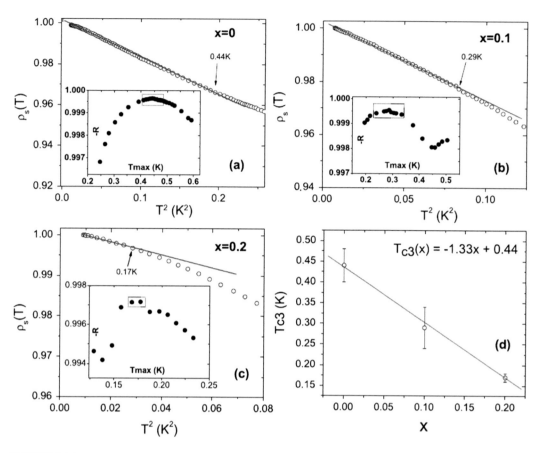

FIGURE 2.32 Low-temperature variation of the density of the superconducting condensate $\rho_s(T)$ plotted as a function of T^2, for (a) pure $PrOs_4Sb_{12}$ and for two low-content solid solutions $Pr(Os_{1-x}Ru_x)_4Sb_{12}$ with (b) $x = 0.1$ and (c) $x = 0.2$. The solid lines are visual guides to determine the temperature where $\rho_s(T)$ deviates from the quadratic dependence. The insets indicate the temperature dependence of the correlation coefficient R of the straight-line fit. Values of $R = +1$ (-1) represent a perfect positive (negative) linear relationship between $\rho_s(T)$ and T^2. Temperature T_{c3} is defined as the point of maximum (absolute) correlation R close to the temperature where $\rho_s(T)$ starts to depart from the T^2-behavior. (d) $T_{c3}(x) = -1.33x + 0.44$, implies that $T_{c3} = 0$ at the Ru content $x = 0.33$. Reprinted from E. E. M. Chia et al., *Journal of Physics: Condensed Matter* **17**, L303 (2005). With permission from the Institute of Physics Publishing.

presented in Figure 2.32, where the main panels of (a)–(c) show the experimental data of $\rho_s(T)$ plotted as a function of T^2 for compositions $x = 0, 0.1,$ and 0.2. The solid line in each plot is a visual aid to assist in identifying the break from the T^2 dependence. The insets provide a measure of correlation R with the quadratic dependence of $\rho_s(T)$. Values of $+1$ (-1) indicate a perfect positive (negative) T^2 behavior. The point of maximum (absolute) R, close to the temperature where $\rho_s(T)$ starts to deviate from the quadratic dependence, is taken as the temperature T_{c3}, which is plotted in Figure 2.32(d) as a function of the content x of Ru. As suggested by Chia et al. (2005), the temperature trend in $\rho_s(T)$ in solid solutions with $x \leq 0.2$ might arise from an active exchange scattering mechanism at high temperatures where the Γ_5 level is populated on the Os-rich end of the phase diagram but is weakened or even suppressed by decreasing temperature or as the rising Ru content increases the Γ_1-$\Gamma_4^{(2)}$ splitting. Aspherical Coulomb scattering, which supports superconductivity, may remain important at lower temperatures and at higher Ru contents. It is interesting to note that the temperatures $T_{c3}(x)$ where the crossover in the density of superconducting condensate is observed are close to 0.6 K, where Cichorek et al. (2005) detected an unexpected enhancement in the lower critical field H_{c1} in $PrOs_4Sb_{12}$, which was interpreted as a sign of yet another phase of the system. The original measurements by Chia et al. (2003) a couple of years earlier on pure $PrOs_4Sb_{12}$ also displayed a small downturn in the penetration depth below 0.62 K and deviated from the T^2-dependence at temperatures above 0.6 K. Thus, it looks like some new and as yet unspecified phase might exist below 0.6 K in $PrOs_4Sb_{12}$ and $Pr(Os_{1-x}Ru_x)_4Sb_{12}$ solid solutions limited to $x \approx 0.2$.

The notable minimum in the transition temperature of $Pr(Os_{1-x}Ru_x)_4Sb_{12}$ solid solutions shown in Figure 2.31(a) drew the attention of Sergienko (2004), who argued that the competition between the s-wave superconductivity of $PrRu_4Sb_{12}$ and the triplet-order parameter of $PrOs_4Sb_{12}$ may give rise to a new superconducting state near $x = 0.6$ composed of both single and triplet components, i.e., a mixed-parity state. However, subsequent zero-field muon spin resonance (ZF-μSR) experiments by Shu et al. (2007) on a series of powder samples of $Pr(Os_{1-x}Ru_x)_4Sb_{12}$ have shown that the small local spontaneous magnetic field, arising as a consequence of the broken TRS in the superconducting state, is suppressed already at Ru concentrations of $x \geq 0.1$. Therefore, the proposed new mixed-parity superconducting state at $x \approx 0.6$ lacks solid grounding.

2.4.10 SUPERCONDUCTING SOLID SOLUTIONS OF $(PR_{1-x}LA_x)OS_4SB_{12}$

Explorations of $(Pr_{1-x}La_x)Os_4Sb_{12}$ solid solutions proved equally important. In this case, the dilution takes place directly on the sites of Pr^{3+} ions, and the average spacing between the cages containing Pr increases. Again, the alloying was performed with an ordinary s-wave superconductor $LaOs_4Sb_{12}$ having the transition temperature $T_c = 0.74$ K and lacking heavy fermions, Sugawara et al. (2002). Particularly revealing were the specific heat measurements of Rotundu et al. (2006) and the already mentioned studies of Sb NQR by Yogi et al. (2006). In both cases, single crystal specimens were used, and the measurements covered the full range of solid solubility. Very small variations, merely 0.4 %, between the lattice constants of the two end members, $PrOs_4Sb_{12}$ and $LaOs_4Sb_{12}$, and the Vegard's law behavior for all x values make $Pr_{1-x}La_xOs_4Sb_{12}$ an ideal system for alloying studies of physical phenomena, including superconductivity, as the microscopic inhomogeneities associated with the lattice mismatch are minimized. Specific heat data from Rotundu et al. (2006) are presented in Figure 2.33, where plots of C/T vs. T, normalized to a mole of Pr, are shown for a wide range of La contents x. Three things are noteworthy: i) the magnitude of the specific heat jump dramatically decreases with the increasing content of La, dropping from some 800 mJK^{-2} per mol of Pr for pure $PrOs_4Sb_{12}$ to about 280 mJK^{-2} per mol of Pr for $x = 0.2$ and further down to 160 mJK^{-2} per mol of Pr for $x = 0.3$. This implies that the electronic specific heat coefficient γ is greatly suppressed by La and the heavy fermion nature is lost. Yet, as shown in Figure 2.34, the composition dependence of T_c is smooth as one goes from the $T_c = 1.85$ K of $PrOs_4Sb_{12}$ to $T_c = 0.74$ K for $LaOs_4Sb_{12}$. ii) While samples with $x = 0$, 0.05 and, to a much lesser degree, $x = 0.1$ show two distinct superconducting transitions, the presence of 20% of La wipes out any sign of a double transition in the structure. Moreover, in very low La content samples, the width of the transition T_{c1}-T_{c2} shrinks rather dramatically as x increases. iii) If one starts with an ordinary s-wave $LaOs_4Sb_{12}$ superconductor with $T_c = 0.74$ K, it is remarkable that the presence of Pr on the sites of La increases the superconducting transition temperature rather than decreases it. Under the usual circumstances, Pr acts as a very efficient pair breaker, yet here it smoothly enhances superconductivity, which, at a full replacement of La by Pr, attains the transition temperature that is a factor of 2.5 larger than in $LaOs_4Sb_{12}$! As I discuss in subsequent paragraphs, the ability of Pr to enhance superconductivity is believed to be due to the quadrupolar interactions that dominate within the crystal field-split energy levels of the Pr^{3+} ions over the magnetic scattering. This has been independently confirmed by Chang et al. (2007) by solving the strong-coupling Eliashberg equations.

Measurements of muon spin rotation (μSR) are one of the most powerful probes with which to study internal magnetic field distributions, they are associated with the vortex lattice or with spontaneous magnetic fields arising as a consequence of the broken TRS in the superconducting state.

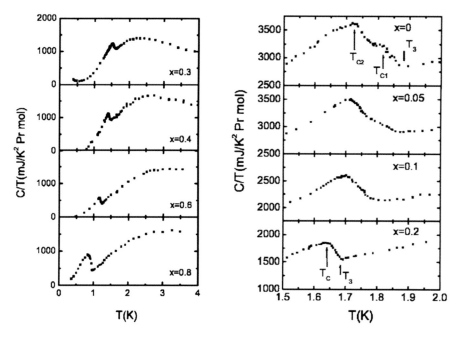

FIGURE 2.33 *C/T* as a function of temperature for $Pr_{1-x}La_xOs_4Sb_{12}$ with different contents *x* of La, near the superconducting transition temperature T_c. The data are normalized to a mole of Pr, and the phonon and normal electron contributions (assumed identical to that for $LaOs_4Sb_{12}$) were subtracted. Arrows indicate the scheme to determine transition temperatures and the width of the transition. Adapted from C. R. Rotundu et al., *Physical Review B* **73**, 014515 (2006). With permission from the American Physical Society.

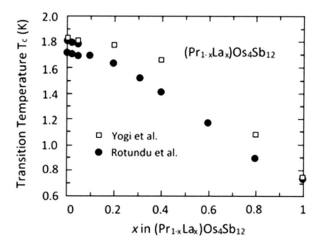

FIGURE 2.34 Superconducting transition temperature of $(Pr_{1-x}La_x)Os_4Sb_{12}$ as a function of the content *x* of La. Drawn from the data of M. Yogi et al., *Journal of the Physical Society of Japan* **75**, 124702 (2006) (open squares) and C. R. Rotundu et al., *Physical Review B* **73**, 014515 (2006) (solid circles). The double transition temperature observed for contents of Pr above about 95% is indicated by two sets of solid circles. With permission from the Physical Society of Japan and the American Physical Society.

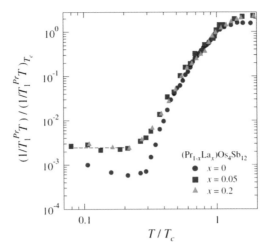

FIGURE 2.35   Plots of $(1/T_1^{Pr}T)/(1/T_1^{Pr}T)_{Tc}$ as a function of reduced temperature T/T_c for three low La content samples of $(Pr_{1-x}La_x)Os_4Sb_{12}$, showing a crossover to $1/T_1T$ = const behavior at temperatures below about $0.2T_c$. Adapted from Yogi et al., *Journal of the Physical Society of Japan* **75**, 124702 (2006). With permission from the Physical Society of Japan.

However, extracting the key superconducting parameters from the μSR spectra is not a trivial task and involves a variety of approaches, including approximations to the London and GL models or using exact GL models. For readers interested in details of such analyses, including also discussions on the validity and accuracy of various approaches, I recommend a paper by Maisuradze et al. (2009b).

Using the zero-field muon spin rotation (ZF-μSR) variant of the technique, Yogi et al. (2006) probed the nuclear spin-lattice relaxation time T_1 as a function of the La content and temperature down to 150 mK. The presence of two distinct icosahedron Sb cages, one containing Pr and the other La, results in two kinds of nuclear spin-lattice relaxation times, T_1^{Pr} and T_1^{La}. In the normal state of the solid solutions, the temperature dependence of the relaxation rate $(T_1^{Pr})^{-1}$ was essentially the same as that of pure $PrOs_4Sb_{12}$ irrespective of the content of La. On the other hand, the relaxation rate of La, $(T_1^{La})^{-1}$, decreased substantially when the La concentration increased. In the superconducting state of the Pr-rich solid solutions with $x = 0.05$ and 0.2, the relaxation rate of Pr decreased exponentially down to $T = 0.7$ K, and no coherence peak (the Hebel–Slichter peak) showed up below T_c, as already noted in the case of pure $PrOs_4Sb_{12}$ measured by Kotegawa et al. (2003). One of the remarkable findings in the measurements of Yogi et al. (2006) was a distinct break in the temperature dependence of the relaxation rate, followed by a constant $1/T_1^{Pr}T$ term, observed only in high Pr-content samples below about 0.3 K, see Figure 2.35. This signaled the presence of the residual DOS at the Fermi level well below T_c, which the authors interpreted as a signature of the multi-band nature of superconductivity. As before, it was assumed that there is a full gap over most of the Fermi surface with a small section of the Fermi surface containing point nodes. With an increasing content of La, the section of the Fermi surface with point nodes is gradually suppressed, and the structure becomes an ordinary *s*-wave superconductor.

2.4.11   SUPERCONDUCTING SOLID SOLUTIONS OF $(Pr_{1-x}Nd_x)Os_4Sb_{12}$

The last solid solution I want to mention combines the unconventional heavy fermion $PrOs_4Sb_{12}$ superconductor with a ferromagnetic compound $NdOs_4Sb_{12}$ that has a Curie temperature $T_{FM} = 0.9$ K, Ho et al. (2005). $(Pr_{1-x}Nd_x)Os_4Sb_{12}$ presents an interesting platform to test which one of the two competing states will form the ground state. Typically, the presence of a magnetic order

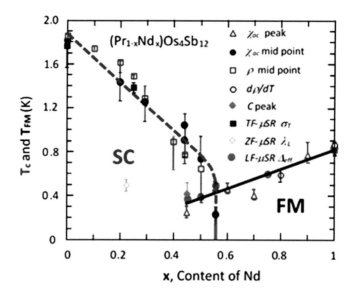

FIGURE 2.36 Superconducting transition temperature T_c and the Curie temperature T_{FM} plotted as a function of the content of Nd in $(Pr_{1-x}Nd_x)Os_4Sb_{12}$. The vertical bars indicate transition widths taken as the difference in temperature associated with the 10% and 90% values of the transition. Drawn from the data of C.-P. Ho et al., *Physical Review B* **83**, 024511 (2011) and D. E. MacLaughlin et al., *Physical Review B* **89**, 144419 (2014). With permission from the American Physical Society.

would very efficiently destroy a superconducting state. However, given reliable reports indicating that Cooper pairing in $PrOs_4Sb_{12}$ is of the triplet type, it is possible that the magnetic moment is somewhat tolerated, and this, indeed, is the case here. Measurements of the electrical resistivity $\rho(T)$, ac susceptibility χ_{ac}, and specific heat C by Maple et al. (2008) and Ho et al. (2008b), shown in Figure 2.36, indicate that while the superconducting transition temperature decreases linearly with the increasing presence of Nd in the structure, the superconductivity persists to quite high concentrations of Nd of about $x = 0.55$. At the same time, the Curie temperature decreases linearly as the content of Pr increases, and the system loses its magnetic moment at about $(1-x) = 0.45$. A further interesting aspect of the measurements was the behavior of the upper critical field $H_{c2}(0)$ extrapolated from measurements of $\rho(T)$ at constant magnetic field and $\rho(H)$ at constant temperature. While the high-field ordered phase is only weakly perturbed by Nd for concentrations up to about $x = 0.5$ (compared with its total destruction by just 10% of Ru), the upper critical field in the presence of Nd initially decreases rapidly for $x < 0.3$ and then more gradually until the superconductivity is destroyed at $x \sim 0.55$, see Figure 2.37. Figure 2.36 also includes the data from measurements of μSR by MacLaughlin et al. (2014) performed on several concentrations of Nd in $(Pr_{1-x}Nd_x)Os_4Sb_{12}$ in three distinct configurations: the transverse field (TF-μSR), longitudinal field (LF-μSR), and zero field (ZF-μSR).

An important outcome of μSR studies was the evidence against any phase separation within the structure. Three sample compositions $x = 0.45$, 0.50, and 0.55, displayed identical damping of the μSR signal characterized by a single exponential, as expected for a homogeneous system. The authors concluded that superconductivity and Nd^{3+} magnetism in $(Pr_{1-x}Nd_x)Os_4Sb_{12}$ coexist on the atomic scale.

All three substitutional studies with skutterudites differ in a major way from the behavior of other heavy fermion systems. The superconducting state in Pr-filled skutterudites tolerates surprisingly large amounts of chemical substitutions, even when they carry a magnetic moment. In contrast, in typical heavy fermion systems, even a minor addition of non-magnetic foreign species has a dramatic effect on their superconductivity and leads to its destruction, Heffner and Norman (1996).

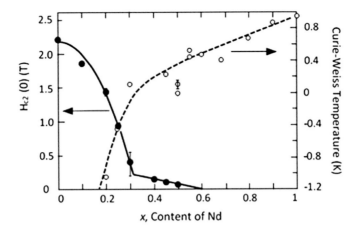

FIGURE 2.37 Upper critical field $H_{c2}(0)$ and the Curie–Weiss temperature of $(Pr_{1-x}Nd_x)Os_4Sb_{12}$ as a function of the content of Nd. The curves drawn are guides to the eye. Adapted from the data of P.-C. Ho et al., *Physical Review B* **83**, 024511 (2011). With permission from the American Physical Society.

2.4.12 OTHER PR-FILLED SKUTTERUDITE SUPERCONDUCTORS

Apart from $PrOs_4Sb_{12}$ with its highly unconventional superconducting state, there are two other superconductors with Pr as the filler, the already mentioned $PrRu_4Sb_{12}$ and $PrRu_4As_{12}$. They both have a fundamentally different nature of superconductivity, closely resembling BCS-type superconductors, and do not share the very heavy electron character, although $PrRu_4Sb_{12}$ ($m^* = 20\ m_e$) and $PrRu_4As_{12}$ ($m^* = 24\ m_e$) have somewhat heavier electrons. Before presenting their fundamental parameters, one might ask what about $PrRu_4P_{12}$, to consider all three pnicogen-type cages with the transition element Ru. The short answer is that $PrRu_4P_{12}$ is not a superconductor under normal conditions. As shown by Sekine et al. (1997), $PrRu_4P_{12}$ undergoes a metal–insulator transition at $T_{MI} = 62$ K, below which its resistivity increases with the decreasing temperature. However, upon application of a high pressure of 11 GPa, the insulating state at low temperatures gives way to a metallic state, and when the pressure exceeds 12 GPa, superconductivity sets in near 2 K, as shown by Miyake et al. (2004). The superconducting state is apparently suppressed by a magnetic field of 2 T.

2.4.12.1 PrRu₄Sb₁₂

In the context of $Pr(Os_{1-x}Ru_x)_4Sb_{12}$ solid solutions, I have mentioned that $PrRu_4Sb_{12}$ is a rather ordinary BCS-type superconductor with the transition temperature $T_c = 1.1$ K, Takeda and Ishikawa (2000). From the initial slope of the upper critical field, $-(dH_{c2}/dT)|_{T_c} = 0.24$ TK^{-1}, and applying Equations 2.3–2.7, one arrives at the effective electron mass $m^* \approx 20\ m_e$. Although this effective mass is somewhat elevated, the value is not exactly in the league of the heavy fermion systems. The BCS superconducting nature of $PrRu_4Sb_{12}$ is further substantiated by the distinct Hebel–Slichter peak just below T_c followed by the exponentially decreasing relaxation rate $1/T_1$ (with a value $\Delta(0)/k_BT_c = 1.5$), observed by Yogi et al. (2003) in Sb-NQR measurements. The penetration depth λ and the superfluid density ρ_s determined by Chia et al. (2004) yielded $\Delta(0)/k_BT_c = 1.9$ and $\lambda(0) = 2900$ Å. In addition, the specific heat measurements by Takeda and Ishikawa (2000) resulted in the specific heat jump *of* $\Delta C/\gamma T_c = 1.87$. All these values are mutually consistent and indicate a moderately coupled and fully gapped superconductor. Moreover, while the superconducting state in $PrOs_4Sb_{12}$ breaks TRS, giving rise to a tiny spontaneous internal magnetic field, similar µSR studies carried out by Androja et al. (2005) on $PrRu_4Sb_{12}$ revealed no temperature dependence in the zero field muon depolarization rate below and above the T_c, documenting the absence of internal spontaneous fields in the superconducting state. Consequently, the superconducting state of $PrRu_4Sb_{12}$ does not break TRS. By applying several modest magnetic fields in their transverse field µSR measurements at 0.05 K and 1.5 K (below and above T_c), the authors were able to extract two important superconducting parameters from muon-spin precession signals: the magnetic penetration depth λ (0.05 K) = 3650 Å and the superconducting coherence length ξ (0.05 K) = 345 Å. The relaxation rate of muon rotation below T_c differs significantly from the rate above T_c on account of the flux-line lattice (vortices), which generates spatial inhomogeneity in magnetic induction in the mixed state of a superconductor, i.e., in the external fields H satisfying $H_{c1} < H < H_{c2}$. The temperature dependence of the muon relaxation rate $\sigma_s(T)$ shown in Figure 2.38 also provided an opportunity for fitting to a phenomenological two-fluid model,

$$\sigma_s(T) = \sigma_s(0)\left[1 - \frac{T}{T_c}\right]^n,\tag{2.26}$$

with $\sigma_s(0)$, T_c, and n as variable parameters. The best fit resulted for T_c = 0.973 K, $\sigma_s(0)$ = 0.452 µs^{-1}, and n = 1.44, the results shown by the solid line in Figure 2.38. A small difference between the fitted T_c and the actual zero field-measured T_c = 1.1 K is accounted for by the applied magnetic field of 0.035 T used in the TF-µSR measurements. However, the discrepancy between the fitted exponent n = 1.44 and the value n = 4 expected for the s-wave isotropic superconducting gap, the dotted curve in Figure 2.38, is rather glaring. The authors offered an explanation for the discrepancy by noting that the muon spin-rotation rate depends not just on the penetration depth λ but also on the coherence length ξ. The relatively low value of $\kappa = \lambda/\xi$ implies an overlap between the flux lines, which decreases the measured muon relaxation rate σ_s. Correcting for the vortex overlap, the authors used a modified London model developed by Brandt (1998) that relates the now field-dependent muon relaxation rate to the penetration depth as depicted by a dash-dot curve in Figure 2.38. The modified penetration depth turned out to be λ (0.05 K) = 3610 Å. The values of ξ and λ calculated from the usual bulk measurements using Equations 2.4 and 2.9, $\xi_0 \approx 400$ Å and $\lambda(0) \approx 3200$ Å are similar to those derived from the TF-µSR measurement. The authors argued that because TF-µSR is a microscopic probe directly sensing the vortex lattice, the values of ξ and λ thus obtained are more reliable

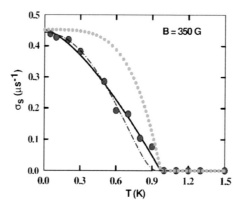

FIGURE 2.38 Temperature dependence of the muon-spin relaxation rate $\sigma_s(T)$ (solid red circles) associated with the flux-line lattice in the mixed state of PrRu$_4$Sb$_{12}$ at 0.035 T transverse field. The solid line is the fit based on the prediction of the phenomenological two-fluid model with n = 1.44. The dotted line is for n = 4 as expected to hold for an isotropic gap of an s-wave BCS superconductor. The dash-dotted line is a fit using the modified London model of Brandt (1998). Reproduced from D. T. Adroja et al., *Physical Review B* **72**, 184503 (2005). With permission from the American Physical Society.

than those resulting from the bulk measurements. Furthermore, although the penetration depth in $PrRu_4Sb_{12}$ is not too different from $PrOs_4Sb_{12}$ ($\lambda(0) = 3440$ Å), the coherence length of $PrRu_4Sb_{12}$ is about three times larger than that of $PrOs_4Sb_{12}$.

The temperature dependence of the specific heat as well as all transport and magnetic measurements indicated a single sharp superconducting transition in $PrRu_4Sb_{12}$. From the specific heat measurements, Takeda and Ishikawa (2000) estimated the Sommerfeld coefficient $\gamma \approx 59$ mJmol^{-1}K^{-2} and the Debye temperature of 232 K. The size of the jump in the specific heat at T_c combined with the value of γ yielded $\Delta C/\gamma T_c = 1.87$, larger than the value of 1.43 expected from the BCS theory, in accordance with the elevated effective mass of electrons. The CEF scheme for the Pr^{3+} ion multiplet was shown by Takeda and Ishikawa (2000) and by Abe et al. (2002) to consist of the Γ_1 singlet ground state and the first excited state $\Gamma_4^{(2)}$ lying some 70 K above.

2.4.12.2 $PrRu_4As_{12}$

Following the discovery of filled arsenide skutterudites by Braun and Jeitschko (1980), the first transport measurements on $PrRu_4As_{12}$ down to low temperatures were performed by Shirotani et al. (1997), who discovered that the electrical resistivity of this compound vanishes below 2.4 K. The bulk nature of the transition was confirmed by the fully diamagnetic response of the magnetic susceptibility cooled through the transition point in zero magnetic field. The sample studied was a polycrystalline structure synthesized from stoichiometric elements at a high pressure of 4 GPa and temperature of 900°C and contained small amounts of impurity phases, most notably $RuAs_2$. Using similar polycrystalline $PrRu_4As_{12}$, Namiki et al. (2007) characterized the superconducting state by performing specific heat and susceptibility measurements as a function of temperature and magnetic field. From the temperature and field sweeps of $C(T, H)$, they extracted the upper critical field $H_{c2}(T)$ with $H_{c2}(0)$ extrapolating to about 0.62 T. The initial slope $-(dH_{c2}(T)/dT)|_{Tc}$ turned out to be equal to 0.365 TK^{-1}. From the specific heat, the estimated value of the Sommerfeld constant was $\gamma \approx 95$ mJmol^{-1}K^{-2}, and the specific heat jump $\Delta C/\gamma T_c \approx 0.83$, a smaller value than 1.43 expected for a weak-coupling BCS superconductor. The analysis of the magnetic susceptibility indicated that the best fit to the CEF structure is made with a Γ_1 singlet as the ground state while the first excited state, separated by about $\Delta_{CEF} \sim 30$ K, is the $\Gamma_4^{(1)}$ triplet (the Γ_4 state in the cubic O_h symmetry), rather than the $\Gamma_4^{(2)}$ triplet as in $PrOs_4Sb_{12}$. The level assignment was confirmed by the ^{75}As-NQR measurements of Shimizu et al. (2008). Moreover, the spin-lattice relaxation rate $1/T_1$ displayed a clear Hebel–Slichter peak just below T_c, followed by exponential temperature dependence, both features revealing the s-wave superconducting nature of this arsenide skutterudite.

As already noted, while single crystals of filled phosphide and antimonide skutterudites can be prepared relatively simply using low-pressure flux techniques, the high vapor pressure of arsenic and its toxicity are major challenges for the flux synthesis of single crystals of arsenide skutterudites. Thus, measurements on single-crystal forms of the structure were not reported on prior to 2008 when Henkie et al. (2008) developed a process of mineralization in a Cd:As flux. With the single crystals available, superconducting measurements by Maple et al. (2008) and Sayles et al. (2010) confirmed the major findings made on polycrystalline structures, but also differed in values for some of the key parameters. The onset of superconductivity was raised to 2.5 K, the fits of C/T vs. T^2 gave a smaller Sommerfeld value of $\gamma \approx 70$ mJmol^{-1}K^{-2}, and from the exponential dependence of the specific heat below T_c, the value of $\Delta C/\gamma T_c = 1.53$, nearly double the value estimated from polycrystalline samples and much closer to the BCS prediction of 1.43, was obtained. The initial slope of the upper critical field $-(dH_{c2}(T)/dT)|_{Tc} = 0.43$ TK^{-1} is about 20% larger than the value obtained for polycrystalline samples. The closeness of the orbital critical field $H_{c2}^{orb}(0) = 0.71$ T calculated using Equation 2.2 to the experimentally extrapolated value of ~ 0.65 T indicated that superconductivity is limited by orbital de-pairing mechanisms rather than being Pauli paramagnetic limited ($H_{c2}^P(0) = 1.84\ T_c \approx 4.4$ T). Using Equations 2.4–2.9, the coherence length $\xi_0 = 178$ Å, the

Fermi velocity $v_F = 3.1 \times 10^4$ ms^{-1}, the effective carrier mass $m^* = 24\ m_e$, and the Sommerfeld coefficient $\gamma \approx 156$ mJmol^{-1}K^{-2} were calculated. The last parameter is about double the value estimated from the normal-state specific heat and $\Delta C(T)$ at T_c, quoted above. The same discrepancy, i.e., much lower γ values extrapolated from the normal-state specific heat, was also observed in the case of PrRu$_4$Sb$_{12}$. Perhaps, the greatest difference between single crystal and polycrystalline data is in the energy spacing of CEF levels, where the best fits to the magnetic susceptibility of single crystalline PrRu$_4$As$_{12}$ returned $\Delta_{CEF} = 95$ K between the Γ_1 ground state and the $\Gamma_4^{(1)}$ first excited state, the energy splitting a factor of 3 larger than the one estimated based on the polycrystalline sample. Moreover, in order to accurately describe the temperature dependence of the specific heat above T_c, it was necessary to include a localized Einstein-like vibration mode to the electronic specific heat, the lattice specific heat, and the specific heat associated with the CEF states.

To summarize, both PrRu$_4$Sb$_{12}$ and PrRu$_4$As$_{12}$ are conventional s-wave BCS superconductors with only moderately enhanced electron masses. Clearly, the cage environment in which the Pr^{3+} ions reside plays an important role, as reflected in the ordering and magnitude of CEF energy splitting that vary greatly from one compound to the next. It is interesting that the highly unconventional nature of superconductivity observed in PrOs$_4$Sb$_{12}$ is associated with the smallest energy spacing of merely $\Delta_{CEF} \approx 8$ K, while the energy spacing Δ_{CEF} between the ground state and the first excited state in the two conventional Pr-filled skutterudites is an order of magnitude larger.

2.5 YT$_4$P$_{12}$, (T = FE, RU, AND OS) SUPERCONDUCTORS

Skutterudites filled with heavier and smaller rare-earth ions (beyond Gd) were not possible to synthesize by the usual solid-state synthesis or using molten fluxes because such small ions could not form bonds and be retained in the oversized skutterudite cage. As demonstrated by Shirotani et al. (2003a), the problem was overcome by the development of a synthesis process utilizing high pressures of 4–5 GPa at temperatures over 1000°C. The technique allowed the synthesis of skutterudites with essentially any rare-earth filler, including Lu and also Y, the latter element being often counted among rare earths even though, technically, it is not. The high-pressure synthesis dramatically expanded the spectrum of filled skutterudites and enhanced chances of discovering new superconducting structures among them. The case in point is Y-filled phosphide skutterudites with all three transition metal elements T = Fe, Ru, and Os.

2.5.1 YF$_{E4}$P$_{12}$

Superconductivity of YFe$_4$P$_{12}$ with the onset of the transition temperature near 7 K was initially observed by Shirotani et al. (2003b). A sharp transition in the electrical resistivity was confirmed as being of bulk nature by essentially 100% flux exclusion observed in magnetic susceptibility measurements. Combining the jump in the specific heat at T_c with the least-squares analysis of the specific heat in the normal state ($C(T) = \gamma T + \beta T^3$ without CEF levels) yielded $\gamma = 2.72$ mJmol^{-1}K^{-2} and the parameter $\Delta C/\gamma T_c \approx 1.33$, the value close to 1.43 expected for the BCS theory. It is remarkable that despite containing iron, YFe$_4$P$_{12}$ is a superconductor and, as a matter of fact, except for the novel iron pnictide superconductors, e.g., Mancini and Citro (2017), a superconductor with the highest T_c of all Fe-containing superconducting structures. Subsequent NMR measurements by Magishi et al. (2005), tracking the nuclear spin-lattice relaxation time T_1 of ^{31}P nuclei in both the normal and superconducting domains, confirmed that YFe$_4$P$_{12}$ is a conventional s-wave superconductor. In the normal state, below 40 K, the metallic regime was characterized by the constant Korringa relation, $(T_1 T)^{-1} = 0.31$ s^{-1}K^{-1}, while in the superconducting state, the relaxation rate T_1^{-1} displayed a small but notable Hebel–Slichter coherence peak just below T_c, followed by an exponentially decreasing relaxation rate at still lower temperatures. The Arrhenius plot of $T_1/T_{1,Tc}$ vs T_c/T returned a value of $2\Delta(0)/k_B T_c = 3.3$, close to the conventional BCS value of 3.5 in the weak coupling limit.

As briefly noted in Section 2.3, YFe_4P_{12} is one of two entries in Table 2.1 (the other one being $LaFe_4P_{12}$) that exhibit a positive pressure derivative $(dT_c/dP)_{P=0}$. Based on band structure calculations, Cheng et al. (2013) and subsequently Nakazima et al. (2014) and Kawamura et al. (2015) pointed out that a filler itself makes a negligible contribution to the DOS at the Fermi level, and therefore, it is unlikely that it could control the slope of the pressure derivative. Rather, positive values of $(dT_c/dP)_{P=0}$ should relate to pressure changes in the physical properties of the normal state of a superconductor, such as a pressure-altered prefactor A and exponent n in the expression for the electrical resistivity $\rho(T) = \rho_o + AT^n$, as a consequence of the pressure-modified electron–phonon coupling constant λ_{e-p}. Following Cheng et al. (2013) and starting with Equation 2.1 of McMillan (1968) for the transition temperature T_c, it is straightforward to take the logarithmic volume derivative of T_c and arrive at the relation

$$-B\frac{dlnT_c}{dP} = -\gamma_{GR} + \Delta\left(\frac{dln\eta}{dlnV} + 2\gamma_{GR}\right). \tag{2.27}$$

Here, B is the bulk modulus, and γ_{GR} is the Grüneisen parameter defined as $\gamma_{GR} \equiv -dln\langle\omega\rangle/dlnV$, where $\langle\omega\rangle$ is the average phonon frequency and V is the volume. The parameter η is called the Hopfield parameter, defined as $\eta \equiv D(E_F)\langle I^2\rangle$, with $D(E_F)$ standing for the DOS at the Fermi energy and $\langle I^2\rangle$ the average squared electronic matrix element that can be calculated theoretically, Hopfield (1971). Because $\gamma_{GR} > 0$ is usually small compared to the second term on the right in Equation 2.27, the sign of the pressure derivative dT_c/dP depends on the relative magnitude of the two terms within the bracket. Δ is clearly always positive, and for simple s- and p-superconductors, $dln\eta/dlnV \approx -1$. With the Grüneisen parameter γ_{GR} typically between +1.5 and +2.5, Equation 2.27 usually leads to a negative dT_c/dP. However, in transition metal-based superconductors, d-electrons result in higher values of $D(E_F)$, and the calculations indicate a larger $dln\eta/dlnV \approx -3$ to -4. In this case, it is possible that the right-hand side of Equation 2.27 will turn negative, resulting in a positive dT_c/dP derivative. From full-potential linearized augmented plane-wave (FPLAPW) calculations of the electronic band structure energy of YFe_4P_{12}, Cheng et al. (2013) established that two conduction bands (the 47th and 48th) cross the Fermi level while in the case of YRu_4P_{12}, only the 48th band does so. Even neglecting the contribution of the 47th band (roughly 25 times smaller contribution to the DOS at the Fermi level compared to the 48th band), the effect of pressure on $D(E_F)$, and therefore on the Hopfield parameter η, is very different for the two Y-filled phosphides. As shown in Figure 2.39, where the angular momentum components of $D(E_F)$ for the 48th band of YFe_4P_{12} and YRu_4P_{12} are plotted at ambient pressure, and at 8 GPa, both the p-derived $D(E_F)_p$ and d-derived $D(E_F)_d$ components increase with pressure in the case of YFe_4P_{12}. Consequently, as indicated by the entries in Table 2.2, the Hopfield parameter η in YFe_4P_{12} is significantly enhanced under pressure. Because the band structure of $LaFe_4P_{12}$ is quite similar to that of YFe_4P_{12}, the same explanation accounts for the positive pressure derivative dT_c/dP in $LaFe_4P_{12}$, as discussed in section 2.3.

2.5.2 YOs_4P_{12} and YRu_4P_{12}

Observations of superconductivity in the sister skutterudites YOs_4P_{12} and YRu_4P_{12} followed shortly. Kihou et al. (2004) noted a sharp transition in the metallic temperature dependence of the electrical resistivity of YOs_4P_{12} near 3 K, and magnetic susceptibility measurements verified the bulk nature of superconductivity. A similar study conducted a few months later by Shirotani et al. (2005a) on YRu_4P_{12} revealed a bulk superconductor with the transition temperature near 8.5 K.

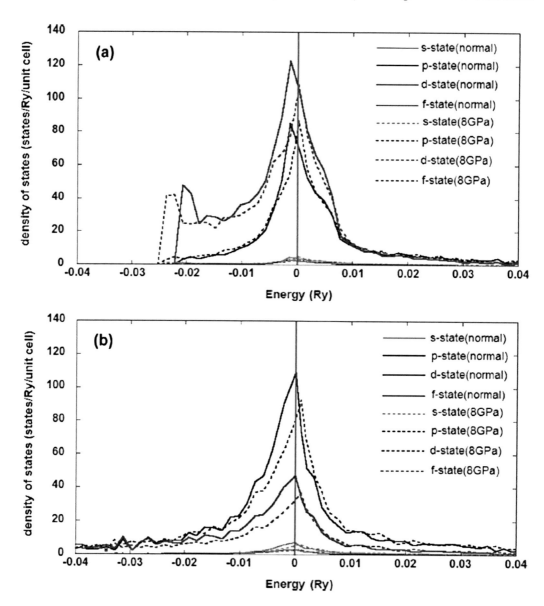

FIGURE 2.39 Angular momentum components of the DOS of the 48th band at ambient pressure (solid lines) and at 8 GPa (broken line) in the vicinity of the Fermi level $E_F = 0$ Ry for (a) YFe_4P_{12} and (b) YRu_4P_{12}. Reproduced from J.-G. Cheng et al., *Physical Review B* **88**, 024514 (2013). With permission from the American Physical Society.

However, the other superconducting features of YOs_4P_{12} and YRu_4P_{12} differ significantly from YFe_4P_{12}. Specifically, because of the decreasing p-derived and d-derived DOS at the Fermi energy with pressure, the Hopfield parameter of YRu_4P_{12} decreases, and thus the transition temperature decreases as the pressure rises, as shown in Table 2.2. Because of the similarity of band structures of YRu_4P_{12} and YOs_4P_{12}, both these Y-filled phosphite skutterudites should show closely related trends. The pressure derivative of the transition temperature for YOs_4P_{12} has been measured by Kawamura et al. (2015), yielding a value of $(dT_c/dP)_{P=0} = -0.11$ K(GPa)$^{-1}$. While the negative pressure derivative of the transition temperature for YRu_4P_{12} has been noted, its exact value has not yet been established.

TABLE 2.2

Contribution of p- and d-states to the DOS at the Fermi Level and the Overall Hopfield Parameter η for YFe$_4$P$_{12}$ and YRu$_4$P$_{12}$ at Ambient Pressure and at 8 GPa

Skutterudite	P (GPa)	T_c (K)	N(E$_F$)$_p$	N(E$_F$)$_d$	η (eVÅ^{-2})
YFe4P12	0	5.6	5.16	7.67	3.68
	8	9.3	6.37	7.98	5.16
YRu4P12	0	8.5	8.03	3.45	4.88
	8	< 8.5	6.64	2.81	4.74

Source: Data from J.-G. Cheng et al., *Physical Review B* **88**, 024514 (2013). With permission from the American Physical Society.

Note: The Angular components of the DOS are in units of States/eV/unit cell. Note substantial increases in the DOS and the Hopfield parameter with increasing pressure in the case of YFe$_4$P$_{12}$, responsible for its positive pressure derivative of the transition temperature. In contrast, for YRu$_4$P$_{12}$, the components of the DOS and the Hopfield parameter decrease with increasing pressure, leading to the usual decrease of T_c with increasing pressure.

2.6 SUPERCONDUCTORS WITH THE [PT$_4$GE$_{12}$] FRAMEWORK

The discovery of skutterudites with Ge replacing the pnicogen-based framework opened new possibilities for observing exciting transport, magnetic, and superconducting properties, especially when the column-8 (Fe, Ru, and Os) and the column-9 (Co, Rh, and Ir) transition metals were replaced with Pt from column 10.

2.6.1 ALKALINE-EARTH-FILLED BaPT$_4$GE$_{12}$ AND SrPT$_4$GE$_{12}$ SUPERCONDUCTORS

The first platinum germanide skutterudites were reported by Bauer et al. (2007), who used Ba and Sr in the voids to stabilize structurally the [Pt$_4$Ge$_{12}$] framework. Superconductivity was revealed immediately upon a rudimentary exploration of the transport properties of BaPt$_4$Ge$_{12}$ and SrPt$_4$Ge$_{12}$ as a sharp drop at T_c = 5.3 K and T_c = 5.1 K, respectively, in the otherwise metallic temperature dependence of the electrical resistivity. The bulk nature of superconductivity in the two skutterudites was confirmed by equally sharp transitions in the magnetic susceptibility, which dropped from a small positive value to a diamagnetic value of $-1/4\pi$ at the transition temperature, revealing the fully developed Meissner effect. Well-formed jumps in the specific heat at T_c of both skutterudites were a further attestation of bulk superconductivity. Because no f-electrons were involved, there was no interference from the CEF contribution, and the specific heat was adequately described by a simple form of Equation 2.19, where only the electronic and lattice terms were present,

$$C(T) = \gamma T + \beta T^3. \tag{2.28}$$

Least-squares fits of the data using Equation 2.28 yielded very similar electronic specific heat coefficients, γ = 42 mJmol^{-1}K^{-2} and γ = 41 mJmol^{-1}K^{-2} for BaPt$_4$Ge$_{12}$ and SrPt$_4$Ge$_{12}$, respectively. From the fitted value of the parameter β, the extrapolated Debye temperatures θ_D = 247 K and θ_D = 220 K, respectively, were somewhat surprising, given that Sr has a much smaller mass than Ba. Bauer et al. (2007) surmised that this is due to the 12% smaller atomic volume of Sr while the volume of the unit cells of the two skutterudites differs by only 1%. In other words, Sr bonding with the framework is weaker than that of Ba, and the weaker force constant results in a smaller value of θ_D. From the size of the jump in the specific heat at T_c, $\Delta C(T_c)/T_c \approx 58$ mJmol^{-1}K^{-2} for both BaPt$_4$Ge$_{12}$ and SrPt$_4$Ge$_{12}$, the parameter $\Delta C/\gamma T_c$ = 1.35 was obtained. This value is very close to the BCS value of 1.43, making

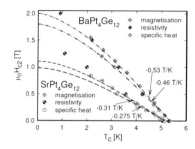

FIGURE 2.40 Temperature-dependent upper critical field $H_{c2}(T)$ of BaPt$_4$Ge$_{12}$ and SrPt$_4$Ge$_{12}$ obtained from measurements of electrical resistivity, magnetization, and specific heat in magnetic field. The dashed and dash-dotted curves represent the orbital pair-breaking limitation, calculated from the WHH (1966) model. Reproduced from E. Bauer et al., *Physical Review Letters* **99**, 217001 (2007). With permission from the American Physical Society.

the two superconducting skutterudites ordinary *s*-wave BCS types. The exponentially falling specific heat in the superconducting state allowed Bauer et al. (2007) to extract a ratio $\Delta(0)/k_B T_c \approx 1.8$, again, very close to the BCS value of the superconducting gap, $\Delta(0) = 1.76\ k_B T_c$. The upper critical field $H_{c2}(0)$ was obtained from extrapolating the magnetic field-dependent transitions in the electrical resistivity, magnetization, and specific heat measurements to absolute zero temperature, as shown in Figure 2.40. Comparing the extrapolated values of $H_{c2}(0) \approx 2$ T for BaPt$_4$Ge$_{12}$ and about 1 T for SrPt$_4$Ge$_{12}$, with predictions of Equation 2.3 for the Clogston–Chandrasekhar limit of nearly 10 T, it is clear that superconductivity in these two platinum germanide skutterudites is limited by orbital pair-breaking. The initial slopes of $dH_{c2}(T)/dT$ are given in Figure 2.40. With the values of the upper critical field, and using Equations 2.4–2.9, one can obtain all important parameters characterizing the superconducting state of the structure. Thus, for instance, the coherence length is $\xi_0 \approx 140$ Å for BaPt$_4$Ge$_{12}$ and ≈ 180 Å for SrPt$_4$Ge$_{12}$. To evaluate the penetration depth, Bauer et al. (2007) used an alternative approach by calculating the thermodynamic critical field from the free energy difference between the superconducting and normal states, $\Delta F(T) = F_n - F_{sc} = H_c^2(T)/2$, where the F_n and F_{sc} are obtained from the specific heat in the normal and superconducting states, respectively. The resulting small values of the thermodynamic critical field $H_c(0) \approx 53$ mT and ≈ 52 mT for BaPt$_4$Ge$_{12}$ and SrPt$_4$Ge$_{12}$, respectively, were then used together with the upper critical field $H_{c2}(0)$ in the Abrikosov's relation for the GL parameter $\kappa_{GL} \equiv \lambda(0)/\xi_0 = H_{c2}(0)/[\sqrt{2}H_c(0)]$, which yielded $\kappa_{GL} = 24$ and 14, respectively. The penetration depth then follows from $\lambda(0) = \kappa_{GL}\xi_0 = 3200$ Å for BaPt$_4$Ge$_{12}$ and 2500 Å for SrPt$_4$Ge$_{12}$. With the transport mean free path of electrons in BaPt$_4$Ge$_{12}$ $l_{tr} \sim 100$ Å and 140 Å in SrPt$_4$Ge$_{12}$, the ratio $l_{tr}/\xi_0 \sim 1$ for both skutterudites implies that BaPt$_4$Ge$_{12}$ and SrPt$_4$Ge$_{12}$ are classified as superconductors near the dirty limit. Fully relativistic calculations of the electronic band structure revealed that Ge p-states dominate the DOS in the vicinity of the Fermi level and that there is some hybridization with Pt 5d-states. On the other hand, the filler species contribute negligibly to the DOS at the Fermi level. The results depicted in Figure 2.41 for both BaPt$_4$Ge$_{12}$ and SrPt$_4$Ge$_{12}$ indicate that the peak in the DOS for BaPt$_4$Ge$_{12}$ is very close to the Fermi level, but not exactly overlapping.

Within a week of submission of the paper by Bauer's et al., Gumeniuk et al. (2008a) submitted their study that included not just the BaPt$_4$Ge$_{12}$ and SrPt$_4$Ge$_{12}$ skutterudites but also expanded the range of platinum germanide skutterudites to include several rare-earth fillers. The two alkaline-earth-filled structures, BaPt$_4$Ge$_{12}$ and SrPt$_4$Ge$_{12}$, became superconducting in a comparable range of temperatures as observed by Bauer et al. (2007), except that the transition temperature of SrPt$_4$Ge$_{12}$ (5.40 K) turned out to be higher than that of BaPt$_4$Ge$_{12}$ (4.98 K). Obviously, there is considerable sensitivity of the transport and superconducting properties to the exact stoichiometry.

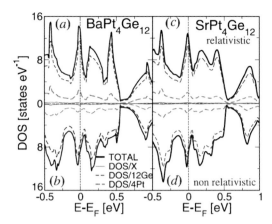

FIGURE 2.41 Calculations of the DOS near the Fermi level for $BaPt_4Ge_{12}$ (panels (a) and (b)) and for $SrPt_4Ge_{12}$ (panels (c) and (d)). The top two panels (a) and (c) show the DOS fully relativistic calculation including the spin-orbit coupling. The bottom two panels (b) and (d) present the DOS of a standard non-relativistic calculation. The relativistic effect is small but important, because it shifts the Fermi level of the non-relativistic DOS almost into the maximum of the relativistic DOS. The solid black line is the total DOS, the blue dashed line is the DOS of 12 Ge atoms, the gray dashed line is the DOS of 4 Pt atoms, and the thin red line is the DOS of the filler. Reproduced from E. Bauer et al., *Physical Review Letters* **99**, 217001 (2007). With permission from the American Physical Society.

Note that the peak in the DOS of $BaPt_4Ge_{12}$ does not fall exactly at the Fermi level, and Gumeniuk et al. (2008b) attempted to adjust its position by chemical substitution. Doping or substitution in superconductors is tricky and must be done with caution as there is always a danger the dopant will act as a pair-breaker and dramatically lower the transition temperature. In some cases, however, such as when the substituting element tends to enhance the DOS at the Fermi level, there is a chance of increasing the transition temperature. This is the case of $BaPt_4Ge_{12}$ where some platinum was replaced with gold, forming $Ba(Pt_{4-x}Au_x)Ge_{12}$ solid solutions. Because Pt $5d$-states lie deep in the valence band and make only a minor contribution to the DOS near the Fermi level, there is a possibility that by replacing Pt with an aliovalent element, such as gold, the density of electrons will be altered, and one may match the Fermi level with the peak in the DOS. In contrast, isovalent substitutions (different size ions of the same valence) would be unlikely to make an impression on T_c because the transition temperature of $BaPt_4Ge_{12}$ was shown by Khan et al. (2008) to be rather little affected by pressure. The effect of pressure on the superconductivity of platinum germanide skutterudites is discussed in Section 2.6.4.

Gumeniuk et al. (2008b) found that Au can replace Pt while the structure remains a single-phase skutterudite for concentrations x up to 1, i.e., 25% substitution. At higher levels of Au, multi-phase structures appeared. The magnetic susceptibility measured in the ZFC and FC modes in a field of 2 mT for samples with the Au content $x = 0, 0.25, 0.5, 0.75$, and 1.0 yielded the transition temperatures 4.98 K, 5.05 K, 5.27 K, 6.27 K, and 7.00 K, respectively. Thus, the initial increase of T_c was slow but became rapid as the content of Au approached $x = 1$. Fits to the specific heat indicated a modestly rising Sommerfeld coefficient γ with the increasing content of Au, which attained a 20% greater value for $x = 1$ than in pure $BaPt_4Ge_{12}$. Concomitantly, while the normalized specific heat jump $\Delta C/\gamma T_c$ increased with the increasing x, both $x = 0$ and $x = 1$ skutterudites showed significantly lower values of 1.03 and 1.37, respectively, compared to the BCS value of 1.43. The authors speculated whether this trend might be hinting at the multi-band nature of superconductivity. On the other hand, within the one-band scenario, the rising value of $\Delta C/\gamma T_c$ with x would suggest a progressively stronger electron–phonon coupling. The upper critical field, determined from the temperature where the electrical resistivity became zero, extrapolated to the same $H_{c2}(0) \approx 2.0$ T for both $BaPt_4Ge_{12}$ and $BaPt_3Au_1Ge_{12}$, implying that the initial slope $-(\partial H_{c2}(T)/\partial T)_{|Tc}$ becomes gradually smaller in magnitude as x increases.

The superconducting state of $BaPt_4Ge_{12}$ and $SrPt_4Ge_{12}$ was further characterized by NQR measurements of Magishi et al. (2011) on ^{73}Ge-enriched samples, which indicated the presence of a small Hebel–Slichter coherence peak in the temperature dependence of the relaxation rate $1/T_1$ just below T_c, another manifestation of the conventional BCS nature of the two superconductors. The ordinary s-wave BCS type of superconductivity in $BaPt_{4-x}Au_xGe_{12}$ was also independently verified by transverse-field muon spin rotation (TF-μSR) measurements of Maisuradze et al. (2012), which probed the state of the superconducting condensate as a function of temperature. The exponentially saturating density of superconducting quasiparticles was an unmistakable proof of the absence of nodes in the superconducting gap function, in complete agreement with the NMR study of Magishi et al. (2011).

The compendium of data on $BaPt_4Ge_{12}$ and $SrPt_4Ge_{12}$ revealed that it is possible to adjust the charge on the $[Pt_4Ge_{12}]$ polyanion by electron transfer *via* either a suitably chosen filler ion or by an aliovalent substitution on the transition metal site. If the latter path were chosen, detailed knowledge of the dependence of the DOS on energy in the proximity of the Fermi level is indispensable in order to select the right level of substitution that places the Fermi level at the maximum of the DOS, as indicated in Figure 2.42 in the case of replacing 25% of Pt atoms with atoms of Au.

Grytsiv et al. (2008) attempted to synthesize a companion alkaline-earth-filled skutterudite, $CaPt_4Ge_{12}$, but with no success. Apparently, only about 20% of Ca can be substituted for Ba in the $[Pt_4Ge_{12}]$ framework to keep the structure intact. The resulting compound $Ba_{0.8}Ca_{0.2}Pt_4Ge_{12}$ maintained very similar superconducting parameters compared to $BaPt_4Ge_{12}$, including the transition temperature $T_c = 5.2$ K, attesting to the conventional nature of superconductivity in APt_4Ge_{12}, $A = Ba$ and Sr, where one would expect a non-magnetic impurity to disturb the superconducting state to only a minor degree.

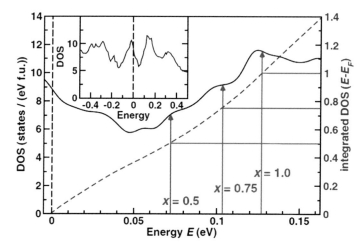

FIGURE 2.42 The inset shows the electronic DOS of $BaPt_4Ge_{12}$ (black curve) and $LaPt_4Ge_{12}$ (gray area). The main panel is a magnified region of the DOS of $BaPt_4Ge_{12}$ very close to the Fermi energy. The blue dotted curve is the integrated DOS between E and E_F. The red arrows indicate the shifting E_F for various substitution levels of Au on the site of Pt. Substitution of Au at $x = 1$ places the Fermi level at the maximum of the DOS, yielding the highest transition temperature $T_c = 7.0$ K, a more than 2 K enhancement over that of pure $BaPt_4Ge_{12}$. Reproduced from R. Gumeniuk et al., *Physical Review B* **78**, 052504 (2008b). With permission from the American Physical Society.

2.6.2 Actinide-Filled ThPt$_4$Ge$_{12}$ Superconductor

Another Ge-based superconducting skutterudite, ThPt$_4$Ge$_{12}$, was discovered by Kaczorowski and Tran (2008) with the transition temperature T_c = 4.62 K and independently by Bauer et al. (2008) with T_c = 4.75 K. Thorium enters the skutterudite structure as a Th^{4+} ion with no f-electrons and is thus expected to behave as a simple system. A weakly diamagnetic susceptibility of ThPt$_4$Ge$_{12}$ in the normal state exhibits a sharp drop at T_c and displays complete flux exclusion. However, in rather low magnetic fields on the order of 0.01 T, the flux exclusion is only partial (~ 20%), revealing strong flux pinning of a type-II superconductor. Using standard approaches to evaluate the superconducting parameters from their experimental data, Kaczorowski and Tran and (Bauer et al.) arrived at similar results that characterize a more or less ordinary s-wave BCS superconductor with γ = 40 mJmol^{-1}K^{-2} (35 mJmol^{-1}K^{-2}); $\Delta C/\gamma T_c$ = 1.7 (1.75); $\Delta(0)/k_B T_c$ = 1.7 (1.8); $H_{c2}(0)$ = 0.29 T (0.21 T); $-(\partial H_{c2}(T)/\partial T)|_{Tc}$ = 0.085 TK^{-1} (0.064 TK^{-1}); ξ_0 = 350 Å (400 Å); and $\lambda(0)$ = 1500 Å (1200 Å). From the McMillan formula in Equation 2.1, it follows that the electron–phonon coupling constant $\lambda_{e\text{-}p} \approx 0.62$ (0.66), which, together with the enhanced value of $\Delta(0)/k_B T_c$, classifies ThPt$_4$Ge$_{12}$ as being strongly coupled. Comparing the transport mean free path $l_{tr} \approx 4350$ Å (6500 Å) with the coherence length, the ratio $l_{tr}/\xi_0 \approx 12$ (16). The values fall close to the regime of clean superconductors, $l_{tr} \gg \xi_0$, the assessment underpinned primarily by the rather low electrical resistivity (the RRR ~ 100) even though the samples are polycrystalline. In an attempt to shed more light on the superconducting state of ThPt$_4$Ge$_{12}$, the same group, Tran et al. (2009), carried out transport studies in the normal state of the skutterudite. Transport properties were analyzed in terms of a two-band conduction model with both electrons and holes participating in the conduction process, a situation different from the dominance of a single carrier type as in most conventional superconductors. Subsequent studies by Tran et al. (2011) probed the superconducting state of ThPt$_4$Ge$_{12}$ via ZT- and TF-μSR measurements. The analysis of the muon spin relaxation rate $\sigma_s(T)$ yielded a power law behavior, which was taken as a sign of the presence of nodes on the superconducing gap function. The authors offered an alternative explanation in terms of a clean superconductor, but the agreement with the data was less satisfactory. The gap parameter $\Delta(0)$ = 1.7 $k_B T_c$ and the penetration depth $\lambda(0)$ = 1560 Å extracted from the measurements agreed well with the group's previous values based on the evaluation of the specific heat. I should also mention that Bauer et al. (2008), in their study of ThPt$_4$Ge$_{12}$, also tested UPt$_4$Ge$_{12}$ but the dominating spin fluctuations prevented the development not only of a superconducting state but also of the magnetic order.

2.6.3 Rare-Earth-Filled [Pt$_4$Ge$_{12}$]-Based Superconductors

Of the five rare-earth-filled skutterudites explored by Gumeniuk et al. (2008a), CePt$_4$Ge$_{12}$, NdPt$_4$Ge$_{12}$, and EuPt$_4$Ge$_{12}$ showed no signs of superconductivity down to 0.48 K. On the other hand, LaPt$_4$Ge$_{12}$ and PrPt$_4$Ge$_{12}$ became superconducting at unexpectedly high transition temperatures of 8.3 K and 7.9 K, respectively. Although a well-developed jump in the specific heat at T_c signaled a bulk form of superconductivity, zero field-cooled curves of the susceptibility gave an order of magnitude smaller Meissner effect than the expected full flux exclusion, apparently the result of strong pinning. Fitting the specific heat data between 3 K and 10 K, the normal state specific heat coefficients γ = 76 mJmol^{-1}K^{-2} and 87 mJmol^{-1}K^{-2} were extracted for LaPt$_4$Ge$_{12}$ and PrPt$_4$Ge$_{12}$, respectively. Making use of an entropy conserving construction, carefully evaluated jumps in the specific heat at T_c resulted in $\Delta C/\gamma T_c \approx 1.49$ for LaPt$_4$Ge$_{12}$ and ≈ 1.56 for PrPt$_4$Ge$_{12}$, both values higher than the 1.43 based on BCS theory. The superconducting energy gap $\Delta(0)$ was evaluated by comparing the specific heat taken at half the superconducting transition temperature T_c under zero magnetic field with the curves of $C_{es}/\gamma T_c$ vs. T_c/T developed for different values of parameter $\alpha \equiv \Delta(0)/k_B T_c$, the so-called α-model of Padamsee et al. (1973). The values of $\Delta(0)/k_B T_c$ = 1.94 and 2.35 were obtained for LaPt$_4$Ge$_{12}$ and PrPt$_4$Ge$_{12}$, respectively, confirming the strong coupling, especially in the case of PrPt$_4$Ge$_{12}$. By the way, throughout the text and in their Table I, the paper of Gumeniuk et al. (2008a) mistakenly used $2\Delta(0)/k_B T_c = 1.76$ for the BCS value instead of the correct $\Delta(0)/k_B T_c = 1.76$, and the reader referring to the paper should be aware of this. The temperature dependence of the upper critical field was determined from the mid-point of the jumps in $C(T)$ and in both $LaPt_4Ge_{12}$ and $PrPt_4Ge_{12}$ gave a nearly T-linear variation that extrapolated to $H_{c2}(0) \approx 1.60$ T for $LaPt_4Ge_{12}$ and $H_{c2}(0) \approx 2.06$ T for $PrPt_4Ge_{12}$.

Because the $4f$-electrons of Pr^{3+} ions are involved, it was of interest to evaluate how the CEF leads to a non-magnetic ground state in $PrPt_4Ge_{12}$. The level splitting was established by two distinct ways. First, the experimentally extracted specific heat contribution $C_{CEF}(T)$ was found by subtracting from the measured $C(T)$ data of $PrPt_4Ge_{12}$ the specific heat of $LaPt_4Ge_{12}$ (no f-electrons) and fitting the difference to a Schottky-like two-level system with an appropriate point symmetry of the T_h group. Second, the paramagnetic susceptibility of $PrPt_4Ge_{12}$ was approximated with the *LLW* energy level model developed by Lea, Leask and Wolf (1962) for the O_h symmetry. Regardless of the method, the ground state of $PrPt_4Ge_{12}$ turned out to be the Γ_1 singlet and the first excited state the $\Gamma_4^{(1)}$ triplet lying at $\Delta E_{CEF}/k_B = 131$ K. The upper energy levels have less certain values, with the Γ_{23} doublet at about 230 K and the triplet $\Gamma_4^{(2)}$ near 300 K. The more than an order of magnitude larger ΔE_{CEF} in $PrPt_4Ge_{12}$ compared to $PrOs_4Sb_{12}$ leaves the ground state substantially isolated; therefore, the CEF has essentially no effect on superconductivity of $PrPt_4Ge_{12}$. Gumeniuk and colleagues also carried out electronic band structure calculations using fully relativistic, LDA-based DFT with the Perdew–Wang exchange correlation potential to obtain the DOS and its atom-resolved contributions near the Fermi level for both $LaPt_4Ge_{12}$ and $PrPt_4Ge_{12}$. The results are reproduced in Figure 2.43 and indicate that the majority of Pt $5d$ states strongly hybridize with Ge $4p$ orbitals in the range between -5.5 eV and -2.5 eV, while at the Fermi level, 80% of electronic states derive from Ge $4p$ bands. The actual values of $D(E_F)$ are 13.4 states eV^{-1} f.u.$^{-1}$ for $LaPt_4Ge_{12}$ and 9.3 states eV^{-1} f.u.$^{-1}$ for $PrPt_4Ge_{12}$. Similar to the case of $PrOs_4Sb_{12}$, the Pr^{3+} ion in $PrPt_4Ge_{12}$ does not play a role as a pair-breaker and maintains the transition temperature comparable to that of $LaPt_4Ge_{12}$.

However, because there is no interference from the magnetic triplet $\Gamma_4^{(1)}$, which lies far above the ground state, the superconductivity is likely due to the intrinsically high transition temperature associated with the $[Pt_4Ge_{12}]$ framework.

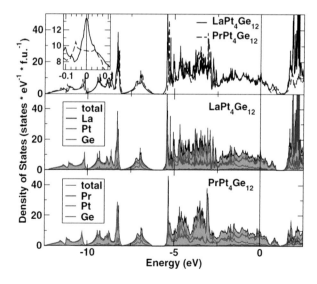

FIGURE 2.43   Total (upper panel) and atom-resolved (lower two panels) electronic DOS for $LaPt_4Ge_{12}$ and $PrPt_4Ge_{12}$. The inset in the upper panel shows the well-pronounced peak in $LaPt_4Ge_{12}$ in a narrow region around the Fermi level. Reproduced from R. Gumeniuk et al., *Physical Review Letters* **100**, 017002 (2008a). With permission from the American Physical Society.

The surprisingly high transition temperatures of $LaPt_4Ge_{12}$ and $PrPt_4Ge_{12}$ were promptly confirmed in measurements by Toda et al. (2008), who also conducted NMR studies on their $LaPt_4Ge_{12}$ sample. Because there are two NMR-active nuclei, ^{195}Pt with the nuclear spin $I = \frac{1}{2}$ and ^{139}La with $I = 7/2$, two sets of NMR spectra are available. As we already know from the discussion of the NMR data of $PrOs_4Sb_{12}$ in Section 2.4, the parameter of interest in the context of superconductivity is the temperature dependence of the nuclear spin-lattice relaxation time T_1. Because the present NMR measurements were performed at rather high magnetic fields of 0.6 T and 1.2 T, some 38% and 75% of $H_{c2}(0)$, respectively, the magnetic field likely suppressed the Hebel–Slichter coherence peak expected to appear just below T_c, and there was no sign of it on the temperature dependence of the relaxation rate of either ^{159}Pt or ^{139}La. Nevertheless, from the Arrhenius plot of $T_1/T_{1,c}$ vs. T_c/T, where $T_{1,c}$ is the relaxation time at T_c, the distinct straight line yielded $\Delta(0)/k_BT_c \approx 1.62$, close to the conventional BCS value of 1.76. However, compared to the estimated value from the specific heat of Gumeniuk et al. (2008a) of 1.94, this is nearly a 20% lower value.

Because the experimentally extracted high values of $\Delta C/\gamma T_c$ and $\Delta(0)/k_BT_c$ suggested that $PrPt_4Ge_{12}$ is a strong coupling superconductor, it was of interest to check whether there was something even more profound regarding the superconducting properties of this skutterudite. The µSR studies of Maisuradze et al. (2009a) showed a continuous increase in the density of the superconducting condensate with decreasing temperature, while the specific heat measurements, extending down to 0.4 K, documented a T^3 power law dependence rather than an exponentially decreasing trend in the electronic specific heat, signaling unusual features of the superconducting state of $PrPt_4Ge_{12}$. From fits to the temperature dependence of the normalized superfluid density $\rho_s/\rho_s(0)$ vs T/T_c, shown in Figure 2.44, and taking into account the experimental estimate by Gumeniuk et al. (2008a) of $\Delta(0)/k_BT_c \approx 2.35$ as a constraint on the gap function, the authors were able to restrict possible forms of the gap function to two model functions C and D, in Table 2.3. Although even smaller least-squares deviation χ^2 was obtained by fitting model F, the much too high value of $\Delta(0)/k_BT_c \approx 3.56$ disqualified it from considerations.

In section 2.4, we have seen that the crucial evidence for the unconventional nature of the superconducting state in $PrOs_4Sb_{12}$ was the detection of a small spontaneous magnetic field due to spin or orbital degrees of freedom of the Cooper pairs observed below T_c in ZF-µSR experiments. Not surprisingly, it did not take long to apply such pivotal experimental measurements also to $PrPt_4Ge_{12}$. Moreover, it made sense to expand measurements to an entire series of solid solutions $(La_{1-x}Pr_x)Pt_4Ge_{12}$ to glean how the nature of the superconducting state changes on going from an ordinary s-wave BCS-type superconductor to a strong-coupling superconductor with point nodes in its gap function. The measurements were performed by Maisuradze et al. (2010) down to 1.5 K and included a carefully reduced magnetic field down to 3×10^{-6} T generated by a set of three orthogonally coupled Helmholtz coils. The depolarization rate of muons obtained from the ZT-µSR time spectra is shown in Figure 2.45 for the end member skutterudites, $LaPt_4Ge_{12}$ and $PrPt_4Ge_{12}$. While the muon depolarization rate in $LaPt_4Ge_{12}$ is small and temperature independent throughout the range of measurements, there is a distinct break at T_c in the much larger depolarization rate of $PrPt_4Ge_{12}$, which increases at still lower temperatures. In fact, the rising muon depolarization rate below T_c is reminiscent of the increasing density of the superconducting condensate $\rho_s(T)$ seen by the same group in their previous study, Maisuradze et al. (2009a), which suggested the gap function of the form $\Delta(0)|k_x \pm ik_y|$ (model C), as being a good description of the superconducting state. Carefully eliminating any spurious reasons for an enhancement in the depolarization rate of muons, the authors concluded that it arises from the spontaneous magnetic field below T_c, documenting that the superconducting state of $PrPt_4Ge_{12}$ breaks TRS. No hint of the broken TRS was seen in the case of $LaPt_4Ge_{12}$. In spite of the notable difference between the superconducting states of the end member compounds, the authors pointed out a smooth and weak variation in T_c with the content of Pr, which they took to mean that the order parameters of $PrPt_4Ge_{12}$ and $LaPt_4Ge_{12}$ are compatible and not separated by any first-order phase transition. To find a common ground between the superconducting states of $PrPt_4Ge_{12}$ and $LaPt_4Ge_{12}$, they speculated that, in spite of its TRS, $PrPt_4Ge_{12}$ might be a spin-singlet superconductor. In this case, however, the gap function would have to belong to a

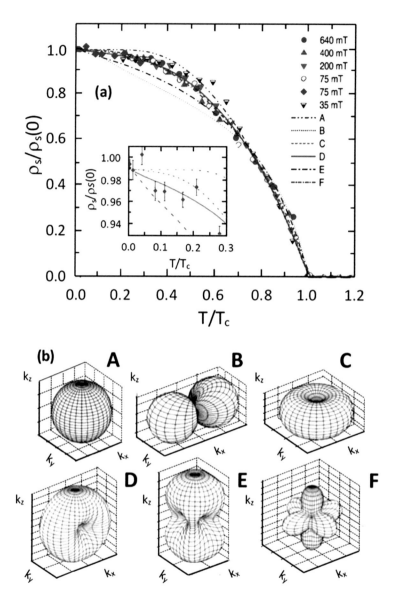

FIGURE 2.44 (a) Temperature dependence of the normalized quasiparticle density $\rho_s/\rho_s(0)$ as a function of the reduced temperature measured at different magnetic fields indicated by symbols. The inset depicts data obtained at 75 mK in the low-temperature region between $0.004 \leq T/Tc \leq 0.3$. The fitted lines refer to models A through F of various gap functions illustrated in (b) and specified by the mathematical formula in Table 2.3. Reproduced from A. Maisuradze et al., *Physical Review Letters* **103**, 147002 (2009). With permission from the American Physical Society.

complex, orbitally degenerate representation leading to an internal orbital moment of the Cooper pairs. The orbital moment could vary and conform to the broken TRS state in $PrPt_4Ge_{12}$, while also satisfying TRS and a smooth transition toward the superconducting state of $LaPt_4Ge_{12}$. The same ZT-μSR technique was used by Zhang et al. (2015a) in their study of the broken TRS in $Pr_{1-x}Ce_xPt_4Ge_{12}$ solid solutions. As we have already seen in Section 2.3, Ce tends to play the role of a pair-breaker and suppresses the superconducting state. The small spontaneous magnetic field decreases linearly with the increasing content of Ce and, at merely $x \approx 0.4$, the field is no longer

TABLE 2.3

Summary of the Analysis of the Superconducting Quasiparticle Density $\rho_s(T)$ for PrPt$_4$Ge$_{12}$

Model	Nodes	Superconducting Gap Function $\Delta(\phi, \theta)$	$\Delta(0)/k_B T_c$	$\chi^2/\chi^2(F)$
A		$\Delta(0)$	1.95	3.99
B	1	$\Delta(0)\lvert k_y \rvert$	4.88	9.32
C	p	$\Delta(0)\lvert k_x \pm i k_y \rvert$	2.68	1.05
D	p	$\Delta(0)\left(1 - k_y^4\right)$	2.29	1.17
E	p	$\Delta(0)\left(1 - k_x^4 - k_y^4\right)$	3.84	3.46
F	p	$\Delta(0)\left[1 - 3\left(k_x^2 k_y^2 + k_x^2 k_z^2 + k_y^2 k_z^2\right)\right]^{1/2}$	3.56	1

Source: Table Reproduced from A. Maisuradze et al., *Physical Review Letters* **103**, 147002 (2009), with Permission from the American Physical Society.

Note: The character of nodes (a line designated by 1 or point designated by p) for the gap functions of models A through F is specified by the functional form of the maximum value of the gap $\Delta(0)$ at $T = 0$ K. Values of the normalized gap $\Delta(0)/k_B T_c$ (the coupling strength) that the functional forms predict and the goodness of fits to the experimental data in Figure 2.44(a) normalized to the goodness of fit of model F are also given.

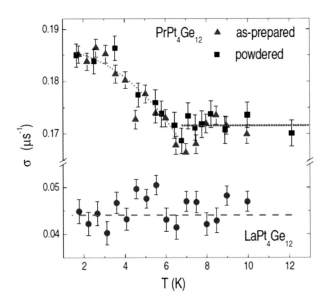

FIGURE 2.45 Temperature dependence of the muon depolarization rate in the as-prepared (solid triangle) and powdered (solid square) samples of PrPt$_4$Ge$_{12}$ and in the as-prepared (solid circle) samples of LaPt$_4$Ge$_{12}$. Reproduced from A. Maisuradze et al., *Physical Review B* **82**, 024524 (2010). With permission from the American Physical Society.

detectable. This is also the critical Ce concentration at which the superconductivity itself ceases to exist with no signs of its re-emergence down to 1 K. Comparing Pr$_{1-x}$Ce$_x$Pt$_4$Ge$_{12}$ solid solutions with the previously discussed Pr$_{1-x}$La$_x$Os$_4$Sb$_{12}$ solid solutions, in both cases, the TRS was broken in high Pr content structures, and in both series, the decrease in the spontaneous magnetic field was proportional to the decreasing amount of Pr. Although it was stated that the filler species in platinum germanide skutterudites contribute negligibly to the DOS at the Fermi energy and, therefore, should not significantly influence the superconducting state, it is clear from the above studies of solid solutions that, as far as breaking the TRS is concerned, it is the Pr–Pr interactions that are responsible.

The effect of Ce substitutions at the site of Pr was also studied by Huang et al. (2014) *via* the temperature dependence of electrical resistivity, magnetic susceptibility, and specific heat down to 50 mK in polycrystalline $Pr_{1-x}Ce_xPt_4Ge_{12}$ samples. The results are in good agreement with the ZT-μSR measurements by Zhang et al. (2015a) and attest to a strong pair-breaking influence of Ce on the superconducting state. Although notable drops in the electrical resistivity are detected even at $x = 0.5$, the transitions are broad and, down to 50 mK, fully completed only for $x \leq 0.2$. Single crystalline forms of $Pr_{1-x}Ce_xPt_4Ge_{12}$ should help to minimize structural disorder and sharpen the transition to pinpoint the concentration of Ce that destroys the superconductivity. The Ce content $x \approx 0.2$ is also where the jump in the specific heat is no longer observed and, from then on, the specific heat decreases smoothly with decreasing temperature. The Sommerfeld coefficient γ, extracted from the specific heat in the normal state, rises from 48 mJmol^{-1}K^{-2} at $x = 0$ to its maximum value of 120 mJmol^{-1}K^{-2} at $x = 0.5$ and then drops down to 86 mJmol^{-1}K^{-2} at $x = 1$. While the initially rising γ with the content of Ce indicates an increased strength of electronic correlations, the enhancement is small compared to values of $\gamma \approx 500$–600 mJmol^{-1}K^{-2} characterizing the heavy fermion state of $PrOs_4Sb_{12}$. The authors detected a crossover in the $C_{es}/\gamma T_c$ vs. T_c/T plot from the power law temperature dependence $\sim T^3$ in $PrPt_4Ge_{12}$ to an exponential dependence $\sim exp(-\Delta/k_BT)$ anytime Ce was present in the samples. The implication is that the gap function with a point-node structure in $PrPt_4Ge_{12}$ switches over to a fully developed gap in solid solutions containing Ce. Alternatively, this could also be viewed as a crossover from the multi-band to a single BCS-type superconducting band upon the presence of Ce, where the superconducting band with a smaller energy gap is suppressed by strong scattering of electrons by substituted Ce ions. In either case, the size of the superconducting gap obtained from the fits of the specific heat data, (Figure 2.46), indicates a strong-coupling superconductivity.

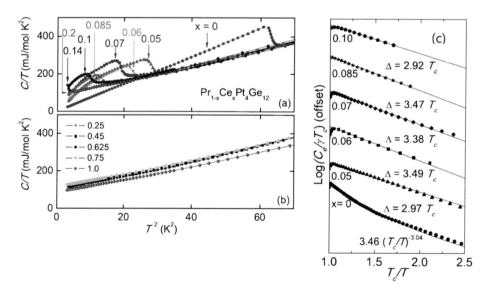

FIGURE 2.46 Specific heat data for $Pr_{1-x}Ce_xPt_4Ge_{12}$ displayed as C/T vs. T^2. (a) Samples with $x < 0.2$ show a jump associated with the transition to the superconducting state. (b) Samples with $x \geq 0.25$ showing no evidence of superconductivity. (c) Electronic contribution to the specific heat plotted as $ln(C_{es}/\gamma T_c$ vs. T_c/T for $Pr_{1-x}Ce_xPt_4Ge_{12}$. The data are offset for clarity. Solid lines are best fits to the data using an exponential temperature dependence for $x > 0$ and the power law expression $3.46 (T_c/T)^{-3.04}$ for $x = 0$. Note that all fitted values of Δ/k_BT_c are significantly larger than the BCS value of 1.76. Reproduced and adapted from K. Huang et al., *Physical Review B* **89**, 035145 (2014). With permission from the American Physical Society.

As in the case of $PrOs_4Sb_{12}$, controversial findings arising from using different experimental probes that test particular aspects of the superconducting state but return results conflicting with other studies took no time to surface. We have already seen it in ZF-μSR results indicating the broken TRS in the superconducting state of $PrPt_4Ge_{12}$, which would often imply a spin triplet state, yet other studies presented $PrPt_4Ge_{12}$ in terms of the anisotropic s-wave gap function.

An even greater challenge has surfaced with NQR measurements by Kanetake et al. (2010) made on ^{73}Ge-enriched $LaPt_4Ge_{12}$ and $PrPt_4Ge_{12}$ skutterudites, where the spin-lattice relaxation rate $1/T_1$ developed a Hebel–Slichter peak in both structures; actually, the peak in $PrPt_4Ge_{12}$ was far more pronounced than in $LaPt_4Ge_{12}$. Because the presence of the Hebel–Slichter peak has always been thought to document an ordinary BCS-type of superconductivity, the result was in clear conflict with the findings of Maisuradze et al. (2009a, 2010), who suggested the presence of point nodes in the gap function. Being aware of such a conflict, Kanetake et al. (2010) fitted their data with equations appropriate for the relaxation rate which are given by

$$\frac{T_1(T_c)}{T_1} = \frac{2}{k_B T_c} \int_0^\infty \frac{N_s(E)^2 + M_s(E)^2}{N_0^2} f(E)\left[1 - f(E)\right] dE \tag{2.29}$$

$$\frac{N_s(E)}{N_0} = \frac{1}{4\pi} \int_0^{2\pi} \int_0^\pi \frac{E}{\sqrt{E^2 - |\Delta(\phi,\theta,T)|^2}} \sin\theta \, d\theta \, d\phi \tag{2.30}$$

$$\frac{M_s(E)}{N_0} = \frac{1}{4\pi} \int_0^{2\pi} \int_0^\pi \frac{\Delta(\phi,\theta,T)}{\sqrt{E^2 - |\Delta(\phi,\theta,T)|^2}} \sin\theta \, d\theta \, d\phi, \tag{2.31}$$

where $N_s(E)$, $M_s(E)$, and N_0 are the DOS for quasiparticles in the superconducting state, the anomalous DOS originating from the coherence effect of the transition probability, and the DOS at the Fermi level in the normal state, respectively. Carrington and Manzano (2003) have shown that the T-dependence of the BCS superconducting gap function can be approximated by

$$\Delta(T) = 1.76 \tanh\left\{ 1.82\left[1.018\left(\frac{T_c}{T} - 1\right)\right]^{0.51} \right\}. \tag{2.32}$$

The overall gap function $\Delta(\phi,\theta,T) = \Delta(\phi,\theta)\Delta(T)$, where $\Delta(\phi,\theta)$ is the angular dependence of the gap function. The usual isotropic s-wave model A is shown by a solid red line in Figure 2.47 and provides quite a good fit, except for the height of the Hebel–Slichter peak that is overestimated. Bringing into play the angular dependence of the gap function, i.e., considering an anisotropic s-wave model, Kanetake et al. (2010) noted that model C, which can also be described as $\Delta(0)|\sin\theta|$, provides an even better fit over the experimentally covered temperatures down to $T \sim 0.3\ T_c$, where the model mimics an exponential-like temperature dependence. At still lower temperatures, on account of the $M_s(E)$ DOS arising from the coherence effect in the transition probability, the relaxation rate should acquire a T^5 power dependence. With the functional form $\Delta(0)|\sin\theta|$, the anisotropic s-wave model C includes a node at $\theta = 0$, bringing the NQR results in line with the specific heat and the temperature dependence of the density of the superconducting condensate seen by Maisuradze et al. (2009a, 2009b). In contrast, the anisotropic s-wave model B, which supports a line node, and the p-wave model, which completely ignores the coherence peak but might possibly be relevant to the broken

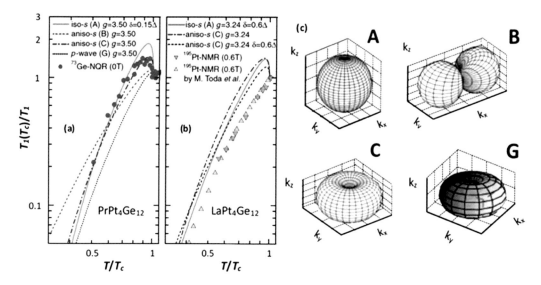

FIGURE 2.47 The left panel shows the temperature dependence of the ^{73}Ge relaxation rate $1/T_1$ for $PrPt_4Ge_{12}$ (solid red circles) fitted with various model functions of the superconducting gap, all calculated with $g \equiv 2\Delta(0)/k_BT_c = 3.5$. Spatial shapes of the gap functions fitted are depicted in the panel on the right. The first three shapes A, B, and C are the same s-wave forms of the gap functions as in Figure 2.44. The shape G represents the p-wave $\Delta_0\exp(i\phi)\sin\theta$. Adapted and redrawn from F. Kanetake et al., *Journal of the Physical Society of Japan* **79**, 063702 (2010). With permission from the Physical Society of Japan.

TRS state observed in $PrPt_4Ge_{12}$, provide distinctly poor fits. All fits were performed assuming the gap size $\Delta(0)/k_BT_c = 1.75$. Extending measurements to well below 1 K would be highly desirable, as the crucial crossover to the power law temperature dependence of $1/T_1$ should be detectable.

In principle, definitive studies of the superconducting gap function of any superconductor should come from high-resolution photoemission measurements, especially when done in a comparative mode with a superconductor that has well characterized properties. An attempt in this direction was made by Nakamura et al. (2012), who compared precision photoemission spectra of $PrPt_4Ge_{12}$ with those of $LaPt_4Ge_{12}$, both above and below T_c. Photoemission spectra were generated by a xenon lamp using its 8.44 eV resonance line and collected with a high resolution of 1.2 meV near the Fermi energy. The task then was to discern the differences in the spectra of the two skutterudites, not an easy job given the apparent close likeness of the intensity of the spectra shown in Figure 2.48, and to try to fit various gap function models. For this purpose, the authors used the Dynes function, Dynes et al. (1978), which is a modified BCS function defined as

$$D(E,\Delta,\Gamma) = \mathrm{Re}\left\{ \frac{(E-i\Gamma)}{\left[(E-i\Gamma)^2 - \Delta^2\right]^{1/2}} \right\}, \tag{2.33}$$

where Δ is the superconducting gap size and Γ is the phenomenological broadening parameter, originally used to represent finite lifetime effects of quasiparticles but, as a fitting parameter, it also includes different factors like the gap anisotropy. To fit the intensity spectra, the Dynes function is multiplied by the Fermi-Dirac function of the measured temperature (4.0 K in this case) and convolved with a Gaussian function corresponding to the experimental energy resolution. Fitting the superconducting spectrum of $LaPt_4Ge_{12}$ to 5 meV above the Fermi level, the intensity was well reproduced by a single Dynes function with the gap size 1.3 meV and $\Gamma = 0.25$ meV. Extrapolating

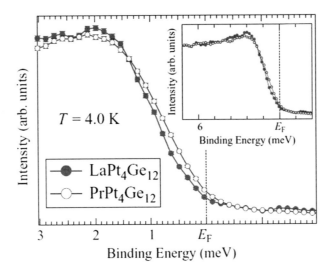

FIGURE 2.48 Comparison of the superconducting spectra of $LaPt_4Ge_{12}$ (solid red circles) and $PrPt_4Ge_{12}$ (open blue circles). The intensities were normalized by setting the average intensity of the normal-state spectra of each compound in the range between 6 and 7 meV to unity. The inset shows the data for $LaPt_4Ge_{12}$ fitted at 4.0 K with a single Dynes function and returning the gap value 1.3 meV. Adapted and redrawn from Y. Nakamura et al., *Physical Review B* **86**, 014521 (2012). With permission from the American Physical Society.

the superconducting gap to absolute zero temperature gave $\Delta(0) = 1.4$ meV, corresponding to $\Delta(0)/k_BT_c = 1.95 \pm 0.25$, somewhat larger but comparable to the BCS value of 1.76 and in excellent agreement with the value of 1.94 extrapolated from the specific heat measurements of Gumeniuk et al. (2008a, 2008b). Attempts to fit the intensity spectrum of $PrPt_4Ge_{12}$ with a single Dynes function with either an isotropic gap or with the allowed anisotropic gaps of models A-F in Figure 2.44(b) were not very successful, as evidenced in Figure 2.49(a). Like before, if no rational explanation for conflicting data of different experiments is on hand, as a last resort, one can always attempt to fit the data assuming that the state consists of more than a single sheet of the Fermi surface. Realizing that no single Dynes function can reasonably fit the photoemission intensity spectra of $PrPt_4Ge_{12}$, Nakamura et al. (2012) showed (Figure 2.49(b)) that two isotropic Dynes functions (model A), appropriately weighted, can fit the data exceptionally well. In this particular case, a Fermi sheet with an isotropic gap of 1.7 meV and 30% weight is augmented by a second isotropic sheet with a smaller gap size of 0.8 meV but a larger weight of 70%. The broadening parameter is common to both bands, and its fitted value is $\Gamma = 0.1$ meV. The authors noted, however, that equally good fits are possible with two Dynes functions with anisotropic gaps. In other words, the result should not be interpreted as a proof of the *s*-wave nature of superconductivity, but merely as an indication of its multi-band character in $PrPt_4Ge_{12}$. After all, multi-band crossings of the Fermi level are well documented in band structure calculations of platinum germanide skutterudites, as shown, for instance, by Sharath Chandra et al. (2016).

The fine details of the superconducting properties of platinum germanide skutterudites may have been affected by the polycrystalline nature of the available samples, which had less than the theoretical density and contained impurities. Moreover, the synthesis of bulk polycrystalline samples inevitably relies on the use of high pressure to compact the powder into ingots, a process that may introduce some degree of structural anisotropy. Such complications were eliminated once extensive recrystallization treatments succeeded in the preparation of ~ 2 mm size single crystals of $PrPt_4Ge_{12}$, tested for the first time by Zhang et al. (2013) in measurements of the penetration depth and specific heat. The advantages of single crystals were brought to light immediately. The analysis of the low-temperature specific heat, previously complicated by the presence of a large nuclear Schottky

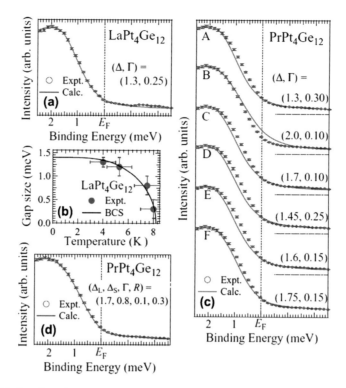

FIGURE 2.49 (a) Results of the fitting analysis of $LaPt_4Ge_{12}$ (solid green curves) compared with the experimental data at 4.0 K (open red circles). (b) Temperature dependence of the superconducting gap of $LaPt_4Ge_{12}$. Solid red circles indicate experimentally obtained temperature dependence of the superconducting gap, and solid curves represent the BCS relation under conditions of $\Delta_0 = 1.4$ meV and $T_c = 8.3$ K. (c) Results of the fitting analysis of $PrPt_4Ge_{12}$ (solid green curve) using various single Dynes functions with models A-F of the gap function given in Table 2.3. The dotted horizontal lines are zero lines for each data set. (d) The same $PrPt_4Ge_{12}$ data fitted using a weighted sum of two isotropic (model A) Dynes functions, corresponding to two sheets of the Fermi surface with different gap sizes of 1.7 meV and 0.8 meV and $R = 0.3$ fraction of the sheet with a larger gap size. Reproduced from Y. Nakamura et al., *Physical Review B* **86**, 014521 (2012). With permission from the American Physical Society.

term, became much simpler as no Schottky-like term was observed in single crystal samples, and C_{es}/T displayed a distinct $T^{2.8}$ power law dependence rather than the previously observed quadratic behavior. Variations of the penetration depth with temperature, proportional to precisely measured shifts of the resonant frequency $\Delta f(T)$ generated by a tunnel diode oscillator used in the self-inductance method, also yielded a large exponent power law dependence $\Delta\lambda = \lambda(T) - \lambda(0) \sim T^{3.2}$, very different from both the BCS-like behavior and the T^2 variation expected for the gap function with a point node. Plotting the density of the superconducting condensate, defined as $\rho_s = [\lambda(0)/\lambda(T)]^2$ *versus* T/T_c, resulted in a similar but more precise plot than the one shown in Figure 2.44(a). Again, neither the specific heat temperature dependence nor the temperature dependence of the superconducting condensate could be adequately fitted with a single gap function, be it isotropic or anisotropic. Even the previously promising model C gap function of the form $\Delta(0)|\sin\theta|$ showed marked differences with the now more precise data. However, when a two-band superconductor was considered with a combined superconducting condensate of the form

$$\tilde{\rho}_s(T) = x\rho_s\left(\Delta_1(0), T\right) + (1-x)\rho_s\left(\Delta_2(0), T\right), \tag{2.34}$$

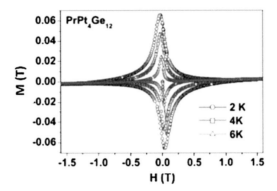

FIGURE 2.50 Isothermal field dependence of magnetization of $PrPt_4Ge_{12}$ at various temperatures below T_c. Note a strongly irreversible behavior reflecting a type-II superconductor. Reproduced from L. S. Sharath Chandra et al., *Philosophical Magazine* **92**, 3866 (2012). With permission from Taylor & Francis Group.

i.e., composed of two distinct BCS-type bands with the weighted gap functions $\Delta_1(0)$ and $\Delta_2(0)$, an excellent fit was obtained. The fitted values turned out to be $\Delta_1(0) = 0.8\,k_B T_c$, $\Delta_2(0) = 2.0\,k_B T_c$, and $x = 0.12$, the weight contributed by the first gap function.

The two-band nature of the superconducting state of $PrPt_4Ge_{12}$ was also strongly supported by detailed measurements of magnetization by Sharath Chandra et al. (2012) carried out in the temperature range of 2–8.5 K and applied fields of 10 mT. To avoid the presence of magnetic flux prior to measuring the isothermal magnetization at various temperatures, the polycrystalline sample was first cooled in zero field from a temperature well above T_c to a desired temperature below T_c and only then was the magnetic field swept between ± 2 T. As expected for a type II superconductor, cycling isothermal magnetization in fields above H_{c1} results in a highly irreversible loop, reflecting strong pinning of vortices on defects in the structure, Figure 2.50. The lower critical field H_{c1} is established as either a value of the applied magnetic field at which the magnetization exceeds 10% of the step size at that field or from a criterion developed by Ren et al. (2008), that H_{c1} is the field in the M vs. H plot where the slope of the linear fit deviates by more than 2% from unity; both estimates agree well with each other. The upper critical field is obtained from both electrical resistivity and magnetization measurements, in the latter case looking for a distinct break in the isothermal magnetization curve as the field is reduced and the sample crosses the phase boundary between the normal and superconducting domains. The most important finding of the study is positive curvatures in both $H_{c1}(T)$ and $H_{c2}(T)$ depicted in Figure 2.51. Attempts to fit the standard single-gap WHH model, Werthamer et al. (1966), to the data close to T_c and at low temperatures (blue and red curves in Figure 2.51) obviously failed due to the presence of positive curvatures in both critical fields. Applying Equation 2.4, the coherence length turned out to be $\xi(0) = 140$ Å, and from H_{c1}, the value of the penetration depth was $\lambda(0) = 1060$ Å. Various estimates of the mean free path yielded a range of values $\ell \approx 100–140$ Å. Hence, $\ell \sim \xi(0)$, and $PrPt_4Ge_{12}$ is classified as being in the dirty limit. With $\xi(0) << \lambda(0)$, $PrPt_4Ge_{12}$ is also in the local limit, and the density of superconducting condensate $\rho_s(T)$ can be written as, Ren et al. (2008),

$$\rho_s(T) = \frac{\lambda^2(0)}{\lambda^2(T)} = \frac{H_{c1}(T)}{H_{c1}(0)}. \tag{2.35}$$

The data for the density of the superconducting condensate could not be fit well with an expression for a single-gap superconductor, but a good fit was obtained with expressions for a superconductor having two gaps, a smaller one $\Delta_S(0) = 0.40\,k_B T_c$ and a larger one $\Delta_L(0) = 1.46\,k_B T_c$, with the weight $c = 0.2$ of the smaller gap. Surprisingly, both gaps are significantly smaller than the BCS

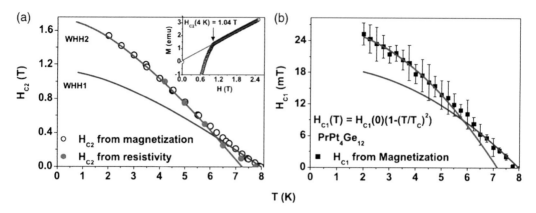

FIGURE 2.51 (a) Temperature dependence of the upper critical field $H_{c2}(T)$ for $PrPt_4Ge_{12}$ obtained from magnetization (open circles) and resistivity (solid green circles) measurements. The solid blue line designated as WHH1 represents the dependence based on the WHH model for the data near T_c. The solid red line marked as WHH2 is the temperature dependence based on the WHH model for data at low temperatures. The inset shows the M vs. H data near H_{c2} at 4 K with a straight line fit to $M(H)$ for $H > H_{c2}$. (b) Temperature dependence of the lower critical field $H_{c1}(T)$ for $PrPt_4Ge_{12}$. The solid blue line represents the temperature dependence $H_{c1}(T) = H_{c1}(0) [1-(T/T_c)^2]$ for data close to T_c. The solid red line is the temperature dependence obtained using the above formula for the data at low temperatures. In both $H_{c2}(T)$ and $H_{c1}(T)$, note a distinct positive curvature in the data brought to light further by the discrepancies between the red and blue curves. Reproduced from L. S. Sharath Chandra et al., *Philosophical Magazine* **92**, 3866 (2012). With permission from Taylor & Francis Group.

value $\Delta(0) = 1.76 k_B T_c$, suggesting that at least the larger gap should be anisotropic. Exploring this possibility, a power low fit of the form $\rho_s(T) = 1 - aT^n$, with a and n constants, was made to the low-temperature data $T/T_c < 0.5$ of the larger gap Δ_L. In a clean limit superconductor, the power law exponents $n = 1$ and $n = 2$ are predicted by Joust and Taillefer (2002) to signal the presence of line and point nodes, respectively. The exponent is modified when electron scattering is enhanced and, in the extreme dirty limit, $n \approx 2$ even though line nodes rather than point nodes are present. The actual fit to the low-temperature values of $\rho_s(T)$ of the large gap Δ_L returned $a = 0.065$ and the exponent $n = 1.29$, taken by the authors to be close to unity and, hence, interpreted as the presence of line nodes in the larger gap. However, in the case of line nodes, the constant parameter $a = 0.065$ is supposed to take the form $2\ln 2/\Delta(0)$, from which the larger gap value should be $\Delta_L(0) = 2.73 k_B T_c$. To confirm the presence of line nodes in the larger gap, Sharath Chandra et al. (2012) estimated the normalized density of the superconducting condensate over the temperature range $0 < T/T_c < 1$ by taking the temperature dependence of the larger gap as

$$\Delta_L(T) \equiv \Delta_{L0}(T)\cos\phi = \Delta_L(0)\tanh\left\{1.82\left[1.018\left(\frac{T_c}{T}-1\right)\right]^{0.51}\right\}\cos\phi, \qquad (2.36)$$

where $\cos(\phi)$ is to account for the presence of line nodes with ϕ being the azimuthal angle, and the hyperbolic tangent function an approximation to the temperature dependence of the gap function. For a two-gap superconductor, the temperature-dependent normalized density of the superconducting condensate is given by

$$\rho_s(T) = 1 + 2\left[c\int_{\Delta_s(T)}^{\infty} \frac{df(E)}{dE} D_s(E)dE + \frac{(1-c)}{4\pi}\int_0^{2\pi} d\phi \int_0^{\pi}\sin\theta d\theta \int_{\Delta_L(T)}^{\infty} \frac{df(E)}{dE} D_L(E)dE\right], \qquad (2.37)$$

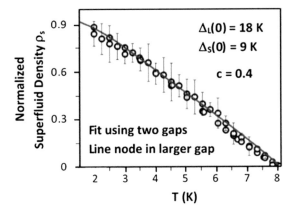

FIGURE 2.52 Comparison of experimental data for normalized density of the superconducting condensate (open circles) with the estimated values based on the model of two superconducting gaps in Equation 2.37 (solid line). The estimate assumes the presence of line nodes in the larger gap. Modified from L. S. Sharath Chandra et al., *Philosophical Magazine* **92**, 3866 (2012). With permission from Taylor & Francis Group.

where c is the fractional contribution of the small gap to the normalized superfluid density, $f(E)$ is the Fermi function, and $D_{L,S}(E) = E / \left[E^2 - \Delta_{L,S}(T)^2 \right]^{1/2}$ for the large (L) and small (S) gaps, respectively.

As documented in Figure 2.52, a very good agreement between the estimated $\rho_s(T)$ and experimental data is obtained when $\Delta_s(0) = 1.15\,k_B T_c$, $\Delta_L(0) = 2.31\,k_B T_c$, and $c = 0.4$. The positive curvature in the upper critical field $H_{c2}(T)$ of PrPt$_4$Ge$_{12}$ was also explained by Sharath Chandra et al. (2012) in terms of a two-band model, this time fitting the experimental data to a simplified and more transparent version of the linearized Eliasberg equations developed by Usadel (1970). For an interested reader, Gurevich (2003) re-derived the Usadel equations and used them in his detailed description of the upper critical field $H_{c2}(T)$ of superconductors in the dirty limit.

Two-band models of superconductivity unify the apparently conflicting experimental findings, as different experimental techniques are sensitive to different aspects of the superconducting state, and the presence of more than one Fermi surface sheet gives a chance to tune to the specific character of one sheet while the other sheet may have entirely different superconducting properties.

Regarding the superconducting properties of LaPt$_4$Ge$_{12}$, the experimental studies described above characterize this skutterudite as a superconductor that does not break TRS, yet it shows an unexpectedly small and perhaps even suppressed Hebel–Slichter peak. The smooth variation of T_c in the series Pr$_{1-x}$La$_x$Pt$_4$Ge$_{12}$ suggests a similar order parameter in the two end members, yet PrPt$_4$Ge$_{12}$ breaks TRS while LaPt$_4$Ge$_{12}$ does not. In other words, LaPt$_4$Ge$_{12}$ is an interesting albeit conflicting structure in its own right as far as its superconducting state is concerned. It is certainly not an ordinary BCS-type reference superconductor against which the unconventional superconducting nature of PrPt$_4$Ge$_{12}$ can be highlighted and compared to, as was the case of LaOs$_4$Sb$_{12}$ *vis-à-vis* PrOs$_4$Sb$_{12}$.

The conflicting characteristics of LaPt$_4$Ge$_{12}$ prompted Zhang et al. (2015b) to reexamine some of the key superconducting parameters, such as the magnetic penetration depth and the density of the superconducting condensate using single crystalline samples of LaPt$_4$Ge$_{12}$ and employing the same sensitive techniques as in their earlier study of single crystals of PrPt$_4$Ge$_{12}$, Zhang et al. (2013). The first surprising result concerning LaPt$_4$Ge$_{12}$ was its linearly varying penetration depth $\lambda(1.5\,K, H)$ with the applied magnetic field, the slope essentially identical to that of PrPt$_4$Ge$_{12}$. In contrast, in alkaline-earth-filled BaPt$_4$Ge$_{12}$ and SrPt$_4$Ge$_{12}$, the magnetic field had no effect on the penetration depth. The field-dependent penetration depth has been predicted for nodal or MSBCs, Amin et al. (2000), while conventional *s*-wave superconductors are expected to show essentially a

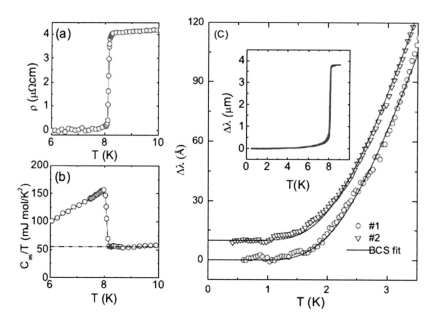

FIGURE 2.53   (a) Temperature dependence of electrical resistivity, (b) the electronic specific heat, and (c) the London penetration depth $\Delta\lambda_L(T)$ for $LaPt_4Ge_{12}$ single crystals. The data for the second sample, indicated by blue triangles, are shifted up by 10 Å for clarity. The solid lines are fits of the BCS model for temperatures $T \ll T_c$, as described in the text. The inset shows $\Delta\lambda_L(T)$ over a wide temperature range. Reproduced from J. L. Zhang et al., *Physical Review B* **92**, 220503(R) (2015). With permission from the American Physical Society.

field-independent penetration depth, Landau and Keller (2007). The temperature dependence of the London penetration depth taken as $\Delta\lambda_L = \lambda_L(T) - \lambda(0)$ for two single crystals of $LaPt_4Ge_{12}$ is shown in Figure 2.53. The value of $\lambda(0) \approx 1200$ Å was used from μSR measurements, and the second sample is shifted up by 10 Å for clarity. At temperatures well below T_c, Prozorov and Giannetta (2006) have shown that the penetration depth within the BCS model can be approximated by

$$\Delta\lambda(T) \approx \lambda(0)\sqrt{\frac{\pi\Delta(0)}{2k_BT}}\, exp\left(-\frac{\Delta(0)}{k_BT}\right). \qquad (2.38)$$

The fit of the penetration depth of the two $LaPt_4Ge_{12}$ single crystals is shown in Figure 2.53 by solid lines, and the fitted values are $\Delta(0) = 1.39\, k_BT_c$ for sample 1 and $\Delta(0) = 1.34\, k_BT_c$ for sample 2. Clearly, the resulting $\Delta(0)$ is much below the BCS value of $1.76\, k_BT_c$, and it looks as if the resonant frequency method, employed to obtain the penetration depth data, is selected to probe a superconducting band with a smaller magnitude of the energy gap. Zhang et al. (2015b) proceeded by calculating the density of the superconducting condensate from $\rho_s(T) = [\lambda(0)/\lambda_L(T)]^2$ for the penetration depth obtained from the resonance frequency measurements, ρ_s^{res}, (in the original paper designated as ρ_s^{TDO} for a tunnel-diode-oscillator) and from TF-μSR studies, $\rho_s\mu^{SR}$, the latter conducted in a magnetic field of 75 mT. The normalized form of the superconducting condensate density for the respective two cases is depicted in Figure 2.54; the ρ_s^{res} data are shown by red open circles and the $\rho_s\mu^{SR}$ data by blue solid triangles. The two sets of data show a distinctly different temperature behavior, with the former depicting a concave rise at temperatures near T_c. It is interesting to note that the derived partial condensate density $\rho_2(T)$ matches quite closely the density measured by TF-μSR, while $\rho_1(T)$ grossly underestimates both ρ_s^{res} and $\rho_s\mu^{SR}$. However, combining $\rho_1(T)$ and $\rho_2(T)$ in the form

$$\rho_s(T) = c\rho_1\big[\Delta_1(0),T\big] + (1-c)\rho_2\big[\Delta_2(0,T)\big], \tag{2.39}$$

i.e., assuming that the overall density of the superconducting condensate is due to contributions of two bands (two sheets of the Fermi surface), ρ_1 and ρ_2, with the parameter c being the weight of sheet 1, an excellent fit to the data of ρ_s^{res} results, as shown by the solid line in Figure 2.54(a).

For a two-band superconductor with the respective densities of states n_1 and n_2 and Fermi velocities v_1 and v_2, the parameter c is determined from

$$c = \frac{n_1\langle v_1^2\rangle}{n_1\langle v_1^2\rangle + n_2\langle v_2^2\rangle}, \tag{2.40}$$

where $\langle v_i^2\rangle$, $i = 1,2$ is the average of the squared Fermi velocity over the corresponding sheet of the Fermi surface. The temperature dependence of the extracted superconducting gaps $\Delta_1(T)$ and $\Delta_2(T)$ is shown in Figure 2.54(b), and their zero temperature values are $\Delta_1(0) = 1.31\,k_BT_c$ and $\Delta_2(0) = 1.80\,k_BT_c$. It is remarkable that the smaller gap value agrees very well with the gap value obtained from the BCS-like fit to the penetration depth in Figure 2.53. The authors surmised that the rapidly diminishing gap $\Delta_1(T)$ with temperature, i.e., its rather non-BCS character, causes the concave shape of the ρ_s^{res} temperature dependence near T_c. Moreover, if the coupling between the bands 1 and 2 is weak, the small value of the gap $\Delta_1(T)$ is likely destroyed by the field of 75 mT used in TF-μSR measurements as the band obviously has a much reduced upper critical field. In fact, the so-called virtual upper critical field above which the vortex cores overlap and drive the majority of electrons in such a band normal is given by

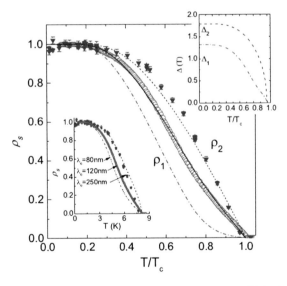

FIGURE 2.54   (a) Normalized density of the superconducting condensate $\rho_s(T)$ in single crystals of LaPt$_4$Ge$_{12}$. Open red circles are experimental data collected using the resonance frequency measurements, ρ_s^{res}, solid blue triangles are data obtained from TF-μSR measurements, $\rho_s\mu^{SR}$, using the field of 75 mT. The solid line is a fit based on a model whereby the density of the superconducting condensate is written in terms of two-band contributions with the weight γ for band 1 and the weight $(1-\gamma)$ for band 2. The respective derived partial densities of the condensate are designated ρ_1 and ρ_2, respectively. (b) The fitted temperature dependences of the gap amplitudes, $\Delta_1(T)$ and $\Delta_2(T)$. Adapted and redrawn from J. L. Zhang et al., *Physical Review B* **92**, 220503(R) (2015). With permission from the American Physical Society.

$$H_{c2}^{vir}(0) \sim H_{c2}(0) \left[\frac{\Delta_1(0) v_2}{\Delta_2(0) v_1} \right]^2.$$ (2.41)

For LaPt$_4$Ge$_{12}$, this virtual upper critical field is about 150 mT, given the upper critical field $H_{c2}(0)$ = 1.6 T. Of course, this value becomes much smaller away from $T = 0$ K. Thus, what likely happens in TF-μSR measurements is the suppression of superconductivity in band 1 by the employed field of 75 mT and $\rho_s \mu^{SR}(T)$ is then governed by a single band 2 contributing the density of superconducting condensate ρ_2 and resulting in the BCS-like behavior with the gap $\Delta_2(T)$. In contrast, in measurements of the penetration depth by the resonant frequency method, the sample is in the Meissner state, and the experiment senses both bands, the large one and the small one, and they contribute their respective condensate densities to the overall ρ_s, resulting in the data described by open red circles in Figure 2.54(a). It seems that at least some of the discrepancies between the different experimental techniques depend on whether the magnetic field is an integral part of the measurements. If so, in situations where multi-band superconductivity is suspected, there is a danger the magnetic field may suppress a band that has a small superconducting gap, and the results will reflect the presence of a band (or bands) with a larger energy gap. The band suppressed by the magnetic field may, however, have a large weight, and in measurements not requiring the presence of a magnetic field would dominate the superconducting state, yielding a very different outcome. The multi-band nature of the Fermi surface of LaPt$_4$Ge$_{12}$ demonstrates such a case.

Single crystals of LaPt$_4$Ge$_{12}$ prepared by the same group using the same synthesis method, Gumeniuk et al. (2010), were also used in the specific heat and thermal conductivity study by Pfau et al. (2016). The crystals had a high RRR of 17 and became superconducting at temperatures below 8.3 K. The specific heat data were overall in good agreement with measurements on polycrystalline samples, except that the jump at the critical temperature was sharp and somewhat higher, and the analysis of the data led to a gap size of $\Delta(0) = 2.03 \, k_B T_c$, comparable with the $\Delta(0) = 1.87 \, k_B T_c$ determined by Humer et al. (2013) on polycrystalline samples and the $\Delta(0) = 1.95 \, k_B T_c$ measured in the photoelectron spectroscopy study by Nakamura et al. (2012). However, it is puzzling why the gap far exceeded the value of $\Delta(0) = 1.39 \, k_B T_c$ obtained from the penetration depth measurements by Zhang et al. (2015b) using the same kind of high quality single crystal (RRR = 16.3). Although Pfau et al. (2016) obtained a convincing fit to the specific heat data with a single superconducting gap, they also made an equally good fit to the data with a two-band model (calling it model α), by writing the electronic contribution to the specific heat as the weighted average of contributions of two individual bands,

$$C_{e,\alpha} = x C_{e,1} + (1-x) C_{e,2}.$$ (2.42)

LaPt$_4$Ge$_{12}$ is likely a strongly coupled superconductor, but whether it contains a single Fermi sheet or more sheets could not be determined with certainty. An interesting aspect of the study was the behavior of the thermal conductivity of LaPt$_4$Ge$_{12}$ crystals at very low temperatures. The overall temperature dependence of the thermal conductivity $\kappa(T)$ from 0.04 K to 100 K is depicted in Figure 2.55(a). A smooth but clearly resolved drop in $\kappa(T)$ is observed at T_c. Below T_c, a hump appears, documenting an enhancement in the mean free path of phonons as the density of quasiparticles (quasiparticles being scattering centers of phonons at these low temperatures) decreases due to the formation of Cooper pairs that do not scatter phonons. At very low temperatures, below 0.5 K, the thermal conductivity can be modeled as consisting of two terms: a residual term κ_0/T and a power law term bT^a,

$$\frac{\kappa(T)}{T} = \frac{\kappa_0}{T} + bT^a.$$ (2.43)

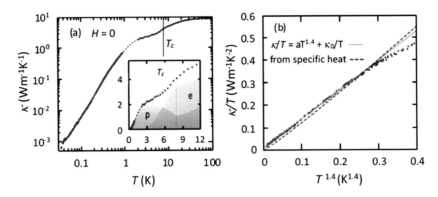

FIGURE 2.55 (a) Temperature dependence of thermal conductivity of LaPt$_4$Ge$_{12}$ from 0.04 K to 100 K. A clear but smooth drop at T_c is marked. A hump below T_c indicates an increased mean free path of phonons. The inset shows schematic contributions of electrons and phonons to the thermal conductivity at low temperatures. (b) Thermal conductivity at temperatures below 0.5 K plotted in the form $\kappa(T)/T$ vs. $T^{1.4}$. The solid yellow line fit returns a residual term $\kappa(T)/T$ $(T \rightarrow 0) = 0.01$ Wm^{-1}K^{-2}, basically the resolution of the experiment and a value far too small to represent any conceivable quasiparticles expected to arise from nodes in the superconducting gap. The dashed red line is a fit using Equation 2.43 with the experimental values of the specific heat. Reprinted from H. Pfau et al., *Physical Review B* **94**, 054523 (2016). With permission from the American Physical Society.

The behavior is illustrated in Figure 2.55(b) in the form of $\kappa(T)/T$ vs. $T^{1.4}$, and the power law is fitted to the data and shown by a solid yellow line. The fit returns a residual value of the thermal conductivity $\kappa(T)/T$ $(T \rightarrow 0) = 0.01$ Wm^{-1}K^{-2}, equivalent to $\kappa_0/\kappa(H_{c2}) \approx 1\%$, and is too tiny to document any presence of quasiparticles arising from nodes in the gap function. For reference, values of $\kappa_0/\kappa(H_{c2}) \approx 20$–35% are typical for unconventional superconductors, e.g., Proust et al. (2002). In s-wave superconductors, well below T_c, essentially all quasiparticles have formed Cooper pairs and the only contribution to heat current is the remaining phonons. Writing the thermal conductivity in a kinetic form

$$\kappa_p = \frac{1}{3} C_p v_p \ell_p, \tag{2.44}$$

where $C_p = \beta T^3$, v_p is the sound velocity and ℓ_p is the mean free path of phonons, one can obtain an estimate of κ_p using the Debye model,

$$\beta = \frac{12\pi^4}{5} \frac{N k_B}{\theta_D^3} \tag{2.45}$$

and

$$v_p = \frac{k_B \theta_D}{\hbar} \left(\frac{V}{6\pi^2 N} \right)^{1/3}. \tag{2.46}$$

Here, V is the volume of the sample and N is the number of atoms it contains. Pfau et al. (2016) found that the cubic temperature dependence of the specific heat is only approximate and a better power law relevant to the data below 4 K is with an exponent 2.5. This is very close to the fitted $T^{2.4}$ temperature dependence of the thermal conductivity (not to be confused with the $T^{1.4}$ dependence of $\kappa(T)/T$). Thus, the negligibly small residual thermal conductivity κ_0, coupled with the dominance of phonons at very low temperatures, implies that the gap is finite at every point of the Fermi surface. The authors also explored and made use of the magnetic field dependence of the thermal conductivity. At very low magnetic fields, below $H_{c1}(0) \approx 14$ mT for $LaPt_4Ge_{12}$, the thermal conductivity is field independent for s-wave superconductors as essentially all quasiparticles have condensed and formed Cooper pairs. The situation is fundamentally different in the mixed state of a superconductor. Here, in addition to quasiparticles thermally excited over the gap, there are quasiparticles associated with vortex cores. As we have discussed in Section 2.4, the vortex core is shielded with circulating currents having velocity v_s, causing the Doppler energy shift in the quasiparticle energy spectrum from $E_k = (\varepsilon_k^2 + \Delta_k^2)^{1/2}$ to $E_k + mv_F \cdot v_s$, where v_F is the Fermi velocity. In turn, the gap size is reduced for directions of the Fermi velocity v_F that have a component parallel to the velocity v_s. The electronic thermal conductivity κ_e^s in the mixed state of a superconductor is therefore enhanced. On the other hand, the presence of vortices, which scatter both electrons and phonons, decreases κ_e^s as well as κ_p^s. It follows that the type of the gap function (nodes or a uniform gap) should be reflected in the behavior of the thermal conductivity measured in externally applied magnetic field. To simplify the analysis, one usually chooses the direction of the magnetic field perpendicular to the direction of the heat current. The data collected by Pfau et al. (2016) at sub-Kelvin temperatures are presented in Figure 2.56(a) as a plot of κ/T vs. the magnetic field H in units of Tesla. Small dots are field-dependent thermal conductivities measured after the sample was zero-field cooled to a desired temperature. At very low temperatures, and starting with zero magnetic field, $\kappa_e^s \approx 0$ as there are no residual quasiparticles, assuming that the superconductor is the s-wave BCS type. The thermal conductivity here is dominated by phonons, κ_p^s, and their contribution obviously decreases as temperature decreases. Applying magnetic field $H > H_{c1}$, phonons are progressively more scattered by the increasing density of vortices and κ_p^s decreases. At the same time, the increasing density of quasiparticles associated with vortex cores leads to a rising contribution of κ_e^s, and the opposite trend of κ_p^s and κ_e^s results in a minimum that shifts to higher fields as the temperature increases. The electronic term κ_e^s eventually takes over and attains a quadratic field dependence ($\kappa(T)/T$ being linear) until, at $H_{c2}(T) \approx 1.4$ T, a sharp break occurs as the superconducting state is destroyed. The shaded areas in Figure 2.56(a) indicate qualitatively how the electronic and phonon contributions change with field. The solid black squares are data extrapolated to zero temperature, and the line through them is a guide to the eye. It is instructive to compare, as done by Pfau et al. (2016), how the low-temperature thermal conductivity of different types of superconductors behaves as a function of magnetic field (Figure 2.56(b)). The field dependence of the thermal conductivity of $LaPt_4Ge_{12}$ at very low temperatures best compares with the data for InBi, representative of a dirty s-wave superconductor. The main difference with a clean s-wave superconductor, such as Nb, is the infinite slope at H_{c2}, which $LaPt_4Ge_{12}$ clearly does not have. The data for InBi were taken at $T/T_c = 0.10$. Another dirty s-wave superconductor, $Ta_{80}Nb_{20}$, was measured at a much higher $T/T_c = 0.35$ and shows a well-developed minimum at intermediate fields similar to the one in $LaPt_4Ge_{12}$ shown in Figure 2.56(a). Because the coherence length $\xi \approx 200$ Å and the carrier mean free path $\ell \approx 600$ Å are comparable in $LaPt_4Ge_{12}$, it too is in the dirty limit.

Continuing with the comparison of the field dependence of thermal conductivity, a typical two-band superconductor, such as MgB_2, develops a plateau at intermediate fields as the smaller of the two bands becomes suppressed. No such plateau is noted in the field dependence of $LaPt_4Ge_{12}$. d-wave superconductors shown in Figure 2.56(b) have a completely different shape of the field-dependent thermal conductivity compared to $LaPt_4Ge_{12}$ and the other s-wave superconducting types. Moreover, superconductors with nodes in the gap function often develop a minimum in $\kappa(H)$, as documented in measurements by Watanabe et al. (2004) and Machida et al. (2012). It would be a stretch to count the minimum seen in $LaPt_4Ge_{12}$ as evidence of its unconventional nature because the minimum arises from the competition between electron and phonon contributions rather than being tied to the structure of the gap function. Furthermore, according to Pfau et al. (2016), $LaPt_4Ge_{12}$

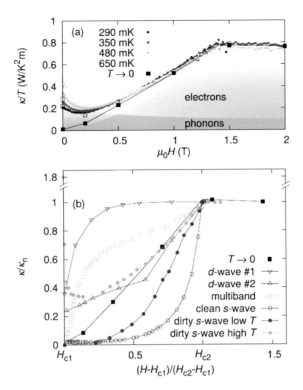

FIGURE 2.56 (a) Magnetic field dependence of the thermal conductivity of $LaPt_4Ge_{12}$ at low temperatures plotted as $\kappa(T)/T$ vs. H. Dots represent field sweeps after zero-field cooling. Open squares are data extracted from temperature sweeps after field cooling; black filled squares are their zero-temperature extrapolations, and the line through them is a guide to the eye. The light and darker shaded areas illustrate qualitatively how the electronic and phonon contributions change with applied field. (b) Comparison of field-dependent thermal conductivity of $LaPt_4Ge_{12}$ as $T \rightarrow 0$ (data from (a)) with superconductors having different gap symmetries: d-wave #1 are data for $Tl_2Ba_2CuO_{6+\delta}$ by Proust et al. (2002); d-wave #2 is for $CeIrIn_5$ by Shakeripour et al. (2009); the multi-band refers to MgB_2 data by Sologubenko et al. (2002); clean s-waves are data for Nb by Lowell and Sousa (1970); dirty s-wave low T stands for InBi ($T/T_c = 0.10$) by Willis and Ginsberg (1976); and dirty s-wave high T is for $Ta_{80}Nb_{20}$ ($T/T_c = 0.35$) by Lowell and Sousa (1970). The scaling of the abscissa is to account for different H_{c1} and H_{c2} of the different compounds. Reproduced from H. Pfau et al., *Physical Review B* **94**, 054523 (2016). With permission from the American Physical Society.

has no residual electronic thermal conductivity, which, of course, should be present in superconductors with nodes. Taking all the experimental findings into account, the measurements of Pfau et al. (2016) provide solid evidence for $LaPt_4Ge_{12}$ being a single-gap BCS s-wave superconductor, although it might be somewhat more strongly coupled.

A picture of $LaPt_4Ge_{12}$ as a single-gap superconductor goes counter to detailed studies of the specific heat made by Sharath Chandra et al. (2016) on polycrystalline samples. The group fit the electronic specific heat measured at various temperatures below T_c in the presence of an externally applied magnetic field and they found statistically smaller deviation from the data when a two-band superconducting model was considered rather than a single-band model. Subtracting first the contributions of phonons, the specific heat due to any Einstein mode associated with the motion of the fillers in the cage, and the Schottky two-level term identified as being dominated by paramagnetic impurities, the remaining electronic specific heat in the superconducting state was written, following the approach of Bouquet et al. (2001), as

$$\frac{C_s(T)}{\gamma_n T_c} = \alpha \frac{C_{s1}(T)}{\gamma_{n1} T_c} + (1 - \alpha) \frac{C_{s2}(T)}{\gamma_{n2} T_c}. \tag{2.47}$$

Here, $C_{s1}(T)$ and $C_{s2}(T)$ are the electronic heat capacities of superconducting bands 1 and 2, respectively, having energy gaps $\Delta_1(T)$ and $\Delta_2(T)$. The Sommerfeld terms γ_{n1} and γ_{n2} combine to give $\gamma_n = \gamma_{n1} + \gamma_{n2}$. The weight factor $\alpha \equiv \gamma_{n1}/\gamma_n$ is a fractional contribution of band 1. Assuming both bands to be isotropic, the fits returned values of $\Delta_1(0) = 0.65\ k_B T_c$, $\Delta_2(0) = 2.05\ k_B T_c$, and $\alpha = 0.15$ for LaPt$_4$Ge$_{12}$ and $\Delta_1(0) = 0.30\ k_B T_c$, $\Delta_2(0) = 1.93\ k_B T_c$, and $\alpha = 0.37$ for PrPt$_4$Ge$_{12}$. It is unfortunate that single crystals were not used in the study as the polycrystalline samples contained significant amounts of the PtGe$_2$ impurity phase, stated by the authors as 2.5% in LaPt$_4$Ge$_{12}$ and a high 9.2% in PrPt$_4$Ge$_{12}$. The latter value can cause ~ 2% error in $C_s(T)/\gamma_n T_c$, which can be significant compared to percentage deviations of the fits with respect to the experimental data of less than ± 10%.

Just as in the case of the thermal conductivity, an important insight into the superconducting state is gained by performing specific heat measurements in a magnetic field. The essential physical aim here is to probe the field dependence of the Sommerfeld parameter γ below T_c, the parameter representing the coefficient of the specific heat of quasiparticles associated with the normal vortex cores in the mixed state of a superconductor. Well below T_c, as the magnetic field increases, the density of vortices increases, which in turn increases the DOS of quasiparticles and, hence, the parameter γ. At the upper critical field, $\gamma = \gamma_n$, the field dependence of $\gamma(H)$ fundamentally depends on the Fermi surface. In single-band s-wave superconductors, one expects $\gamma(H)$ to approach γ_n as a linear function of field because the DOS of quasiparticles is proportional to the number of vortices. For d-wave superconductors, there is a prediction by Volovik (1993) that $\gamma(H) \sim H^{1/2}$. In two-band s-wave superconductors, the DOS of quasiparticles should grow linearly at low fields, but flatten somewhat at higher fields as the superconductivity is suppressed in the band with a smaller gap and the density of quasiparticle states is thus saturated in that band. Consequently, two linear regions are anticipated, the low field one being steeper and the crossover indicating the upper critical field of the first band. Because both LaPt$_4$Ge$_{12}$ and PrPt$_4$Ge$_{12}$ are classified as being in the dirty limit, one can employ the formalism developed by Vieland (1965) and write the specific heat in the superconducting state subjected to a magnetic field in terms of the specific heat in zero field plus a term expressing the effect of the magnetic field *via* the Sommerfeld parameter $\gamma(H)$ as

$$\frac{C_s^H(T)}{\gamma_n T_c} = \frac{1}{\beta'} \frac{C_s(T)}{\gamma_n T_c} + \gamma \frac{T}{T_c}. \tag{2.48}$$

Here, $\beta' = \beta(2\kappa^2 - 1)/2\kappa^2$, with $\beta = 1.18$ and κ being the GL parameter $\kappa = \lambda/\xi$. In the case of a two-band superconductor, $C_s(T)/\gamma_n T_c$ should be taken as given in Equation 2.47. Figure 2.57 depicts the field dependence of $C_s(H)/\gamma_n T_c$ and $\gamma(H)/\gamma_n$ obtained at 2 K for both LaPt$_4$Ge$_{12}$ and PrPt$_4$Ge$_{12}$. Sharath Chandra et al. (2016) discerned two linear regions in their γ/γ_n data for both LaPt$_4$Ge$_{12}$ and PrPt$_4$Ge$_{12}$, with a larger slope at low magnetic fields. They interpret such a behavior as strong support for the two-band nature of superconductivity in both skutterudites.

They also point out a rather large value of the γ/γ_n ratio as an indication of anisotropy in either one of the two superconducting gaps. The multi-band nature of superconductivity is also supported by their DFT studies of the electronic band structure, which reveal several bands crossing the Fermi level and distinct anisotropy of the Fermi surface. Because both LaPt$_4$Ge$_{12}$ and PrPt$_4$Ge$_{12}$ displayed multi-band superconductivity, the authors surmised that this may be intrinsic to all skutterudite superconductors.

Multi-band superconductivity is, indeed, a favorite interpretation of many recent studies of superconducting skutterudites and was further supported by specific heat studies by Singh et al. (2016) of

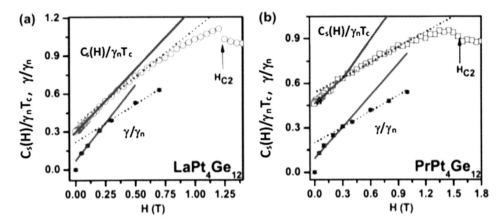

FIGURE 2.57 Magnetic field dependence of $C_s(H)/\gamma_n T_c$ and γ/γ_n at 2 K for (a) LaPt$_4$Ge$_{12}$ and (b) PrPt$_4$Ge$_{12}$. Note two linear regions in γ/γ_n depicted by red solid lines and blue dotted lines. Reprinted from L. S. Sharath Chandra et al., *Philosophical Magazine* **96**, 2161 (2016). With permission from Taylor & Francis Group.

superconducting solid solutions of Pr$_{1-x}$Ce$_x$Pt$_4$Ge$_{12}$ (superconductivity up to $x \approx 0.2$) that specifically addressed the nature of the gap function as a strong pair-breaking element Ce was introduced into the structure. Measurements were carried out in zero magnetic field but extended down to 0.5 K. After subtracting the lattice and Schottky contributions (due to a small amount of impurity phases present in the polycrystalline samples), the best fits of the electronic specific heat $C_s/\gamma_n T_c$ were obtained with a power law dependence T^m, $m \approx 3.5$–4 for $x \leq 0.04$ while an exponential dependence applied to compositions with $x > 0.04$. The power law temperature dependence at low temperatures for low-Ce content solid solutions was interpreted as the evidence of nodes in the gap functions. In contrast, the exponential behavior observed for all samples with $x > 0.04$ indicated the dominance of the nodeless gap. Because the presence of the nodes and the nodeless gap were intrinsic to these solid solutions, Singh et al. (2016) decided to fit their data using a two-band superconducting model along the lines outlined above in the measurements of Pfau et al. (2016) and Sharath Chandra et al. (2016). However, in this case, at least one gap function was not BCS-like, but contained nodes. The experimental specific heat data and the fitted two gap functions are shown in Figure 2.58 for pure PrPt$_4$Ge$_{12}$ and for $x = 0.07$, both combining nodal and nodeless gaps, and for $x = 0.085$ where a single gap with no nodes is present. The authors pointed out that for all samples with $0 \leq x \leq 0.07$, the zero-field electronic specific heat could be fit very well with a larger nodal gap and a smaller nodeless gap. The size of both gaps decreases with the increasing content of Ce, but the rate of decrease is larger for the nodal gap, so that at $x = 0.085$, the nodal gap has disappeared and only a nodeless gap of 0.55 meV supports superconductivity. Because a similar situation was observed in the case of Pr(Os$_{1-x}$Ru$_x$)$_4$Sb$_{12}$ by Chia et al. (2005), perhaps it is a general feature of skutterudites.

Solid solutions are a prolific medium with which to extract information on all sorts of physical properties, including superconductivity and its crossover to other ordered states. Here, I wish to mention briefly the effect of divalent Eu on superconducting properties of Pr$_{1-x}$Eu$_x$Pt$_4$Ge$_{12}$ solid solutions. The effect of Eu on the magnetic properties of Pr$_{1-x}$Eu$_x$Pt$_4$Ge$_{12}$ solid solutions is described in section 3.3.4.13.4. Superconductivity of Pr$_{1-x}$Eu$_x$Pt$_4$Ge$_{12}$ was studied by Jeon et al. (2017) for polycrystalline samples. The range of superconductivity, as detected in the resistive and susceptibility transitions, extends to at least $x = 0.5$ and may overlap with the antiferromagnetic state that weakens with the increasing content of Pr. Clear jumps in the specific heat were resolved up to $x = 0.38$. The phase diagram, based on a collection of superconducting and magnetic studies, is illustrated in Figure 2.59. In the superconducting state, the specific heat data exhibit a crossover from a nodal to a nodeless superconducting energy gap as x increases or, alternatively, might possibly be explained in

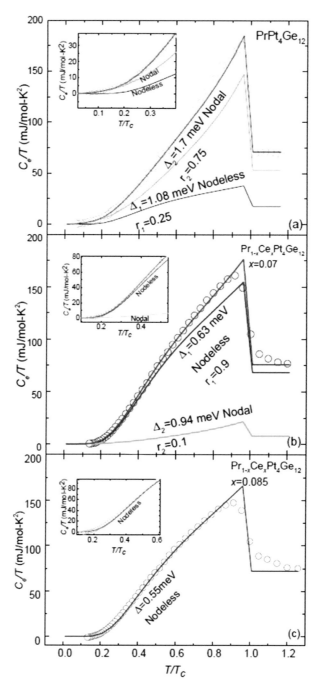

FIGURE 2.58 Electronic specific heat C_e/T plotted *vs.* T/T_c for $Pr_{1-x}Ce_xPt_4Ge_{12}$ with (a) $x = 0$ and (b) $x = 0.07$ using nodal and nodeless gaps and (c) $x = 0.085$ maintaining a single gap without nodes. The red solid lines indicate the overall fit to the data. Green and blue lines are the individual contributions of the nodal and nodeless gaps, respectively. The insets show the low-temperature region of the data in the main panel. Gap values in units of meV together with the weight of each superconducting band are also given. Reproduced from Y. P. Singh et al., *Physical Review B* **94**, 144502 (2016). With permission from the American Physical Society.

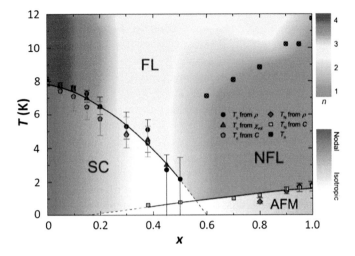

FIGURE 2.59   Phase diagram of $Pr_{1-x}Eu_xPt_4Ge_{12}$. FL and NFL stand for the Fermi liquid and non-Fermi liquid, respectively. The vertical bars represent the width of the superconducting transitions. The suppression of T_c has a negative curvature and extrapolates to 0 K near $x = 0.6$. Superconductivity may coexist with an antiferromagnetic order in the range between $x = 0.2$ and $x = 0.6$. The blue gradient-filled area under the T_c vs. x curve indicates the change of the temperature dependence of the low-temperature specific heat $C_e/\gamma_n T_c$ from the power law variation of the form $b(T_c/T)^{-m}$ to an exponential variation as x increases. Reproduced from I. Jeon et al., *Physical Review B* **95**, 134517 (2017). With permission from the American Physical Society.

terms of a multi-band superconducting state. Compared to Ce, paramagnetic Eu is far less effective as a pair-breaker in $PrPt_4Ge_{12}$.

In Section 2.6.1, I discussed the influence of substituted gold at the site of Pt in the alkaline-earth-filled $Ba(Pt_{4-x}Au_x)Ge_{12}$ solid solution skutterudites, where the gold led to a significant increase (by some 2 K) in the superconducting transition temperature at 25% replacement of Pt. Interestingly, a similar study performed by Li et al. (2010) with La-filled solid solutions $La(Pt_{1-x}Au_x)_4Ge_{12}$ (note a different designation of the content of Au compared to the chemical formula used previously) resulted in a completely different outcome; the high $T_c = 8.3$ K of $LaPt_4Ge_{12}$ was suppressed by the presence of a comparable 20% of Au at the site of Pt down to 6.3 K. While the lattice parameter increases with the increasing content of Au in both solid solutions at a very similar rate, the presence of Au in the $[Pt_4Ge_{12}]$ framework filled with a divalent filler seems to tune the DOS peak to coincide with the Fermi level but apparently detunes the DOS peak away from the Fermi energy in the same framework filled with trivalent La, even though La has no f-electrons. It would be interesting to confirm this trend by exploring what happens when gold is introduced into the $[Pt_4Ge_{12}]$ framework filled with different alkaline-earth and rare-earth fillers.

Counting electrons of $LaPt_4Ge_{12}$, it is obvious that the structure is a distinct metal with five hole carriers. Replacing Ge with Sb, the holes should be gradually compensated until the composition $LaPt_4Ge_7Sb_5$ is reached, at which point the structure should be perfectly compensated and become an insulator. But what happens to superconductivity as Sb is substituted for Ge? To answer this question, Humer et al. (2013) synthesized a series of $LaPt_4Ge_{12-x}Sb_x$ skutterudites and found that the solubility range ends at $x = 5$, beyond which the lattice parameter stops increasing (the radius of Sb is larger than that of Ge, hence the increasing lattice parameter) as Sb segregates at the grain boundaries. The composition $LaPt_4Ge_7Sb_5$ is not only a crossover point from the metallic to insulating state, but it is also the terminal point of the structural stability. From resistivity, magnetic susceptibility, and specific heat measurements in zero and applied magnetic field, the authors established that the superconducting state is depressed rather quickly as Sb is brought into the structure and vanishes entirely at a critical concentration of Sb $x_{Sb}^{crit} \approx 1.3$, i.e., well before the expected metal–insulator transition takes place at $x = 5$. It should be noted, however, that even for $x = 5.1$, i.e., slightly above the solubility range of Sb, the resistivity maintains its metallic character at all temperatures but with its magnitude an order of magnitude larger compared to pure $LaPt_4Ge_{12}$. Apparently, intrinsic disorder and inhomogeneities close to the solubility limit prevent the manifestation of the insulating state. A similar set of measurements was carried out on $PrPt_4Ge_{12-x}Sb_x$ by Jeon et al. (2016). From XRD measurements, it was established that the occupancy of Pr in the voids stays at 100% up to about $x = 3$, but then it decreases rather quickly and, at $x = 5$, the occupancy is down to 70%. The structures possess a similar metallic character of transport typified by the rising resistivity with increasing temperature up to $x = 5$ as in the case of $LaPt_4Ge_{12}$. However, superconductivity is maintained to higher contents of Sb and is suppressed only above $x = 4$. In the context of the void occupancy by Pr and its effect on the transition temperature, I should mention a paper by Venkateshwarlu et al. (2014) where the authors intentionally prepared a platinum germanide with nominally 50% of the voids filled with Pr, i.e., a structure $Pr_{0.5}Pt_4Ge_{12}$. Remarkably, the transition temperature stayed essentially intact at $T_c = 7.66$ K. The only significant difference with respect to fully filled $PrPt_4Ge_{12}$ was a much smaller gap size $\Delta(0) = 1.13\, k_B T_c$, which implied a large weakening of the coupling constant. XRD data indicated a considerable level of impurities in the half-filled material, but their composition and volume fraction were not analyzed to extract the actual filling fraction of Pr.

Figure 2.60 shows the temperature dependence of the upper critical field determined by Jeon et al. (2016) from transition temperatures taken as a 50% value of the normal state resistivity just above the transition at fixed values of the magnetic field. Interestingly, the initial slopes of the temperature dependence of $H_{c2}(T)$ close to T_c are similar for all concentrations with a value of about -0.39 T/K. The bulk nature of superconductivity was confirmed by magnetic susceptibility and specific heat measurements. As before, the specific heat in the superconducting state of $PrPt_4Ge_{12}$ obeyed the power law dependence, suggesting nodes in the gap or multi-band superconductivity, while a switch over to the exponential temperature dependence, implying a nodeless gap function or a suppression of a superconducting gap in a MSBC, was noted when Sb was present in the structure. Overall, there are many similarities in the effect of Sb replacing some Ge in $PrPt_4Ge_{12}$ and in $LaPt_4Ge_{12}$. Perhaps,

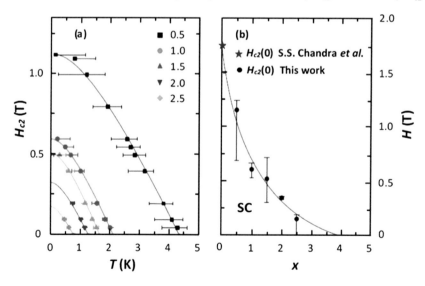

FIGURE 2.60 (a) Temperature dependence of the upper critical field $H_{c2}(T)$ for several samples of $PrPt_4Ge_{12-x}Sb_x$ determined from 50% values of the normal state resistivity just above the transition. The legend numbers in (a) indicate the concentration x of Sb. Horizontal bars indicate the transition width taken at 10% and 90% values of the normal state resistivity. Reprinted from I. Jeon et al., *Physical Review B* **93**, 104507 (2016). With permission from the American Physical Society.

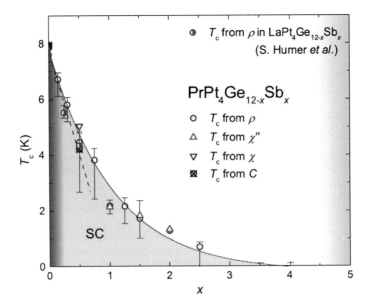

FIGURE 2.61 Phase diagram showing the superconducting transition temperature T_c as a function of Sb concentration x based on measurements of the electrical resistivity, ac and dc magnetic susceptibility, and specific heat in $PrPt_4Ge_{12-x}Sb_x$. The vertical bars indicate the width of the superconducting transition derived from measurements of the electrical resistivity taken at 10% and 90% values of the transition. For Sb concentrations $x \geq 3.5$, only the onset of the transition was detected down to 140 mK. The red dashed line shows the suppression of superconductivity in $LaPt_4Ge_{12-x}Sb_x$ measured by Humer et al. (2013). Reprinted from I. Jeon et al., *Physical Review B* **93**, 104507 (2016). With permission from the American Physical Society.

the only notable difference is the faster rate of suppression of superconductivity with the content of Sb in $LaPt_4Ge_{12}$, as shown by a red dashed line in Figure 2.61.

2.6.4 Effect of Pressure on Platinum Germanide Superconductors

As we have seen with pnicogen-based superconducting skutterudites, pressure-dependent studies of the superconducting transition temperature have potential to reveal much useful information about the superconducting state. However, due to the complexities of combining the high-pressure environment with low temperatures, such studies are performed rather sporadically. Moreover, the pressure-transmitting media (frozen fluids and powders) rarely facilitate truly hydrostatic pressure conditions, and the outcome of experiments might be affected by anisotropic strains that develop in a superconducting specimen.

The first attempt to study the pressure dependence of the transition temperature in the $[Pt_4Ge_{12}]$ skutterudite framework was made with $BaPt_4Ge_{12}$ and $SrPt_4Ge_{12}$ in the measurements by Khan et al. (2008), who used a piston-and-cylinder apparatus with daphne oil to generate pressures up to 20 kbar (2 GPa). As indicated by the data in Figure 2.62, the effect of pressure on T_c is rather small, but it shows an interesting non-linear behavior. In $BaPt_4Ge_{12}$, the transition temperature initially rises with the applied pressure, reaches a maximum near 14 kbar with about 0.14 K higher T_c than at ambient pressure, and then decreases. In contrast, the transition temperature of $SrPt_4Ge_{12}$ initially falls a bit but then rises and gains about 0.27 K over the ambient pressure value. Using DFT calculations, the authors estimated changes in the DOS at the Fermi level for a set of decreasing lattice parameters that mimicked the effect of hydrostatic pressure up to 20 kbar. In the case of $SrPt_4Ge_{12}$, such changes matched closely the pressure variation of the transition temperature. Of course, there is definitely

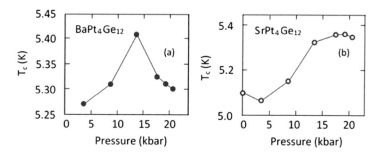

FIGURE 2.62 Pressure dependence of the transition temperature in (a) $BaPt_4Ge_{12}$ and (b) $SrPt_4Ge_{12}$. Adapted and redrawn from data by T. Khan et al., *Journal of the Physical Society of Japan* **77**, 350 (2008). With permission from the Physical Society of Japan.

some uncertainty about the extent the actual applied pressure was hydrostatic given that the pressure-transmitting medium was frozen oil. It is interesting to compare what happens when the pressure is applied to rare-earth-filled skutterudites, and $PrPt_4Ge_{12}$, with its unconventional superconducting nature is, of course, the prime candidate for such a study. In measurements performed by Foroozani et al. (2013), the effect of purely hydrostatic pressure up to 0.6 GPa on the superconducting transition temperature of $PrPt_4Ge_{12}$ was monitored by changes in the real part of the ac magnetic susceptibility. To ensure that the pressure was truly hydrostatic, the authors used a three-stage He-gas compressor connected by a flexible capillary tube to a CuBe pressure cell mounted at the tip of a liquid helium cryostat. Although much higher pressures are available *via* a piston-and-cylinder apparatus and diamond anvil cells, the emphasis here was on the hydrostatic pressure condition to avoid strains in the sample that might mask the intrinsic effect of pressure. As the pressure increased, the transition temperature decreased linearly, with the pressure derivative $dT_c/dP = -0.19 \pm 0.03$ KGPa^{-1}. The data are displayed in Figure 2.63. The numbers associated with the data points indicate the sequence with which the data were collected. Given that the lattice parameter of $PrPt_4Ge_{12}$ is about 0.14% smaller than the lattice parameter of $LaPt_4Ge_{12}$, it has often been speculated whether the marginally smaller transition temperature of $PrPt_4Ge_{12}$ (7.91 K) compared to that of $LaPt_4Ge_{12}$ (8.27 K) is a consequence of the larger internal pressure acting on Pr. Based on the pressure effect they measured, the authors

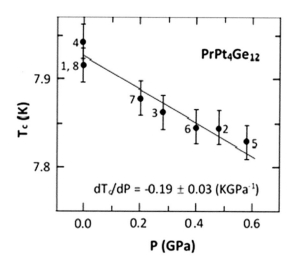

FIGURE 2.63 Superconducting critical temperature plotted as a function of external hydrostatic pressure for $PrPt_4Ge_{12}$. The numbers associated with each point indicate the sequence in which the data were taken. Reproduced from N. Foroozani et al., *Physica C* **485**, 160 (2013). With permission from Elsevier.

estimated that the 0.14% smaller lattice parameter of $PrPt_4Ge_{12}$ would cause only about 0.164 K change in the transition temperature, less than the $\Delta T_c = 0.36$ K between the transition temperatures of the two skutterudites. Consequently, the small difference in the transition temperatures is not due to the size of the cage but more likely arises from subtle changes in the DOS of the two materials at the Fermi energy. An interesting point is that the rate with which the pressure suppresses superconductivity in $PrPt_4Ge_{12}$ (-0.19 KGPa^{-1}) is very close to the rate observed in $PrOs_4Sb_{12}$ (-0.15 KGPa^{-1}) by Maple et al. (2002). Subsequently, Tayama et al. (2006) attributed the pressure suppression in $PrOs_4Sb_{12}$ to a competition between the superconducting state and the high-field AFQ ordered phase. However, given the similar rates with which the superconductivity is suppressed under pressure in $PrPt_4Ge_{12}$ and $LaPt_4Ge_{12}$, yet the vastly different CEF environments in which the Pr^{3+} ions reside (only about 7 K splitting between the ground state and the first excited state in $PrPt_4Ge_{12}$ compared to some 130 K splitting in $LaPt_4Ge_{12}$), it is unlikely that the CEF levels play an important role here.

2.7 CRITICAL CURRENT DENSITY OF SUPERCONDUCTORS

In any practical applications of superconductors, and especially in superconducting magnets and power transmission, an important issue is the maximum electric current a superconductor can support before it reverts to the normal state. The value of the electric current density that destroys the superconducting state is called the critical current density J_c, a parameter that depends on the shape of a specimen and is closely related to the degree to which flux lines (vortices) are effectively immobilized (pinned) in the body of a superconductor. Unlike the intrinsic superconducting parameters, such as the superconducting transition temperature and the lower and upper critical fields, the critical current density, because of its dependence on the pinning centers that can be altered for example by cold working, is not an intrinsic superconducting parameter. Critical current density J_c, or critical current $I_c = A\,J_c$, where A is a cross-section of the superconducting wire, can be determined by two main approaches. The first one is a resistive measurement where one measures the current at which the first measurable voltage is detected. Typically, the above rather vague criterion is made more meaningful by specifying some low level of the threshold voltage. The technique is straightforward but requires attaching electrical contacts to the sample.

A contactless method of determining the critical current density that also does not rely on an arbitrarily specified threshold value is based on measurements of an M-H hysteresis loop. The drawback of the technique is the theoretical modeling of the dependence of the critical current on the magnetic induction inside a superconductor. The simplest and often used model was developed by Bean (1962, 1964). Before I introduce the Bean model, I advise the readers to review the distinction between the vectors of the magnetic field $\boldsymbol{H}$, magnetic induction $\boldsymbol{B}$, and magnetization $\boldsymbol{M}$ discussed in section 2.2, and the units in which they are measured.

2.7.1 FIELDS AND CURRENTS INSIDE TYPE-I SUPERCONDUCTORS

Let us assume a long type-I superconducting wire (pure metal) of radius a placed in an applied magnetic field $B_0 = \mu_0 H_0$ directed parallel to its length. Please note that I am using here the proper form of the applied field $\mu_0 H_0$ in units of Tesla and not just H_0. The superconductor will react to the presence of the applied field by generating a shielding current J_{sh}, which circles around the wire. The value of the shielding current at the surface is J_0, and it decays exponentially inside the wire as

$$J_{sh}(r) \approx J_0\, exp\left[-\frac{(a-r)}{\lambda}\right], \tag{2.49}$$

where λ is the penetration depth. The purpose of the shielding current is to suppress the interior B_{in} field that exponentially decays from its surface value B_0 toward the center of the wire,

$$B_{in}(r) \approx B_0 \exp\left[-\frac{(a-r)}{\lambda}\right].$$

(2.50)

The shielding current is induced by magnetization M *via* the equation

$$\boldsymbol{J}_{sh} = \Delta \times \boldsymbol{M}.$$

(2.51)

The magnetization rises from zero at the surface of the wire and exponentially approaches the field H_0 as

$$M(r) \approx -H_0\left[1 - exp\left(-\frac{(a-r)}{\lambda}\right)\right].$$

(2.52)

Note the negative sign of M, as dictated by the Lenz law for an induced entity. The magnetization cancels the internal field B_{in} *via* an equation

$$B_{in}(r) = \mu_0\left[H_0 + M(r)\right].$$

(2.53)

Using a one-dimensional form of Equation 2.51, $J_{sh} = dM/dr$, it is straightforward to arrive at a relation

$$B_0 = \mu_0 H_0 = \mu_0 \lambda J_0.$$

(2.54)

2.7.2 CRITICAL STATE MODEL OF BEAN

The situation in type-II superconductors is more complicated by the presence of vortices. As the applied field exceeds H_{c1}, vortices start to enter the body of a superconductor. However, they do not spread uniformly throughout the superconductor because they pile up close to the surface, being held there by pinning forces. The resultant gradient in the flux density inside the superconductor, according to the Maxwell equation

$$\nabla \times \boldsymbol{B} = \mu_0 \boldsymbol{J}(\boldsymbol{B}),$$

(2.55)

gives rise to a current flowing perpendicular to the applied field. The essence of the Bean model, Bean (1962, 1964), is an assumption that vortex pinning yields the maximum possible gradient and, therefore, generates the maximum current density J_c the superconductor can sustain. This is referred to as a critical state of a superconductor. A common assumption of all critical state models is that, in low applied fields or currents, the outer part of the superconductor is in the critical state with each region that carries the current doing so at the critical value of the current density J_c, while the rest of the superconductor is well shielded from the influence of these fields and currents. Many critical state models have been developed, and the difference among them is the form the critical current density takes as a function of the magnetic field. Bean, in his critical state model, made two simplifying assumptions: (a) he neglected the lower critical field H_{c1}, which is not a big problem as

H_{c1} is typically quite small and (b) he assumed that the critical current density is independent of the magnetic field, i.e.,

$$J(B) = J_c. \tag{2.56}$$

This major simplification makes the problem of sorting out the dependence of the internal magnetic field and the current density of the condensate on the external magnetic field relatively easy. In spite of its simplicity, the model is still useful in providing a reasonable estimate of the critical current density. The other model frequently used is due to Kim et al. (1962) but is more complicated by relaxing the condition on J_c being constant.

The one-dimensional form of Equation 2.55 relevant in the model of Bean is

$$\frac{dB_z(x)}{dx} = \mu_0 J_y(x). \tag{2.57}$$

We assume a sample in the form of a slab of width $2a$ in the x-direction, with y and z dimensions very much larger, as depicted in Figure 2.64. The assumption is that the slab has not been previously exposed to any external magnetic field and is cooled in zero magnetic field to a desired temperature below T_c. Applying a small magnetic field B_0 parallel to the slab, i.e., along the z-direction, we inquire about the magnetic induction inside the slab and the density of supercurrent the applied field B_0 generates.

The reader will note that, apart from the permeability μ_0, the slope of $B_z(x)$ in Equation 2.57 is equal to the current density J_c. Thus, for the assumed small B_0, the profile of the internal field and the supercurrent that flows is sketched in Figure 2.65(a). The magnetic flux inside of the slab is

$$
\begin{aligned}
B_z(x) &= 0 & a' \le x \le a'' \\
&= B_0 - \mu_0 J_c x & 0 \le x \le a' \\
&= B_0 + \mu_0 J_c(x - 2a) & a'' \le x \le 2a
\end{aligned}
\tag{2.58}
$$

The extent to which the field and supercurrent penetrate the slab follows from the boundary condition by setting $B_0 - \mu_0 J_c a' = 0$, which yields

$$a' = \frac{B_0}{\mu_0 J_c}. \tag{2.59}$$

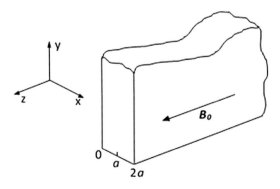

FIGURE 2.64 A sketch of a slab of a superconductor of width $2a$ in applied magnetic field B_0.

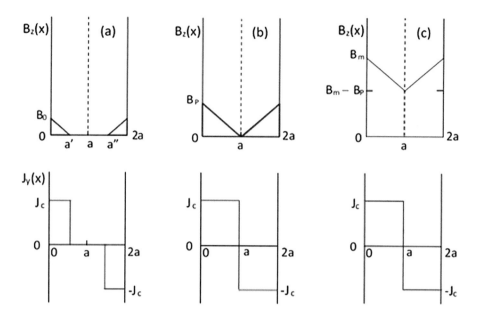

FIGURE 2.65 Profiles of the interior magnetic field and supercurrent density for various strengths of applied field B_0. (a) Low-level applied field B_0. (b) Applied field equal to the penetrating field B_P. (c) Applied field B_m larger than the penetration field B_P. The bottom halves of the schematics indicate the current density in the slab.

As the applied field increases, the penetration of the flux and supercurrent increases until an applied field reaches a value, called the penetration field $B_P = \mu_0 J_c a$, at which the magnetic flux from the sides of the slab meet at $x = a$, Figure 2.65(b). The average magnetic flux in the slab depicted in Figure 2.65(a) is

$$
\begin{aligned}
B_{av}^0 \equiv B^0 &= \frac{1}{2a} \int_0^{2a} B_z(x)\,dx = \frac{1}{2a} \times \text{shaded area} \\
&= 2 \times \frac{1}{2a} \frac{B_0}{2} a' = \frac{B_0}{2a} \frac{B_0}{\mu_0 J_c} = \frac{B_0^2}{2B_P}.
\end{aligned}
\tag{2.60}
$$

The magnetization when the field B_0 is applied to the slab becomes

$$
-M(B_0) = \frac{B_0}{\mu_0} - \frac{\langle B^0 \rangle}{\mu_0} = \frac{B_0}{\mu_0} - \frac{B_0^2}{2\mu_0 B_P}.
\tag{2.61}
$$

The average flux in the slab at full penetration is

$$
B_{av}^P = \langle B^P \rangle = \frac{B_P}{2}.
\tag{2.62}
$$

Consequently, the magnetization at full penetration becomes

$$
-M(B_P) = \frac{B_P}{\mu_0} - \frac{\langle B^P \rangle}{\mu_0} = \frac{B_P}{\mu_0} - \frac{B_P}{2\mu_0} = \frac{B_P}{2\mu_0}
\tag{2.63}
$$

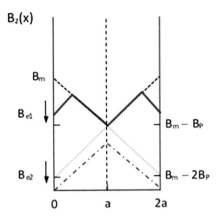

FIGURE 2.66 Profile of the magnetic flux in the interior of the slab after the applied field is reversed and reduced to B_e (blue solid line) on its way to zero from the maximum applied field B_m (dashed black line). The dash-dotted blue curve represents the remanent magnetization retained in the slab when the applied field is returned to zero.

If now the applied field exceeds the value of the penetration field B_P and reaches its maximum value B_m, the flux inside the slab is as shown in Figure 2.65(c). The average flux inside the slab is then obviously

$$B_{av}^m = \left\langle B^m \right\rangle = \frac{2B_m - B_P}{2}, \tag{2.64}$$

and the corresponding magnetization is

$$-M(B_m) = \frac{B_m}{\mu_0} - \frac{(2B_m - B_P)}{2\mu_0} = \frac{B_P}{2\mu_0}. \tag{2.65}$$

Thus, in the increasing magnetic field, for applied fields below the penetration field B_P, the magnetization varies with the applied field according to Equation 2.61. However, once the applied field attains the value of B_P, and for any higher fields (up to presumably H_{c2}), the magnetization stays constant and has the value $B_P/2\mu_0$.

The situation becomes a bit more complicated when, after reaching the highest applied field, the field is now reduced on its way toward zero. The different stages of magnetization as the applied field is gradually reduced are displayed in Figure 2.66. At the highest applied field B_m, the profile of the flux in the interior of the slab is indicated by a dashed black curve (the same as in Figure 2.65(c)). Reversing and reducing the applied field to B_{e1}, the interior magnetic flux assumes a profile shown by the blue solid line. As the field is reduced further to B_{e2}, where $B_{e2} = B_m - 2B_P$, the profile becomes a perfect mirror image of the profile at the applied field B_m. Consequently, $-M = -B_P/2\mu_0$, and the magnetization is now clearly positive, as if the superconductor became a paramagnet. The magnetization stays constant when the applied field is reduced further and maintains the value of $-B_P/2\mu_0$ even when the applied field is fully reduced to zero. The flux retained in the slab at zero applied field is called the remanent magnetization and is indicated by a dash-dotted blue curve in Figure 2.66. One cannot get rid of the remanent magnetization by any manipulation of the magnetic field. Rather, to do so, one must warm up the superconductor above its transition temperature, Although tedious, it is not difficult to show that on its return sweep from B_m, the magnetization for any applied field B_e follows the functional form

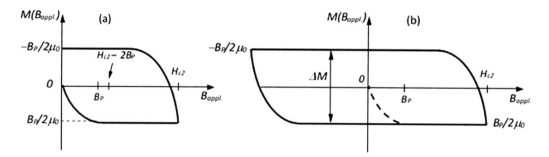

FIGURE 2.67 (a) Magnetization of an originally non-magnetized superconducting slab during a half-cycle starting from zero applied field, increasing the field to H_{c2} and back to zero. (b) A full hysteretic cycle for the slab. The magnetization is derived under Bean's assumption of constant critical field J_c. Note that the difference in magnetization ΔM between the first down-sweeping and the second up-sweeping branches is used in estimations of the critical current density, $\Delta M = B_P/\mu_0 = J_c\, a$. If one were to use only a half-cycle loop in (a) to determine the critical current, the magnetization difference ΔM must be taken at applied fields satisfying $B_P < B_{appl.} < H_{c2} - 2B_p$, which may be a very narrow interval, depending on values of the penetration field and the upper critical field.

$$-M\left(B_e\right)= \frac{B_P}{2\mu_0} - \frac{B_m}{\mu_0} + \frac{B_m^2}{4\mu_0 B_P} + \left(1 - \frac{B_m}{2B_P}\right)\frac{B_e}{\mu_0} + \frac{B_e^2}{4\mu_0 B_P}, \qquad (2.66)$$

and the values of $-M(B_m)$ and $-M(B_{c2})$ are, indeed, $B_P/2\mu_0$ and $-B_p/2\mu_0$, respectively. Tracing the value of magnetization $-M$ during a half-cycle (starting with a non-magnetized superconductor) by sweeping the applied field from zero to some maximum value, such as H_{c2}, and back to zero, we get the profile sketched in Figure 2.67a.

The whole point of going through the analysis of the Bean model is to establish a simple relationship between the sample magnetization and the superconducting critical current density J_c. As follows from Figure 2.67 and the relation $B_p = \mu_0 J_c a$, for a sample shaped as a slab with width $2a$, the critical current density is simply

$$J_c = \frac{\Delta M}{a}. \qquad (2.67)$$

For superconductors with a shape different from a slab, Equation 2.67 is modified. For instance, for a long superconducting bar having a rectangular cross-section with dimensions a and b where $a < b$, the critical current is

$$J_c = \frac{\Delta M}{a\left(1 - \dfrac{a}{3b}\right)}. \qquad (2.68)$$

For a long cylinder (wire) with radius R, the critical current density is

$$J_c = \frac{3}{2}\frac{\Delta M}{R}. \qquad (2.69)$$

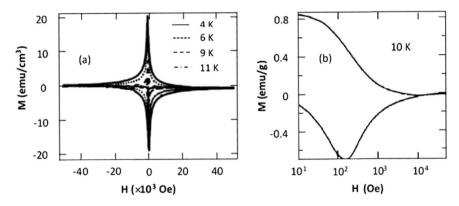

FIGURE 2.68 (a) M-H loop of $La_{0.8}Rh_4P_{12}$ for various temperatures. (b) An expanded view of the M-H half-loop for $La_{0.8}Rh_4P_{12}$ measured at 10 K. Note that the magnetization in (a) is in volume units emu/cm³, while the one in (b) is in mass units emu/g. Adapted from M. Imai et al., *Superconducting Science and Technology* **20**, 832 (2007b) and *Physical Review B* **75**, 184535 (2007). With permission from the Institute of Physics and the American Physical Society, respectively.

The above expressions assume SI units, i.e., M being in Amperes per meter and R in meters, so that the current density J_c comes out in Amperes per square meter. I remind the reader that if the magnetization is measured in cgs units and M is taken as volume magnetization, i.e., as emu/cm³, the conversion of such magnetization to SI units involves a factor of 10^3 A/m.

There are only a couple of measurements of the critical current density for superconducting skutterudites described in the literature. Perhaps it is not so surprising given that skutterudites are borderline cases between ductile and brittle structures, which would make it challenging to fabricate long thin wires from skutterudites necessary for most practical applications of superconductors. Moreover, in the case of platinum germanide skutterudites, the cost of such wires would be prohibitive. Nevertheless, critical current measurements provide information on important structural features of the material *via* the pinning force the vortices experience as they are held in place against the Lorentz force by metallurgical defects. The defects can be introduced by various forms of cold working, by precipitation of secondary phases, and by irradiation, and their primary function is to alter the mean free path.

The first attempt to measure the critical current density of superconducting skutterudites was made by Imai et al. (2007a) on $La_{0.8}Rh_4P_{12}$ prepared by the HPHT growth. This synthesis dramatically enhanced the La occupancy of voids and, as noted in section 2.3, resulted in a skutterudite with the highest T_c of 17 K, as measured by Shirotani et al. (2005a, 2005b). The samples of Imai et al. (2007a) had a somewhat smaller T_c of about 14.9 K, but still much higher than that of other skutterudites. The hysteresis loops collected for $La_{0.8}Rh_4P_{12}$ at various temperatures are shown in Figure 2.68(a). An expanded view of an *M-H* half-loop taken at 10 K is depicted in Figure 2.68(b).

Following the Bean model described above, the loops served to determine the critical current density J_c, displayed in Figure 2.69 as a function of applied magnetic field. The authors modeled their sample as a rectangular parallelepiped. Notable values at 4.2 K are $J_c (H = 0) = 1.84 \times 10^8$ Am^{-2} and $J_c (H = 2000 \text{ Oe}) = 6 \times 10^7$ Am^{-2}. For comparison, the critical current density J_c of $La_{0.8}Rh_4P_{12}$ is smaller than $J_c (4.2 \text{ K}, H = 2000 \text{ Oe}) = 1 \times 10^9$ Am^{-2} of Nb_3Sn, measured in a classical superconductor often used in superconducting magnets, Swartz (1962).

The critical current density of platinum germanide skutterudites was measured by Sharath Chandra et al. (2012) for a polycrystalline sample of $PrPt_4Ge_{12}$. Again, magnetization measurements were used, and the generated *M-H* loops were very similar to the loops shown in Figure 2.68(a). The critical current density determined *via* the model of Bean is depicted for various temperatures between 2 K and 7 K in Figure 2.70(a). At zero magnetic field, the critical current density at 4.2 K

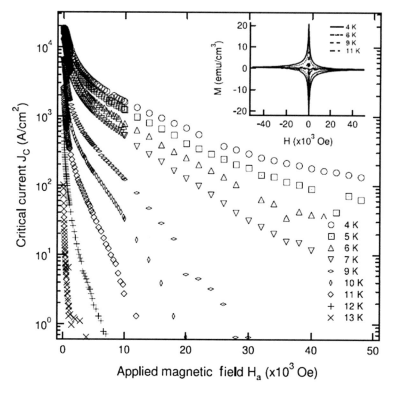

FIGURE 2.69 Critical current density J_c *vs.* applied magnetic field H for $La_{0.8}Rh_4P_{12}$ at various temperatures. Note an initial rapid decrease with the field followed by a more modest one at higher fields. At temperatures above 10 K, the critical current density falls precipitously with the applied field. Adapted from M. Imai et al., *Superconducting Science and Technology* **20**, 832 (2007b). With permission from the Institute of Physics.

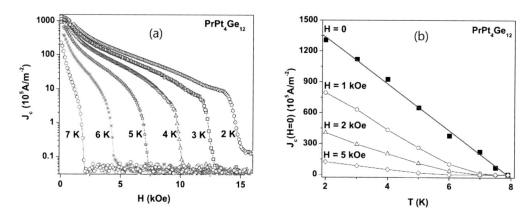

FIGURE 2.70 (a) Field dependence of the critical current density of $PrPt_4Ge_{12}$ at various temperatures. (b) Critical current density as a function of temperature at different values of applied field derived from the data in (a). Note that the units of J_c in both panels are in Am^{-2}. Reprinted from L. S. Sharath Chandra et al., *Superconducting Science and Technology* **25**, 105009 (2012). With permission from the Institute of Physics.

is about 10^8 Am^{-2}, comparable to that of La$_{0.8}$Rh$_4$Sb$_{12}$ under the same conditions, but an order of magnitude smaller than that for Nb$_3$Sn. Overall, the decrease of the J_c is exponential with respect to the applied magnetic field, plausibly related to thermally activated flux creep, but with different rates at low and high magnetic fields. In Figure 2.70(b) is shown the critical current density as a function of temperature for several applied magnetic fields. In particular, the temperature dependence of the critical current density at zero applied field shows a distinctly linear behavior. The authors argued that this is symptomatic of the presence of two superconducting bands, trying to further augment their experimental results and strengthen their arguments presented in section 2.6.3. Although it is very likely that PrPt$_4$Ge$_{12}$ is a MSBC, using as supporting evidence the linear dependence of the critical current density on temperature is not a strong argument. The reason is that the linear temperature depedence, rather than, say, the GL-predicted $J_c \propto (1-T/T_c)^{3/2}$ power law, is not a unique discriminator. Indeed, all granular superconductors are predicted by Ambegaokar and Baratoff (1963) to follow $J_c \propto (1-T/T_c)$, as the structures are essentially weakly Josephson coupled superconductors. The polycrystalline form of PrPt$_4$Ge$_{12}$ is not that far from a granular material.

2.8 CONCLUDING REMARKS

So, what has been learned about the superconducting state of skutterudites from the wealth of experimental studies performed? It is clear that PrOs$_4$Sb$_{12}$ is a 1.85 K chiral superconductor (breaks TRS) with exotic properties in the highly correlated realm of unconventional superconductors. There is no dispute that the conduction electrons have heavy mass, which gives PrOs$_4$Sb$_{12}$ the status of a heavy fermion superconductor. Based on a variety of measurements, the superconductivity of PrOs$_4$Sb$_{12}$ is of strong-coupling nature with $\Delta/k_B T_c \sim 3$, and the invariance of spin susceptibility with temperature, extracted from measurements of the Knight shift, documents the spin-triplet pairing. It is also well established that the ground state of Pr^{3+} ions is non-magnetic, and the relevant energy landscape is provided by the CEF with an energy spacing between the ground state and the first excited state of 0.7 meV (equivalent to 8.1 K). A surprising and pivotal finding is the existence of the HFOP that develops at low temperatures in magnetic fields between about 4.5 T and 14 T. Finally, there is strong evidence that the superconducting state is supported by more than a single sheet at the Fermi surface. Except for the large positive pressure derivative in YFe$_4$P$_{12}$ and LaFe$_4$P$_{12}$, the remaining superconducting pnicogen-based skutterudites show properties compatible with the usual BCS-type s-wave superconductors.

Regarding superconductivity in the [Pt$_4$Ge$_{12}$] framework, here too, the Pr-filled structure shows unusual properties starting with the breaking of TRS in the superconducting state, the strong-coupling nature of superconductivity, and multi-band features that may possibly extend to other platinum germanide skutterudites, such as LaPt$_4$Ge$_{12}$. However, there is a major difference between PrOs$_4$Sb$_{12}$ and PrPt$_4$Ge$_{12}$ in the size of the CEF energy splitting between the ground state and the first excited state, which is only about 8 K in PrOs$_4$Sb$_{12}$ but some 130 K in PrPt$_4$Ge$_{12}$. Accordingly, the superconductivity of platinum germanide skutterudites is more closely related to the properties of the [Pt$_4$Ge$_{12}$] polyanion than with the CEF levels. This is also consistent with observations that platinum germanide skutterudites have only mildly enhanced carrier effective masses that are nowhere near the values characterizing PrOs$_4$Sb$_{12}$ as a heavy fermion superconductor.

Indeed, an impressive array of experimental studies has been carried out to provide a deeper insight into the superconducting state of skutterudites and much has been learned. However, it was also amply documented that even the usually reliable experimental techniques found so useful and successful in revealing the unconventional superconducting properties of other heavy fermion systems often generate conflicting results when applied to superconducting skutterudites and PrOs$_4$Sb$_{12}$

in particular. Thus, while μSR measurements of the penetration depth by MacLaughlin et al. (2002) and Shu et al. (2006) indicate a fully open gap, many studies imply nodal features in the gap function, such as radiofrequency measurements of the surface penetration depth by Chia et al. (2003), an observation of flux line lattice distortion at very low temperatures and small magnetic fields by Huxley et al. (2004), point-contact spectroscopy by Turel et al. (2008), and especially the magnetic field dependence of the thermal conductivity by Izawa et al. (2003). Moreover, some observations likely arise from the inherent crystal inhomogeneity, such as the two superconducting transition temperatures seen in many $PrOs_4Sb_{12}$ crystals, Méasson et al. (2008), or two distinctly different kinds of images (one with a well-defined gap and the other showing finite DOS in the gap) obtained by Suderow et al. (2004) with high-resolution scanning tunneling microscopy on the same crystal. In one particular case, oscillations in the angular dependence of thermal conductivity that change symmetry with decreasing temperature were taken as proof of a two-phase nature of the superconducting condensate, yet were never seen in other studies where they should have been prominently manifested. There are also speculations, Cichorek et al. (2005), concerning a possible transition to yet another superconducting phase at around $T/T_c \sim 0.3$, based on an unexpected enhancement of the lower critical field H_{c1} and increased critical current $I_c(T)$ observed below about 0.6 K. It is true that there really is no experimental technique that couples directly and in a simple way to the microscopic order parameter of $PrOs_4Sb_{12}$. Combined with the fact that the well-understood mechanism of pairing and mass renormalization based on magnetic fluctuations has no relevance to $PrOs_4Sb_{12}$ and $PrPt_4Ge_{12}$ on account of their non-magnetic ground state, it is no surprise that in spite of a vast compendium of experimental data, it has been challenging to come up with a viable scenario of pairing in $PrOs_4Sb_{12}$ and $PrPt_4Ge_{12}$ and large mass renormalization in the former skutterudite.

Theorists have tried to make sense of the experimental data and attempted to explain the highly unconventional interactions at the heart of the superconducting state of $PrOs_4Sb_{12}$ and $PrPt_4Ge_{12}$ and the presence of heavy electrons in $PrOs_4Sb_{12}$. An obvious starting point for interaction theories in Pr-filled skutterudites seemed to be the hybridization between the p-orbitals of the pnicogen and the f-orbitals of Pr, as used by Shiina and Shiba (2010) to describe the charge ordered state in $PrRu_4P_{12}$ at 63 K. However, Pr does not behave the same way in all cages of Pr-filled skutterudites and, in the case of $PrOs_4Sb_{12}$, such hybridization is much weaker and is swamped by the quadrupolar interactions realized by the singlet ground state and the triplet first excited states of the crystal field of the Pr^{3+} ion. Likewise, quadrupolar interactions predominate over the hybridization between d-bands associated with the transition metal T and the f-electrons of Pr^{3+} ions, as shown by Shiina (2012). Consequently, interest has shifted toward two important features that might provide the best hope for identifying the interaction mechanism imparting heavy mass to electrons and leading to the formation of Cooper pairs in $PrOs_4Sb_{12}$: the CEF energy levels of Pr^{3+} ions that harbor multi-polar moments and the existence of the magnetic field-induced AFQ ordered phase (also referred to as FIOP or HFOP).

Electrons acquire a mass heavier than their free space mass m_e by interacting with other elementary excitations. The most familiar case is the electron–phonon interaction that underpins the ordinary BCS superconductivity. In this case, the effective mass becomes

$$m^* = \left(1 + \lambda_{ep}\right)m_e, \tag{2.70}$$

where λ_{ep} is the coupling constant of the electron–phonon interaction. In the usual harmonic oscillations of the lattice, the coupling constant λ_{ep} is limited to values less than 2, making only a modest enhancement in the effective mass of electrons. To attain effective masses of several tens or even hundreds of m_e, a much stronger interaction with electrons is needed. In heavy fermion systems, such interactions arise from hybridization of the conduction electron states with the f-electrons, the so-called c-f hybridization, which leads to Kondo screening in the lattice of magnetic ions that strongly renormalizes electrons. As I have already mentioned, interactions of this kind are not applicable to

$PrOs_4Sb_{12}$ because the f-electrons are localized and the ground state is non-magnetic. We are thus left with the quadrupolar moments formed *via* excitations within the CEF-split states of the Pr^{3+} ion. The concept of mass renormalization by inelastic scattering of conduction electrons by low-energy CEF transitions that possess quadrupolar electric moments has been known for some time, since the work of Fulde and Jensen (1983). In solids containing impurities, such as rare-earth ions with CEF splitting on the order of $k_B T$, conduction electrons undergo basically two main inelastic scattering processes: exchange scattering, in which the spin of an electron is altered (flipped), and inelastic charge scattering, often called aspherical Coulomb scattering, which involves a transition between the crystal field levels. Formally, the mass enhancement is expressed as

$$\frac{m^*}{m_e} = 1 + \left(g_J - 1\right)^2 J_{sf}^2 N\left(0\right)\frac{2\left|\langle i|\hat{J}|j\rangle\right|^2}{\Delta}, \tag{2.71}$$

where g_J is the Landé factor, J_{sf} is the exchange integral coupling the conduction electrons to the $4f$-electrons, $N(0)$ is the bare density of conduction electron states at the Fermi level, and $\langle i|\hat{J}|j\rangle$ is the magnetic dipole matrix element appropriate for the states of the CEF separated by energy Δ. With typical values for the praseodymium metal, Goremychkin et al. (2004) obtained the mass enhancement of about 20, which falls nicely between the estimates of Maple et al. (2002) based on the specific heat of $PrOs_4Sb_{12}$ (~50) and the value of 7.6 obtained for this skutterudite from dHvA measurements by Sugawara et al. (2002). Evidently, the CEF energy levels greatly facilitate mass renormalization.

In the case of the crystal field levels of $PrOs_4Sb_{12}$, the strongest dipole matrix elements arise from $\Gamma_1 \to \Gamma_4^{(1)}$ ($\Gamma_1 \to \Gamma_4$ in O_h representation) transitions at an energy of 11 meV, equivalent to the temperature of about 130 K. However, just as such magnetic dipole interactions enhance the effective mass, they would surely act as strong pair breakers and suppress superconductivity. Fortunately, the $\Gamma_1 \to \Gamma_4^{(2)}$ transition ($\Gamma_1 \to \Gamma_5$ in O_h), with its energy spacing of 0.7 meV (8.2 K) relevant to superconductivity, has strong quadrupolar matrix elements and only a very weak magnetic dipole. Consequently, as shown by Zwicknagl et al. (2009), the main scattering mechanism is the aspherical Coulomb scattering rather than magnetic exchange scattering. This particular situation was considered quantitatively by Chang et al. (2007), who made use of the strongly coupled Eliashberg equations to calculate both mass renormalization and the enhancement of the superconducting transition temperature as the content of Pr increased in $(Pr_{1-x}La_x)Os_4Sb_{12}$. At large Pr concentrations, the RKKY interaction between the $4f$ ions gives rise to the formation of a quadrupolar excitonic band, observed in experiments by Kuwahara et al. (2005) and Kanako et al. (2007). Excitons are basically low-level bosonic excitations, which, *via* interactions with conduction electrons, may also contribute to Cooper pairing. As pointed out by Chang et al. (2007), while mass renormalization benefits from both the magnetic and quadrupolar scattering of conduction electrons, the formation of Cooper pairs depends critically on the relative strength of the two. In the case of $(Pr_{1-x}La_x)Os_4Sb_{12}$, the favorable CEF structure of the Pr^{3+} ion makes quadrupolar (aspherical) scattering stronger, thus explaining why replacing La with Pr leads to an enhancement in T_c. Large amplitude motion of Pr^{3+} ions in the oversized cage of the skutterudite structure referred to as rattling has also been proposed by Hattori et al. (2005) and Hotta (2008) as a mechanism that enhances electron–phonon interaction and contributes to mass renormalization. Summarizing all experimental evidence and theoretical evidence, a broad consensus has emerged that quadrupolar electric interactions facilitated by a favorable structure of the crystal electric field of Pr^{3+} ions is the origin of the heavy fermion nature of electrons as well as of Cooper pair formation in $PrOs_4Sb_{12}$.

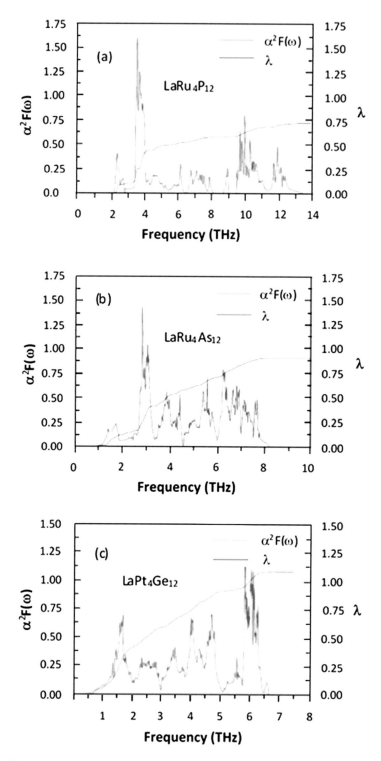

FIGURE 2.71 Calculated electron–phonon spectral function $\alpha^2 F(\omega)$ (red line) and the frequency accumulated electron–phonon coupling constant λ_{e-p} (blue line) for (a) LaRu$_4$P$_{12}$, (b) LaRu$_4$As$_{12}$, and (c) LaPt$_4$Ge$_{12}$ superconductors. Adapted from H. M. Tütüncü et al., *Physical Review B* **95**, 214514 (2017). With permission from the American Physical Society.

Ab initio DFT-based electronic energy band structure calculations, particularly those utilizing a generalized gradient approximation to estimate the exchange-correlation energy and carried out within the FP-LAPW formalism, have reached the level of accuracy one can rely on to provide essential information on the position and ordering of energy bands in the all-important range at and near the Fermi level. All such studies indicate, *via* the partial DOS, that in pnicogen-based superconducting skutterudites other than $PrOs_4Sb_{12}$, the dominant contribution to the DOS at the Fermi energy comes from pnicogen *p*-states and transition metal *d*-states, with the filler providing a very minor (~ 2%) contribution, Takegahara and Harima (2007), Gumeniuk et al. (2008a, 2008b), and Sharath Chandra et al. (2016). The conclusion regarding the lack of weight of the filler in the DOS at the Fermi level holds true whether or not the calculations include spin-orbit interaction, Ram et al. (2014). Consequently, superconductivity here originates in the $[T_4X_{12}]$ polyanion. The much wider separation (~ 130 K) of the first excited state from the ground state in CEF of $PrPt_4Ge_{12}$ likely renders the quadrupolar interaction ineffective, and the origin of superconductivity in all platinum germanide skutterudites rests in the band structure of the $[Pt_4Ge_{12}]$ complex, where Ge atoms contribute as much as 80% of electronic states at the Fermi energy, most of them being p-states. One must marvel at the close agreement achieved in recent DFT calculations by Tütüncü et al. (2017) between the calculated and measured critical temperatures T_c of La-filled $LaRu_4P_{12}$, $LaRu_4As_{12}$, and $LaPt_4Ge_{12}$ with values of 6.95 K (7.2 K), 11.56 K (10.45 K), and 8.32 K (8.23 K), respectively, with the experimental values in parentheses. Making use of the Migdal–Eliashberg approach, Migdal (1958) and Eliashberg (1960), the authors determined the Eliashberg spectral function $\alpha^2F(\omega)$ and, by integrating it, arrived at an estimate of the average electron–phonon coupling parameter λ_{e-p}, which turned out to be 0.74, 1.03, and 1.08, respectively, for the three La-filled skutterudites, see Figure 2.71. The results indicate that electron–phonon coupling is the essential mechanism in all three La-filled skutterudites, with $LaRu_4P_{12}$ being a medium coupled superconductor, while $LaRu_4As_{12}$ and $LaPt_4Ge_{12}$ are somewhat more strongly coupled superconductors.

Of course, to truly understand the peculiar superconducting state of $PrOs_4Sb_{12}$ and $PrPt_4Ge_{12}$, it is essential to know the symmetry of the superconducting order parameter. Here, progress has been slow, and the issue has not been settled even today, in spite of numerous attempts to learn more about the exact shape and form of the superconducting gap function. Theoretical approaches have often focused on attempting to explain some particularly striking experimental result without taking into account the totality of experimental evidence. This is the case of the A and B phases of the superconducting condensate extracted from the apparent symmetry change in the oscillations of the angular dependence of magneto-thermal conductivity in $PrOs_4Sb_{12}$, Izawa et al. (2003). A half dozen reports, Maki et al. (2003), Miyake et al. (2003), Goryo (2003), Ichioka et al. (2003), and Maki et al. (2004), trying to assign the order parameters to both phases not only did not agree with each other but also addressed an experimental finding that has never been reproduced and, in fact, was contradicted by several subsequent measurements where it failed to manifest, Seyfarth et al. (2005, 2006), Custers et al. (2006), Hill et al. (2008), and Sakakibara et al. (2008). Moreover, some of the theories, Miyake et al. (2003), assumed an assignment of the CEF ground state (doublet Γ_{23} in T_h) that was subsequently shown to be incorrect.

Short of putting a finger on the exact form of the superconducting order parameter in $PrOs_4Sb_{12}$, some progress has been made by a careful symmetry screening of the space groups that return gap functions consistent with the structure and properties of $PrOs_4Sb_{12}$. The traditional classification scheme for the order parameter summarized by Sigrist and Ueda (1991) is based on point group symmetry. Sergienko and Curnoe (2004) used Landau theory to classify all superconducting phases within the T_h point group symmetry, including those that may be reached by breaking additional symmetries within the superconducting state. They concluded that the most faithful description of the superconducting state of $PrOs_4Sb_{12}$ should come from the three-dimensional representations of the T_h point group that encompass translational symmetry, i.e., those that are based on the space group symmetry rather than just the point group symmetry. This has been done most recently by Sumita and Yanase (2018), who applied a new scheme to a variety of unconventional superconductors, including $PrOs_4Sb_{12}$. The new scheme importantly includes point nodes that depend on the z-component of the angular momentum of the Bloch state and that may emerge on the threefold rotation axis in the Brillouin zone. Because the body-centered-cubic space group relevant to skutterudites includes two threefold axes, the new classification allows for the emergence of point nodes in $PrOs_4Sb_{12}$, provided the order parameter belongs to a specific E_u representation, something that could be tested for. In the same spirit are calculations by Bohloul and Curnoe (2016), who derived general formulae for point-contact conductance from the metal tip into a superconductor using the Blonder–Tinkham–Klapwijk (1982) theory of Andreev reflection, and applied them to all symmetry-allowed gap functions for superconductors with tetrahedral symmetry, having in mind future tests of the order parameter of $PrOs_4Sb_{12}$.

Although the exhilarating days of frantic research activity trying to shed light on the mysterious nature of superconductivity in $PrOs_4Sb_{12}$ during the first ten years of this century have passed, the exotic properties of this skutterudite continue to generate scientific interest and I have no doubt that, in time, even the most challenging issue, the exact form of the superconducting order parameter, will be successfully resolved.

REFERENCES

Abe, K., H. Sato, T. D. Matsuda, T. Namiki, H. Sugawara, and Y. Aoki, *J. Phys.: Condens. Matter* **14**, 11757 (2002).

Affleck, I., arXiv:0911.2209v2 (2010).

Ambegaokar, V. and A. Baratoff, *Phys. Rev. Lett.* **10**, 4861 (1963).

Amin, M. H. S., M. Franz, and I. Affleck, *Phys. Rev. Lett.* **84**, 5864 (2000).

Andraka, B. and K. Pocsy, *J. Appl. Phys.* **111**, 07E115 (2012).

Androja, D. T., A. D. Hillier, J.-G. Park, E. A. Goremychkin, K. A. McEwen, N. Takeda, R. Osborn, B. D. Rainford, and R. M. Ibberson, *Phys. Rev. B* **72**, 184503 (2005).

Aoki, Y., T. Namiki, S. Ohsaki, S. R. Saha, H. Sugawara, and H. Sato, *J. Phys. Soc. Jpn.* **71**, 2098 (2002).

Aoki, Y., T. Namiki, S. Ohsaki, S. R. Saha, H. Sugawara, and H. Sato, *Physica C* **388–389**, 557 (2003a).

Aoki, Y., A. Tsuchiya, T. Kanayama, S. R. Saha, H. Sugawara, H. Sato, W. Higemoto, A. Koda, K. Ohishi, K. Nishiyama, and R. Kadono, *Phys. Rev. Lett.* **91**, 067003 (2003b).

Aoki, Y., W. Higemoto, S. Sanada, K. Ohishi, S. R. Saha, A. Koda, K. Nishiyama, R. Kadono, H. Sugawara, and H. Sato, *Physica B* **359–361**, 895 (2005).

Aoki, Y., S. Sanada, H. Aoki, D. Kikuchi, H. Sugawara, and H. Sato, *Physica B* **378–380**, 54 (2006).

Aoki, Y., T. Tayama, T. Sakakibara, K. Kuwahara, K. Iwasa, M. Kohgi, W. Higemoto, D. E. MacLaughlin, H. Sugawara, and H. Sato, *J. Phys. Soc. Jpn.* **76**, 051006 (2007).

Arii, K., K. Igawa, H. Takahashi, M. Imai, M. Akaishi, and I. Shirotani, *J. Phys.: Conf. Ser.* **121**, 052014 (2008).

Asano, Y., Y. Tanaka, Y. Matsuda, and S. Kashiwaya, *Phys. Rev. B* **68**, 184506 (2003).

Bauer, E., N. A. Frederick, P.-C. Ho, V. S. Zapf, and M. B. Maple, *Phys. Rev. B* **65**, 100506(R) (2002).

Bauer, E., A. Grytsiv, X.-Q. Chen, N. Melnychenko-Koblyuk, G. Hilscher, H. Kaldarar, H. Michor, E. Royanian, G. Giester, M. Rotter, R. Podloucky, and P. Rogl, *Phys. Rev. B* **99**, 217001 (2007).

Bauer, E., X.-Q. Chen, P. Rogl, G. Hilscher, H. Michor, E. Royanian, R. Podloucky, G. Giester, O. Sologub, and A. P. Goncalves, *Phys. Rev. B* **78**, 064516 (2008).

Bean, C. P., *Phys. Rev. Lett.* **8**, 250 (1962).

Bean, C. P., *Rev. Mod. Phys.* **36**, 31 (1964).

Blonder, G. E., M. Tinckham, and T. M. Klapwijk, *Phys. Rev. B* **25**, 4515 (1982).

Bochenek, L., R. Wawryk, Z. Henkie, and T. Cichorek, *Phys. Rev. B* **86**, 060511(R) (2012).

Bohloul, S. and S. H. Curnoe, *J. Phys.: Condens. Matter* **28**, 045701 (2016).

Bouquet, F., Y. Wang, R. A. Fisher, D. G. Hinks, J. D. Jorgensen, A. Junod, and N. E. Phillips, *Europhys. Lett.* **56**, 856 (2001).

Brandt, E. H., *Phys. Rev. B* **37**, R2349 (1998).

Briarty, M. E., P. Kumar, G. R. Stewart, and B. Andraka, *J. Phys.: Condens. Matter* **21**, 385701 (2009).

Braun, D. J. and W. Jeitschko, *J. Solid State Chem.* **32**, 35 (1980).

Carrington, A. and F. Manzano, *Physica C* **385**, 205 (2003).

Chang, J., I. Eremin, P. Thalmeier, and P. Fulde, *Phys. Rev. B* **76**, 220510(R) (2007).

Cheng, J.-G., J.-S. Zhou, K. Matsubayashi, P. P. Kong, Y. Kubo, Y. Kawamura, C. Sekine, C. Q. Jin, J. B. Goodenough, and Y. Uwatoko, *Phys. Rev. B* **88**, 024514 (2013).

Chia, E. E. M., M. B. Salamon, H. Sugawara, and H. Sato, *Phys. Rev. Lett.* **91**, 247003 (2003).

Chia, E. E. M., M. B. Salamon, H. Sugawara, and H. Sato, *Phys. Rev. B* **69**, 180509(R) (2004).

Chia, E. E. M., D. Vandervelde, M. B. Salamon, D. Kikuchi, H. Sugawara, and H. Sato, *J. Phys.: Condens. Matter* **17**, L303 (2005).

Cho, W. and S. A. Kivelson, *Phys. Rev. Lett.* **116**, 093903 (2016).

Cichorek, T., A. C. Mota, F. Steglich, N. A. Frederick, W. M. Yuhasz, and M. B. Maple, *Phys. Rev. Lett.* **94**, 107002 (2005).

Cox, D. L. and A. Zawadowski, *Adv. Phys.* **47**, 599 (1998).

Custers, J., Y. Namai, T. Tayama, T. Sakakibara, H. Sugawara, Y. Aoki, and H. Sato, *Physica B* **378–380**, 179 (2006).

DeLong, L. E. and G. P. Meisner, *Solid State Commun.* **53**, 119 (1985).

Deminami, S., Y. Kawamura, Y.-Q. Chen, M. Kanazawa, J. Hayashi, T. Kuzuya, K. Takeda, M. Matsuda, and C. Sekine, *J. Phys.: Conf. Ser.* **950**, 042032 (2017).

Dynes, R. C., V. Narayanamurti, and J. P. Garno, *Phys. Rev. Lett.* **41**, 1509 (1978).

Eliashberg, G. M., *Sov. Phys. JETP* **11**, 696 (1960).

Fisher, R. A., S. Kim, B. F. Woodfield, N. E. Phillips, L. Taillefer, K. Hasselbach, J. Flouquet, A. L. Giorgi, and J. L. Smith, *Phys. Rev. Lett.* **62**, 1411 (1989).

Foroozani, N., J. J. Hamlin, J. S. Schilling, R. E. Baumbach, I. K. Lum, L. Shu, K. Huang, and M. B. Maple, *Physica C* **485**, 160 (2013).

Frederick, N. A., T. D. Do, P.-C. Ho, N. P. Butch, V. S. Zapf, and M. B. Maple, *Phys. Rev. B* **69**, 024523 (2004).

Frederick, N. A., T. A. Sayles, and M. B. Maple, *Phys. Rev. B* **71**, 064508 (2005).

Frederick, N. A., T. A. Sayles, S. K. Kim, and M. B. Maple, *J. Low Temp. Phys.* **147**, 321 (2007).

Fulde, P. and J. Jensen, *Phys. Rev. B* **27**, 4085 (1983).

Gardner, W. E. and T. F. Smith, *Phys. Rev.* **138**, A 484 (1965).

Gopal, E. S. R., in *Specific Heat*, Plenum Press, New York (1966).

Goremychkin, E. A., R. Osborn, E. D. Bauer, M. B. Maple, N. A. Frederick, W. M. Yuhasz, F. M. Woodward, and J. W. Lynn, *Phys. Rev. Lett.* **93**, 157003 (2004).

Goryo, J., *Phys. Rev. B* **67**, 184511 (2003).

Goto, T., Y. Nemoto, K. Onuki, K. Sakai, T. Yamaguchi, M. Akatsu, T. Yanagisawa, H. Sugawara, and H. Sato, *J. Phys. Soc. Jpn.* **74**, 263 (2005).

Gross, F., B. S. Chandrasekhar, D. Einzel, K. Andres, P. J. Hirschfeld, H. R. Ott, J. Beuers, Z. Fisk, and J. L. Smith, *Z. Phys. B-Condens. Matter* **64**, 175 (1986).

Grytsiv, A., X.-Q. Chen, N. Melnychenko-Koblyuk, P. Rogl, E. Bauer, G. Hilscher, H. Kaldarar, H. Michor, E. Royanian, R. Podloucky, M. Rotter, and G. Giester, *J. Phys. Soc. Jpn.* **77**, 121 (2008).

Gumeniuk, R., W. Schnelle, H. Rosner, M. Nicklas, A. Leithe-Jasper, and Y. Grin, *Phys. Rev. Lett.* **100**, 017002 (2008a).

Gumeniuk, R., H. Rosner, W. Schnelle, M. Nicklas, A. Leithe-Jasper, and Y. Grin, *Phys. Rev. B* **78**, 052504 (2008b).

Gumeniuk, R., H. Borrmann, A. Ormeci, H. Rosner, W. Schnelle, M. Nicklas, Y. Grin, and A. Leithe-Jasper, *Z. Kristallogr.* **225**, 531 (2010).

Gurevich, A., *Phys. Rev. B* **67**, 184515 (2003).

Harima, H. and K. Takegahara, *Physica C* **388–389**, 555 (2003).

Harima, H. and K. Takegahara, *Physica B* **359–361**, 920 (2005).

Hattori, K., Y. Hirayama, and K. Miyake, *J. Phys. Soc. Jpn.* **74**, 3306 (2005).

Heffner, R. H. and M. R. Norman, *Comments on Cond. Matter Phys.* **17**, 361 (1996).

Henkie, Z., M. B. Maple, A. Pietraszko, R. Wawryk, T. Cichorek, R. E. Baumbach, W. M. Yuhasz, and P.-C. Ho, *J. Phys. Soc. Jpn.* **77**, Suppl. A, 128 (2008).

Higemoto, W., S. R. Saha, A. Koda, K. Ohishi, R. Kadono, Y. Aoki, H. Sugawara, and H. Sato, *Phys. Rev. B* **75**, 020510(R) (2007).

Hill, R. W., S. Li, M. B. Maple, and L. Taillefer, *Phys. Rev. Lett.* **101**, 237005 (2008).

Ho, P.-C., V. S. Zapf, E. D. Bauer, N. A. Frederick, M. B. Maple, G. Giester, P. Rogl, S. T. Berger, C. H. Paul, and E. Bauer, *Int. J. Mod. Phys. B* **16**, 3008 (2002).

Ho, P.-C., W. M. Yuhasz, N. P. Butch, N. A. Frederick, T. A. Sayles, J. R. Jeffries, M. B. Maple, J. B. Betts, and A. H. Lacerda, *Phys. Rev. B* **72**, 094410 (2005).

Ho, P.-C., N. P. Butch, V. S. Zapf, T. Yanagisawa, N. A. Frederick, S. K. Kim, W. M. Yuhasz, M. B. Maple, J. B. Betts, and A. H. Lacerda, *J. Phys.: Condens. Matter* **20**, 215226 (2008a).

Ho, P.-C., T. Yanagisawa, N. P. Butch, W. M. Yuhasz, C. C. Robinson, A. A. Dooraghi, and M. B. Maple, *Physica B* **403**, 1038 (2008b).

Hopfield, J. J., *Physica* **55**, 41 (1971).

Hotta, T., *J. Phys. Soc. Jpn.* **77**, 103711 (2008).

Huang, K., L. Shu, I. K. Lum, B. D. White, M. Janoschek, D. Yazici, J. J. Hamlin, D. A. Zocco, P.-C. Ho, R. E. Baumbach, and M. B. Maple, *Phys. Rev. B* **89**, 035145 (2014).

Humer, S., E. Royanian, H. Michor, E. Bauer, A. Grytsiv, M. X. Chen, R. Podloucky, and P. Rogl, in *New Materials for Thermoelectric Applications: Theory and Experiment*, NATO Science for Peace and Security Series B: Physics and Biophysics, edited by V. Zlatic and A. Hewson, Springer, Netherlands, pp. 115–127 (2013).

Huxley, A. D., M.-A. Méasson, K. Izawa, C. D. Dewhurst, R. Cubitt, B. Grenier, H. Sugawara, J. Flouquet, Y. Matsuda, and H. Sato, *Phys. Rev. Lett.* **93**, 187005 (2004).

Ichioka, M., N. Nakai, and K. Machida, *J. Phys. Soc. Jpn.* **72**, 1322 (2003).

Imai, M., M. Akaishi, E. D. Sadki, T. Aoyagi, T. Kimura, and I. Shirotani, *Phys. Rev. B* **75**, 184535 (2007a).

Imai, M., M. Akaishi, and I. Shirotani, *Supercond. Sci. Technol.* **20**, 832 (2007b).

Ishida, K., H. Mukuda, Y. Kitaoka, K. Asayama, Z. Q. Mao, Y. Mori, and Y. Maeno, *Nature* **396**, 658 (1998).

Ishida, K., D. Ozaki, T. Kamatsuka, H. Tou, M. Kyogaku, Y. Kitaoka, N. Tateiwa, N. K. Sato, N. Aso, C. Geibel, and F. Steglich, *Phys. Rev. Lett.* **89**, 037002 (2002).

Izawa, K., Y. Nakajima, J. Goryo, Y. Matsuda, S. Osaki, H. Sugawara, H. Sato, P. Thalmeier, and K. Maki, *Phys. Rev. Lett.* **90**, 117001 (2003).

Jeon, I., K. Huang, D. Yazici, N. Kanchanavatee, B. D. White, P.-C. Ho, S. Jang, N. Pouse, and M. B. Maple, *Phys. Rev. B* **93**, 104507 (2016).

Jeon, I., S. Ran, A. J. Breindel, P.-C. Ho, R. B. Adhikari, C. C. Almasan, B. Luong, and M. B. Maple, *Phys. Rev. B* **95**, 134517 (2017).

Joust, R. and L. Taillefer, *Rev. Mod. Phys.* **74**, 235 (2002).

Juraszek, J., Z. Henkie, and T. Cichorek, *Acta Phys. Polon.* **130**, 597 (2016).

Kaczorowski, D. and V. H. Tran, *Phys. Rev. B* **77**, 180504(R) (2008).

Kamerlingh Onnes, H., *Comm. Phys. Lab. Univ. Leiden* No. 120b (1911).

Kanako, K., N. Metoki, R. Shiina, T. D. Matsuda, M. Koghi, K. Kuwahawa, and N. Bernhoeft, *Phys. Rev. B* **75**, 094408 (2007).

Kanetake, F., H. Mukuda, Y. Kitaoka, K. Magishi, H. Sugawara, K. M. Itoh, and E. E. Haller, *J. Phys. Soc. Jpn.* **79**, 063702 (2010).

Kapitulnik, A., *Physica B* **460**, 151 (2015).

Kawamura, Y., M. Hayato, Y. Q. Chen, J. Hayashi, C. Sekine, H. Gotou, and Z. Hiroi, *Phys. Procedia* **75**, 200 (2015).

Kawamura, Y., T. Kawaai, J. Hayashi, C. Sekine, H. Gotou, J. Cheng, K. Matsubayashi, and Y. Uwatoko, *J. Phys. Soc. Jpn.* **82**, 114702 (2013).

Kawamura, Y., S. Deminami, L. Salamakha, A. Sidorenko, P. Heinrich, H. Michor, E. Bauer, and C. Sekine, *Phys. Rev. B* **98**, 024513 (2018).

Khan, R. T., E. Bauer, X.-Q. Chen, R. Podloucky, and P. Rogl, *J. Phys. Soc. Jpn.* **77**, Suppl. A, 350 (2008).

Kihou, K., I. Shirotani, Y. Shimaya, C. Sekine, and T. Yagi, *Mater. Res. Bull.* **39**, 317 (2004).

Kim, Y. B., C. F. Hempstead, and A. R. Strnad, *Phys. Rev. Lett.* **9**, 306 (1962).

Kohgi, M., K. Iwasa, M. Nakajima, N. Metoki, S. Araki, N. Bernhoeft, J. M. Mignot, A. Gukasov, H. Sato, Y. Aoki, and H. Sugawara, *J. Phys. Soc. Jpn.* **72**, 1002 (2003).

Kotegawa, H., M. Yogi, Y. Imamura, Y. Kawasaki, G.-Q. Zheng, Y. Kitaoka, S. Ohsaki, H. Sugawara, Y. Aoki, and H. Sato, *Phys. Rev. Lett.* **90**, 027001 (2003).

Kozii, V., J. W. F. Venderbos, and L. Fu, *Sci. Adv.* **2**, e1601835 (2016).

Kuwahara, K., K. Iwasa, M. Kohgi, K. Kaneko, S. Araki, N. Metoki, H. Sugawara, Y. Aoki, and H. Sato, *J. Phys. Soc. Jpn.* **73**, 1438 (2004).

Kuwahara, K., K. Iwasa, M. Kohgi, K. Kaneko, N. Metoki, S. Raymond, M.-A. Méasson, J. Flouquet, H. Sugawara, Y. Aoki, and H. Sato, *Phys. Rev. Lett.* **95**, 107003 (2005).

Landau, I. L. and H. Keller, *Physica C* **466**, 131 (2007).

Lea, K. R., M. J. M. Leask, and W. P. Wolf, *J. Phys. Chem. Solids* **23**, 1381 (1962).

Levenson-Falk, E. M., E. R. Schemm, M. B. Maple, and A. Kapitulnik, *Phys. Rev. Lett.* **120**, 187004 (2018).

Li, L.-W., E. Sakada, and K. Nishimura, *Mater. Trans.* **51**, 227 (2010).

Lowell, J. and J. Sousa, *J. Low Temp. Phys.* **3**, 65 (1970).

Machida, K., T. Nishira, and T. Ohmi, *J. Phys. Soc. Jpn.* **68**, 3364 (1999).

Machida, Y., A. Itoh, Y. So, K. Izawa, Y. Haga, E. Yamamoto, N. Kimura, Y. Onuki, Y. Tsutsumi, and K. Machida, *Phys. Rev. Lett.* **108**, 157002 (2012).

MacLaughlin, D. E., J. E. Sonier, R. H. Heffner, O. O. Bernal, B.-L. Young, M. S. Rose, G. D. Morris, E. D. Bauer, T. D. Do, and M. B. Maple, *Phys. Rev. Lett.* **89**, 157001 (2002).

MacLaughlin, D. E., P.-C. Ho, L. Shu, O. O. Bergal, S. Zhao, A. A. Dooraghi, T. Yanagisawa, M. B. Maple, and R. H. Fukuda, *Phys. Rev. B* **89**, 144419 (2014).

Magishi, K., T. Saito, K. Koyama, I. Shiratoni, Y. Shimaya, K. Kihou, C. Sekine, N. Takeda, M. Ishikawa, and T. Yagi, *Physica B* **359–361**, 883 (2005).

Magishi, K., H. Sugawara, T. Saito, K. Koyama, F. Kanetake, H. Mukuda, Y. Kitaoka, K. M. Itoh, and E. E. Haller, *J. Phys. Soc. Jpn.* **80**, SA028 (2011).

Maisuradze, A., M. Nicklas, R. Gumeniuk, C. Baines, W. Schelle, H. Rosner, A. Leithe-Jasper, Y. Grin, and R. Khasanov, *Phys. Rev. Lett.* **103**, 147002 (2009a).

Maisuradze, A., R. Khasanov, A. Shengelaya, and H. Keller, *J. Phys.: Condens. Matter* **21**, 075701 (2009b).

Maisuradze, A., W. Schelle, R. Khasanov, R. Gumeniuk, M. Nicklas, H. Rosner, A. Leithe-Jasper, Y. Grin, A. Amato, and P. Thalmeier, *Phys. Rev. B* **82**, 024524 (2010).

Maisuradze, A., R. Gumeniuk, W. Schnelle, M. Nicklas, C. Baines, R. Khasanov, A. Amato, and A. Leithe-Jasper, *Phys. Rev. B* **86**, 174513 (2012).

Majorana, E., *Nuovo Cimento* **14**, 171 (1937).

Maki, K., H. Won, P. Thalmeier, Q. Yuan, K. Izawa, and Y. Matsuda, *Europhys. Lett.* **64**, 496 (2003).

Maki, K., S. Haas, D. Parker, H. Won, K. Izawa, and Y. Matsuda, *Europhys. Lett.* **68**, 720 (2004).

Mancini, F. and R. Citro, eds., in *Iron Pnictide Superconductors*, Springer Series in Solid-State Sciences, Vol. 186, Springer International Publishers (2017).

Maple, M. B., P.-C. Ho, V. S. Zapf, N. A. Frederick, E. D. Bauer, W. M. Yuhasz, F. M. Woodward, and J. W. Lynn, Proc. Int. Conf. Strongly Correlated Electrons with Orbital Degrees of Freedom (ORBITAL 2001), *J. Phys. Soc. Jpn.* **71**, Suppl, 23 (2002).

Maple, M. B., P.-C. Ho, V. S. Zapf, W. M. Yuhasz, N. A. Frederick, and E. D. Bauer, *Physica C* **388–389**, 549 (2003).

Maple, M. B., Z. Henkie, R. E. Baumbach, T. A. Sayles, N. P. Butch, P.-C. Ho, T. Yanagisawa, W. M. Yuhasz, R. Wawryk, T. Cichorek, and A. Petraszko, *J. Phys. Soc. Jpn.* **77** Suppl. 1, 7 (2008).

McBriarty, M. E., P. Kumar, G. R. Stewart, and B. Andraka, *J. Phys.: Condens. Matter* **21**, 285701 (2009).

McCollam, A., B. Andraka, and S. R. Julian, *Phys. Rev. B* **88**, 075102 (2013).

McMillan, W. L., *Phys. Rev.* **167**, 331 (1968).

Migdal, A. B., *Sov. Phys. JETP* **34**, 996 (1958).

Méasson, M.-A., D. Braithwaite, J. Flouquet, G. Seyfarth, J.-P. Brison, E. Lhotel, C. Paulsen, H. Sugawara, and H. Sato, *Phys. Rev. B* **70**, 064516 (2004).

Méasson, M.-A., D. Braithwaite, B. Salce, J. Flouquet, G. Lapertot, H. Sugawara, H. Sato, and Y. Onuki, *J. Magn. Magn. Mater.* **310**, 626 (2007).

Méasson, M.-A., D. Braithwaite, G. Lapertot, J.-P. Brison, J. Flouquet, P. Bordet, H. Sugawara, and P. C. Canfield, *Phys. Rev. B* **77**, 134517 (2008).

Meisner, G. P., *Physica B* **108**, 763 (1981).

Meisner, G. P., Ph. D. thesis, University of California, San Diego (1982).

Meissner, W. and R. Ochsenfeld, *Naturwissenschaften* **21**, 787 (1933).

Miyake, K., H. Kohno, and H. Harima, *J. Phys.: Condens. Matter* **15**, L275 (2003).

Miyake, A., K. Shimizu, C. Sekine, K. Kihou, and I. Shirotani, *J. Phys. Soc. Jpn.* **73**, 2370 (2004).

Nakai, N., P. Miranovic, M. Ichioka, and K. Machida, *Phys. Rev. B* **70**, 100503(R) (2004).

Nakai, Y., K. Ishida, D. Kikuchi, H. Sugawara, and H. Sato, *J. Phys. Soc. Jpn.* **74**, 3370 (2005).

Nakamura, Y., H. Okazaki, R. Yoshida, T. Wakita, H. Takeya, K. Hirata, M. Hirai, Y. Muraoka, T. Yokoya, *Phys. Rev. B* **86**, 014521 (2012).

Nakazima, M., S. Arai, and Y. Kubo, *J. Phys. Soc. Jpn.* **83**, 065003 (2014).

Namiki, T., Y. Aoki, H. Sato, C. Sekine, I. Shirotani, T. D. Masuda, Y. Haga, and T. Yagi, *J. Phys. Soc. Jpn.* **76**, 093704 (2007).

Namiki, T., C. Sekine, K. Matsuhira, M. Wakeshima, and I. Shirotani, *J. Phys. Soc. Jpn.* **A 77**, 336 (2008).

Oeschler, N., P. Gegenwart, F. Weickert, I. Zerec, P. Thalmeier, F. Steglich, E. D. Bauer, N. A. Frederick, and M. B. Maple, *Phys. Rev. B* **69**, 235108 (2004).

Padamsee, H., J. E. Neighbor, and C. A. Shiffman, *J. Low Temp. Phys.* **12**, 387 (1973).

Pfau, H., M. Nicklas, U. Stockert, R. Gumeniuk, W. Schnelle, A. Leithe-Jasper, Y. Grin, and F. Steglich, *Phys. Rev. B* **94**, 054523 (2016).

Proust, C., E. Boaknin, R. W. Hill, L. Taillefer, and A. P. Mackenzie, *Phys. Rev. Lett.* **89**, 147003 (2002).

Prozorov, R. and R. W. Giannetta, *Supercond. Sci. Technol.* **19**, R41 (2006).

Qi, Y. P., H. C. Lei, J. G. Guo, W. J. Shi, B. H. Yan, C. Felser, and H. Hosono, *J. Am. Chem. Soc.* **139**, 8106 (2017).

Ram, S., V. Kanchara, and M. C. Valsakumar, *J. Appl. Phys.* **115**, 093903 (2014).

Ren, C., Z. S. Wang, H. Q. Luo, H. Yang, L. Shan and H. H. Wen, *Phys. Rev. Lett.* **101**, 257006 (2008).

Rotundu, C. R., H. Tsujii, Y. Takano, B. Andraka, H. Sugawara, Y. Aoki, and H. Sato, *Phys. Rev. Lett.* **92**, 037203 (2004).

Rotundu, C. R., P. Kumar, and B. Andraka, *Phys. Rev. B* **73**, 014515 (2006).

Sakakibara, T., J. Custers, K. Yano, A. Yamada, T. Tayama, Y. Aoki, H. Sato, H. Sugwara, H. Amitsuka, and M. Yokoyama, *Physica B* **403**, 990 (2008).

Sayles, T. A., R. E. Baumbach, W. M. Yuhasz, M. B. Maple, L. Bochenek, R. Wawryk, T. Cichorek, A. Pietraszko, Z. Henkie, and P.-C. Ho, *Phys. Rev. B* **82**, 104513 (2010).

Sekine, C., T. Uchiumi, I. Shirotani, and T. Yagi, *Phys. Rev. Lett.* **79**, 3218 (1997).

Sergienko, I. A., *Phys. Rev. B* **69**, 174502 (2004).

Sergienko, I. A. and S. H. Curnoe, *Phys. Rev. B* **70**, 144522 (2004).

Setty, C., Y. Wang, and P. W. Phillips, *Phys. Rev. B* **96**, 054508 (2017).

Seyfarth, G., J. P. Brison, M.-A. Méasson, J. Flouquet, K. Izawa, Y. Matsuda, H. Sugawara, and H. Sato, *Phys. Rev. Lett.* **95**, 107004 (2005).

Seyfarth, G., J. P. Brison, M.-A. Méasson, D. Braithwaite, G. Lapertot, and J. Flouquet, *Phys. Rev. Lett.* **97**, 236403 (2006).

Shakeripour, H., C. Petrovic, and L. Taillefer, *New J. Phys.* **11**, 055065 (2009).

Sharath Chandra, L. S., M. K. Chattopadhyay, and S. B. Roy, *Phil. Mag.* **92**, 3866 (2012).

Sharath Chandra, L. S., M. K. Chattopadhyay, S. B. Roy, and S. K. Pandey, *Phil. Mag.* **96**, 2161 (2016).

Shenoy, G. K., D. R. Noakes, and G. P. Meisner, *J. Appl. Phys.* **53**, 2628 (1982).

Shiina, R. and Y. Aoki, *J. Phys. Soc. Jpn.* **73**, 541 (2004).

Shiina, R. and H. Shiba, *J. Phys. Soc. Jpn.* **79**, 044704 (2010).

Shiina, R., *J. Phys. Soc. Jpn.* **81**, 024706 (2012).

Shimizu, M., H. Amanuma, K. Hachitani, H. Fukazawa, Y. Kohori, T. Namiki, C. Sekine, and I. Shirotani, *J. Phys. Soc. Jpn.* **77**, Suppl. A, 229 (2008).

Shirotani, I., T. Adachi, K. Tachi, S. Todo, Y. Nakazawa, T. Yagi, and M. Kinoshita, *J. Phys. Chem. Solids* **57**, 211 (1996).

Shirotani, I., T. Uchiumi, T. Ohno, C. Sekine, Y. Nakazawa, K. Kanoda, S. Todo, and T. Yagi, *Phys. Rev. B* **56**, 7866 (1997).

Shirotani, I., K. Ohno, C. Sekine, T. Yagi, T. Kawakami, T. Nakanishi, H. Takahashi, J. Tang, A. Matsushita, and T. Matsumoto, *Physica B* **281–282**, 1021 (2000).

Shirotani, I., Y. Shimaya, K. Kihou, C. Sekine, and T. Yagi, *J. Solid State Chem.* **174**, 32 (2003a)

Shirotani, I., Y. Shimaya, K. Kihou, C. Sekine, N. Takeda, M. Ishikawa, and T. Yagi, *J. Phys.: Condens. Matter* **15**, S2201 (2003b).

Shirotani, I., N. Araseki, Y. Shimaya, R. Nakata, K. Kihou, C. Sekine, and T. Yagi, *J. Phys.: Condens. Matter* **17**, 4383 (2005a).

Shirotani, I., S. Sato, C. Sekine, K. Takeda, I. Inagawa, and T. Yagi, *J. Phys.: Condens. Matter* **17**, 7353 (2005b).

Shu, L., D. E. MacLaughlin, R. H. Heffner, F. D. Callaghan, J. E. Sonier, G. D. Morris, O. O. Bernal, A. Bosse, J. E. Anderson, W. M. Yuhasz, N. A. Frederick, and M. B. Maple, *Physica B* **374**, 247 (2006).

Shu, L., W. Higemoto, Y. Aoki, N. A. Frederick, W. M. Yuhasz, R. H. Heffner, K. Ohishi, K. Ishida, R. Kadono, A. Koda, D. Kikuchi, H. Sato, H. Sugawara, T. U. Ito, S. Sanada, Y. Tunashima, Y. Yonezawa, M. B. Maple, and D. E. MacLaughlin, *J. Magn. Magn. Matter.* **310**, 551 (2007).

Sigrist, M. and K. Ueda, *Rev. Mod. Phys.* **63**, 239 (1991).

Singh, Y. P., R. B. Adhikari, S. Zhang, K. Huang, D. Yazici, I. Jeon, M. B. Maple, M. Dzero, and C. C. Almasan, *Phys. Rev. B* **94**, 144502 (2016).

Smith, T. F. and H. L. Luo, *J. Phys. Chem. Solids* **28**, 596 (1967).

Sologubenko, A. V., J. Jun, S. M. Kazakov, J. Karpinski, and H. R. Ott, *Phys. Rev. B* **66**, 014504 (2002).

Suderow, H., S. Vieira, J. D. Strand, S. Bud'ko, and P. C. Canfield, *Phys. Rev. B* **69**, 060504 (2004).

Sugawara, H., S. Osaki, S. R. Saha, Y. Aoki, H. Sato, Y. Inada, H. Shishido, R. Settai, Y. Onuki, H. Harima, and K. Oikawa, *Phys. Rev. B* **66**, 220504(R) (2002).

Sugawara, H., M. Kobayashi, S. Osaki, S. R. Saha, T. Namiki, Y. Aoki, and H. Sato, *Phys. Rev. B* **72**, 014519 (2005).

Sumita, S. and Y. Yanase, *Phys. Rev. B* **97**, 134512 (2018).

Swartz, P. S., *Phys. Rev. Lett.* **9**, 448 (1962).

Takeda, N. and M. Ishikawa, *J. Phys. Soc. Jpn.* **69**, 868 (2000).

Takegahara, K., H. Harima, and A. Yanase, *J. Phys. Soc. Jpn.* **70**, 1190 (2001).

Takegahara, K. and H. Harima, *J. Magn. Magn. Mater.* **310**, 861 (2007).

Tanaka, K., T. Namiki, A. Imamura, M. Ueda, T. Saito, S. Tatsuoka, R. Miyazaki, K. Kuwahara, Y. Aoki, and H. Sato, *J. Phys. Soc. Jpn.* **78**, 063701 (2009).

Tayama, T., T. Sakakibara, H. Sugawara, Y. Aoki, and H. Sato, *J. Phys. Soc. Jpn.* **72**, 1516 (2003).

Tayama, T., T. Sakakibara, H. Sugawara, and H. Sato, *J. Phys. Soc. Jpn.* **75**, 043707 (2006).

Tee, X. Y., H. G. Luo, T. Xiang, D. Vandervelde, M. B. Salamon, H. Sugawara, H. Sato, C. Panagopoulos, and E. E. M. Chia, *Phys. Rev. B* **86**, 064518 (2012).

Tenya, K., N. Oeschler, P. Gegenwart, F. Steglich, N. A. Frederick, E. D. Bauer, and M. B. Maple, *Acta Phys. Polon. B* **34**, 995 (2003).

Thalmeier, P., K. Maki, and Q. S. Yuan, *Physica C* **408–410**, 177 (2004).

Tinkham, M., in *Introduction to Superconductivity*, McGraw-Hill, New York (1983).

Toda, M., H. Sugawara, K. Magishi, T. Saito, K. Koyama, Y. Aoki, and H. Sato, *J. Phys. Soc. Jpn.* **77**, 124702 (2008).

Tou, H., Y. Kitaoka, K. Asayama, N. Kimura, Y. Ōnuki, E. Yamamoto, and K. Maezawa, *Phys. Rev. Lett.* **77**, 1374 (1996).

Tou, H., Y. Inaoka, M. Doi, M. Sera, K. Asaki, H. Kotegawa, H. Sugawara, and H. Sato, *J. Phys. Soc. Jpn.* **80**, 074703 (2011).

Tran, V. H., D. Kaczorowski, W. Miller, and A. Jezierski, *Phys. Rev. B* **79**, 054520 (2009).

Tran, V. H., A. D. Hillier, D. T. Adroja, and D. Kaczorowski, *J. Phys. Soc. Jpn.* **80**, SA030 (2011).

Turel, C. S., J. Y. T. Wei, W. M. Yuhasz, R. Baumbach, and M. B. Maple, *J. Phys. Soc. Jpn.* **77**, Suppl. A, 21 (2008).

Tütüncü, H. M., E. Karaca, and G. P. Srivastava, *Phys. Rev. B* **95**, 214514 (2017).

Uchiumi, T., I. Shirotani, C. Sekine, S. Todo, T. Yagi, Y. Nakazawa, and K. Kanoda, *J. Phys. Chem. Solids* **60**, 689 (1999).

Usadel, K., *Phys. Rev. Lett.* **25**, 507 (1970).

Venkateshwarlu, D., S. S. Samatham, M. Gangrade, and V. Ganesan, *Proc. Int. Conf. on Recent Trends in Physics, J. Phys. Conf. Ser.* **534**, 012040 (2014).

Vieland, L. J., *Phys. Lett.* **15**, 23 (1965).

Vollmer, R., A. Faisst, C. Pfleiderer, H. V. Löhneysen, E. D. Bauer, P.-C. Ho, V. Zapf, and M. B. Maple, *Phys. Rev. Lett.* **90**, 057001 (2003).

Volovik, G. E., *JETP Lett.* **58**, 469 (1993).

Watanabe, T., K. Izawa, Y. Kasahara, Y. Haga, Y. Onuki, P. Thalmeier, K. Maki, and Y. Matsuda, *Phys. Rev. B* **70**, 184502 (2004).

Werthamer, N. R., E. Helfand, and P. C. Hohenberg, *Phys. Rev.* **147**, 295 (1966).

Willis, J. O. and D. M. Ginsberg, *Phys. Rev. B* **14**, 1916 (1976).

Xia, J., Y. Maeno, P. T. Beyersdorf, M. M. Fejer, and A. Kapitulnik, *Phys. Rev. Lett.* **97**, 167002 (2006).

Yogi, M., H. Kotegawa, Y. Imamura, G.-Q. Zheng, Y. Kitaoka, H. Sugawara, and H. Sato, *Phys. Rev. B* **67**, 180501(R) (2003).

Yogi, M., T. Nagai, Y. Imamura, H. Mukuda, Y. Kitaoka, D. Kikuchi, H. Sugawara, Y. Aoki, H. Sato, and H. Harima, *J. Phys. Soc. Jpn.* **75**, 124702 (2006).

Yosida, K., *Phys. Rev.* **110**, 769 (1958).

Zhang, J. L., Y. Chen, L. Jiao, R. Gumeniuk, M. Nicklas, Y. H. Chen, L. Yang, B. H. Fu, W. Schnelle, H. Rosner, A. Leithe-Jasper, Y. Grin, F. Steglich, and H. Q. Yuan, *Phys. Rev. B* **87**, 064502 (2013).

Zhang, J. L., D. E. MacLaughlin, A. D. Hillier, Z. F. Ding, K. Huang, M. B. Maple, and L. Shu, *Phys. Rev. B* **91**, 104523 (2015a).

Zhang, J. L., G. M. Pang, L. Jiao, M. Nicklas, Y. Chen, Z. F. Weng, M. Smidman, W. Schnelle, A. Leithe-Jasper, A. Maisuradze, C. Baines, R. Khasanov, A. Amato, F. Steglich, R. Gumeniuk, and H. Q. Yuan, *Phys. Rev. B* **92**, 220503(R) (2015b).

Zwicknagl, G., P. Thalmeier, and P. Fulde, *Phys. Rev. B* **79**, 115132 (2009).

3 Magnetic Properties of Skutterudites

3.1 INTRODUCTION

As we have already seen, skutterudites are solids that encompass a rich variety of physical properties. Among them, one of the most fascinating and important characteristics is magnetism. Studies of the magnetic properties have greatly aided our understanding of the transport behavior in skutterudites, and the magnetic response of the structure provided a pivotal input to rationalize bonding in these materials. It is fair to say that without the insight gained from magnetic measurements, our knowledge of skutterudites would have been very incomplete.

All substances have some magnetism associated with them, and what kind of magnetism the substance possesses depends on the nature of interactions among its electrons. Each electron reveals a dual form of a tiny magnet. Orbital motion of an electron around its nucleus can be viewed as a tiny loop of current generating an orbital magnetic moment μ_{orb}. An electron can also be naively considered as spinning about its own axis, giving rise to its intrinsic spin magnetic moment μ_s. The orbital and spin angular momenta of electrons of a given atom combine by the rules of quantum mechanics to yield the overall magnetic moment of the atom μ that relates to the total atomic angular momentum J. The magnetic moments are usually expressed in units of Bohr magneton, $\mu_B = e\hbar/2m_e = 9.274 \times 10^{-24}$ Am^2, which is a magnetic moment generated by an electron of mass m_e as it orbits around a nucleus of a hydrogen atom. A solid substance contains a vast number of electrons, and they not only interact but may actually cooperate to find a particular harmonious order we categorize into basically one of five major classes of magnetic materials: diamagnetic, paramagnetic, ferromagnetic (FM), antiferromagnetic (AFM), and ferrimagnetic materials. There are other forms of magnetism, such as spiral and helical magnetic structures, in which the orientation of the magnetic moment precesses around a cone or rotates around a circle as one advances from site to site, as well as situations where the magnetic moment is frozen at a lattice site in random orientation, called spin glasses. In skutterudites, these other forms of magnetism are rarely encountered, and we shall not be concerned about them. Each of the above forms of magnetism has its specific properties that reflect the overall magnetic moment of the substance. One usually normalizes this magnetic moment per unit volume and arrives at the quantity referred to as magnetization M. In the study of magnetism and superconductivity, one must carefully distinguish between the vectors of applied magnetic field H and magnetic induction B. In free space, using the SI units, the two vectors are merely a scaled version of each other, related by

$$B = \mu_0 H, \tag{3.1}$$

where $\mu_0 = 4\pi \times 10^{-7}$ Hm^{-1} is the permeability of free space. In a magnetic solid, however, the two vectors may have not only very different magnitudes but also different directions, and the vector relationship between the two is

$$B = \mu_0 \left(H + M \right), \tag{3.2}$$

where the magnetization M expresses the property of the magnetic medium. As long as the magnetization M is linearly related to the magnetic field H, i.e.,

$$M = \chi H, \tag{3.3}$$

where χ is the magnetic susceptibility, the linearity between B and H is preserved as we can write

DOI: 10.1201/9781003225898-3

113

$$B = \mu_0 \left(1 + \chi\right) H = \mu_0 \mu_r H. \tag{3.4}$$

Here, $\mu_r = 1 + \chi$ is called the relative permeability. Magnetic induction B is measured in units of Tesla, while magnetization M is measured in the same units as the magnetic field H, i.e., Am^{-1}. This makes the magnetic susceptibility χ a dimensionless quantity. It is the magnetic susceptibility that conveniently differentiates between different forms of magnetism.

Before I discuss the magnetic susceptibility, it is important to understand how the usual non-relativistic Hamiltonian $\hat{\mathcal{H}}_0$ for an atom consisting of Z electrons that interact among themselves and with the nucleus,

$$\hat{\mathcal{H}}_0 = \sum_{i=1}^{Z} \left(\frac{p_i^2}{2m_e} + V_i \right) + \sum_{i \neq j} \frac{e^2}{\left| r_i - r_j \right|}, \tag{3.5}$$

gets modified when placed in a uniform magnetic field B given in terms of the magnetic vector potential $A(r)$ as $\nabla \times A(r) = B$. Choosing a symmetric gauge,

$$A\left(r\right) = \frac{B \times r}{2}, \tag{3.6}$$

the momentum of an electron p_i is altered to $p_i + eA(r_i)$, and the energy of all electron spins in the magnetic field B, amounting to

$$E = g\mu_B B \cdot S, \tag{3.7}$$

must be added to the Hamiltonian. Here, S is the total spin operator (in units of $\hbar$) given by $S = \sum s_i$. A simple manipulation of Equation 3.5 with the condition expressed in Equation 3.6 and including Equation 3.7 yields the (perturbed) Hamiltonian

$$\hat{\mathcal{H}} = \hat{\mathcal{H}}_0 + \mu_B \left(L + gS\right) \cdot B + \frac{e^2}{8m_e} \sum_{i=1}^{Z} \left(B \times r_i\right)^2. \tag{3.8}$$

Here, the vector L is the orbital angular momentum of all Z electrons given by

$$L = \frac{1}{\hbar} \sum_{i=1}^{Z} \left(r_i \times p_i\right). \tag{3.9}$$

In Equation 3.8, the orbital angular momentum L is combined with the total spin operator S. The parameter g is the Landé factor, approximately equal to 2. The second term on the RHS of Equation 3.8 represents a contribution of paramagnetic moments arising from the orbital and spin angular momentum of all electrons of an atom, while the third term is the diamagnetic contribution. In general, the energy change that the magnetic field imparts to the unperturbed Hamiltonian is quite small for all magnetic fields attainable in a laboratory. Consequently, the effect of the magnetic field on the atom can be treated with the usual tools of the perturbation theory. In spite of a small change the magnetic field imparts to the energy of the atom, the effect of the field is easily measured in terms of the magnetization it gives rise to. I briefly consider each term, starting with the diamagnetic contribution.

3.1.1 DIAMAGNETISM

Diamagnetic materials are typified by the absence of unpaired electrons and, consequently, the lack of magnetic moments in zero applied magnetic field. Diamagnets acquire a small magnetic moment upon an application of the field, and this induced moment is directed opposite to the direction of the applied field. Thus, diamagnetic materials have a small and negative susceptibility. In other words, the external magnetic field tends to repel diamagnetic objects. Among diamagnetic substances are counted pure binary skutterudites. We can get a better sense of the diamagnetic contribution by assuming that the substance has all its electron shells fully occupied. Assuming that the magnetic field B is directed along the z-axis, $B \equiv (0, 0, B)$, the vector product $B \times r_i$ in Equation 3.8 thus has components $B(-y_i, x_i, 0)$. The square of the vector product is then $(B \times r_i)^2 = B^2(x_i^2 + y_i^2)$. Assuming that the ground state of the Hamiltonian $\hat{\mathcal{H}}_0$ is nondegenerate, the ion in question will have zero spin and zero angular momentum in its ground state $|0\rangle$,

$$J|0\rangle = L|0\rangle = S|0\rangle = 0, \tag{3.10}$$

and the paramagnetic term is zero[1]. In this case, the energy shift due to the magnetic field is given purely by the diamagnetic term $\Delta E_{o,D}$ equal to

$$\Delta E_{0,D} = \frac{e^2 B^2}{12 m_e} \sum_{i=1}^{Z} \langle 0 | r_i^2 | 0 \rangle, \tag{3.11}$$

where it is assumed that closed shell atoms or ions are spherically symmetric, i.e., $\langle x_i^2 \rangle = \langle y_i^2 \rangle = \langle r_i^2 \rangle / 3$. The magnetization M follows from the differential of the Helmholtz free energy $dF = -SdT - pdV - MdB$ as

$$M(T, B) = -\frac{\partial F}{\partial B}\bigg|_{T,V}. \tag{3.12}$$

At $T = 0$, and for N atoms or ions, all having Z electrons and occupying volume V, the Helmholtz free energy is just the energy in Equation 3.11, and therefore

$$M(0, B) = -\frac{N}{V} \frac{\partial \Delta E_{0,D}}{\partial B} = -\frac{e^2 B}{6 m_e} \frac{N}{V} \sum_{i=1}^{Z} \langle 0 | r_i^2 | 0 \rangle = -\frac{N}{V} \frac{e^2 Z}{6 m_e} B \langle r^2 \rangle, \tag{3.13}$$

where $\langle r^2 \rangle$ stands for a mean square of radii of electron orbits or, if you like, a mean square atomic or ionic radius,

$$\langle r^2 \rangle = \frac{1}{Z} \sum_{i=1}^{Z} \langle 0 | r_i^2 | 0 \rangle. \tag{3.14}$$

The diamagnetic susceptibility χ_D is then

$$\chi_D = \frac{M}{H} \approx \frac{\mu_0 M}{B} = -\frac{N}{V} \frac{e^2 \mu_0 Z}{6 m_e} \langle r^2 \rangle. \tag{3.15}$$

Typical values of the diamagnetic (volume) susceptibility in SI units are in the range of 10^{-4} to 10^{-6}, and the diamagnetic susceptibility is only very weakly temperature-dependent. Diamagnetism is present in all substances but, because it is quite weak, in many cases, it is masked by a much stronger paramagnetism or perhaps even some form of an ordered magnetic state, such as ferromagnetism.

3.1.2 Paramagnetism of Localized Magnetic Moments and Curie Law

Paramagnetic materials have a positive susceptibility as a consequence of the presence of magnetic moments associated with unpaired electron spins. In the absence of an external magnetic field, the spins are randomly oriented because the interaction between the neighboring magnetic moments is very weak. However, the applied magnetic field tends to orient the spins in the direction of the field, and the alignment gets better and better as the field strength increases. It is easier to align spins at low temperatures because thermal energy is smaller and the tendency to randomize spins is weaker. It is thus clear that the paramagnetic susceptibility is expected to be rather strongly temperature- and field-dependent. Unpaired spins introduced by doping or substitution at the transition metal site of binary skutterudites will make such skutterudites paramagnetic substances. The same outcome is expected upon filling the voids of the skutterudite structure.

Let us briefly inspect the paramagnetic (second) term in Equation 3.8. The first thing to notice is how the orbital L and spin S angular momenta (both in units of $\hbar$) of electrons of a given atom or ion combine. Unless one deals with heavy mass atoms or ions, the spin-orbit interaction (interaction between the orbital and spin angular momenta) is weak, and one relies on the so-called Russell-Saunders coupling scheme, whereby one first works out separately the total orbital angular momentum L and the total spin angular momentum S (again, in units of $\hbar$) and only then combines the two to arrive at the total angular momentum J of the atom or ion by setting $J = L + S$. However, when the substance contains heavy elements (e.g., Pb, Bi, etc), the spin-orbit interaction may dominate the energy landscape, and, in this case, it must be taken into account *a priory* by coupling the angular and spin angular momenta of each electron to form its j number and only then the weaker electrostatic interaction couples all individual j's to arrive at the total angular momentum J. In the jargon of magnetism, this is referred to as the j-j coupling.

Staying within the weak spin-orbit interaction, i.e., in the regime where the Russell–Saunders coupling applies, let us assume that an atom or ion has an incomplete electronic shell corresponding to an orbital angular momentum ℓ and, hence, it possesses a permanent magnetic moment. We would like to know how large this magnetic moment is. To answer what appears to be a simple question requires an understanding of the quantum mechanics of atoms, the topic exhaustingly covered in a number of authoritative texts. My aim here is to outline the essential issues so that the readers can reasonably orient themselves when I discuss magnetic properties of skutterudites.

Quantum mechanics states that, for a given angular momentum ℓ, there are $(2\ell + 1)$ possible orientations of the angular momentum, i.e., $2\ell + 1$ states. The Pauli Exclusion Principle dictates that no two electrons can occupy the same energy state unless they have opposite spins, Pauli (1925). Thus, for a given ℓ, there are $2(2\ell + 1)$ states that can be occupied by electrons. As an anachronism from the days past when optical transitions between electronic levels were studied, it is customary to label various ℓ values by letters according to the scheme:

$$\ell = 0 \quad \text{s shell} \quad n_{max} = 2(2\ell + 1) = 2 \text{ electrons}$$
$$\ell = 1 \quad \text{p shell} \quad\quad\quad\quad\quad\quad\quad = 6 \text{ electrons}$$
$$\ell = 2 \quad \text{d shell} \quad\quad\quad\quad\quad\quad\quad = 10 \text{ electrons}$$
$$\ell = 3 \quad \text{f shell} \quad\quad\quad\quad\quad\quad\quad = 14 \text{ electrons}$$
$$\ell = 4 \quad \text{g shell} \quad\quad\quad\quad\quad\quad\quad = 18 \text{ electrons}$$
$$\ell = 5 \quad \text{h shell} \quad\quad\quad\quad\quad\quad\quad = 22 \text{ electrons}$$

Provided the spin-orbit coupling is weak, the total orbital angular momentum $L = \sum \ell_i$, the total spin angular momentum $S = \sum s_i$, and the total angular momentum $J = L + S$ of the atom are "good" quantum numbers in the sense that they commute with the Hamiltonian of the atom or ion. In that case, the states of the atom or ion can be specified by quantum numbers L, L_z, S, S_z, J, and J_z, which are eigenstates of the quantum mechanical operators L^2, L_z, S^2, S_z, J^2, and J_z having eigenvalues $L(L + 1)$, L_z, $S(S + 1)$, S_z, $J(J + 1)$, and J_z, respectively. Thus, given the total orbital and spin angular momenta, they combine to form $(2L + 1)(2S + 1)$ different configurations, reflecting the $(2L + 1)$ choices of the z-component of the orbital angular momentum multiplied by $(2S + 1)$ choices of the z-component of the spin angular momentum. Clearly, this yields a large number of possible configurations for fitting electrons into an incomplete shell. Luckily, we do not have to worry about all configurations, but only those that minimize the energy as these are the ones that have the decisive influence on the magnetic properties. How such configurations of spin and orbital angular momenta are selected is conveniently answered by Hund's rules, which are simple instructions in place of challenging quantum mechanical calculations of the exchange/correlation energies.

The first Hund's rule states that the configuration should be chosen so as to maximize the spin angular momentum S but, at the same time, not to violate the Pauli Exclusion Principle. Physically, this means that the energy of an atom will be lowered by preventing electrons with the same spin to come close to each other, and thus, their Coulomb repulsion is minimized. Obviously, if the number of electrons n to be accommodated on an ℓ-shell is equal to one half of all available states, $n = 2\ell + 1$, all electrons can have their spins in the same direction, say "up", giving the largest possible spin angular momentum $S = n/2 = \ell + \frac{1}{2}$. If the number of electrons is smaller than one half of the available states, $n < 2\ell + 1$, there is no problem and all spins can be parallel, giving again $S = n/2$, except that the spin angular momentum S is now smaller because n is smaller. If the number of electrons is larger than one half of the available states on the shell, $n > 2\ell + 1$, some electrons will have to have their spins in the opposite direction ("down") to avoid violating the Pauli principle. For each electron above the $(2\ell + 1)$-th, the spin angular momentum will be reduced by $\frac{1}{2}$. A fully occupied shell will have $S = 0$, because for every electron with the spin "up" there is an electron with the spin "down".

The second Hund's rule addresses the total orbital angular momentum L and states that the lowest lying state should have maximum L consistent with the Pauli Exclusion Principle and with the first Hund's rule. Again, the physics behind this rule is to minimize Coulomb repulsion by preventing electrons coming close together by forcing them to spread out in space and orbit in the same direction. To achieve the largest L, the first electron will go to a level with the largest z-component of the orbital angular momentum ℓ_z, i.e., the maximum ℓ value. By rule 1, the second electron should have the same spin as the first electron, but the Pauli principle prevents it to go into a level with the same value of ℓ_z. It will thus go into the next best option, which is a state with $|\ell_z| = \ell - 1$. Placing the first two electrons, the total orbital angular momentum is $L = \ell + (\ell - 1) = 2\ell - 1$. Continuing filling the states in the same fashion, for a shell less than half filled, the orbital angular momentum will be $L = \ell + (\ell - 1) + (\ell - 2) + \ldots + [\ell - (n - 1)]$. At exactly half-filled shell, all available ℓ_z values have been used, half of them are positive and half of them are negative, resulting in $L = 0$. With more electrons than the number of ℓ_z states, electrons must have opposite spin and the filling will generate the same sequence of L values as when filling the first half of states.

Rules 1 and 2 provide a procedure to work out values of L and S for the lowest energy states. However, there are many other states, and they are covered by the third Hund's rule. The rule governs the total angular momentum J and attempts to minimize the spin-orbit interaction. It states that J is given by

$$J = |L - S| \text{ for shells less than half filled, i.e., } n \leq (2\ell + 1)$$
$$\text{and} \tag{3.16}$$
$$J = |L + S| \text{ for shells more than half filled, i.e., } n \geq (2\ell + 1).$$

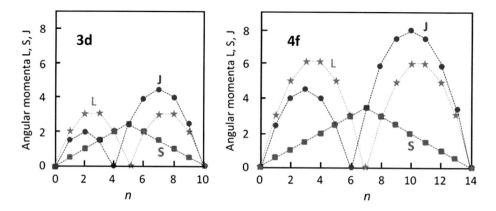

FIGURE 3.1 Angular momenta L, S, and J (in units of $\hbar$) for $3d$ and $4f$ ions calculated using Hund's rules. Integer n indicates the number of electrons in the $3d$ and $4f$ shell, respectively.

L, S, and J values of $3d$ and $4f$ ions (ions relevant to skutterudites) calculated using Hund's rules are presented in Figure 3.1. The integer n indicates the number of electrons in the $3d$ and $4f$ shell, respectively.

To illustrate applications of the rules, I choose a rare-earth ion Pr^{3+} that in $PrOs_4Sb_{12}$ and $PrPt_4Ge_{12}$-filled skutterudites is responsible for their exotic superconducting (SC) nature, as discussed in Chapter 2. The Pr^{3+} ion has the valence shell $4f^2$, i.e., the shell $\ell = 3$ contains two electrons.

With $\ell = 3$, the shell can accommodate up to $2(2\ell + 1) = 14$ electrons, 7 with the spin up and 7 with the spin down. Obviously, with just two electrons, the shell is less than half-filled. Rule 1 instructs us to place the two electrons to maximize S, i.e., we take $S = \frac{1}{2} + \frac{1}{2} = 1$. The spin degeneracy is thus $2S + 1 = 3$. Rule 2 commands us to maximize orbital angular momentum L. We do so by placing one electron at $\ell_z = 3$ and the other one at $\ell_z = 2$, i.e., $L = 3 + 2 = 5$. With less than half-filled shell, Rule 3 implies that we take the total angular momentum as $J = |L\text{-}S| = 5 - 1 = 4$. Consequently, the Pr^{3+} ion is designated as $^3H_4$ where H corresponds to $L = 5$ in the spectroscopic designation above, the upper script 3 stands for the spin degeneracy $2S + 1$ (not the spin S itself), and 4 indicates the total angular momentum J.

According to quantum mechanics, a permanent magnetic moment of an atom or ion cannot rotate arbitrarily in response to the applied magnetic field but, rather, takes on a discrete set of angles with respect to the field. The rotations are specified by the z-component of the magnetic moment $m_J g \mu_J$, where m_J is the z-component of the total angular momentum and takes on any one of the $2J + 1$ values given by

$$m_J = -J, -J + 1, \ldots, J - 1, J. \tag{3.17}$$

With the partition function Z,

$$Z = \sum_{m_J=-J}^{+J} \exp\left(\frac{m_J g_J \mu_B B}{k_B T}\right) = \sum_{m_J=-J}^{+J} \exp m_J x, \tag{3.18}$$

where a substitution

$$x = \frac{g_J \mu_B B}{k_B T} \tag{3.19}$$

was made, the average value of the magnetic quantum number m_J is

$$\langle m_J \rangle = \frac{\sum_{m_J=-J}^{+J} m_J e^{m_J x}}{\sum_{m_J=-J}^{+J} e^{m_J x}} = \frac{1}{Z} \frac{\partial Z}{\partial x}. \tag{3.20}$$

The reader recognizes the partition function Z as a geometrical series of the form

$$Z = e^{-Jx} + e^{(-J+1)x} + \ldots + e^{(J-1)x} + e^{Jx} \tag{3.21}$$

with the first term e^{-Jx} and the ratio of the terms e^x. Such a series can be evaluated to yield

$$Z = \frac{e^{-Jx} - e^{(J+1)x}}{1 - e^x}. \tag{3.22}$$

Multiplying both the numerator and denominator by $e^{-x/2}$, the partition function becomes

$$z = \frac{e^{-x\left(J+\frac{1}{2}\right)} - e^{x\left(J+\frac{1}{2}\right)}}{e^{-\frac{x}{2}} - e^{\frac{x}{2}}} = \frac{\sinh\left[(2J+1)\frac{x}{2}\right]}{\sinh\frac{x}{2}}. \tag{3.23}$$

For n identical atoms or ions in a unit volume, the magnetization is then

$$M = n g_J \mu_B \langle m_J \rangle = n g_J \mu_B \frac{1}{Z} \frac{\partial Z}{\partial x} = n k_B T \frac{1}{Z} \frac{\partial Z}{\partial B} = n k_B T \frac{\partial \ln Z}{\partial B}. \tag{3.24}$$

Taking the logarithm of Equation 3.23 and differentiating it with respect to B (remember, x is a function of B given by Equation 3.19), the magnetization becomes

$$M = n g_J \mu_B \left\{ \frac{2J+1}{2} \coth(2J+1)\frac{x}{2} - \frac{1}{2}\coth\frac{x}{2} \right\}. \tag{3.25}$$

The largest possible magnetization, called the saturation magnetization M_s, is achieved with the largest J, i.e.,

$$M_s = n g_J \mu_B J. \tag{3.26}$$

In this case, all magnetic moments are in the direction of the applied magnetic field. Using Equation 3.26 and substituting

$$y = xJ \equiv \frac{g_J \mu_B}{k_B T} JB, \tag{3.27}$$

Equation 3.25 takes the form

$$M = M_s \left\{ \frac{2J+1}{2J} \coth \frac{(2J+1)}{2J} y - \frac{1}{2J} \coth \frac{y}{2J} \right\} \equiv M_s B_J(y). \qquad (3.28)$$

The function in the curly brackets is known as the Brillouin function $B_J(y)$, and Equation 3.28 applies to any magnetic field strength. The plot of the Brillouin function for several values of the total angular momentum J is shown in Figure 3.2. I should point out that the Brillouin function reduces to a Langevin function (classical treatment of magnetism where the magnetic moment can be rotated by any angle in the presence of the external magnetic field) when $J = \infty$ and becomes a hyperbolic tangent function when $J = \frac{1}{2}$, the restricted case often treated in introductory textbooks.

As is easily checked by assuming room temperature and taking a typical magnetic field of 1 Tesla, the parameter y in Equation 3.27 is much smaller than unity. Thus, unless one deals with very low temperatures and extreme magnetic fields, the approximation $y \ll 1$ is appropriate and the Brillouin function reduces to

$$B_J(y) \approx \frac{J+1}{3J} y + O(y^3). \qquad (3.29)$$

In this approximation, the paramagnetic susceptibility χ_{par} becomes

$$\chi_{par} = \frac{M}{H} \cong \frac{\mu_0 M}{B} = \frac{n\mu_0 g_J^2 \mu_B^2}{k_B T} \frac{J(J+1)}{3} = \frac{n\mu_0 \mu_{eff}^2}{3k_B T} \equiv \frac{C_{Curie}}{T}, \qquad (3.30)$$

where

$$\mu_{eff} = g_J \mu_B \sqrt{J(J+1)}. \qquad (3.31)$$

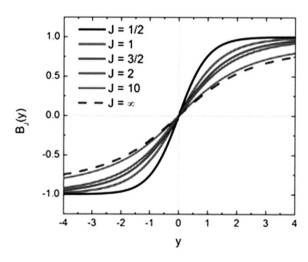

FIGURE 3.2   Brillouin function plotted for several values of the total angular momentum J, including the case $J = \frac{1}{2}$, in which case it is just a tanh y function, and for $J = \infty$ (dashed curve), where it becomes a Langevin function $L(y) = \coth y - 1/y$.

Equation 3.30 is the famous Curie law of paramagnetism, which can be used to find the experimental value of the effective magnetic moment of an atom or ion and compare it with the theoretical value given by Equation 3.31. One often speaks in terms of the effective number of Bohr magnetons $p = g_J [(J(J+1)]^{1/2}$. In practice, one plots the inverse susceptibility $1/\chi_{par}$ vs. T to obtain a straight line with the slope $1/C_{Curie}$. To make use of Equation 3.31, we also need the Landé g_J-factor, which is given by

$$g_J = \frac{3}{2} + \frac{S(S+1) - L(L+1)}{2J(J+1)}. \tag{3.32}$$

It is useful to note that when $J = L$, the Landé g-factor is $g_J = 1$, while when $J = S$, it is $g_J = 2$.

Taking again the Pr^{3+} ion as an example, its theoretical effective magnetic moment according to Equations 3.31 and 3.32 is 3.58 μ_B. The experimental value is 3.56 μ_B, a remarkable agreement that holds true for most of the rare-earth ions, i.e., ions of the form $^{2S+1}L_J$. The reason why the experimental and theoretical magnetic moments (or, equivalently, the number of experimental and theoretical Bohr magnetons) agree so well in the case of rare-earth ions is because their $4f$ levels lie inside the outermost $5s^2 5p^6$ orbitals that very effectively screen the effect of the crystalline electric field (CEF). For $3d$ ions (ions of transition metals starting with Ti^{3+}or V^{4+} ($3d^1$) and ending with Zn^{2+} ($3d^{10}$)), calculations of the effective magnetic moment μ_{eff} by the Hund's rules generally do not agree with the experimental effective magnetic moments. Rather, the measured effective magnetic moments are in much better agreement with calculations that totally neglect to include the orbital angular momentum, and the total magnetic moment is given purely by its spin angular momentum,

$$\mu_{eff} = 2\mu_B \sqrt{S(S+1)}. \tag{3.33}$$

This situation is referred to as quenching of the angular momentum. Unlike the $4f$ orbitals of rare-earth ions, the $3d$ orbitals of the transition metals of the iron group are the outermost orbitals and the CEF has a direct effect on them, resulting in quenching of the angular momentum.

3.1.3 Paramagnetism in Metals and Pauli Susceptibility

In metals, all states up to the Fermi level are occupied by electrons. Each energy level can accommodate two electrons, one with the spin "up" and one with the spin "down". In other words, there is the density of "up" spin states and the density of the "down" spin states in the respective subbands, schematically illustrated in Figure 3.3(a). Applied magnetic field splits the two subbands, raising the energy of the spin down subband and decreasing the energy of the spin up subband by the same amount. Consequently, the subbands are split in energy by $g\mu_B B = 2\mu_B B$, Figure 3.3(b). Here, we assume that the effect of magnetic field affects only the spin part of the angular momentum and, hence, $g = 2$. Because the Fermi energy must be the same throughout the metal, some electrons in the spin-down subband must flip their spins to equate the Fermi energy between subbands, Figure 3.3(c). This leaves an excess of spin-up electrons $n_\uparrow$ over the spin-down electrons $n_\downarrow$, giving rise to magnetization and, consequently, the magnetic susceptibility.

Let us assume first that the metal is degenerate and we consider the situation at T = 0 K. The number of extra electrons per unit volume with spin up is (showing a step-by-step treatment)

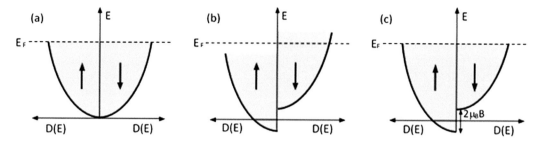

FIGURE 3.3 (a) Spin up and spin down bands in zero magnetic field. (b) Applied magnetic field along the z-direction rises the energy of the spin down band by $\mu_B B$ and decreases the energy of the spin up band by the same amount $\mu_B B$, creating two spin subbands separated by $g\mu_B B = 2\mu_B B$, taking the spin only g value of 2. (c) Fermi energy of the metal must be the same everywhere; hence some down spin electrons must flip their spin to equate the Fermi energy in both subbands. The metal finishes with an excess of spin up electrons, giving rise to magnetization and, consequently, paramagnetic susceptibility known as Pauli susceptibility, Pauli (1927).

$$n_\uparrow = \frac{1}{2}\int_0^{E_F} D(E + \mu_B B) f(E) dE = \frac{1}{2}\int_0^\infty D(E + \mu_B B) f(E) dE = \frac{\mu_B B}{2}\int_0^\infty \frac{dD(E)}{dE} f(E) dE$$

$$= \frac{\mu_B B}{2}\left[\left(D(E) f(E) \right)_0^\infty - \int_0^\infty D(E)\frac{df(E)}{dE} dE \right] \tag{3.34}$$

$$= \frac{\mu_B B}{2}\left[0 + \int_0^\infty D(E)\delta(E - E_F) dE \right] = \frac{\mu_B}{2} D(E_F) B.$$

The product of the density of states (DOS) and the Fermi function, $D(E)f(E)$, is obviously zero in limits of $E = 0$ and $E = \infty$. Because T = 0 K, $df(E)/dE$ is the delta function $-\delta(E-E_F)$, which brings the final result. Following the same steps for the number of electrons with spin down, we obtain

$$n_\downarrow = \frac{1}{2}\int_0^{E_F} D(E - \mu_B B) f(E) dE = -\frac{\mu_B}{2} D(E_F) B. \tag{3.35}$$

The spin imbalance $n_\uparrow - n_\downarrow$ leads to a magnetization $M = \mu_B(n_\uparrow - n_\downarrow)$ equal to

$$M = \mu_B^2 D(E_F) B, \tag{3.36}$$

and the Pauli susceptibility in this degenerate case becomes

$$\chi_{Pauli}^{deg} = \frac{M}{H} \approx \frac{\mu_0 M}{B} = \mu_0 \mu_B^2 D(E_F). \tag{3.37}$$

The magnitude of the Pauli susceptibility is quite small, $\chi_{Pauli} \ll 1$, and essentially temperature independent, unlike the case of the paramagnetism of localized magnetic moments. In metals, the degenerate approximation is perfectly good up to the melting point. Note that measurements of the Pauli susceptibility give directly the value of the DOS at the Fermi energy, which is very useful information.

In lower carrier systems, such as weakly doped semiconductors and at elevated temperatures, the Fermi–Dirac distribution function can be replaced by the Maxwell–Boltzmann function $f(E) \approx \exp[-(E-\mu)/k_BT]$, which has a derivative $df/dT = -f/k_BT$, and the spin imbalance becomes

$$n_\uparrow - n_\downarrow = \frac{\mu_B B}{k_B T} \int_0^\infty D(E) f(E) dT = \frac{n\mu_B}{k_B T} B. \tag{3.38}$$

In Equation 3.38, the carrier concentration n is the integral over the DOS multiplied by the distribution function, hence the last equity. The Pauli susceptibility for the non-degenerate system is then

$$\chi_{Pauli}^{nondeg} = \frac{M}{H} \approx \frac{\mu_0 M}{B} = \frac{n\mu_0 \mu_B^2}{k_B T}. \tag{3.39}$$

Equation 3.39 is, apart from the factor of 3 missing in the denominator, the susceptibility of n localized magnetic moments μ_B per unit volume given by Equation 3.30.

In filled skutterudites, the electrons donated by the filler ions raise the carrier concentration to as much as 10^{20}–10^{21} cm^{-3}, which makes the system degenerate, and Equation 3.37 is a reasonable approximation of the paramagnetic state of filled skutterudites. In undoped binary skutterudites, the carrier concentration is significantly lower, on the order of 10^{18} cm^{-3}, and the system is non-degenerate with the Curie-like paramagnetic dependence of Equation 3.39 frequently observed.

3.1.4 MAGNETICALLY ORDERED STRUCTURES

We have seen that randomly oriented permanent magnetic moments in a paramagnetic material can be aligned upon application of a strong magnetic field. After the field is removed, thermal fluctuations at any temperature randomize the direction of the magnetic moments, resulting in net zero magnetization. There are, however, materials in which the magnetic moments align (order) spontaneously along a particular direction without any assistance of the applied magnetic field. Such a spontaneously ordered state of magnetic moments, revealed *via* its spontaneous magnetization, persists against the tendency to thermally disorder it up to a certain critical temperature at which, eventually, the order is destroyed, leaving behind randomly oriented magnetic moments, i.e., a paramagnetic state. Materials where all magnetic moments spontaneously align along one direction are called ferromagnets, and they possess large positive susceptibility. The above statement should be clarified in the following sense: many FM materials contain magnetic domains, which are small regions within the sample where all magnetic moments are aligned, but the magnetization of different domains has a random orientation. In this case, the overall magnetic moment may be very small or even zero. On the other hand, it is possible to align the magnetization of all domains in the direction of the applied field using a rather small magnetic field because all the hard work of aligning magnetic moments within the domains has already been done and all the magnetic field has to do now is to displace or rotate the domains, the process requiring much less energy.

There are also materials, actually more of them, where one-half of magnetic moments is spontaneously aligned along one direction and the other half of the moments is aligned in an exactly opposite direction, resulting in net zero magnetization. Such structures are referred to as antiferromagnets, and they are typified by two interpenetrating sublattices, each with magnetic moments of the same magnitude but of opposite orientation, an example provided in Figure 3.4.

Ferrimagnetic materials can be viewed as a special case of antiferromagnets where the magnetic moments on the two sublattices are not of the same magnitude and therefore do not compensate each other. Such structures have a net magnetization. Ferrimagnetism is often encountered in ferrites and garnets and, as we shall see, can be realized also in skutterudites. A schematic illustration of linear

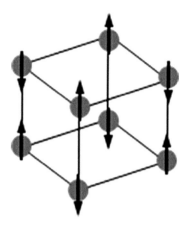

FIGURE 3.4 AFM ordering on a simple cubic lattice viewed as two interpenetrating face-centered cubic sublattices with mutually opposing magnetic moments (spins).

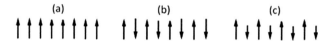

FIGURE 3.5 Linear array of (a) ferromagnetically, (b) antiferromagnetically, and (c) ferrimagnetically ordered magnetic moments (spins).

arrays of ferromagnetically, antiferromagnetically, and ferrimagnetically aligned magnetic moments (spins) is depicted in Figure 3.5.

All the above magnetically ordered states rely on the ability of magnetic moments to align spontaneously with no assistance of an applied magnetic field, as if they communicated among themselves what to do. The communication channel between the magnetic moments is interactions among electrons constituting the magnetic material. The first obvious interaction mode that comes to mind is the magnetic dipolar interaction. Magnetic dipoles arise naturally when the north and south magnetic poles are separated by an infinitesimal distance or, alternatively, are realized as a tiny closed loop of the circulating electric current. In either case, the interaction between two magnetic moments μ_1 and μ_2 separated by a distance r gives rise to energy of magnitude

$$E = \frac{\mu_0}{4\pi r^3}\left[\mu_1 \cdot \mu_2 - \frac{3}{r^2}(\mu_1 \cdot r)(\mu_2 \cdot r)\right]. \tag{3.40}$$

A numerical estimate of this energy is obtained by assuming that the magnetic moments are Bohr magnetons (9.27×10^{-24} JT^{-1}) situated at a typical lattice spacing ~ 1Å $= 10^{-10}$ m apart. With the permeability of the free space $\mu_0 = 4\pi \times 10^{-7}$ Hm^{-1}, the magnetic dipolar interaction energy comes out as $E \approx 8.6 \times 10^{-24}$ J, equivalent to the temperature of about 0.6 K. In view of the high temperatures, where the FM order is still sustained (e.g., up to 1043 K in the case of Fe and even higher 1394 K in Co, Blundell (2001)), the dipolar magnetic interaction is hopelessly inadequate to cause the alignment of magnetic moments at ordinary temperatures. Indeed, the interaction that gives rise to spontaneous magnetization is not of magnetic nature but, rather, has its origin in electrostatic interactions, whereby the system tries to minimize its energy by keeping like electric charges as far apart as possible. This is generally referred to as the *exchange interaction*. It follows from the Pauli Exclusion Principle that the overall wavefunction describing electrons must be antisymmetric. Because the wavefunction is a product of the space part and the spin part, considering just two electrons, one each associated with two protons separated some distance apart, the electrons must form a symmetric singlet state $|\uparrow\downarrow\rangle$ when the space part of the wavefunction is antisymmetric and an anti-symmetric triplet state $|\uparrow\uparrow\rangle$ when the space part is symmetric. There is an energy difference between the singlet and triplet states $(E_{singlet} - E_{triplet})S_1 \cdot S_2$ that depends on the orientation of spins, expressed through the respective spin operators S_1 and S_2 in the form of $2JS_1 \cdot S_2$, where J is the exchange constant. Because $S_1 \cdot S_2 = -3/4$ for a singlet state and $S_1 \cdot S_2 = 1/4$ for the triplet state, if J > 0, $E_{singlet} > E_{triplet}$, and the triplet state S = 1 is a more probable state. In contrast, if J < 0, $E_{singlet} < E_{triplet}$, and the singlet S = 0 is favored. Extending the concept to multi-electron systems (not a trivial matter), a very complicated and essentially intractable multi-electron Hamiltonian reduces to a relatively simple Hamiltonian containing only spin operators, which is further simplified by assuming that the most important interactions are between the nearest neighbor spins taken at constant strength with all other interactions neglected,

$$\hat{\mathcal{H}} = -\sum_{ij} J_{ij} S_i \cdot S_j = -2\sum_{i>j} J_{ij} S_i \cdot S_j \approx -2J\sum_{n.n} S_i \cdot S_j, \qquad (3.41)$$

where the last sum in Equation 3.41 is taken over the nearest neighbors only. Note that the coupling in the above equation depends only on the relative orientation of the two spins. Equation 3.41 is a formal expression of the famous Heisenberg (1926) model, which is the starting point of most of the more sophisticated theories of magnetism. The above interaction is referred to as the *direct exchange* because it is a consequence of direct Coulomb interaction among electrons of the two ions. There are other forms of exchange interactions, and I just mention them without much elaboration.

It frequently happens that the two magnetic ions are separated by an ion that has all electronic shells filled, i.e., a nonmagnetic ion. The interaction between the magnetic ions is then not direct but, rather, mediated by electrons of the nonmagnetic ion, and this type of interaction is called *superexchange*.

Another kind of interaction is the *indirect exchange*. This occurs often with rare-earth metals where their 4f shells are inside the 5s and 5p orbitals and do not spread far from the nucleus. In other words, the direct coupling is limited, and the f-electrons prefer to couple to conduction electrons, making this coupling stronger than the direct coupling.

Exchange interactions, generally referred to as itinerant exchange, also pertain to conduction electrons in metals. The area generated much interest in the past, particularly in the context of the Hartree–Fock approximation, when Felix Bloch (1929) predicted that the gas of electrons could couple *via* their mutual Coulomb interactions to form a FM state. The theory requires a rather dilute electron density for the transition to a FM state to take place, and, at such low densities, the electrons tend to crystallize to form the Wigner crystal, Wigner (1938). The free electron model here seems to be hopelessly inadequate to serve as a viable model for ordered magnetic phases in metals.

3.1.4.1   Weiss Molecular Field Model and Mean Field Theory

In 1907, attempting to explain the cooperative behavior of magnetic moments characterized by their ability to spontaneously align along a particular direction, Pierre Weiss (1907) came up with an idea that on every magnetic moment is acting a magnetic field of the form $B + \lambda M$, where B is the external magnetic field and the term λM is the "molecular" field with the molecular (Weiss) constant λ that represents the cooperative interaction of all other magnetic moments. Predating the development of the quantum mechanics by nearly 20 years, Weiss used the classical treatment based on Langevin theory and showed that his molecular field approach, indeed, leads to spontaneous magnetization and the existence of the Curie (since then Curie–Weiss) critical temperature, above which the spontaneous magnetization is destroyed. Some years later, Weiss's ideas were recast in terms of the quantum mechanics. Specifically, the Weiss molecular field was ascribed to an exchange interaction between electron spins on neighboring atoms. Here, I outline the main points, following the treatment in a book by Blundell (2001). For simplicity, we assume that the system has no orbital angular momentum ($L = 0$), and thus the total angular momentum is purely due to the spin angular momentum S.

3.1.4.1.1 *The Case of a Ferromagnet*

In the applied magnetic field B, Heisenberg and Dirac have shown that the magnetic Hamiltonian can be written as

$$\hat{\mathcal{H}} = -\sum_{ij} J_{ij} S_i \cdot S_j + g\mu_B \sum_j S_j \cdot B, \tag{3.42}$$

where S_i stands for the total spin operator for the atom at the i-th site, and J_{ij} is the exchange integral that is a function of the relative positions of the sites i and j. In reality, the exchange integral is large only between the nearest neighbor sites. The first term in Equation 3.42 is called the Heisenberg exchange energy and the second term the Zeeman energy. Because we are considering a FM state, the exchange constants J_{ij} between the nearest neighbors will all be positive. The Weiss' effective molecular field B_{wmf}, the field generated by all magnetic moments at sites j acting at the site i, is taken as

$$B_{wmf} = -\frac{2}{g\mu_B} \sum_j J_{ij} S_j. \tag{3.43}$$

The Heisenberg exchange term in Equation 3.42 includes double counting as the sum is over all i and j sites. This can be rewritten instead as twice the sum over just j sites, i.e.,

$$-\sum_{ij} J_{ij} S_i \cdot S_j = -2S_i \sum_j J_{ij} S_j = -2S_i \frac{-g\mu_B B_{wmf}}{2} = g\mu_B S_i \cdot B_{wmf}, \tag{3.44}$$

where a substitution for the sum over $J_{ij}S_j$ from Equation 3.43 was made. The Hamiltonian in Equation 3.42 can then be written as

$$\hat{\mathcal{H}} = g\mu_B \sum_i S_i \left(B + B_{wmf} \right). \tag{3.45}$$

Equation 3.45 has the form of the paramagnetic term in Equation 3.8, except that it is written assuming $L = 0$, considering spins on all sites i, and the magnetic field is now $B + B_{wmf}$. Consequently, we will treat the situation as that of a paramagnet placed in the field $B + B_{wmf}$. Obviously, due to the presence of the molecular field B_{wmf} in Equation 3.45, it follows that the spins can be aligned even in the absence of the external magnetic field B. Of course, as the temperature increases, thermal fluctuations will tend to destroy the alignment and, at some critical temperature, the alignment (ordered magnetic state) will be lost. This is the essence of the Weiss picture of ferromagnetism.

We now specifically write the Weiss molecular field B_{wmf} as λM, where the Weiss constant λ is positive (we are considering a ferromagnet) and "measures" the strength of the molecular field in terms of the magnetization the field causes. As in the case of a paramagnet described in Equations 3.27 and 3.28, we must solve simultaneously equations

$$M = M_s B_J \left(y \right) \tag{3.46}$$

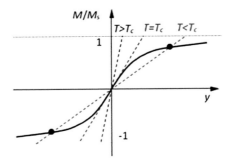

FIGURE 3.6 Graphical solution to Equations 3.46 and 3.48 (zero external magnetic field). The black solid curve is the Brillouin function. The intercept of the straight line with the Brillouin function represents a non-zero spontaneous magnetization for $T < T_c$, which is indicated by solid black points.

and

$$y = \frac{g_J \mu_B J (B + \lambda M)}{k_B T}.$$ (3.47)

This can be done graphically. Let us first assume that there is no applied magnetic field, i.e., $B = 0$. From Equation 3.47, we can write for the magnetization M

$$M = \frac{k_B T}{g_J \mu_B \lambda J} y.$$ (3.48)

Referring to Figure 3.6, the plot of M/M_s vs. y shows the Brillouin function in Equation 3.46. The figure also shows straight lines that represent the behavior of the magnetization as a function of y according to Equation 3.48. Obviously, a plot of M vs. y is a straight line with the slope proportional to temperature T. At high temperatures, the slope will be large and the only common point with the Brillouin function is at the origin $y = 0$, where magnetization is zero. As the temperature decreases, the slope becomes smaller and, at some particular temperature T_C, the slope of the straight line will be a tangent to the Brillouin function at the origin. For any smaller temperature than this, the straight line will intersect the Brillouin function at two points marked on the graph; in other words, below T_C the system develops non-zero spontaneous magnetization, which grows as the temperature falls further.

The transition temperature, known as the Curie temperature T_C, is obtained by equating the slopes of the straight line in Equation 3.48 with the slope of the Brillouin function at the origin. Because for small y, the Brillouin function can be approximated to the leading term as $(J+1)y/3J$, see Equation 3.29, we set

$$\frac{M}{M_s} \equiv \frac{k_B T}{g_J \mu_B \lambda J M_s} y = B_J(y) \approx \frac{J+1}{3J} y.$$ (3.49)

From Equation 3.49, the transition temperature T_C follows after substituting for the saturation magnetization M_s from Equation 3.26 and using the definition of the effective magnetic moment μ_{eff} in Equation 3.31,

$$T_C = \frac{(J+1) g_j \mu_B \lambda J M_s}{3 J k_B} = \frac{(J+1) g_j \mu_B \lambda}{3 k_B} n g_J \mu_B J = \frac{n \lambda \mu_{eff}^2}{3 k_B}.$$ (3.50)

Often, important exchange interactions take place between the nearest neighbor magnetic moments only. Assuming that there are z such nearest neighbors and making a further simplification by assuming that all interactions are of the same strength, $J_{ij} = J$, it is easy to relate the Weiss constant λ and the exchange parameter J by making use of Equations 3.42 and 3.45,

$$\lambda = \frac{2zJ}{n g_J^2 \mu_B^2}.$$ (3.51)

Substituting this Weiss parameter λ into Equation 3.50, the critical Curie temperature becomes

$$T_C = \frac{2zJJ(J+1)}{3k_B}.$$ (3.52)

The strength of the Weiss molecular field B_{wmf} (taking the largest magnetization, i.e., setting $M = M_s$) is

$$B_{wmf} = \lambda M_s = \frac{3k_B T_C}{(J+1)g_J \mu_B}.$$ (3.53)

For a ferromagnet with $T_C \sim 10^3$ K and the typical $J = \frac{1}{2}$, the magnitude of the molecular field is surprisingly large, having a value of 1500 T! This reflects how strong the exchange interaction governing the spontaneous alignment of magnetic moments is.

To determine magnetic susceptibility, we need to apply a small magnetic field at temperatures $T \geq T_C$. Because small magnetic fields result in small magnetization, the parameter $y \ll 1$ and the Brillouin function can be approximated as in Equation 3.29. The magnetization then becomes

$$\frac{M}{M_s} \approx \frac{g_J \mu_B (J+1)}{3k_B} \left(\frac{B + \lambda M}{T} \right).$$ (3.54)

With the aid of Equations 3.31 and 3.50, this can be rearranged to read

$$M \approx \frac{T_C}{\lambda} \left(\frac{B + \lambda M}{T} \right) = \frac{1}{T - T_C} \frac{T_C}{\lambda} B.$$ (3.55)

Magnetic susceptibility χ is then

$$\chi = \frac{M}{H} = \frac{\mu_0 M}{B} \approx \frac{\mu_0 T_C}{\lambda} \frac{1}{T - T_C} = \frac{n \mu_0 \mu_{eff}^2}{3k_B} \frac{1}{T - T_C}.$$ (3.56)

As the temperature approaches the critical point from the high-temperature side, the susceptibility in the Weiss model diverges as $(T - T_C)^{-1}$, reflecting the onset of the phase transition. Equation 3.56 is known as the Curie–Weiss law.

From the intercepts between the straight line and the Brillouin function obtained at different temperatures below T_C, one can construct a graph (phase diagram) of the spontaneous magnetization M/M_s as a function of the reduced temperature T/T_C, shown in Figure 3.7. While the graph depends somewhat on the particular value of the total angular momentum J appropriate for a given ferromagnet, the general features of the dependence are common to all values of J. The most important aspect is to recognize that the magnetization is continuous at $T = T_C$, while its gradient is not. This classifies the phase transition between the FM and paramagnetic regimes as the second-order phase transition.

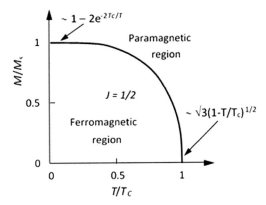

FIGURE 3.7 Magnetization of a ferromagnet as a function of reduced temperature (phase diagram) constructed from intercepts of the straight lines with the Brillouin function at temperatures $T < T_C$ and plotted for a ferromagnet with $J = \frac{1}{2}$. Temperature dependences for limiting cases of $T << T_c$ and $T \approx T_c$ follow from relevant approximations of the Brillouin function and are shown in the figure. However, neither one agrees with the experiment. The critical exponent near the Curie temperature T_c is 1/3 rather than the shown 1/2, the result of an oversimplified treatment of the molecular field, where each spin was replaced by the same average value. The exponential temperature dependence at very low temperatures is not seen in experiments because of the presence of spin waves, which are the lowest energy excitations in the FM state and their limiting temperature variations is $T^{3/2}$, the Bloch law, as discussed in section 3.1.4.2.

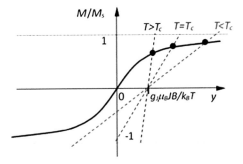

FIGURE 3.8 Graphical solution of Equations 3.46 and 3.55 for non-zero applied field B. Note that an intercept with the Brillouin function happens at any temperature and thus there is no phase transition.

When an external field is applied, the situation is somewhat changed. The presence of the magnetic field increases the parameter y in Equation 3.47 and shifts the origin in Figure 3.6 by $g_J\mu_B JB/k_BT$ to the right to a new point, see Figure 3.8. The magnetization must again satisfy Equation 3.46, but instead of Equation 3.48, it now must also satisfy

$$M = \frac{1}{\lambda}\left(\frac{yk_BT}{g_J\mu_BJ} - B \right). \tag{3.57}$$

In this case, clearly, there is always an intercept of a straight line M vs. y with the Brillouin function at whatever temperature. This means that in non-zero applied magnetic field, the phase transition does not happen. The theory here places no restriction on the direction of the applied field; the magnetization will rotate and follow whatever field direction. In reality, this is not so because the samples possess magnetic anisotropies, which determine the easy and hard directions of magnetization.

While the Weiss molecular field model accounts for the spontaneous magnetization and is a very simple and convenient approach to understand the ordered magnetic state, it does not do well in predicting the detailed behavior of magnetization. For instance, let us consider what the model predicts for the behavior of spontaneous magnetization as very low temperatures, $T \ll T_C$, and near the critical point, $T \approx T_C$. Near the absolute zero temperature, the Brillouin function decays exponentially, which is in discord with the experimentally observed $T^{3/2}$ behavior, sometimes referred to as the Bloch $T^{3/2}$ law. Likewise, as the critical point is approached, the Weiss model predicts the magnetization to vanish proportionally to $(T_C - T)^{1/2}$, in disagreement with the experimentally determined exponent of 1/3. The failure to predict the asymptotic and critical exponents is a notorious shortcoming of all mean-field theories, and the Weiss molecular field model is one such example.

3.1.4.1.2 The Case of an Antiferromagnet

When the exchange interaction constant is negative, $J < 0$, an antiparallel arrangement of the neighboring magnetic moments is favored. The simplest formulation of an AFM state is to assume that the molecular field acting on one sublattice is due to the magnetization of the other sublattice and *vice versa*. Again, we assume the form of this molecular field to be proportional to the Weiss constant λ, except now this parameter is negative. The two respective sublattices are designated as "up" and "down", reflecting the orientation of their magnetization. The molecular fields acting on the sublattice "up" and the sublattice "down" are then

$$B_{\text{up}} = -\left|\lambda\right| M_{\text{down}}$$

and

$$B_{\text{down}} = -\left|\lambda\right| M_{\text{up}}. \tag{3.58}$$

Assuming no applied magnetic field, we retrace the steps we took when considering the FM state. The magnetization of each sublattice is written as

$$\frac{M_{\text{up}}}{M_s} = B_J\left(y_{\text{up}}\right), \text{where } y_{\text{up}} = -\frac{g_J \mu_B J \left|\lambda\right| M_{\text{up}}}{k_B T} \tag{3.59}$$

and

$$\frac{M_{\text{down}}}{M_s} = B_J\left(y_{\text{down}}\right), \text{where } y_{\text{down}} = -\frac{g_J \mu_B J \left|\lambda\right| M_{\text{down}}}{k_B T}. \tag{3.60}$$

The two sublattices are equivalent in all aspects, except that the magnetization of one points "up" while that of the other points "down". We thus have

$$\left|M_{\text{up}}\right| = \left|M_{\text{down}}\right| \equiv M, \tag{3.61}$$

and write

$$\frac{M}{M_s} = B_J\left(y\right) \text{with } y = -\frac{g_J \mu_B J \left|\lambda\right| M}{k_B T}. \tag{3.62}$$

Equation 3.62 is similar to Equations 3.46 and 3.47, and thus, the magnetization of each sublattice as a function of temperature will have a similar shape (for the same $J = \frac{1}{2}$) as shown by the curve in Figure 3.7. This is understandable, because the ordered magnetic state on each sublattice arises due to precisely the same mechanism, i.e., the action of the Weiss molecular field. Specifically, the spontaneous magnetization (the ordered state) will disappear at a particular critical temperature, which, in this case, is called the Néel temperature T_N. Similarly to Equation 3.50 that describes the FM case, the Néel temperature is given by

$$T_N = \frac{g_J \mu_B (J+1)|\lambda| M_s}{3k_B} = \frac{n|\lambda| \mu_{eff}^2}{3k_B}. \tag{3.63}$$

Clearly, because the magnetizations M_{up} and M_{down} are of equal magnitude but pointing in opposite directions, the overall magnetization $M_{up} + M_{down} = 0$.

In the presence of a small applied field at $T \geq T_N$, proceeding exactly the same way as when developing the expression for the susceptibility of a ferromagnet, the magnetization of each sublattice will be small and, hence, $y \ll 1$. This justifies expanding the Brillouin function as in Equation 3.49,

$$M = M_s B_J(y) \approx M_s \frac{J+1}{3J} y = M_s \frac{(J+1)}{3k_B T} g_J \mu_B (B - |\lambda| M). \tag{3.64}$$

Isolating the magnetization M and substituting the Néel temperature from Equation 3.63, we obtain

$$M = \frac{T_N B}{|\lambda|(T + T_N)}. \tag{3.65}$$

The susceptibility then follows from

$$\chi = \frac{M}{H} \approx \frac{\mu_0 M}{B} = \frac{\mu_0 T_N}{|\lambda|} \frac{1}{T + T_N}. \tag{3.66}$$

This is the Curie–Weiss law, but with T_N replacing $-T_C$. Introducing the Weiss temperature θ in place of T_C and T_N in Equations 3.56 and 3.66, the susceptibility becomes

$$\chi \propto \frac{1}{T - \theta}. \tag{3.67}$$

We recognize the case $\theta > 0$ as corresponding to a FM structure, $T_C = \theta$, while

$\theta = 0$ corresponds to a paramagnet, and

$\theta < 0$ indicates an AFM material, $T_N = -\theta$.

Figure 3.9(a) sketches the behavior of susceptibility of a paramagnet, ferromagnet, and antiferromagnet as a function of temperature. In practice, it is convenient to plot the experimental data as $1/\chi$ vs. T, as the slopes of the expected straight lines are the Curie and Curie–Weiss constants, from which the effective number of Bohr magnetons is easily determined *via* Equations 3.50 and 3.63. However, a word of caution is in order regarding the value of the Néel temperature expected based on Equation 3.63. This often differs significantly from the experimentally determined Weiss temperature θ, and it is the consequence of a rather crude treatment where the Weiss molecular field was assumed to be the field exerted by one sublattice on the other, totally disregarding any field contribution from their own sublattice.

Unlike in a ferromagnet where the magnetic susceptibility diverges as the temperature approaches the critical temperature T_C and is infinite below T_C, the susceptibility of an antiferromagnet stays finite even below T_N and displays a characteristic feature depending on the direction of the applied magnetic field, depicted in Figure 3.8(b). Because at $T = 0$ K both sublattices have saturated magnetization, a modest magnetic field applied parallel to one of the sublattices (antiparallel to the

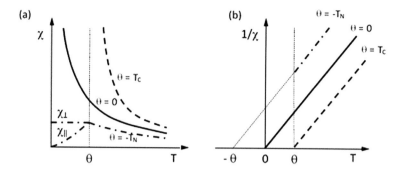

FIGURE 3.9 (a) A sketch of the Curie–Weiss law for a ferromagnet (dashed curve), paramagnet (solid curve), and antiferromagnet (dash-dot curve). Parallel $\chi_{//}$ and perpendicular $\chi_\perp$ susceptibilities of an antiferromagnet below the Néel temperature are also shown. (b) Plots of $1/\chi$ vs. T result in straight lines with the Weiss temperature θ as an intercept on the temperature axis.

other) will not have any effect and $\chi_\parallel$ will be zero. If, however, the field is applied perpendicular to the direction of magnetization of the sublattices, the field will turn slightly the magnetization of both sublattices toward the direction of the applied field, and both sublattices will generate a component of magnetization in the field direction, i.e., a non-zero susceptibility $\chi_\perp$. As the temperature increases, but is maintained below T_N, the magnetization of each sublattice is no longer saturated as the thermal energy tends to disrupt the perfect alignment of magnetic moments on both sublattices. The applied field now has an opportunity to improve the alignment of magnetic moments on the sublattice with the magnetization parallel to the applied field while it further weakens the alignment on the sublattice where the magnetization is opposite to the direction of the field. This creates a greater and greater imbalance in the magnetization of the two sublattices as the temperature increases, and the parallel susceptibility $\chi_\parallel$ increases up to T_N. In contrast, with the applied field perpendicular to the magnetization of sublattices, progressively stronger thermal fluctuations disrupt the alignment of the tilted magnetic moments on both sublattices equally, and the non-zero perpendicular susceptibility $\chi_\perp$ stays essentially constant up to T_N.

3.1.4.1.3   The Case of a Ferrimagnet

In the case of an antiferromagnet, the two sublattices were assumed to possess equal but opposite magnetizations and therefore cancel each other. It often happens that the two sublattices are not identical, and although they have opposite magnetic moments, their strength is not equal, resulting in a net spontaneous magnetization. This situation is called ferrimagnetism and is schematically depicted in Figure 3.5(c). Typical ferrimagnetic materials are oxides, such as ferrites, spinels, and garnets with different magnetic ions occupying both tetrahedral and octahedral sites, which may have not only opposing magnetic moments but also the number of the respective sites may not be the same. Because the vast majority of ferrimagnets are insulators, they are used in high-frequency applications requiring materials with spontaneous magnetization, such as ferrite cores, where ordinary metallic ferromagnets are not suitable because of induced Eddy currents and energy losses they cause.

Beyond the above three major categories of materials possessing spontaneous magnetization, there are also magnetic structures with a helical order where the magnetic moments are aligned ferromagnetically in a plane, but the magnetization rotates between successive planes. Finally, one should also mention spin glasses where the magnetic moments are dispersed randomly but their "freezing", as the temperature decreases, is done cooperatively at a well-defined temperature. Although interesting, these topics are beyond the purpose introducing the magnetic behavior of skutterudites.

3.1.4.2　Magnetic Excitations and Spin Waves

At absolute zero temperature, a FM material has all its magnetic moments (spins) perfectly parallel along a particular direction. As the temperature increases, fluctuations can disturb this perfect order by flipping spins at some lattice sites. A question is how it happens. Naively, one might assume that a certain minimum thermal energy, equivalent to the difference in energies of the final and initial spin states, i.e., the energy of exchange interaction, is required to do so. In other words, this suggests that there is an energy gap to overcome in order for a spin to flip its orientation. In reality, this is not so. Rather than a single electron at some particular lattice site fully flipping its spin, there are many electrons that cooperatively "turn" their spins, one by one, slightly away from the z-direction. Thus, the flipped spin is "smeared out" over all lattice sites, as sketched in Figure 3.10 for a one-dimensional system of spins. This is the essence of spin waves, which are the lowest energy magnetic excitations.

A formal quantum mechanical derivation of the dispersion relation of spin waves is somewhat cumbersome and involves defining spin reversal operators related to Pauli spin matrices and is beyond the needs of a brief review of magnetic properties. Nevertheless, the results are interesting and worth to summarize here.

Since by flipping a spin, the total spin of the system changes by $1/2 - (-1/2) = 1$, the magnetic excitations have an integer spin and are thus boson particles called magnons. The energy of excitation (energy of a magnon) turns out to be

$$E_{magnon} = \hbar\omega = 4JS\left(1 - \cos qa\right), \tag{3.68}$$

where J is the usual exchange integral, q is the wavevector ($q = 2\pi/\lambda_{sw}$) of the spin wave, and a is the lattice spacing. For small q, i.e., for a long wavelength λ_{sw} of the spin waves, $\cos(qa)$ can be approximated ($\cos\varphi \approx 1 - \varphi^2/2$) and the magnon energy becomes

$$\hbar\omega \approx 2JSq^2a^2. \tag{3.69}$$

Equation 3.69 implies that at $q = 0$, the frequency $\omega = 0$. In other words, it takes a vanishingly small energy to excite a magnon, and there is no need to overcome any energy gap. The situation is similar as with acoustic phonons, except that the frequency of long-wavelength phonons is linearly, rather than quadratically, dependent on the wavevector, $\omega = v\,q$, where v is the speed of sound.

Because the DOS in three dimensions is $g(q)dq \propto q^2dq$, in terms of the frequency, the DOS becomes

$$g(\omega)d\omega \propto \omega^{1/2}d\omega. \tag{3.70}$$

Being bosons, the number of magnons is given by the Bose–Einstein distribution function as

$$n_m = \int_0^\infty \frac{g(\omega)d\omega}{\exp\left(\hbar\omega/k_BT\right) - 1} = \left(\frac{k_BT}{\hbar}\right)^{3/2} \int_0^\infty \frac{x^{1/2}}{e^x - 1}dx \propto T^{3/2}, \tag{3.71}$$

where a substitution $x = \hbar\omega/k_BT$ was made in the last equality.

Wavelength of the spin wave λ_{sw}

FIGURE 3.10　(a) A sketch of a spin at a lattice site flipping its orientation. (b) Lowest energy magnetic excitations are spin waves where the deviation of a spin from the z-direction is shared cooperatively by many neighboring spins as a spin wave.

As noted above, each thermally excited magnon reduces the total magnetic moment by $S = 1$. Consequently, the spontaneous magnetization at very low temperatures will deviate from the value of magnetization at absolute zero temperature by

$$\frac{M(T=0) - M(T)}{M(T=0)} \propto T^{3/2}. \tag{3.72}$$

Equation 3.72 expresses the so-called Bloch $T^{3/2}$ law, which is in excellent agreement with the temperature dependence of magnetization observed at the lowest temperatures. It is the existence of spin waves, as the low energy excitations of the FM state, which causes the departure from the expected exponential temperature dependence in Figure 3.7 to the 3/2 power law of Bloch.

3.1.4.3 Magnetic Anisotropy

In the above discussion, we have assumed that the magnetic properties are independent of any particular crystallographic direction. But, this is generally not so. The dependence of magnetic properties on direction is referred to as magnetic anisotropy, and it describes the energetics that determines the preferred direction of magnetization of a material possessing spontaneous magnetization. I will mention the four most important sources of anisotropy without going into details.

Shape anisotropy has its roots in long-range dipole–dipole interactions between free poles at surfaces and is particularly important in thin-film magnetic structures where stray field energy is minimized when the magnetization lies in plane and is maximized when it is oriented perpendicular to the surface. In thin films, this is usually the most important contribution to the anisotropy, and the energy associated with it is written as

$$E_{shape} = 2\pi M_s^2 \cos^2 \theta, \tag{3.73}$$

with M_s being the saturation magnetization and θ the angle between M_s and the film normal.

Magnetocrystalline anisotropy arises from spin-orbit coupling and reflects how the atomic orbitals respond to their local environment. It shows up as the tendency of the magnetic moment to align along a particular crystal axis. The effect is most notable in measurements of single crystalline samples where at certain crystallographic directions it is easy to magnetize the materials (understood as needing a smaller magnetic field to achieve magnetic saturation) while at other directions, called directions of hard magnetization, larger applied fields are needed to do the same. In polycrystalline specimens, the magnetocrystalline anisotropy is substantially smaller but may still be present on account of preferred grain orientation. The form of the energy associated with magnetocrystalline anisotropy reflects the underlying crystalline structure and typical magnitudes at room temperature are on the order of 10^5 Jm^{-3} (10^6 erg per cm^3).

Stress-related anisotropy, sometimes called magnetostriction anisotropy or magnetoelastic anisotropy, is often present as a consequence of coupling between the magnetization direction and the mechanical strain due to spin-orbit interaction. It describes the change in the size of the sample as it is magnetized along different crystal axes. This anisotropy can be very large and even dominant in heavily strained thin film structures.

Surface anisotropy is one of the earliest proposed anisotropies and arises because of the reduced symmetry the surface atoms experience. However, it often also includes a component from magnetocrystalline anisotropy. The magnitude of surface anisotropy was estimated by Néel as 10^{-3} Jm^{-2} (1 erg per cm^2), Néel (1954). The surface anisotropy is usually expressed through its leading order term

$$E_{surf} = K_s \cos^2 \theta, \tag{3.74}$$

where θ is the angle between the magnetic moment and the surface normal, with the constant K_s known as the surface anisotropy constant. Because of the interference of other anisotropies, it is difficult to isolate and provide a reliable estimate of K_s.

3.1.4.4 Effect of Crystalline Electric Field (CEF)

Rare-earth filler ions in a skutterudite cage "feel" the CEF potential of 12 surrounding pnicogen or germanium (in the case of APt_4Ge_{12} skutterudites) ions that constitute a slightly distorted icosahedron and, at a somewhat greater distance, eight transition metal ions at the vertices of a cube. The resulting non-spherical charge distribution around each rare-earth site causes splitting of the f-electron Hund's rule multiplet, giving rise to what is referred to as a CEF energy scheme. In particular, the energy separation between the ground state and the first excited state of the CEF-split energy levels governs the low-temperature transport and magnetic properties and is a topic discussed in numerous papers devoted to physics of rare earth-filled skutterudites. To appreciate the importance of the CEF effect, I sketch here its most important aspects.

In cubic structures, the effect of electrostatic field of the neighboring charges on the $4f$ magnetic ions of rare earths has been worked out long time ago, first by Bethe (1929) using group-theoretical methods and subsequently, in a more amenable way, by Lea, Leask and Wolf (1962). Instrumental in the latter treatment was the development of an operator equivalent technique by Stevens (1952), the essence of which is a replacement of successive terms in the expansion of the electrostatic potential by appropriately chosen angular momentum operators that share the transformation characteristics with that of the electrostatic potential. Because the number of expansion terms required is rather limited, in the case of f-electron rare-earth ions no more than the sixth order is required, the process is easily tractable.

The most general operator equivalent potential with cubic point symmetry is specified by a Hamiltonian

$$\hat{H}_{CEF} = A_4\left(O_4^0 + 5O_4^4\right) + A_6\left(O_6^0 - 21O_6^4\right), \tag{3.75}$$

where the Stevens operators expressed *via* the total angular momentum J and its z-component J_z are

$$O_4^0 = 35J_z^4 - \left[30J\left(J+1\right) - 25\right]J_z^2 - 6J\left(J+1\right) + 3J^2\left(J+1\right)^2, \tag{3.76}$$

$$O_4^4 = \frac{1}{2}\left(J_+^4 + J_-^4\right), \tag{3.77}$$

$$O_6^0 = 231J_z^6 - 105\left[3J\left(J+1\right) - 7\right]J_z^4 + \left[105J^2\left(J+1\right)^2 - 525J\left(J+1\right) + 294\right]J_z^2 \\ - 5J^3\left(J+1\right)^3 + 40J^2\left(J+1\right)^2 - 60J\left(J+1\right), \tag{3.78}$$

$$O_6^4 = \frac{1}{4}\left[11J_z^2 - J\left(J+1\right) - 38\right]\left(J_+^4 + J_-^4\right) + \frac{1}{4}\left(J_+^4 + J_-^4\right)\left[11J_z^2 - J\left(J+1\right) - 38\right]. \tag{3.79}$$

The coefficients A_4 and A_6 determine the scale of the CEF splitting and are linear functions of mean fourth $\langle r^4 \rangle$ and sixth $\langle r^6 \rangle$ powers of the radii of the magnetic electrons. As they are difficult to calculate, they are usually left as fitting parameters. With substitutions $O_4 = \left[O_4^0 + 5O_4^4\right]$ and

$O_6 = \left[O_6^0 \right] - 21 O_6^4$, and capturing all possible values of the ratio between the fourth and sixth order terms by setting

$$A_4 F_4 = Wx \tag{3.80}$$

and

$$A_6 F_6 = W \left(1 - |x| \right), \tag{3.81}$$

with $-1 < x < +1$, it immediately follows that

$$\frac{A_4}{A_6} = \frac{F_6}{F_4} \frac{x}{1 - |x|}. \tag{3.82}$$

Hence, for $x = 0$, $A_4 / A_6 = 0$ and for $x = \pm 1$, $A_4 / A_6 = \pm \infty$.

Substituting from Equations 3.79 and 3.80 into Equation 3.74, the Hamiltonian can be written as

$$\hat{H}_{CEF} = W \left[x \frac{O_4}{F_4} + \left(1 - |x| \right) \frac{O_6}{F_6} \right]. \tag{3.83}$$

The term in brackets in Equation 3.83 is a matrix with eigenvectors corresponding to the most general combination of fourth- and sixth-order crystal field and the eigenvalues related to the crystal field energy levels scaled by a factor W defined in Equations 3.80 and 3.81. By diagonalizing the matrix, one obtains the CEF energy level spectrum including the associated eigenvectors. Results for all J-manifolds between 2 and 8 in half-integer steps can be found in the study by Lea, Leask, and Wolf (1962).

The above approach is well established for structures with the cubic O_h point symmetry. Although skutterudites are cubic structures, a small distortion of the icosahedron due to unequal distances d_1 and d_2 on the pnicogen ring locally lowers the cubic symmetry O_h to T_h symmetry, the chief distinction between the two being the lack of symmetry operations C_4 (rotation by $\pi/2$ about a four-fold symmetry axis) and C_2' (rotation through π perpendicular to the principal rotation axis) in T_h. In a vast majority of situations, such local symmetry lowering is irrelevant as it does not alter the ordering of the CEF-split states or their degeneracy. However, it somewhat alters the wavefunctions and their mixing and may lead to a different response *vis-à-vis* dipole and quadrupole fields. As we saw in discussions of the SC state in PrOs$_4$Sb$_{12}$ in Chapter 2, the treatment within the T_h symmetry was essential in understanding the nature of exotic superconductivity in this filled skutterudite.

Takegahara et al. (2001) realized that there is a distinction between the O_h and T_h point symmetries and showed that the CEF Hamiltonian appropriate for the T_h symmetry is given by

$$\hat{H}_{CEF} = A_4 \left(O_4^0 + 5 O_4^4 \right) + A_6 \left(O_6^0 - 21 O_6^4 \right) + A_6' \left(O_6^2 - O_6^6 \right), \tag{3.84}$$

or, in a form equivalent to Equation 3.83, as

$$\hat{H}_{CEF} = W \left[x \frac{O_4}{F_4} + \left(1 - |x| \right) \frac{O_6}{F_6} + y \frac{O_6'}{F_6} \right]. \tag{3.85}$$

Here, $O_6' = \left[O_6^2 - O_6^6 \right]$, and the parameter y represents the effect of T_h symmetry. Clearly, for $y = 0$, the Hamiltonian for the O_h point symmetry is recovered.

Let us briefly consider the most important case of a skutterudite where it is absolutely essential to take into account the lowered T_h symmetry, i.e., Pr^{3+} ions in $PrOs_4Sb_{12}$. Assuming O_h point symmetry at the site of the Pr^{3+} ion, its nine-fold degenerate ground state $^3H_4$ manifold would split under the O_h symmetry into a singlet Γ_1, a non-magnetic doublet Γ_3, and two magnetic triplets Γ_4 and Γ_5. When the proper T_h symmetry is used, we obtain again four states with the same degeneracies, now labeled a Γ_1 singlet, a Γ_{23} non-magnetic doublet, and two magnetic triplets $\Gamma_4^{(1)}$ and $\Gamma_4^{(2)}$. The extra term in the Hamiltonian in Equation 3.83 mixes the wavefunctions of states Γ_4 and Γ_5 and yields the states $\Gamma_4^{(1)}$ and $\Gamma_4^{(2)}$. In the process of mixing, the usual crystal field selection rules are altered, most notably the dipole matrix element between the Γ_1 singlet and the Γ_5 triplet is no longer zero. Proper application of these selection rules in the analysis of inelastic neutron scattering (INS) experiments by Goremychkin et al. (2004) was the key factor in identifying the non-magnetic singlet Γ_1 as the ground state of $PrOs_4Sb_{12}$, as we discussed in Chapter 2. I wish to stress that the CEF energy scheme is not a fixed entity identical in all skutterudites filled with the same rare-earth element. Slight compositional modifications, be it a different transition metal or different pnicogen atom, may lead to entirely different CEF energy levels with different ground states. The case in point is $PrOs_4As_{12}$ where, as shown by Chi et al. (2008), the ground state is a magnetic triplet $\Gamma_4^{(2)}$ in contrast to the non-magnetic Γ_1 singlet in $PrOs_4Sb_{12}$. Likewise, the ground state of $SmFe_4P_{12}$ is the Γ_5 doublet while in $SmOs_4Sb_{12}$, the ground state is the Γ_{67} quartet. In my discussions of skutterudites filled with rare-earth elements, I will mostly label states using the T_h point group symmetry and, where appropriate, will also include their designation within the O_h symmetry.

3.2 TECHNIQUES OF MEASURING MAGNETIC PROPERTIES

Although there are a number of ways how one can assess the magnetic state of a substance, three techniques are considered essential to shed light on the magnetic properties of solids: measurements of the magnetic moment, Mössbauer effect, and neutron scattering measurements. Magnetic moment, or magnetization when expressed per unit volume, informs about the global magnetic character of a solid, and its magnitude and sign determine what kind of magnetic behavior one deals with. When divided by the applied field, we talk in terms of magnetic susceptibility and, from its temperature dependence, one can extract important magnetic parameters and any hints the structure is developing a long-range order. The relative ease with which one can measure magnetic susceptibility, especially given a widespread use of very sensitive superconducting quantum interference device (SQUID)-based magnetometers, makes this a favorite parameter to assess magnetic properties. The Mössbauer effect, on the other hand, is a rather specialized spectroscopic technique that probes the magnitude of the magnetic field acting on a specific nucleus (Mössbauer nucleus) or the presence of the gradient of the electric field at the nucleus. Neutron scattering is a sharp and most direct technique informing about an arrangement of magnetic moments inside solids and, as such, is indispensable in determining the magnetic structure. A combination of the three techniques is a powerful armory to tackle often very complicated magnetic structures the nature presents. In the following three sections, I briefly describe the above techniques.

3.2.1 MEASUREMENTS OF MAGNETIC MOMENTS AND MAGNETIC SUSCEPTIBILITY

Instruments that measure magnetic moments or magnetization are generally called magnetometers. While they are of various designs, the principle is substantially the same; they rely on Faraday law where the motion of a magnetized sample in a pick up coil gives rise to an induced voltage that is proportional to the magnetic moment. Nowadays, the most popular magnetometers are vibrating sample magnetometers (VSMs), magnetometers that use SQUIDs) in their detection circuit, torque magnetometers, Faraday-type magnetometers, and optical magnetometers.

The VSM relies on a vibrational motion of the sample in proximity of the pickup coil that generates an induced voltage. The sample is attached to a rod and initially placed in the center of a highly homogeneous electromagnet (fields up to 10^6 A/m) to induce magnetization in the sample. The rod is then vibrated (typically 60 Hz with 1 mm amplitude produced by a piezoelectric actuator), and the change in the magnetic flux in the pickup coils attached to the pole pieces of the electromagnet induces voltage that is processed using a lock-in amplifier with a piezoelectric signal serving as a reference signal. VSMs can typically operate in the range down to about 2 K (limited by heat generated by vibrations) and up to above 1200 K. The sensitivity of VSMs is about an order of magnitude inferior to the SQUID-based magnetometers, but the instrument is versatile, sensitive enough, and convenient for collecting hysteresis loops and performing demagnetization measurements.

SQUID-based magnetometers are arguably the most sensitive devices to detect faint magnetic fields, but are also more expensive to operate as they require a liquid helium environment for their superconducting quantum interference detector circuit and SC solenoids operating in their persistent mode and producing large magnetic fields. A step-like motion of a sample inside a SC loop generates a flux change that is inductively transformed to a SQUID circuit that acts as an extremely sensitive flux-to-voltage converter. Depending on whether one uses RF or DC SQUID circuits, there is one or two Josephson junctions as part of the circuit. The SQUID is located outside of the magnetic solenoid and is well shielded from the effect of magnetic fields. To minimize noise current that may arise in the detector circuit due to (tiny) relative motion of the detection coil with respect to the stationary magnetic field of a SC solenoid, one uses a second-order gradiometer configuration, sketched in Figure 3.11. Commercially available SQUID-based magnetometers can operate from about 400 mK up to about 400 K and use magnetic fields as high as 14 T. The sensitivity of the instruments is on the order of 10^{-14}THz$^{-1/2}$.

Torque magnetometers measure the torque $\tau = \mu \times B$ exerted on the magnetic moment by the applied magnetic field and, in principle, can achieve comparable sensitivity to SQUID-based magnetometers. Sample magnetization must be extracted from the measured torque, and a reliable relation between the two is essential. A torque magnetometer is an ideal instrument to measure magnetic phase transitions rather than the absolute values of the magnetic moment.

Excellent sensitivity of the instrument relies on mounting a sample on a cantilever the displacement of which is measured *via* changes in capacitance between a fixed reference and the cantilever.

A Faraday-type magnetometer measures the force that a magnetic field gradient exerts on a magnetic sample, $(M.\nabla)B$. The force is measured by a sensitive balance or by a compression of a fine spring. Often, a cantilevel arrangement is set up, and the displacement is measured capacitively. Typical sensitivity is an order of magnitude smaller than in the case of SQUID-based magnetometers. The technique requires not only magnetic field but also the field gradient that can be achieved by suitably shaped magnet pole faces or by using gradient coils.

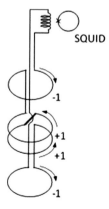

FIGURE 3.11   Superconducting pickup coil in a second-order gradiometer configuration and a SC transformer coupled to a SQUID.

Optical magnetometers make use of the magneto-optic Kerr effect and rely on a rotation of polarization of light impinging on the surface of a magnetic sample. The polarization of the reflected light is measured by a suitable detector. This is primarily a surface technique and is extensively used in the study of thin-film magnetic structures.

As a practical note, the susceptibility is often measured and reported as a molar susceptibility χ_{mol}. This is simply the volume susceptibility multiplied by the molar volume of the substance $\chi_{mol} = \chi V_{mol}$, where V_{mol} is the volume occupied by 1 mole (amounting to 6.022×10^{23} formula units) of the substance, i.e., $V_{mol} = M_{mol}/\rho$, with M_{mol} the molar mass and ρ the density. In case one is interested in mass susceptibility χ_g, it is obtained by dividing volume susceptibility by the density of the substance, $\chi_g = \chi/\rho$. Conversion between the SI units of the susceptibility and the cgs units is accomplished by using a relation $\chi^{SI} = 4\pi\chi^{cgs} = 4\pi \, \rho^{cgs}\chi_g^{cgs}$.

3.2.2 Mössbauer Spectroscopy

The Mössbauer effect, Mössbauer (1958), is a remarkable realization of a recoil-free emission of a γ-ray photon as the source nucleus undergoes a transition from its excited nuclear state to the ground state, and this photon is absorbed, again with no recoil, by a target nucleus of the same nuclear species. In gases, the Mössbauer effect is not possible because the respective nuclei would necessarily recoil, leaving the emitted γ-ray with less energy than the energy difference between the two nuclear energy states. In solids, where the nucleus is tightly bonded to other nuclei *via* the crystal lattice, the recoil is taken up by very many atoms (the recoiling mass is essentially the entire sample and thus enormous in comparison to a single atom), and the energy shift of the emitted γ-ray is very small. Specifically, as long as the energy lost in recoil is smaller than one-half of the linewidth of the nuclear transition, the γ-ray photon can be resonantly absorbed by the nucleus of the same species as the source nucleus (energy lost in the absorption recoil must, again, be less than one-half of the linewidth). Rudolf Mössbauer received the Nobel Prize in Physics for the discovery of the effect now carrying his name in 1961. The Mössbauer effect has proved to be an extremely sensitive probe of the magnetic field and electric field gradients in the vicinity of nuclei under study. It should be noted that not every nucleus is suitable for Mössbauer studies. In fact, only few nuclei have sufficiently low-lying excited states and long enough lifetimes and can serve as Mössbauer nuclei. The most prominent among them is a ^{57}Fe nucleus that emits a 14.41 keV γ-ray photon (the frequency equivalent of this energy is 3.5×10^{18} Hz) as the nucleus transits from the $I = 3/2$ nuclear state to $I = 1/2$ ground state. For studies of skutterudites, apart from this ^{57}Fe nucleus, it is advantageous that ^{119}Sn and ^{121}Sb are also Mössbauer nuclei, so three independent Mössbauer effect assessments can be made. A sketch of a typical experimental setup to measure Mössbauer spectra is indicated in Figure 3.12.

There are three main features of a Mössbauer experiment that reveal useful complementary information: a position of the resonance called the isomer shift, the quadrupolar splitting, and further

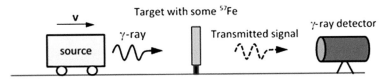

FIGURE 3.12 Schematic picture of the Mössbauer resonance experiment using ^{57}Fe nuclei. The source undergoes a nuclear transition from the $I^{3/2}$ excited state to the $I^{1/2}$ ground state, emitting a recoilless γ-ray photon. The source is moving with the velocity v (typically several mm /s) to fine tune the energy of the emitted photon by the Doppler effect to exactly match the energy for its resonant absorption in ^{57}Fe contained in the fixed target. The transmitted signal is detected by a γ-ray detector. At resonance, there will be a dip in the transmitted signal.

splitting due to the action of magnetic field (both internal and external) generally called fine structure Zeeman splitting.

The isomer shift reflects the fact that the source and absorber nuclei, although they are of identical species, might not be situated in exactly the same environment, and this will slightly alter the energy of nuclear transitions. To ensure that the target nucleus can resonantly absorb the emitted γ-ray photon, this slight energy level difference is compensated for by moving the source at various velocities back and front with respect to a stationary target until the resonance condition is satisfied. Here, one relies on the Doppler effect, which results in a small frequency shift that compensates for the small energy difference. A motion of the source with respect to the target at 1 mm/s is equivalent to an energy shift of 4.8075×10^{-8} eV or, in terms of the frequency, to a shift of 1.2×10^{7} Hz, Fultz (2011). Because the isomer shift is not an absolute measure, it is usually specified with respect to a reference, most often an iron foil of 40 μm thickness. The physical basis of the isomer shift is a finite size of the nucleus and the electronic charge density of *s*-electrons that slightly change the nuclear energy levels. When the *s*-electronic environments of the source and target nuclei differ, either a positive or a negative shift in the position of the resonance will be observed. The isomer shift is useful in determining valency of the states, electron shielding, and ligand states.

Quadrupolar splitting arises as due to the electric field gradient perturbing the quadrupolar electric field. Nuclei in states with non-spherical charge distribution, i.e., those with the nuclear angular momentum quantum number larger than ½, possess a nuclear quadrupole moment. Any asymmetric electric charge distribution then splits the nuclear energy levels. In the case of ^{57}Fe and ^{119}Sn, the nuclear state is split into two sublevels with $m_I = \pm 1/2$ and $m_I = \pm 3/2$, separated in energy by the quadrupolar splitting $\Delta E_Q = eV_{zz}Q/2$, where e is an elementary electric charge, Q is the quadrupole moment of the ^{57}Fe (or ^{119}Sn) in the excited state, $V_{zz} = \partial^2 V/\partial z^2$, with z as the principal axis of the electric field gradient tensor, and V is the electric potential due to all electric charges surrounding the ^{57}Fe nucleus. Thus, instead of a single resonance line, there will be a doublet with the separation reflecting the electric field at the nucleus and related to the oxidation state, spin state, and site symmetry, Figure 3.13.

Magnetic (Zeeman) splitting arises from the presence of many possible internal sources of magnetic field as well as an applied magnetic field that split a nuclear level with spin I into $2I + 1$ sublevels. Quantum mechanics dictates that the allowed transitions between the ground state and the excited states are only those where m_I changes by 0 or 1. In the case of ^{57}Fe, this yields six allowed transitions depicted in Figure 3.13.

3.2.3 NEUTRON SPECTROSCOPY

Neutrons are spin ½ particles (fermions) possessing no electric charge but having a magnetic moment associated with the spin angular momentum. The lack of the electric charge allows neutrons to penetrate deep into the interior of a structure and, unlike X-rays or electron diffraction, the information neutron scattering provides is not limited to the state of a surface. On account of the neutron's much larger mass compared to the mass of an electron, the neutron magnetic magneton $\mu_N = e\hbar/2m_N = 5.05 \times 10^{-27}$ JT^{-1} is some three orders of magnitude smaller than the Bohr magneton μ_B.

Neutrons are produced in large numbers in nuclear reactors as a result of the fission process, and their energy is moderated by interactions (collisions) with substances having low atomic mass, such as graphite or water, typically at room temperature. Alternatively, neutrons are generated at spallation sources where synchrotron-accelerated energetic protons are directed at heavy metal targets, such as uranium, tantalum, or mercury, producing fast neutrons that are, again, moderated by collisions with CH_4, H_2O, or D_2O. Neutrons emerge with a modified Maxwellian distribution of velocities of the form

$$n(v) \propto v^3 e^{-\frac{\frac{1}{2}m_N v^2}{k_B T}},$$

(3.86)

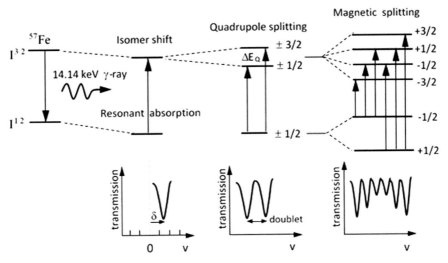

FIGURE 3.13 Upper panels indicate nuclear transitions involved in Mössbauer measurements using ^{57}Fe nuclei. Lower panels depict the signal transmitted through an absorber and detected by a suitable detector. The emitted 14.41 keV γ-ray photon is resonantly absorbed with an isomer shift δ indicated. Asymmetric electrical charge distribution splits the nuclear level by ΔE_Q, resulting in a doublet observed in the transmitted signal. Magnetic field (either of internal origin or an applied field) splits the magnetic quantum levels as indicated, resulting in six transitions of unequal intensity, indicated in the transmitted signal. The ratio of intensities of the transmitted sextet is 3: $4\sin^2\theta/(1+\cos^2\theta)$: 1 where the θ is the angle between the direction of Mössbauer γ-ray photon momentum and the nuclear spin momentum. The ratio refers to outside transmission minima with respect to middle transmission minima and with respect to inner transmission minima. Note that the outside to inside transmission minima ratio is fixed at 3. It is the middle minima, which depend on the mutual orientation of the momentum of γ-ray photon and the nuclear spin momentum.

where $n(v)$ represents the number of neutrons passing through a unit area per unit time. In comparison to the usual Maxwellian velocity distribution with a quadratic velocity term, Equation 3.86 picks up an additional velocity factor v due to effusion of neutrons through a hole. Typical de Broglie wavelength of moderated (thermalized) neutrons is $\lambda_N \sim 10^{-10}$ m, of the same order of magnitude as the lattice spacing, and thus strong scattering on atomic nuclei is expected. On the other hand, the wavelength is several orders of magnitude larger than the extremely short range of the strong force ($\sim 10^{-15}$ m), and thus neutrons are unable to probe the internal structure of the nucleus. Consequently, scattering of neutrons on nuclei is isotropic. The neutron wavelength can be tuned to some degree by performing neutron scattering experiments at different temperatures. The fact that neutrons do not care much about the atomic number of the scattering target, unlike X-rays and electrons, which are scattered by the electron clouds and are thus substantially insensitive to light elements, neutrons can easily detect and resolve the presence of light species, such as hydrogen, and distinguish between elements sitting in the nearby positions in the periodic table. Apart from scattering on nuclei, neutrons possess a magnetic moment and are thus also sensitive to the presence of magnetic moments in the structure, resulting in magnetic scattering. In fact, neutron magnetic scattering and its dependence on the direction of the magnetic moment provide a unique and revealing insight into the structural arrangement of magnetic moments in the crystalline lattice and are the most trustworthy input toward analyzing and understanding often-complicated magnetically ordered structures. Although Néel proposed the existence of AFM ordering back in 1932, Néel (1932), it was not until Shull and Smart 1949) observed additional (magnetic) diffraction peaks, apart from the Bragg peaks, in the spectrum of MnO at low temperatures, that the existence of the AFM order was so vividly demonstrated, Figure 3.14. Together with Brockhouse, who contributed

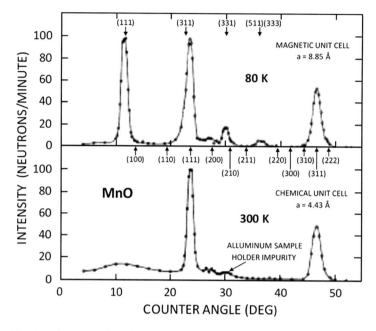

FIGURE 3.14 The first demonstration of the AFM ordering in the neutron diffraction spectrum. The upper panel shows the spectrum of MnO obtained at 80 K, i.e., below the Néel transition temperature, where an additional (magnetic) peak is seen compared to the spectrum obtained in the paramagnetic range at 300 K (bottom panel) where only the usual Bragg peaks are shown. The data of C. G. Shull and J. S. Smart, *Physical Review* **76**, 1256 (1949). With permission from the American Physical Society.

greatly to the development of neutron spectroscopy and invented a triple-axis spectrometer, Clifford Shull and Bertram Brockhouse were awarded the 1994 Nobel Prize in Physics for the development of the neutron scattering and its application to magnetic structure determination.

Both nuclear and magnetic scattering are examples of elastically scattered neutrons whereby the incoming beam with a wavevector k is scattered in the k' direction with the scattering vector $Q = k - k'$, but with no loss of neutron energy. Neutrons can also be scattered inelastically, i.e., with their energy altered, and this kind of scattering is used in the study of magnetic excitations, such as spin waves. The quantized unit of spin waves (magnon) has energy in the range of 10^{-3}–10^{-2} eV, the range accessible by neutrons. Just as in the case of phonons, neutron scattering of this type is a powerful probe of dispersion of spin waves and is nowadays studied using triple-axis spectrometers to provide maximum flexibility in the selection of neutron energy, sample orientation, and neutron detection.

3.3 MAGNETIC PROPERTIES OF SKUTTERUDITES

3.3.1 MÖSSBAUER STUDIES OF SKUTTERUDITES

The first Mössbauer measurements for skutterudites were performed by Kjekshus and colleagues, Kjekshus et al. (1973) and Kjekshus and Rakke (1974), who studied the ^{121}Sb and ^{57}Fe Mössbauer spectra in the binary antimonide skutterudites $CoSb_3$, $RhSb_3$, $IrSb_3$, and $Fe_{0.5}Ni_{0.5}Sb_3$. The observed isomer shifts of the ^{121}Sb transition were comparable, and their magnitude (-9 mm/s to -10.1 mm/s) was consistent with the covalent bonding. The authors were particularly interested in the relation between the magnitude of the quadrupole interaction and the difference in the bond lengths d_1 and d_2 within the rectangular pnicogen ring. Because the difference between d_1 and d_2 is inevitably the result of an electron imbalance, this should be reflected in the magnitude of the quadrupole split. As shown in Figure 3.15, they found an approximately linear relationship between the quadrupole

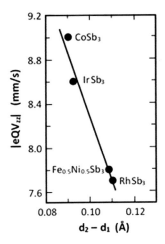

FIGURE 3.15 Quadrupole splitting as a function of the deviation $d_2 - d_1$ of the pnicogen ring from a perfect square measured for binary antimonide skutterudites and for $Fe_{0.5}Ni_{0.5}Sb_3$. Redrawn from the data of A. Kjekshus and T. Rakke et al., *Acta Chemica Scandinavica* **A28**, 99 (1974).

interaction and $(d_2 - d_1)$. From the isomer shift of ^{57}Fe, measured with respect to an iron foil, the authors concluded that the Fe atom in $Fe_{0.5}Ni_{0.5}Sb_3$, indeed, has a formal $3d^6$ configuration, implying the transfer of one electron from the Ni atom to the nonbonding $3d$ orbital of Fe. The authors pointed out that the lack of a resolvable quadrupole interaction attests to a high crystallographic symmetry at the Fe site.

The early Mössbauer spectrum of the filled skutterudites was recorded by Shenoy et al. (1982), who hoped to determine the magnetic moment of Fe in $LaFe_4P_{12}$ using the ^{57}Fe resonance. The results indicated a well-defined quadrupole splitting that was fitted by superposing appropriate Lorenzians. Repeating the measurement in the presence of an external magnetic field of 6.1 T applied parallel to the direction of gamma ray resonance absorption, the data analysis of the spectrum returned a hyperfine magnetic field equal to the applied field with a maximum error of ± 0.1 T. This important result indicated that Fe in $LaFe_4P_{12}$ has a magnetic moment of less than 0.01 μ_B. In other words, the polyanion $[Fe_4P_{12}]^{4-}$ does not carry a magnetic moment! On the one hand, the result is in accordance with the fact that $LaFe_4P_{12}$ is a superconductor, as discussed in Chapter 2. On the other hand, the result is surprising as one would expect the polyanion $[Fe_4P_{12}]^{4-}$ complex to have one of the four Fe atoms in the d^5 configuration with a spin-only moment of 1.73 μ_B. The Mössbauer measurement showed no presence of such atoms.

Gerard et al. (1983) and Grandjean et al. (1984) reexamined the Mössbauer spectra of $LaFe_4P_{12}$ and extended the study to include $CeFe_4P_{12}$ and $EuFe_4P_{12}$. Regarding $LaFe_4P_{12}$, the authors confirmed the non-magnetic state of Fe in this skutterudite and found zero magnetic moment also on the polyanionic framework of $CeFe_4P_{12}$ and $EuFe_4P_{12}$. They concluded that Fe must have a very similar electronic configuration in all three compounds and, whatever it is, its magnetic moment is zero, Figure 3.16. ^{57}Fe Mössbauer parameters for all three compounds are presented in Table 3.1. Based on this result, they suggested that the non-bonding Fe $3d$ states ($\sim t_{2g}$ block bands) are filled in all three compounds and are located well below the Fermi level. According to the authors, these bands cannot be responsible for the magnetic properties, at least in the filled phosphide skutterudites.

Obviously, it was of considerable interest to check the magnetic state of Fe in other skutterudites and, most notably, in those with an iron-antimonide framework. The first attempt was made by Kitagawa et al. (1998), who measured the Mössbauer spectra of La- and Ce-filled structures. The spectra showed virtually the same isomer shift for the two compounds but different quadrupole splitting, 0.374 mm/s for $LaFe_4Sb_{12}$ and 0.406 mm/s for $CeFe_4Sb_{12}$. The authors related the quadrupole splitting to the DOS at the Fermi level and made a statement that the Fe atoms are in a non-magnetic

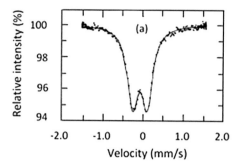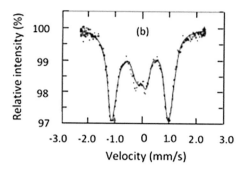

FIGURE 3.16   Room-temperature ^{57}Fe Mössbauer spectra of $LaFe_4P_{12}$ (A), $CeFe_4P_{12}$ (B), and $EuFe_4P_{12}$ (C). Solid lines are the least-squares fits with symmetrical doublets. Reproduced from F. Grandjean et al., *Journal of Physics and Chemistry of Solids* **45**, 877 (1984). With permission from Elsevier.

TABLE 3.1

^{57}Fe Mössbauer Hyperfine Parameters for RFe_4P_{12} Compounds, R = La, Ce, and Eu at Temperatures of 300 K, 77 K, and 4.2 K

Skutterudite	ε	δ	Γ
		300 K	
$LaFe_4P_{12}$	0.30	0.05	0.30
$CeFe_4P_{12}$	0.38	0.02	0.30
$EuFe_4P_{12}$	0.50	0.06	0.24
		77 K	
$LaFe_4P_{12}$	0.33	0.14	0.26
$CeFe_4P_{12}$	0.37	0.15	0.28
$EuFe_4P_{12}$	0.57	0.16	0.42
		4.2 K	
$LaFe_4P_{12}$	0.36	0.17	0.30
$CeFe_4P_{12}$	0.41	0.14	0.29

Source: Adapted from F. Grandjean et al., *Journal of Physics and Chemistry of Solids* **45**, 877 (1984). With permission from Elsevier.

Note: The Quadrupole Splitting is designated by ε, the Isomer Shift with respect to Fe is indicated by δ, and the Linewidth Γ is taken at Half-maximum. All parameters are in units of mm/s.

---

state at room temperature. However, the crucial test to decide the magnetic state of Fe—to carry out the measurements in an external magnetic field—has not been performed, and thus, the measurements were inconclusive.

Interpretation of the Mössbauer spectra of filled skutterudites with the $[Fe_4X_{12}]$ framework is a bit tricky in the sense that it depends on the level of occupancy of the skutterudite voids, which is rarely at 100%. The point is that, with less than full occupancy, the number of the filler atoms in the next-nearest positions to an atom of Fe may not be the expected two (full void occupancy), see Figure 3.17, but, rather, zero, one or two, depending on the statistical distribution of the fillers.

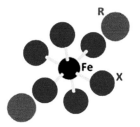

FIGURE 3.17 The nearest and the next-nearest atoms of an atom of Fe (black circle) in the skutterudite structure. Fe is octahedrally coordinated by six pnicogen atoms X (blue spheres), and in the next-nearest positions are 12 other pnicogen atoms (not shown) and two filler atoms (large red circles). This is a configuration corresponding to 100% occupancy of the voids. However, if the void occupancy falls below 100%, the number of fillers in the next-nearest positions to Fe could be zero, one or two, depending on the particular statistical distribution of the fillers in the skutterudite structure.

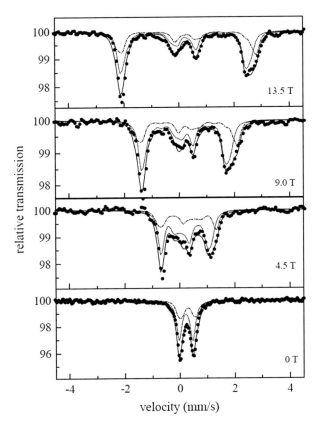

FIGURE 3.18 Mössbauer spectra of $Pr_{0.73}Fe_4Sb_{12}$ at zero field (the lowest panel) and at various magnetic fields indicated. Reproduced from M. Reissner et al., *Journal of Magnetism and Magnetic Materials* **272–276**, 813 (2004). With permission from Elsevier.

This, in turn, modifies the charge density and the electric field gradient at the Fe nucleus, making the Fe sites non-equivalent, in contrast to the uniqueness of the Fe site (8c in Wyckoff's scheme) in the $Im\bar{3}$ space group structure in which skutterudites crystallize. The resulting asymmetry of the Mössbauer outer spectral lines then necessitates fitting of the spectral intensity data with more than a single spectrum. An example of asymmetry is illustrated in Figure 3.18, where the Mössbauer spectra of incompletely filled $Pr_{0.73}Fe_4Sb_{12}$ obtained by Reissner et al. (2004) are shown at zero applied field and at a field of 13.5 T. Two subspectra were needed to fit the intensity data as shown in Figure 3.18; the subspectrum with a larger intensity was assigned to Fe having completely filled Pr positions (i.e., two) at the next-nearest distance from Fe, while the other subspectrum was allocated to Fe where, at the next-nearest distance, at least one Pr atom was missing (it was not possible to determine if one or none Pr atom was in the next-nearest position). The authors claimed that the resulting intensity ratio of the two sublattices was in good agreement with the X-ray analysis. The Pr-filled iron antimonide skutterudite was synthesized several times by the usual route consisting of grinding, heating, and quenching under ambient pressure, with the void occupancy no better than 87%, Danebrock et al. (1996), Bauer et al. (2002a), and Butch et al. (2005). All studies indicated some kind of magnetic ordering setting in at around 5 K, but each with different characteristics: FM ordering (no occupancy reported), AFM ordering for the void occupancy of 73%, and ferrimagnetism for an occupancy of 87%. A surprise came when a near 100% void occupancy was attained by synthesizing $PrFe_4Sb_{12}$ under a pressure of 4 GPa, Tanaka et al. (2007). Not only was the magnetic order missing down to at least 0.15 K, but also the entire CEF level scheme of the Pr^{3+} ion has changed, with the ground state now being a singlet instead of a triplet, and the first excited state a triplet lying merely 22 K above the ground state, as opposed to a singlet as the first excited state at 28 K. It is remarkable that the filler occupancy change of 20% leads to such drastic changes in the magnetic properties. Just as in the case of $PrOs_4Sb_{12}$ with its unconventional SC properties discussed in Chapter 2, it seems that a small crystal electric field splitting characterizing the environment of the $4f$ electron states of Pr^{3+} ions in $PrFe_4Sb_{12}$ gives rise to such unexpected magnetic behavior.

The asymmetric shape of the Mössbauer spectra is seen in all skutterudites of the form RFe_4Sb_{12} with fillers R = La, Pr, Nd, and Eu that do not fully occupy the voids of the skutterudite structure. Because the estimated void filling fractions of the above fillers were comparable (0.80, 0.73, 0.72, and 0.88, respectively), the measured asymmetries of the Mössbauer spectra were not too different. Also, except for Eu that I will consider separately, the remaining filled skutterudites displayed similar isomer shifts and quadrupole splits of around 0.25 mm/s, but the values were not the same for the two different Fe environments.

The only ^{57}Fe Mössbauer measurement where the intensity of the spectrum was apparently adequately fitted with just a single component was reported by Leithe-Jasper et al. (1999) for $YbFe_4Sb_{12}$. However, no actual spectrum was shown to judge whether the absorption peaks of the quadrupole doublet were symmetric. Moreover, as the analysis of the magnetic susceptibility revealed an intermediate valence state of Yb, there should have been two subspectra present corresponding to Yb^{2+} and Yb^{3+} states. Because $YbFe_4Sb_{12}$ is one of the most prospective filled skutterudites for thermoelectric applications, its physical properties, including magnetism, have attracted much interest, and several Mössbauer studies were performed with this structure. Apart from the one mentioned above, all of them required more than one component to fit the spectrum. The intent of Mössbauer measurements on $YbFe_4Sb_{12}$ was to shed light on the often-contradictory reports regarding not only the relative magnetic contributions of Yb ions and of Fe in the $[Fe_4Sb_{12}]$ polyanion framework, but also the overall assessment of magnetism in this skutterudite. Depending on a particular Yb void-filling fraction, some reports indicated FM ordering at low temperatures, Dilley et al. (1998), Bauer et al. (2000), and Ikeno et al. (2007), while others claimed the persistence of paramagnetism down to low temperatures, Leithe-Jasper et al. (1999). I have already mentioned that rarely is the void occupancy at 100%, and thus, it was suspected that the susceptibility and other magnetic parameters, including the Mössbauer spectra, might reflect the actual degree of filling. With ytterbium, the problem is compounded by the fact that Yb is known to be either divalent or trivalent and the relative fraction of the two ions makes a big difference because Yb^{2+} is a non-magnetic ion with $4f^{14}$ configuration (closed shell) while any presence of Yb^{3+} ions ($4f^{13}$ configuration) would give rise to a magnetic moment. With divalent Yb, $YbFe_4Sb_{12}$ resembles the alkaline-earth skutterudites, and it is of interest to probe the magnetic state of the $[Fe_4Sb_{12}]^{2-}$ polyanion. This was done first by Leithe-Jasper et al. (1999) who noted that the measured hyperfine field always matched the value of the applied magnetic field. Based on this observation, they concluded that Fe atoms carry no magnetic moment. The conclusion was surprising because the $[Fe_4Sb_{12}]^{2-}$ polyanion should have had a net moment as there were two Fe atoms in the nominally d^5 configuration. In an extensive study of $Ce_yFe_{4-x}Co_xSb_{12}$ solid solutions, Long et al. (1999) also pointed out that Fe in the structure is non-magnetic. However, in skutterudites where Co enters in a sub-stoichiometric amount, a possibility exists that a solid solution consists of a completely filled Fe-based skutterudite ($CeFe_4Sb_{12}$) and pure $CoSb_3$, Meissner et al. (1998), and the magnetic properties of such a fully filled $[Fe_4Sb_{12}]$ framework might be different from those of a framework filled only partially. In their Mössbauer measurements for two samples of $YbFe_4Sb_{12}$, labeled #1 and #2, Tamura et al. (2007) noted that the quadrupole splitting was slightly larger for their sample #2 than for sample #1, and the difference between the two was temperature independent. Using the experimental data of Ikeno et al. (2007) to relate the transition temperature of the weakly FM $Yb_xFe_4Sb_{12}$ to the content x of Yb (sample #1 with $x \approx 0.89$ and sample #2 with $x \approx 0.94$), they established that the vacancies in the voids tend to reduce the quadrupolar splitting by reducing the electric field gradient at the Fe nucleus. Temperature dependence of the quadrupole splitting for samples #1 and #2, together with the quadrupole splitting for $La_xFe_4Sb_{12}$ with large but unspecified x, is shown in Figure 3.19. While the temperature dependence of $La_xFe_4Sb_{12}$ can be fitted to the usual behavior of the quadrupole splitting given by Verma and Rao (1983) as

$$\Delta E_Q(T) = \Delta E_Q(0)\left[1 - \beta T^\gamma\right] \tag{3.87}$$

with $\gamma = 3/2$, the temperature dependence of both $YbFe_4Sb_{12}$ samples shows a distinct plateau below about 60 K and a linear temperature dependence at higher temperatures. The onset of the plateau coincides with the departure of the magnetic susceptibility from the Curie–Weiss behavior depicted in the inset of Figure 3.19. Because the electric field gradient at the Fe nucleus consists of two main contributions,

$$eq = (1 - R)eq_{ce} + (1 - \gamma_\infty)e(q_{Yb} + q_{Sb}). \tag{3.88}$$

Here the first term represents a contribution of the conduction electrons and the second term is a combined contribution of Yb and Sb atoms. The two terms in the brackets, $(1-R)$ and $(1-\gamma\infty)$, are

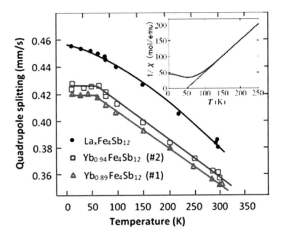

FIGURE 3.19 Temperature dependence of the quadrupole splitting for $La_xFe_4Sb_{12}$ (large but unspecified content x) and samples #1 and #2 of $YbFe_4Sb_{12}$ with Yb contents of 0.89 and 0.94, respectively. The inset shows the inverse magnetic susceptibility of $YbFe_4Sb_{12}$ plotted as a function of temperature. The straight line in the inset represents $C^{-1}(T\text{-}50)$. Adapted from I. Tamura et al., *Journal of the Physical Society of Japan* **81**, 074703 (2012) with permission from the Japanese Physical Society.

the so-called Steinheimer factors, Steinheimer (1967). Because the second term in Equation 3.88 is substantially temperature independent, the temperature variation comes from the q_{ce}. Takahashi and Moriya (1978) related q_{ce} to the magnetic susceptibility *via*

$$q_{ce}(T) = \frac{q'}{\chi(T)^{-1} + d'}, \tag{3.89}$$

where q' is related to the electric field gradient in the absence of electron–electron interaction, and d' involves the effect of exchange enhancement. The observed plateau in the quadrupole splitting below 60 K thus arises due to the T-dependent inverse susceptibility, specifically its departure from the Curie–Weiss law observed at low temperatures. Such departure has also an impact on the asymmetry of the absorption double peak, which varies with temperature below about 50 K. In their subsequent paper, Tamura et al. (2012) explained this variation as a result of the increasing valence of Yb ions located in voids that are adjacent to a void where Yb is missing. In their study, they also relied on the reports of Yamaoka et al. (2011) and Möchel et al. (2011) that indicated an increase in the average valence of Yb ions as the temperature decreased below 50 K, respectively 20 K. Saito et al. (2011) succeeded in growing single crystals of $YbFe_4Sb_{12}$ from flux under a high pressure of 4 GPa that achieved a near full filling of $x = 0.991$. The crystals were used in [121,123]Sb nuclear quadrupole resonance (NQR) measurements by Magishi et al. (2014), and the results indicated no sign of a phase transition, in contrast to measurements of Yb-deficient samples. The results confirm a strong dependence of the magnetic state of $YbFe_4Sb_{12}$ on the void occupancy, and the distinct weakening of magnetism as the full occupancy of skutterudite voids is approached. It would be of interest to see to what extent the Mössbauer spectral parameters obtained for such high-quality single crystals are modified.

The only other rare earth with a potentially variable valence is Eu. Indeed, this filler has generated considerable interest as far as the magnetic properties are concerned, and several Mössbauer spectra have been collected on a variety of Eu-filled skutterudites. Eu has mostly a divalent character, but such ions are often accompanied by a fraction of trivalent ions, especially when Eu is in contact with oxygen. First Mössbauer spectra of Eu-filled skutterudites were recorded by Gèrard et al. (1983). $EuFe_4P_{12}$ becomes FM below $T_c \sim 100$ K and has an effective magnetic moment of 6.2 μ_B/f.u., based on measurements of susceptibility. As a Mössbauer nucleus served [151]Eu and the spectra were taken from room temperature down to 4.2 K. A single absorption line above 100 K became split below the Curie temperature, and an asymmetric doublet, similar to the one shown in the lower panel of Figure 3.18, was observed. The measured isomer shift was unusually large at −6.0 ± 0.2 mm/s and reflected a high s-electron charge density at the nucleus. The low-temperature quadrupole splitting amounted to 1.2 ± 0.3 mm/s. Upon an application of an external magnetic field, the [151]Eu hyperfine field was reduced by the value of the external field, showing that the hyperfine field is oriented antiparallel to the Eu^{2+} magnetic moment. The low-temperature saturation value of the hyperfine field of −67 ± 1 T was the largest ever measured for a compound containing Eu^{2+}. The obtained effective magnetic moment of 6.2 μ_B/f.u. was low compared to the theoretical value of 7.94 μ_B expected for the Eu^{2+} ion and might be a consequence of incompletely filled voids, as also indirectly suggested by the asymmetry of the Mössbauer spectra. The fact that the effective moment is smaller than the theoretical value, and taking into account that the hyperfine field was reduced by exactly the value of the applied field, i.e., no hyperfine field was induced by the applied field, clearly indicated that the polyanion $[Fe_4P_{12}]^{2-}$ had no magnetic moment. Using, again, phosphide skutterudites, but now with the $[Ru_4P_{12}]$ framework, the Mössbauer spectra of [151]Eu were measured by Grandjean et al. (1983) and by Indoh et al. (2002). The spectra obtained by the latter authors are shown both above and below the FM transition temperature of $T_c = 17.8$ K in Figure 3.20. The spectral intensity could be fitted only by invoking three subspectra: one for a minor fraction of Eu^{3+} and two for Eu^{2+} designated as $Eu^{2+}(A)$ and $Eu^{2+}(B)$. The respective isomer shifts of these states of europium turned out to be $\delta = -13.4$ mm/s for $Eu^{2+}(A)$, $\delta = -9.9$ mm/s for $Eu^{2+}(B)$, and $\delta = +0.3$ mm/s for Eu^{3+}. Taking the

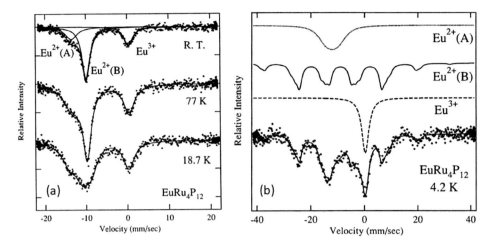

FIGURE 3.20 (a) Mössbauer spectrum of $EuRu_4P_{12}$ above T_c. Black circles are experimental data and solid lines are calculated spectra. Three subspectra are shown by solid lines in the spectrum taken at room temperature, which are used to fit the intensity of the experimental spectra. They represent an Eu^{3+} state and two kinds of Eu^{2+} states, $Eu^{2+}(A)$ and $Eu^{2+}(B)$. (b) Mössbauer spectrum at 4. 2K, again, fitted by three sets of subspectra. Because Eu^{3+} is a non-magnetic ion, there is no Zeeman splitting. Reproduced from K. Indoh et al., *Journal of the Physical Society of Japan* **71**, 243 (2002). With permission from the Physical Society of Japan.

ratio of the relative areas of the respective absorption peaks as an indication of the ratio of Eu ions in the three states, the authors obtained $Eu^{2+}(A)$: $Eu^{2+}(B)$: Eu^{3+} = 20: 56: 24. Thus, the Mössbauer spectra suggested that the overall fraction of Eu^{2+} ions is 76% while Eu^{3+} accounted for 24%. From Figure 3.20(b), it follows that $Eu^{2+}(B)$ ions are the carriers of the magnetic moment because their subspectrum clearly displays the Zeeman splitting, while no splitting is observed in the subspectra of $Eu^{2+}(A)$ and Eu^{3+}. The Eu^{3+} ion is, of course, not expected to show any Zeeman effect as its ground state is nonmagnetic ($J = 0$). Based on the above observations, one can surmise that the subspectrum of $Eu^{2+}(B)$ ions is the one that reflects intrinsic magnetic properties of $EuRu_4P_{12}$, while both $Eu^{2+}(A)$ and Eu^{3+} ions reside in the impurity phases. Because the measured saturation magnetic moment is about 4.3 μ_B at 5 T and 5 K, while theoretically one would expect $g\mu_B J = 7\ \mu_B$, this suggests that no more than $4.3/7 \approx 60\%$ of Eu ions order magnetically, close to the fraction of Eu ions (56%) that are in the $Eu^{2+}(B)$ state. The hyperfine field measured in $EuRu_4P_{12}$ at 4.2 K was -39.8 T, some 40% lower than the hyperfine field in $EuFe_4P_{12}$. As in $EuFe_4P_{12}$, the effective magnetic moment obtained from magnetic susceptibility $\mu_{eff} = 6.57\ \mu_B/f.u.$ is below the theoretical value of 7.94 μ_B for Eu^{2+}, and thus, the $[Ru_4P_{12}]$ polyanion has no magnetic moment, as expected for non-magnetic Ru.

Shifting from phosphide-based skutterudites filled with Eu to their antimonide skutterudite cousins results in an increased lattice constant while the structure is maintained. From studies of the magnetic susceptibility, $EuFe_4Sb_{12}$ orders ferromagnetically at about 84 K, but the saturated magnetic moment of 4.26 μ_B at 2 K and 6 T is inconsistent with the theoretical value of the saturated moment of 7 μ_B expected to correspond to complete alignment of magnetic moments of Eu^{2+}. This would mean that either filling is quite low or that a significant fraction of Eu ions is trivalent. To clarify this issue, Reissner et al. (2006) prepared two samples of $EuFe_4Sb_{12}$ with void filling fractions of $x = 0.88$ and $x = 0.98$, calling them samples A and B, respectively, and collected Mössbauer spectra on both between 4.2 K and room temperature. Again, the occupancy of voids by Eu is an important factor as different local environments influence the electric field gradient at the Fe nucleus and the hyperfine interaction parameters. Although the Mössbauer spectra collected for samples A and B were similar, both required two subspectra to fit the intensity profile, reflecting two different Eu configurations (one with two Eu atoms and one with less than two Eu atoms) in the next-nearest

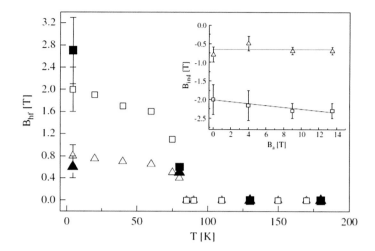

FIGURE 3.21   Temperature dependence of the hyperfine field for sample A (open symbols) and sample B (full symbols). Triangles indicate a situation where Fe has both Eu^{2+} ions as its next-nearest neighbors, while squares represent an environment of Fe where less than two Eu^{2+} ions are located at the next-nearest position. The inset displays the dependence of the induced hyperfine field B_{ind} as a function of the applied field B_a measured at 4.2 K. Triangles indicate the case where Fe has both Eu^{2+} ions at its next-nearest position, while squares designate the case where less than two Eu^{2+} ions are the next-nearest neighbors of Fe. Adapted and redrawn from M. Reissner et al., *Physica B* **378–380**, 232 (2006). With permission from Elsevier.

distance from Fe. The onset of the magnetic order below 84 K was reflected in the loss of an absorption double peak and the distinct change in the temperature dependence of the induced hyperfine field B_{ind}, see Figure 3.21, calculated as the difference between the measured hyperfine field B_{hf} and the applied field B_a. For both A and B samples, B_{ind} is negative, as expected for ferromagnetically ordered structures. However, as indicated in the inset in Figure 3.21, there is a significant difference in the magnitude of B_{ind} depending on whether Fe has two or fewer Eu^{2+} ions as its next-nearest neighbors. At full occupancy of voids (triangles in the inset), the magnitude of the induced hyperfine field is independent of the applied field and a factor of more than 2 smaller than in the case where some voids are not occupied (open squares), and the difference increases with the strength of the applied field. The existence of the induced hyperfine field suggests that Fe carries a magnetic moment. How large such a moment is depends on what one believes is the appropriate value of the magnetic susceptibility.

Taking the value of the effective magnetic moment of Danebrock et al. (1996), $\mu_{exp} = 8.4$ μ_B/f.u., implies a large moment associated with the $[Fe_4Sb_{12}]^{2-}$ polyanion of 2.6 μ_B/Fe^{3+}. On the other hand, using a smaller effective moment of 7.3 μ_B/f.u. measured by Bauer et al. (2001a), leaves nothing for the magnetic moment of Fe, unless the filling is very low or a large fraction of ions is Eu^{3+}. Thus, although Mössbauer data indicate the presence of a magnetic moment on Fe, it cannot be too large given how small the induced hyperfine field is, especially the one corresponding to the full occupancy of voids. Because the induced hyperfine field is negative and small, Reissner et al. (2008) argued that the negative sign of B_{ind} observed in the Mössbauer studies of $EuFe_4Sb_{12}$ indicates that the contributions of valence and core electrons to the hyperfine field have comparable magnitudes but opposite signs, giving rise to a small value of B_{ind}. They pointed out that rather than taking the magnetic moment of Fe^{3+} as due to a single unpaired spin in the low spin configuration of d^5 electrons (1.73 μ_B), a more realistic picture emerges when one assumes itinerant ferromagnetism with small ordered moments. Because the measured values of the saturation magnetization are low ($\mu_s = 5.1$ μ_B/Eu, Bauer et al. (2004)) compared to the expected fully saturated spins of $2[s(s+1)]^{1/2} = 7$ μ_B/Eu, it was suggested that rather than simple FM ordering their might be an element of canting of spins or a ferrimagnetic order between the Eu and Fe moments. The latter suggestion was subsequently confirmed by Krishnamurthy et al. (2007), based on a combined study of X-ray magnetic circular dichroism (XMCD) spectroscopy at Fe $L_{2,3}$ edges and Eu $M_{4,5}$ edges, X-ray absorption spectroscopy (XAS) to reveal the actual valence of Eu, and spin density calculations to show that, indeed, a ferrimagnetic state arises as the Fe $3d$ moment and Eu^{2+} $4f$ moments order magnetically with AFM coupling at high $T_c = 84$ K.

The last skutterudite I want to mention in the context of its Mössbauer effect is a Tl-filled skutterudite, first synthesized by Sales et al. (2000a). With its nominal +1 valence, Tl filling mimics that of the alkali metal-filled Na and K skutterudites, Leithe-Jasper et al. (2004). Its two versions, $Tl_xCo_{4-y}Fe_ySb_{12}$ and $Tl_xCo_4Sb_{12-z}Sn_z$, were used by Long et al. (2002) to assess the microscopic state of magnetism following some Co and Sb replaced by Fe and Sn in order to maintain the overall charge neutrality because of the low +1 valence of Tl. The first set of samples, $Tl_xCo_{4-y}Fe_ySb_{12}$, was studied with ^{57}Fe nuclei, while for the sample with the composition $Tl_xCo_4Sb_{12-z}Sn_z$, the ^{119}Sn nucleus served as the source. In the $CoSb_3$ matrix, the Tl occupancy is limited to no more than about 21% of voids. The void occupancy increases rapidly by replacing some Co with Fe. In the actual ^{57}Fe Mössbauer measurements, two Tl filling fractions were used: $Tl_{0.5}Co_{3.5}Fe_{0.5}Sb_{12}$ and $Tl_{0.8}Co_3Fe_1Sb_{12}$. The spectra were measured at 90 K and 295 K and were fit with two symmetric Lorenzian quadrupole doublets. The major subspectrum was assigned to Fe on the Co lattice while the origin of the minor subspectrum was less certain, but was not due to $FeSb_2$ or $FeSb$ impurities as these two compounds have very different hyperfine parameters. The authors speculated that the minor subspectrum might originate due to Tl vacancies in the next-nearest position to the Fe nucleus on the Co lattice. The respective probabilities that in the next-nearest position to the Fe nucleus for two, one, and none Tl atoms are 0.25, 0.50, and 0.25 for the compounds with the Tl content $x = 0.5$; and 0.64, 0.32, and 0.04 for the compound with $x = 0.8$. As always, the quadrupole splitting in the ^{57}Fe Mössbauer spectrum suggests a distortion from the cubic symmetry, and the obvious measure of distortion is a deviation of the pnicogen rings from a perfect square, i.e., the difference in edge lengths d_1 and d_2 of the ring structure. The quadrupole splitting for the Tl-filled skutterudite turned out to be considerably smaller than that expected for the actual lattice distortion. In the case of $Ce_xCo_{4-y}Fe_ySb_{12}$, based on Mössbauer measurements, Long et al. (1999) estimated the lattice and conduction electron contribution to the quadrupole splitting of 0.16 mm/s and 0.25 mm/s, respectively. Assuming that the lattice term in $Tl_xCo_{4-y}Fe_ySb_{12}$ is the same as in $Ce_xCo_{4-y}Fe_ySb_{12}$, the electronic contribution in the Tl-filled skutterudite should be much smaller at merely 0.06 mm/s. Such a large difference in the electronic contribution might possibly be due to a very different nature of conduction in the two skutterudite series, p-type in $Ce_xCo_{4-y}Fe_ySb_{12}$ and n-type in $Tl_xCo_{4-y}Fe_ySb_{12}$.

With ^{119}Sn and ^{121,123}Sb nuclei, Mössbauer spectroscopy offers an additional quadrupole splitting of about 1.15 mm/s for Sn substituted for Sb at the 24g site. Combined with a nuclear quadrupole moment of 0.064 barn, given in the Mössbauer Effect Data Index (1975), (1 barn = 10^{-28} m², and expresses a cross section for a scattering process), Long et al. (2002) estimated the electric field gradient at the Sn site of 2.8×10^{22} V/m². From the value of the quadrupole interaction of 9 mm/s measured by Kjekshus et al. (1973) on $CoSb_3$, and using a ground-state nuclear quadrupole moment of Sb of 0.28 barn, Mössbauer Effect Data Index (1975), the electric field at the Sb nuclear site of 3×10^{22} V/m² comes very close to the value for Sn, providing undisputable evidence that Sn, in this case, substitutes on the Sb sublattice.

In pure $TlFe_4Sb_{12}$, Leithe-Jasper et al. (2008) reported the presence of magnetic order below 80 K and their spin-polarized linear discriminant analysis (LDA) calculations gave a FM ground state with a moment on Fe of 0.82 μ_B. Reissner et al. (2008) measured Mössbauer spectra for a structure with a quite high Tl filling fraction of $x = 0.982$, as determined by chemical and X-ray analyses. However, the Mössbauer spectrum shown in Figure 3.22, again, consisted of two subspectra and, from the area ratio of the two components, at least 10% of voids were not occupied, in considerable discord with the chemical composition. The isomer shift for the two subspectra measured at 4.2 K was 0.22 mm/s and 0.33 mm/s, respectively. In external fields above 4.5 T, the spectra became fully

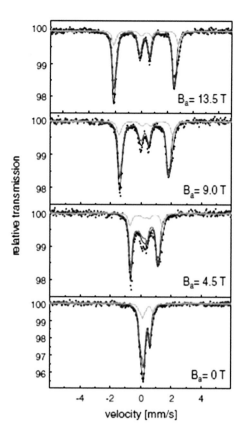

FIGURE 3.22 Mössbauer spectra of nearly fully filled $TlFe_4Sb_{12}$ measured at 4.2 K with the applied magnetic fields indicated. Reproduced from M. Reissner et al., *Hyperfine Interactions* **182**, 15 (2008). With permission from Springer Nature.

polarized (the intensity of $\Delta m = 0$ transition vanished). Similar to the Eu-filled $[Fe_4Sb_{12}]$ framework, the induced hyperfine field B_{ind} was negative, suggesting that the contribution of valence and core electrons to the hyperfine field is of comparable magnitude. However, in both cases, the induced hyperfine fields were rather small and unlikely to cause directly the large effective moments assigned to the Fe site based on measurements of magnetic susceptibility. Rather, similarly as in the case of Na- and K-filled $[Fe_4Sb_{12}]$ skutterudite frameworks, it is likely that itinerant ferromagnetism with small ordered moments is a more viable mechanism of magnetism than counting the unpaired spins of Fe^{3+} in a low spin configuration, as done originally by Danebrock et al. (1996).

To summarize Mössbauer measurements for skutterudites, they provide a very useful insight into the magnetic state of the Mössbauer nucleus, from which important conclusions can be made about the magnetic properties of a structure. However, as we are going to see in the following sections discussing magnetic susceptibility of skutterudites, not all Mössbauer results can be taken at a face value. This is especially the case of a magnetic moment residing on the $[Fe_4Sb_{12}]$ polyanion, where Mössbauer studies indicate zero moment, yet susceptibility measurements clearly document the presence of a nonzero magnetic moment as, indeed, it should be when one or two Fe atoms are in the d^5 configuration.

Readers interested in more details concerning Mössbauer measurements are referred to several excellent review articles and monographs, e.g., *Mössbauer Spectroscopy and Its Applications* by T. E. Cranshaw et al., Cambridge University Press (1985), *Mössbauer Spectroscopy Applied to Magnetism and Materials Science*, Vol.1, eds. G. J. Long and F. Grandjean, Plenum Press, N.Y. (1993), *Mössbauer Spectroscopy in Characterization of Materials* by B. Fultz, ed. E. Kaufmann, John Wiley, N.Y. (2011).

3.3.2 MAGNETIC SUSCEPTIBILITY OF BINARY SKUTTERUDITES

Table 3.2 lists room-temperature magnetic susceptibilities of most of the binary skutterudites. With the exception of NiP_3, all other binary skutterudites show a small, negative, and weakly temperature-dependent magnetic susceptibility χ. The data are uncorrected for the core diamagnetism of the constituent species that represents the dominant contribution to the diamagnetic signal. The core susceptibility is independent of the temperature, and for $CoSb_3$, its value is $\chi_{core} = -0.212 \times 10^{-6}$ emu/g. By subtracting the core diamagnetism, one obtains considerably less than one Bohr magneton ($1 \mu_B = 9.27 \times 10^{-24}$ J/T). Such small values confirm the essentially covalent bonding in the binary skutterudites. Had the bonds been of ionic character, just the "spin only" contribution to the magnetic susceptibility would yield a value of several μ_B.

Anno et al. (1998) measured the magnetic susceptibility of $CoSb_3$, including both polycrystalline and single-crystal specimens. The temperature dependence of the susceptibility indicated a gradual decrease in the magnitude of the diamagnetic signal down to about 50 K. At lower temperatures, the susceptibility started to be dominated by the presence of paramagnetic impurities, and the susceptibility was rapidly increasing, evidencing the Curie law behavior. As might be expected, polycrystalline samples of $CoSb_3$ showed a stronger paramagnetic signal than the single crystal. The data were fitted assuming three distinct contributions: the temperature independent diamagnetic susceptibility of the

TABLE 3.2
Room-Temperature Susceptibilities of Binary Skutterudites

Binary Skutterudite	Magnetic Susceptibility χ (in units of 10^{-6} emu/g)	Reference
CoP_3	−0.092	Ackermann and Wold (1977)
$CoAs_3$	−0.158	Pleass and Heyding (1962)
	−0.106	Ackermann and Wold (1977)
$CoSb_3$	−0.165	Hulliger (1961)
	−0.111	Ackermann and Wold (1977)
	−0.189	Mandrus et al. (1995)
	−0.121	Morelli et al. (1995)
	−0.200	Morelli et al. (1997)
RhP_3	−0.281	Hulliger (1961)
NiP_3	+280	Shirotani et al. (1996)[a]
$RhAs_3$	−0.214	Hulliger (1961)
	−0.219	Pleass and Heyding (1962)
$RhSb_3$	−0.203	Hulliger (1961)
$IrAs_3$	−0.251	Hulliger (1961)
	−0.253	Pleass and Heyding (1962)
	−0.210	Kjekshus and Pedersen (1961)
$IrSb_3$	−0.260	Hulliger (1961)
	−0.160	Kjekshus and Pedersen (1961)

[a] Prepared under a pressure of 4 GPa.

ion cores χ_{core}, the paramagnetic susceptibility of charge carriers χ_e (assumed proportional to their number density), and the paramagnetic susceptibility of magnetic impurities χ_{imp} (given by the Curie law)

$$\chi = \chi_{core} + \chi_e + \chi_{imp} = \chi_{core} + N_0 e^{-\frac{E_g}{2k_B T}} + C\!\!\Big/\!\!_T. \qquad (3.90)$$

Here, N_o is a constant, E_g is the band gap, and C is the Curie constant. The band gap value from the fit came out as $E_g \approx 70$–80 meV, in good agreement with the theoretical estimate of 50 meV, Singh and Pickett (1994). The fitted Curie constants for the two polycrystalline samples turned out to be 9.66×10^{-8} emuK/g and 2.36×10^{-7} emuK/g, while for the single crystal, the value was 5.19×10^{-8} emuK/g. Had impurities been identified (i.e., had their effective magnetic moments been known), one could, given the values of the Curie constant, calculate the respective concentrations of the magnetic impurities in the samples.

It is possible to partially replace Co by its immediate neighbors in the periodic table, Fe and Ni, and form limited ranges of substitutional solid solutions $Fe_x Co_{1-x} X_3$ and $Ni_x Co_{1-x} X_3$. In each case, such a substitution alters the electron count and, because Fe and Ni are magnetic, the skutterudite acquires a magnetic moment. The first study of magnetism in substituted binary skutterudites was performed by Pleass and Heyding (1962) using $CoAs_3$, where Fe can replace up to 16% of Co and Ni can substitute for 65% of Co. In a series of $Ni_x Co_{1-x} As_3$ samples, the authors observed a crossover from p-type to n-type conduction at just 1% of Ni replacing Co. Diamagnetism at room temperature persisted up to $x = 0.03$, and only at larger Ni contents, the paramagnetic response prevailed. The Fe-substituted skutterudites, $Fe_{1-x} Co_x As_3$, because of the much lower solubility limit of Fe and weaker paramagnetism, required a correction for the diamagnetic contribution in order to display clearly the paramagnetic temperature dependence. This was done simply by assuming that the diamagnetic term in the substituted skutterudite is the same as in $CoAs_3$. The effective moments per Ni atom and the Curie–Weiss temperatures θ_{CW} for $Ni_x Co_{1-x} As_3$ samples and the corresponding effective moments per Fe atom and the Curie–Weiss temperatures for $Fe_{1-x} Co_x As_3$ samples noted in the text of the paper presented in Table 3.3.

Turning to cobalt triantimonide, a skutterudite that holds considerably more appeal than $CoAs_3$ as the base thermoelectric material, there has always been keen interest to reduce its large thermal conductivity by introducing various dopant species, including Fe and Ni. Although the void radius of $CoSb_3$ is somewhat greater than that of $CoAs_3$, one might thus expect higher solubility limits; this is so only in the case of Fe but not of Ni. Already in the late 1950s, Dudkin and Abrikosov (1957),

TABLE 3.3

Effective Magnetic Moments and Curie–Weiss Temperatures per atom of Ni and Fe Substituting for Co in $CoSb_3$

NixCo$_{1-x}$As$_3$	p_{eff} (μ_B per atom)	θ_{CW} (K)
$x = 0.20$	0.804	−290
$x = 0.40$	1.24	−1200
FexCo$_{1-x}$As$_3$		
$x = 0.06$	2.06	−215
$x = 0.11$	2.71	−290
$x = 0.15$	1.88	−250

Note: Table created from the data in the text of the paper by Pleass and Heyding (1962).

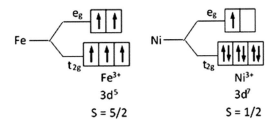

FIGURE 3.23 Trivalent electronic states of Fe and Ni in the octahedral crystalline field.

studying the effect of Fe and Ni on the transport properties of $CoSb_3$, established that Co can be replaced with up to 25% of Fe, while merely 10% of Ni can substitute for Co. The effect of Fe and Ni on the magnetic properties of $CoSb_3$ was not looked at seriously until Anno et al. (1999) reported on the temperature dependence of magnetic susceptibility in hot-pressed $Fe_xCo_{1-x}Sb_3$ and $Ni_xCo_{1-x}Sb_3$ with Fe and Ni concentrations $x = 0.03, 0.05$, and 0.1. While the primary motivation was to ascertain the effect of Fe and Ni doping on the thermal conductivity, the authors analyzed the magnetic susceptibility in the context of the Curie–Weiss law and reported the effective magnetic moments per Fe atom in the range of 1–6 μ_B, while the moments per Ni atom were close to 1.7 μ_B. Based on these results, they concluded that the electronic states of Fe and Ni in $CoSb_3$ are trivalent high spin states, d^5 for Fe with $S = 5/2$ and d^7 for Ni with $S = 1/2$, as schematically shown in Figure 3.23.

However, at least in the case of Ni, there is a major problem with a purely trivalent state of Ni. As shown in detailed transport and magnetic studies of Dyck et al. (2002), even very minor amounts of Ni in $CoSb_3$ (~ 0.1 %) dramatically increase the charge carrier concentration and alter the transport properties, implying that Ni is not a neutral impurity but donates its valence electron to the conduction band. In this scenario, Ni atoms must attain the d^6 state and become Ni^{4+}. However, just as Anno et al. (1999), Dyck et al. (2002) too observed the Curie–Weiss behavior of the magnetic susceptibility to yield the effective magnetic moment per Ni atom of around 1.73 μ_B, expected for a single unpaired electron spin of Ni, in effect implying the Ni^{3+} state. So how can one get a consistent picture of both transport and magnetic properties of Ni-doped $CoSb_3$?

Dyck et al. (2002) proposed a picture that accounts for both the transport and magnetism behavior in $Ni_xCo_{1-x}Sb_3$. It rests on a premise that, unlike the electron transport, the magnetic susceptibility is not sensitive to whether the valence electron of Ni is in the impurity state or in the conduction band. At ordinary temperatures, Ni donates its valence electron to the conduction band (Ni^{4+}, d^6 configuration). This electron is thus delocalized and drives not only a dramatic increase in the carrier concentration, but it also gives rise to "band" (Pauli) paramagnetism. At low temperatures, this conduction electron is recaptured by the Ni impurity as a localized spin with $s = \frac{1}{2}$, consistent with the effective magnetic moment per Ni atom of 1.73 μ_B obtained from the analysis of the Curie–Weiss law. The model provides a consistent explanation of both the transport properties and the Curie–Weiss behavior of the magnetic susceptibility in $Ni_xCo_{1-x}Sb_3$. The donor character of Ni is essential for tuning the carrier concentration in order to achieve optimally doped skutterudite structures.

The magnetic state of Fe in $Fe_xCo_{1-x}Sb_3$, although less controversial than that of Ni, required further clarification regarding the valence state of Fe by exploring a broader range of Fe concentrations and its effect on the transport and magnetism. This was undertaken by Yang et al. (2000) who used polycrystalline samples with the Fe content $x = 0, 0.005, 0.01, 0.02, 0.05$, and 0.1. The state of the magnetism depends on whether the Fe ion attains the zero-spin Fe^{2+} (d^6) valence or the Fe^{3+} (d^5) valence state, the latter being in either a high-spin $S = 5/2$ configuration as assumed by Anno et al. (1999) or in a low-spin $S = \frac{1}{2}$ configuration. The total magnetization measured by Yang et al. (2000) comprised three terms,

$$M_{total} = M_d + M_{PM} + M_{FM},$$ (3.91)

where M_d, M_{PM}, and M_{FM} are the diamagnetic, paramagnetic, and FM moments, respectively. The FM component was separated by performing differential magnetic susceptibility measurements, $\chi_{total} = \partial M_{total}/\partial H$ at each temperature in the field of 4 T, high enough for M_{FM} to saturate and, hence, $\partial M_{FM}/\partial H = 0$. The FM component was consistent with a very small fraction of Fe in the samples being α-Fe. The fraction was not detected in X-rays primarily because it was quite small and because its strongest X-ray diffraction peak would overlap with the strong skutterudite peak at $2\theta \approx 45°$. The authors estimated the percentage of this α-Fe phase from $M_{MF}/(2.22\,\mu_B/\text{Fe})$ where M_{MF} is the measured FM moment per Fe atom, and the magnetic saturation moment of Fe atoms with metallic bonds is 2.22 μ_B/Fe. It was found that the percentage of α-Fe decreases with the increasing Fe doping content. For instance, in the nominal $x = 0.005$ sample, α-Fe amounted to 10.5% of Fe dopants, while in the nominal $x = 0.1$ sample, the amount of α-Fe dropped down to 3.6%. The actual amount of Fe in the skutterudite phase was corrected using

$$x_{corr} = x\left(1 - \frac{M_{FM}}{2.22\,\mu_B\,/\,Fe}\right). \tag{3.92}$$

Subtracting the diamagnetic background susceptibility (taken as the susceptibility of pure $CoSb_3$ χ_{pure}) from the total measured susceptibility χ_{total}, the temperature-dependent magnetic susceptibility per Fe atom is

$$\chi = \frac{\chi_{total} - \chi_{pure}}{x_{corr}} = \chi_0' + \frac{C'}{T + \theta_{CW}'}. \tag{3.93}$$

The second equality represents the form of fitting the experimental data, where χ_0' is any remaining temperature independent susceptibility per Fe atom, C' is the Curie constant per Fe atom, and θ_{CW}' is the Curie–Weiss temperature. Isolating the Curie–Weiss term and plotting the inverse susceptibility $1/(\chi_0 - \chi_0')$ as a function of temperature, Figure 3.24, one can obtain the effective magnetic

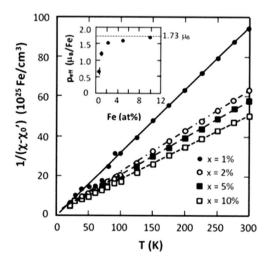

FIGURE 3.24 Inverse susceptibility of $Fe_xCo_{1-x}Sb_3$ as a function of temperature. The lines are fits to the data according to Equation 3.79. The fitted parameters are presented in Table 3.4. The inset shows the effective number of Bohr magnetons as a function of the content of Fe. The dashed line indicates the effective magnetic moment per Fe atom of 1.73 μ_B expected for Fe in a low-spin d^5 electron configuration. Adapted from the data of J. Yang et al., *Physical Review B* **63**, 014410 (2000). With permission from the American Physical Society.

TABLE 3.4

Fitting Parameters of $Fe_xCo_{1-x}Sb_3$ Skutterudites from the Plots of the Inverse Susceptibility *vs.* Temperature Displayed in Figure 3.28

$Fe_xCo_{1-x}Sb_3$	$\chi_0'E$ (10^{-27} cm³/Fe)	C' (10^{-25} cm³K/Fe)	θ_{CW} (K)	p_{eff} (μ_B/Fe)
$x = 0.005$	4.06	0.888	0	0.654
$x = 0.01$	6.79	3.14	0	1.23
$x = 0.02$	0.34	4.80	2.52	1.52
$x = 0.05$	1.65	5.25	3.54	1.59
$x = 0.1$	1.24	6.03	7.37	1.70

Source: The data from J. Yang et al., *Physical Review B* **63**, 014410 (2000). With permission from the American Physical Society.

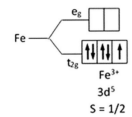

FIGURE 3.25 Trivalent electronic state of Fe in the low-spin electron configuration.

moments from the slopes and the Curie–Weiss temperature from the intercepts, as given in Table 3.4. The inset in Figure 3.24 depicts the development of the magnetic moment (in terms of the effective number of Bohr magnetons) as the concentration of Fe increases. The data indicate that the effective number of Bohr magnetons asymptotically approaches 1.73 μ_B, the number expected for a low spin d^5 configuration of electrons in Fe^{3+}, shown in Figure 3.25.

A very interesting situation is encountered in the coupled replacement of Co with Fe and Ni, i.e., in the alloys $Fe_xNi_{1-x}X_3$ with $x \approx 0.5$. The model of Dudkin predicts such structure to be a strong paramagnet because of the unpaired spin on Fe and, at the same time, a good metallic conductor because of the conduction electron on Ni. But this is not the case. Not only is such a compound semiconducting, but the magnetic susceptibility is very small. The magnetic moment per metal atom is much smaller than the values obtained with the $Fe_xCo_{1-x}X_3$ systems and, in fact, independent measurements by Nickel (1969) indicated that $Fe_{0.5}Ni_{0.5}Sb_3$ is a diamagnetic solid. It appears that an electron transfer occurs between Fe and Ni resulting in the formation of (Fe^-Ni^+) pairs effectively equivalent to Co. Mössbauer measurements discussed in section 3.3.1 confirmed that Fe has the electron configuration d^6 and thus acquired an electron from Ni.

3.3.3 Magnetic Properties of Partially Filled $CoSb_3$

By inserting filler species into the voids of the skutterudite structure, one effectively dopes the system. In the case of rare earth, the ions such as La^{3+} or Ce^{3+} donate three electrons, and this should be immediately evident on the behavior of magnetic susceptibility. From the data in Figure 3.26, the reader will appreciate a rapid development of a paramagnetic signal in Ce_yCoSb_3 that becomes

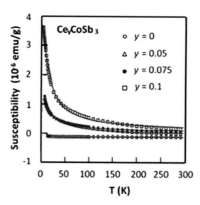

FIGURE 3.26 Temperature dependence of magnetic susceptibility of Ce_yCoSb_3. Solid lines are fits using Equation 9.94 with the following parameter values: $y = 0.05$, $C = 2.43 \times 10^{-5}$ emuK/g, $\theta = 30.2$ K, $\chi_0 = -1.57 \times 10^{-7}$ emu/g; $y = 0.075$, $C = 1.85 \times 10^{-5}$ emuK/g, $\theta = 0.2$ K, $\chi_0 = -3.64 \times 10^{-8}$ emu/g; $y = 0.1$, $C = 4.50 \times 10^{-5}$ emuK/g, $\theta = 10.6$ K, and $\chi_0 = -2.00 \times 10^{-8}$ emu/g. Adapted from C. Uher et al., *Materials Research Society Symposia Proceedings* **478**, 315 (1997). With permission of the Materials Research Society.

progressively stronger with the increasing amount of Ce ($y \approx 0.1$ is a limiting Ce occupancy of voids in $CoSb_3$). Fitting the susceptibility using the Curie–Weiss form (solid lines through the data points in Figure 3.26),

$$\chi = {C}/{(T - \theta)} + \chi_0, \tag{3.94}$$

Uher et al. (1997) determined the Curie constant for each compound, which, in turn, allowed them to deduce the effective magnetic moment per Ce atom of 2.58 μ_B, 1.84 μ_B, and 2.48 μ_B for samples with $y = 0.05$, 0.075, and 0.1, respectively. The average effective moment of 2.30 μ_B is consistent with the theoretical value of 2.54 μ_B for trivalent cerium.

3.3.4 Magnetic Properties of Filled Skutterudites with the $[T_4X_{12}]$ Framework

Turning our attention to filled skutterudites with the $[T_4X_{12}]$ framework, there have been numerous investigations of mostly phosphide- and antimonide-based compounds. Arsenides, because of difficulties with their synthesis, have been explored to a lesser extent. The study of filled skutterudites greatly benefited from the development of high-pressure/high-temperature synthesis, which facilitated not only greater filling fractions of the skutterudite voids but, even more important, was instrumental in the synthesis of skutterudites with heavier and smaller rare-earth ions (generally Gd through Lu) that could not form sufficiently strong bonding within the icosahedral pnicogen cage during the synthesis under ambient pressures.

As we have seen throughout the book, the CEF plays an important role when considering electronic levels, and therefore, magnetic properties of various rare-earth ions as they occupy the voids of the skutterudite structure. The effect of CEF is to split Hund's rule multiplet of f-electron rare-earth ions into a particular set of levels, with the ground state and the first excited state playing the most important role. As discussed in section 3.1.4.4, although skutterudites are cubic structures, the local environment of the fillers is actually tetrahedral and, instead of the O_h symmetry that is often used to describe the CEF-split states, it is more appropriate to use the T_h symmetry to identify the states. Where appropriate, I will point out the particular CEF energy scheme.

To provide an overview of the magnetic properties of a vast number of filled skutterudites, one has to decide how to group them for a presentation. I have decided to categorize them by the type of the pnicogen atom and divide them into three groups: phosphites, arsenides, and antimonides. Any important cross-relations will be pointed out.

In general, magnetism in filled skutterudites can be associated with both the $[T_4X_{12}]$ framework and magnetic filler ions. The issue is particularly relevant in structures where the framework contains Fe atoms. A question arises regarding the magnetic state of Fe and what fraction, if any, of the total magnetic moment it might account for. On the one hand, there is strong evidence for the Fe atoms in most of the phosphide and antimonide skutterudites to have a zero or vanishingly small magnetic moment. On the other hand, in arsenides there are reasons to believe that the $[Fe_4As_{12}]$ polyanion carries a magnetic moment. As far as the filler species are concerned, all rare earths, except for La, carry a magnetic moment and are usually the dominant source of magnetism in filled skutterudites.

The evidence for a non-magnetic state of Fe in phosphide and antimonide skutterudites is quite compelling. Perhaps, the most persuasive experimental evidence is the occurrence of superconductivity with a surprisingly high transition temperature of 4.6 K in $LaFe_4P_{12}$, see Table 2.1. Had Fe carried a magnetic moment, it would have surely broken the Cooper pairs *via* exchange coupling. Moreover, an essentially zero moment of Fe is also heralded by the Mössbauer measurements discussed in section 3.3.1. The bottom line is that, in rare earth-filled iron phosphide and antimonide skutterudites, there is not much evidence for a contribution of Fe to the magnetic susceptibility of the structure.

Table 3.5 lists all phosphide-based skutterudites and gives the key magnetic properties observed for each structure. Tables 3.6 and 3.7 follow with arsenide- and antimonide-based skutterudites. Skutterudites based on the $[Pt_4Ge_{12}]$ framework are presented later. Discussion of more interesting

TABLE 3.5
Relevant Parameters of *Phosphide-Based* Filled Skutterudites Collected from the Literature

Compound	a (Å)	Ground State	$T_{sc}/T_C/T_N$ (K)	μ_{eff} (μ_B/ion)	θ_{CW} (K)	γ (mJ/moleK^2)	Reference
$LaFe_4P_{12}$	7.8316	SC	4.1–4.6	–	–	52–57	1–9
$CeFe_4P_{12}$	7.7920	Semicond	–	–	–	–	1,4,5,10,11
$PrFe_4P_{12}$	8.0420	AFM	6.5[a]	3.63	−16.5	465–2700	1,3,5,12–17
$NdFe_4P_{12}$	7.8079	FM	~2	3.53	–	–	1–3,19,20
$SmFe_4P_{12}$	7.8029	FM	1.6	0.79	0.1	370	1,8,18,21–23
$EuFe_4P_{12}$	7.8055	FM	100	6.2	–	–	1,10
$GdFe_4P_{12}$	7.7950	FM	23	7.9	17.5	–	24–26
$TbFe_4P_{12}$	7.7926	FM	10	9.48	10	–	25,26,28,29
$DyFe_4P_{12}$	7.7891	FM	4, 10	10.7	0.6	–	26,30
$HoFe_4P_{12}$	7.7854	FM	5	10.4	–	–	31
$ErFe_4P_{12}$	7.7832	Metal	–	9.59	–	–	31
$TmFe_4P_{12}$	7.7802	Metal	–	6.58	–	–	31
$YbFe_4P_{12}$	7.7832	metal[b]	–	3.58	−67	200–300	31–33
$LuFe_4P_{12}$	7.7771	–	–	–	–	–	30
YFe_4P_{12}	7.7896	SC	~7	–	–	27	30,34,35
UFe_4P_{12}	7.7729	FM,semicond	3.15	2.25	–	–	4,11,36–38
$ThFe_4P_{12}$	7.7999	Metal	–	–	–	–	3,11,39–41
$NpFe_4P_{12}$	7.7702	FM	23	–	–	10	42,43
$LaRu_4P_{12}$	8.0561	SC	7.2	–	–	26–44.4	1,2,7–9,44–47

(Continued)

TABLE 3.5 (Continued)

Compound	a (Å)	Ground State	$T_{sc}/T_C/T_N$ (K)	μ_{eff} (μ_B/ion)	θ_{CW} (K)	γ (mJ/moleK^2)	Reference
$CeRu_4P_{12}$	8.0376	Semicond	–	–	–	–	1,48
$PrRu_4P_{12}$	8.0420	Semicond	–	3.84	–7	< 60	1,49–52
$NdRu_4P_{12}$	8.0364	FM	1.7	3.68	–	–	1,2,53,54
$SmRu_4P_{12}$	8.0397	AFM	16.5	–	–	13.5–28.8	8,47,53,55–57
$EuRu_4P_{12}$	8.0406	FM	17.8	7.75	18	–	1,58
$GdRu_4P_{12}$	8.0375	AFM	22	8.04	23	–	25,53,59
$TbRu_4P_{12}$	8.0338	AFM	20	9.76	8	–	26,59
$DyRu_4P_{12}$	8.0294	AFM	15	12.23	–0.1	–	26
YRu_4P_{12}	8.0298	SC	8.5	–	–	–	31
$ThRu_4P_{12}$	8.0461	–	–	–	–	–	39
$CaOs_4P_{12}$	8.084	SC	2.5	–	–	21	60
$BaOs_4P_{12}$	8.124	SC	1.8	–	–	–	61
$LaOs_4P_{12}$	8.0844	SC	1.82	–	–	20–21.6	1,7–9,11,62,63
$CeOs_4P_{12}$	8.0626	Semicond	–	–	–	–	1,48
$PrOs_4P_{12}$	8.0710	Metal	–	3.63	–17	26–56	1,62–65
$NdOs_4P_{12}$	8.0638	FM	1.15	3.03	–23		1
$SmOs_4P_{12}$	8.0752	AFM	4.5	–	–	13.5–20.3	8,18,25
$EuOs_4P_{12}$	8.0792	FM	15	7.64	–	–	25
$GdOs_4P_{12}$	8.0657	FM	5	8.54	2.9	–	25
$TbOs_4P_{12}$	8.0631	Metal	–	10.86	3	–	25,26
$DyOs_4P_{12}$	8.0601	FM	2	13.55	4.4	–	25,26
$HoOs_4P_{12}$	8.0579						25
YOs_4P_{12}	8.0615	SC	3	–	–	–	25,32

Note: SC stands for a superconducting structure, and AFM and FM for an antiferromagnetic and ferromagnetic material, respectively. T_{sc}, T_C, and T_N indicate the SC transition temperature, the Curie temperature, and the Néel temperature, respectively. θ_{CW} is the Curie–Weiss temperature, and γ is the coefficient of the electronic specific heat (Sommerfeld coefficient).

[a] Originally associated with AFM transition. Later viewed as an AFQ-ordered phase, and more recently thought of as antiferro-multipole order.

[b] Develops a minimum in resistivity near 45 K.

1. Jeitschko and Braun (1977), 2. Meisner (1981), 3. Torikachvili et al. (1987), 4. Dordevic et al. (1999), 5. Sato et al. (2000), 6. Sugawara et al. (2000), 7. DeLong and Meisner (1985), 8. Matsuhira et al. (2005a), 9. Matsuhira et al. (2009), 10. Grandjean et al. (1984), 11. Meisner et al. (1985), 12. Sugawara et al. (2002b), 13. Namiki et al. (2003a), 14. Tayama et al. (2004), 15. Sugiyama et al. (2005), 16. Aoki et al. (2002a), 17. Aoki et al. (2005b), 18. Giri et al. (2003), 19. Keller et al. (2001), 20. Nakanishi et al. (2004), 21. Takeda and Ishikawa (2003), 22. Takeda et al. (2008), 23. Konno et al. (2015), 24. Jeitschko et al. (2000), 25. Kihou et al. (2004), 26. Sekine et al. (2008a), 27. Matsunami et al. (2005), 28. Kihou et al. (2005), 29. Shirotani et al. (2006), 30. Shirotani et al. (2003b), 31. Shirotani et al. (2005), 32. Shirotani et al. (1997), 33. Yamamoto et al. (2006), 34. Cheng et al. (2013), 35. Shirotani et al. (2003a), 36. Guertin et al. (1987), 37. Nakotte et al. (1999), 38. Torikachvili et al. (1986), 39. Jeitschko and Braun (1980), 40. Takegahara and Harima (2003), 41. Khenata et al. (2007), 42. Aoki , D. et al. (2006a), 43. Tokunaga et al. (2009), 44. Shirotani et al. (1996), 45. Uchiumi et al. (1999), 46. Tsuda et al. (2006), 47. Aoki et al. (2007), 48. Shirotani et al. (1999), 49. Sekine et al. (1997), 50. Matsuhira et al. (2002a), 51. Iwasa et al. (2005a), 52. Iwasa et al. (2005b), 53. Sekine et al. (1998), 54. Masaki et al. (2008b), 55. Matsuhira et al. (2002b), 56. Yoshizawa et al. (2005), 57. Kikuchi et al. (2007a), 58. Sekine et al. (2000b), 59. Sekine et al. (2000c), 60. Kawamura et al. (2018), 61. Deminami et al. (2017), 62. Matsuhira et al. (2005b), 63. Sugawara et al. (2009), 64. Yuhasz et al. (2007), 65. Sugawara et al. (2008a).

TABLE 3.6

Relevant Parameters of *Arsenide-Based* Filled Skutterudites Collected from the Literature

Compound	a (Å)	Ground State	$T_{sc}/T_C/T_N$ (K)	μ_{eff} (μ_B/ion)	θ_{CW} (K)	γ (mJ/moleK^2)	Reference
SrFe$_4$As$_{12}$	8.351	metal	–	1.36	36	58	66
BaFe$_4$As$_{12}$	8.3975	metal	–	1.46	−57	62	67
LaFe$_4$As$_{12}$	8.3252	FM	5.2	0.96	−70 to −48	78–170	9,68–70
CeFe$_4$As$_{12}$	8.2959	semicond	–	–	–	3.3–5	68,71–73
PrFe$_4$As$_{12}$	8.3125	FM	18	3.52–3.98	–	340	68,74
NdFe$_4$As$_{12}$	8.3090	FM	14.6	4.43–4.7	−35	134	24,75
SmFe$_4$As$_{12}$	8.3003	FM[a]	39	–	–	< 170	76–78
EuFe$_4$As$_{12}$	8.3374	FM	152	6.93	46	–	79,80
GdFe$_4$As$_{12}$	8.3024	FM	56	8.09	23	–	81
TbFe$_4$As$_{12}$	8.2961	FM	38	9.76	16	–	81
SrRu$_4$As$_{12}$	8.521	metal	–	–	–	–	66
LaRu$_4$As$_{12}$	8.5081	SC	10.3	–	–	58–73	9,32,68,82,83
CeRu$_4$As$_{12}$	8.4908	semicond	–	–	–	20–26	68,71,84–86
PrRu$_4$As$_{12}$	8.4963	SC	2.4	3.30	−11	70–95	9,32,68,71, 82,83,87
NdRu$_4$As$_{12}$	8.4941	FM	2.3	3.58	−37	240	88
EuRu$_4$As$_{12}$	8.5120	metal	–	8.31	−7.4	–	79
SrOs$_4$As$_{12}$	8.561	SC	4.8	–		–	66
LaOs$_4$As$_{12}$	8.5437	SC	3.2			49	9,68,84,89–91
CeOs$_4$As$_{12}$	8.5249	semicond	–	–	–	–	68,71,92
PrOs$_4$As$_{12}$	8.5311	AFM	2.3	3.81	−26	50–1000[b]	68,90,93–95
NdOs$_4$As$_{12}$	8.5386	FM	1.1	3.66	−15	259	68,96
EuOs$_4$As$_{12}$	8.5504	FM	25	7.29	9.7	–	79
ThOs$_4$As$_{12}$	8.5183	–	–	–	–	–	39

[a] Namiki et al. (2010) claimed ferrimagnetic ordering with the itinerant magnetism of [Fe$_4$As$_{12}$] opposing the 4f-moments of Sm^{3+}.

[b] An enormous range of the electronic specific heat reflects the temperature range where it was measured. In zero field and over 10 K < T < 18 K, γ = 211 mJmol^{-1}K^{-2} (75,) while below 1.6 K, $\gamma \approx$ 1000 mJmol^{-1}K^{-2} was measured (74,75).

66. Nishine et al. (2017), 67. Sekine et al. (2015), 68. Braun and Jeitschko (1980a), 69. Tatsuoka et al. (2008), 70. Nowak et al. (2009), 71. Maple et al. (2008), 72. Wawryk et al. (2011), 73. Ogawa et al. (2014), 74. Sayles et al. (2008), 75. Higashinaka et al. (2013), 76. Kikuchi et al. (2008a), 77. Kikuchi et al. (2008b), 78. Namiki et al. (2010), 79. Sekine et al. (2009b), 80. Kawamura et al. (2015), 81. Sekine et al. (2011a), 82. Shimizu et al. (2007), 83. Namiki et al. (2007), 84. Shirotani et al. (2000), 85. Sekine et al. (2007a), 86. Baumbach et al. (2008), 87. Sayles et al. (2010), 88. Rudenko et al. (2016), 89. Ho et al. (2007), 90. Henkie et al. (2008), 91. Wawryk et al. (2008), 92. Sekine et al. (2008b), 93. Chi et al. (2008), 94. Maple et al. (2006), 95. Yuhasz et al. (2006), 96. Cichorek et al. (2014).

aspects of magnetism in various filled skutterudites is presented subsequently. Several structures have been studied multiple times, and I indicate the key references as well as a range of parameters obtained. The lattice constants quoted are mostly those determined originally by Prof. Jeitschko and his group. The lattice parameters measured subsequently by various researchers may differ slightly from the quoted values.

TABLE 3.7

Relevant Parameters of *Antimonide-Based* Filled Skutterudites Collected from the Literature

Compound	a (Å)	Ground State	$T_{sc}/T_C/T_N$ (K)	μ_{eff} (μ_B/ion)	θ_{CW} (K)	γ (mJ/moleK²)	Reference
NaFe₄Sb₁₂	9.1767	FM	85	1.6–1.8	76–88	116–145	97–99
KFe₄Sb₁₂	9.1994	FM	85	1.6–1.8	85	113–116	97–99
TlFe₄Sb₁₂	9.1973	FM	80	1.69	88	104–127	100
CaFe₄Sb₁₂	9.162	metal	–	1.52–2.6	54	109–118	97,99,101–104
SrFe₄Sb₁₂	9.178	metal	–	1.47–2.7	53	87–115	99,101–105
BaFe₄Sb₁₂	9.202	metal	–	1.5–2.8	31	98–104	99,101–105
SrRu₄Sb₁₂	9.289	metal	–	–	–	11–16	101–104
BaRu₄Sb₁₂	9.315	metal	–	–	–	9–10	101–104
SrOs₄Sb₁₂	9.322	metal	–	–	–	44–48	101–104,106
BaOs₄Sb₁₂	9.340	metal	–	–	–	43–46	101–104,106
LaFe₄Sb₁₂	9.1395	metal, s.f.	$T_{sf} \sim 50$ K	1.2–3.0	−55 to +3.4	122–195	9,45,102, 104,107–115
CeFe₄Sb₁₂	9.1350	semicond.		2.0–4.15	−124 to 1	63.8	8,102,108, 110–112, 116,117
PrFe₄Sb₁₂	9.1362	AFM, ferri	4.1–4.6	4.2–4.6	−22	300–1000	102,108,109, 118,119
NdFe₄Sb₁₂	9.1300	FM	8.6–16.5	3.86–4.5	−15 to 36	–	102,120–122
Sm$_x$Fe₄Sb₁₂	9.1300	FM	43–45	–	–	–	102,121,123
Eu$_x$Fe₄Sb₁₂	9.1650	FM or ferri	82–87	7.7–8.4	−18 to 19	85	102,121, 124–127
Yb$_x$Fe₄Sb₁₂	9.1580	metal		1.85–4.6[a]	14 to 70	100–175	99, 125, 128–133
LaRu₄Sb₁₂	9.2700	SC	3.58			37–48	9,45,108,131, 134–138
CeRu₄Sb₁₂	9.2657	non–Fermi	–	2.26–2.35	−37 to −26	187–377	108,134–136, 139
PrRu₄Sb₁₂	9.2648	SC	1.03–1.3	3.58	−11	59	108,135,136, 138–140
NdRu₄Sb₁₂	9.2642	FM	1.3	3.45	−28	–	7,135,136
SmRu₄Sb₁₂	9.2590	–	–	–	–	–	121
EuRu₄Sb₁₂	9.2824	FM	3.3–4.0	7.2–8.0	3 to 6.1	13–73	108,126,135, 136,141
LaOs₄Sb₁₂	9.3029	SC	0.74	–	–	36–57	9,108,131, 142–148
CeOs₄Sb₁₂	9.3011	semicond.	–	1.88	6	92–180	108,144,149
PrOs₄Sb₁₂	9.2994	SC	1.85	2.97	−16	310–750	17,107,108,142, 150–154
NdOs₄Sb₁₂	9.2989	FM	0.9	3.84	−43	520	108,148,155
SmOs₄Sb₁₂	9.3009	FM	3	0.63	−0.99	820–880	108,156,157

EuOs$_4$Sb$_{12}$	9.3187	FM	9	7.3	8	135	108,126
YbOs$_4$Sb$_{12}$	9.3160	metal	–	–	–	37	158,159

Note: SC stands for a superconducting structure, and AFM and FM for an antiferromagnetic and ferromagnetic material, respectively. T_{sc}, T_C and T_N indicate the SC transition temperature, the Curie temperature, and the Néel temperature, respectively. Spin fluctuations are designated as s.f. θ_{CW} is the Curie–Weiss temperature, and γ is the coefficient of the electronic specific heat (Sommerfeld coefficient).

[a] Wide range of μ_{eff} and θ_{CW} values reflects incomplete filling by Yb and its intermittent valence. In this case, the effective moments μ_{eff} should be taken as per formula unit rather than per ion.

97. Leithe-Jasper et al. (2003), 98. Leithe-Jasper et al. (2004), 99. Schnelle et al. (2008), 100. Leithe-Jasper et al. (2008), 101. Evers et al. (1994), 102. Danebrock et al. (1996), 103. Matsuoka et al. (2005), 104. Takabatake et al. (2006), 105. Matsuoka et al. (2006), 106. Narazu et al. (2008), 107. Bauer et al. (2002c), 108. Braun and Jeitschko (1980b), 109. Tanaka et al. (2007), 110. Mori et al. (2007), 111. Viennois et al. (2003), 112. Viennois et al. (2004), 113. Viennois et al. (2005), 114. Yamada et al. (2007), 115. Ravot et al. (2001), 116. Gajewski et al. (1998), 117. Morelli & Meisner (1995), 118. Bauer et al. (2002a), 119. Butch et al. (2005), 120. Bauer et al. (2002b), 121. Evers et al. (1995), 122. Ikeno et al. (2008), 123. Ueda et al. (2008), 124. Bauer et al. (2001b), 125. Dilley et al. (1998), 126. Bauer et al. (2004), 127. Krishnamurthy et al. (2007), 128. Ikeno et al. (2007), 129. Leithe-Jasper et al. (1999), 130. Schnelle et al. (2005), 131. Tamura et al. (2006), 132. Saito et al. (2011), 133. Möchel et al. (2011), 134. Abe et al. (2002), 135. Takeda & Ishikawa (2000a), 136. Takeda & Ishikawa (2001), 137. Yogi et al. (2003), 138. Adroja et al. (2005), 139. Bauer et al. (2001c), 140. Matsuda et al. (2002), 141. Sugawara et al. (2008b), 142. Sugawara et al. (2002c), 143. Aoki et al. (2005a), 144. Bauer et al. (2001d), 145. Maple et al. (2003a), 146. Maple et al. (2003b), 147. Nakai et al. (2008), 148. Sugawara et al. (2005a), 149. Namiki et al. (2003b), 150. Aoki et al. (2002b), 151. Aoki et al. (2003), 152. Tayama et al. (2003), 153. Vollmer et al. (2003), 154. Kuwahara et al. (2005), 155. Ho et al. (2005), 156. Sanada et al. (2005), 157. Yuhasz et al. (2005), 158. Kaiser & Jeitschko (1999), 159. Kunitoshi et al. (2016).

A comment is in order regarding values of the effective magnetic moment μ_{eff} entered in the tables in units of μ_B per ion. In non-magnetic frameworks of [Ru$_4$X$_{12}$] and [Os$_4$X$_{12}$], μ_{eff} indicates the effective moment of the filler ion. In frameworks containing Fe, i.e., [Fe$_4$X$_{12}$], the magnetic moment of the filled skutterudite can, in principle, comprise contributions of both the filler ion (provided it is magnetic) and the moment associated with Fe. In that case, the entry under μ_B/ion should be understood as μ_B/f.u., because rarely it is specified how large are the individual moments due to the filler and due to Fe in the framework. However, it should be clear that in the case of trivalent fillers, there is only one of the four Fe atoms uncompensated, while in the case of divalent or monovalent fillers there are two or three Fe atoms, respectively, left uncompensated thus contributing to the overall magnetic moment.

In assembling relevant references, I found very useful an excellent review by Sato et al. (2009) on magnetic properties of filled skutterudites. In particular, this concerns the work of Japanese scientists who, throughout the mid-2000s, were exceptionally prolific and contributed greatly to the understanding of SC and magnetic properties of skutterudites.

Most of the early rare earth-filled skutterudites were prepared as single crystals by a growth from the melt. All skutterudites with heavier and smaller fillers beyond Gd were synthesized by the high-pressure/high-temperature technique.

3.3.4.1 Lanthanum-Filled Skutterudites

Lanthanum with its electron configuration [Xe]5d^{1}6s^2 has no f-electrons, which makes its band structure less complex. La enters the skutterudite void as a trivalent ion La^{3+} and thus cannot saturate the [T$_4$X$_{12}$]$^{4-}$ framework. Consequently, LaT$_4$X$_{12}$-filled skutterudites are metals with positive Hall and Seebeck coefficients. As discussed in Section 3.3.1, Mössbauer studies indicated that the framework [Fe$_4$P$_{12}$] carries no magnetic moment and LaFe$_4$P$_{12}$ is thus a Pauli paramagnet. With no magnetic moment in the structure, it is no surprise that LaFe$_4$P$_{12}$ becomes a rather ordinary BCS-type s-wave superconductor, see Chapter 2. The transition temperature of 4.1 K measured originally by Meisner (1981) was subsequently raised to 4.6 K in measurements of Sato et al. (2000). In both cases, the crystals were flux-grown from the melt of Sn. To have no magnetic moment, in spite of having a large Fe content (24 at%) in the structure, Fe must be in a divalent spin-paired d^6 configuration, as documented by the value of the isomer shift, Shenoy et al. (1982). In any case, according to calculations of Harima and Takegahara (2003), Fe $3d$ orbitals in $LaFe_4P_{12}$ make only a minor contribution to the DOS at the Fermi level. As indicated by the data in Tables 3.5–3.7, with the exception of $LaFe_4Sb_{12}$ and $LaFe_4As_{12}$, all other La-filled skutterudites turn SC at low enough temperatures. In $LaFe_4Sb_{12}$, Long et al. (1999) documented by their Mössbauer measurements that the polyanion $[Fe_4Sb_{12}]$, just like the $[Fe_4P_{12}]$ framework, has no magnetic moment. It is interesting to note that of all La-filled skutterudites, only one, $LaFe_4As_{12}$, is able to develop a full-fledged ordering, albeit not a very strong one, Tatsuoka et al. (2008). A FM transition is clearly revealed as a kink near 5.2 K in a temperature-dependent resistivity as well as in a rapidly rising magnetization measured in a low field of 0.01 T, as shown in Figure 3.27. With a rather modest effective magnetic moment $\mu_{eff} =$ 0.96 μ_B/Fe, but a very small saturated moment of $\mu_s = 0.03$ μ_B/Fe, the Rhodes–Wohlfarth ratio μ_{eff}/μ_s, which expresses the extent to which the magnetism has either an itinerant character (the ratio much larger than 1) or localized character (the ratio equal or close to 1), is large at $\mu_{eff}/\mu_s = 16$, suggesting weak itinerant electron ferromagnetism, Rhodes and Wohlfarth (1963).

Subsequent measurements by Nowak et al. (2009) placed the FM transition at a somewhat lower temperature of 3.8 K with the same effective moment of 0.96 μ_B/Fe, but a smaller saturated moment of 0.015 μ_B/Fe, increasing the Rhodes–Wohlfarth ratio to 32. Because, according to Harima and Takegahara (2003), the DOS at the Fermi level increases on going from $LaFe_4P_{12}$ (13.97 states/eV) to $LaFe_4As_{12}$ (18.16 states/eV) to $LaFe_4Sb_{12}$ (25.21 states/eV), it is surprising that, following ferromagnetism in $LaFe_4As_{12}$, the La-filled antimonide structure does not develop a magnetically ordered state. A possible clue why this is so comes from the temperature dependence of the resistivity. While $LaFe_4P_{12}$ shows a T^2 temperature variation, indicating an ordinary Fermi liquid behavior, both $LaFe_4As_{12}$ and $LaFe_4Sb_{12}$ follow the $T^{5/3}$ dependence, reflecting a contribution (and perhaps even dominance) of spin fluctuations. In $LaFe_4As_{12}$, this particular power-law dependence is not strong enough to destroy the FM ordering. However, in $LaFe_4Sb_{12}$, the coefficient of the power-law dependence in the electrical resistivity is several times larger, and such strong spin fluctuations prevent the development of a robust magnetic order. Whether the spin fluctuations in $LaFe_4Sb_{12}$ are of the FM or AFM nature has been discussed for some time. As pointed out by Nowak et al. (2009), negative values of θ_{CW} cannot be used as decisive evidence for the presence

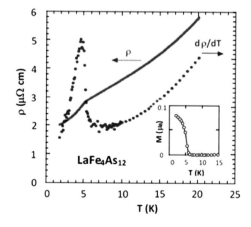

FIGURE 3.27 Temperature dependence of the electrical resistivity $\rho(T)$ and its derivative $d\rho/dT$ for a polycrystalline sample of $LaFe_4As_{12}$. The inset shows the temperature dependence of magnetization taken in the field of 0.01 T with a sharp upturn near the FM transition at 5.2 K. Adapted from S. Tatsuoka et al., *Physica B* **404**, 2912 (2009). With permission from Elsevier.

of AFM fluctuations. In fact, their ^{139}La Knight shift and ^{75}As relaxation rate $1/T_1$ do not follow a $T^{1/2}$ dependence expected for AFM spin fluctuations. Rather, the temperature trend in these two microscopic parameters is more in line with the self-consistent renormalization (SCR) theory of spin fluctuations for weak itinerant ferromagnets developed by Moriya and Takimoto (1995). According to this theory, in systems, such as LaFe$_4$Sb$_{12}$, where large departures from linearity are observed in the temperature dependence of inverse susceptibility on account of strong spin fluctuations, the usual meaning of the θ_{CW} as the ordering temperature is inappropriate because the system, after all, never orders magnetically. Instead, the extracted value of θ_{CW} is more symptomatic of the spin fluctuation temperature T_{sf}, in this case equal to about 50 K. At temperatures below T_{sf}, the susceptibility should scale as $T^{-\eta}$ with an exponent that reflects the dimensionality of the system and the nature of spin fluctuations, FM or AFM. For bulk structures ($d = 3$) and FM fluctuations, $\eta = 4/3$, while for AFM fluctuations, $\eta = 3/2$. A least-squares fit to the low-temperature susceptibility of LaFe$_4$Sb$_{12}$ by Viennois et al. (2005) yielded an exponent $\eta = 1.35 \approx 4/3$, clearly distinct from 3/2. Thus, this measurement, too, favored more FM fluctuations over AFM ones. In comparison to LaFe$_4$P$_{12}$, specific heat measurements of LaFe$_4$As$_{12}$, and even more so of LaFe$_4$Sb$_{12}$, show enhanced Sommerfeld constant γ (coefficient of the electronic specific heat). This means that both LaFe$_4$As$_{12}$ and LaFe$_4$Sb$_{12}$ skutterudites possess aspects of non-Fermi liquid (NFL) nature, which classify them as moderately heavy fermion systems. In conjunction with the presence of strong spin fluctuations, Viennois et al. (2005) placed them into a category of structures being close to the FM quantum critical point (QCP). Magishi et al. (2012) made ^{121,123}Sb-nuclear quadrupole resonance measuements on LaFe$_4$Sb$_{12}$ samples synthesized under high pressure and compared them to measurements on samples of the same chemical composition but synthesized at ambient pressure. The idea was to investigate the effect of void filling on the magnetic properties. The data revealed a significantly sharper NQR spectra on samples prepared under high pressure, where the filling was close to 100% and the electronic states around Sb nuclei became uniform. High-pressure samples also possessed a smaller Weiss temperature, suggesting that as the filling increases, the system becomes close to the ferromagnetic instability.

3.3.4.2 Ce-Filled Skutterudites

As a filler in the skutterudite structure, Ce can, in principle, pose as a nonmagnetic Ce^{4+} ion with no $4f$ electrons or as a magnetic Ce^{3+} ion possessing one $4f$ electron. Nearly all Ce-filled skutterudites are semiconductors, a rather rare case among filled skutterudites. The fact that the Ce-filled skutterudites, and phosphides and arsenides in particular, have notably smaller lattice parameters than expected from trivalent lanthanide contraction led to early speculations that Ce enters the voids as a tetravalent rather than trivalent ion. Experimental support for the Ce^{4+} state of cerium came from magnetization measurements of CeRu$_4$P$_{12}$ by Shirotani et al. (1999) that yielded a saturated moment of only 0.15 μ_B/Ce at 5 T. This was nowhere near the value of 2.54 μ_B/Ce expected for the Ce^{3+}-free ion. On the other hand, the magnetic susceptibility was weakly paramagnetic, which conflicted with the strictly tetravalent, and therefore non-magnetic Ce ion that should perfectly saturate [Fe$_4$P$_{12}$]$^{4-}$ and [Ru$_4$P$_{12}$]$^{4-}$ and result in the diamagnetic response, just like the isovalent CeP$_3$ binary skutterudite. Indirect evidence for the tetravalent Ce came also from two sets of ^{31}P-nuclear magnetic resonance (NMR) measurements made by Fujiwara et al. (2000) and Magishi et al. (2006) on CeFe$_4$P$_{12}$ and CeRu$_4$P$_{12}$, where the Knight shift was observed essentially temperature independent and with a near zero magnitude, implying that Ce is non-magnetic and, hence, in the form of Ce^{4+}. The two sets of data also agreed on the fact that the nuclear spin-lattice relaxation time T_1 at 300 K is much longer in both CeFe$_4$P$_{12}$ and CeRu$_4$P$_{12}$ than in LaFe$_4$P$_{12}$, the latter structure having no f-electrons. Consequently, it is unlikely that spin fluctuations are an important relaxation mechanism in either CeFe$_4$P$_{12}$ or CeRu$_4$P$_{12}$. However, attempting to resolve the valence state of Ce in CeFe$_4$P$_{12}$, Xue et al. (1994), and in CeRu$_4$P$_{12}$, Lee et al. (1999), X-ray absorption near edge structure spectroscopy (XANES) measurements indicated the dominance of the trivalent Ce state with some contribution of Ce^{4+}. It thus appears that Ce has a dynamically intermediate valence in these skutterudites, sometimes showing up in a more trivalent form while at other times the Ce^{4+} character prevails.

For a trivalent Ce ion, Lea et al. (1962) prescribed that the six-fold degenerate $J = 5/2$ Hunds' rule multiplet is split into a Γ_7 (Γ_5^-) doublet and a Γ_8 (Γ_{67}^-) quartet, with the states in the brackets expressed in the local T_h symmetry. The band gap, a necessary requirement to underpin the semiconducting nature of conduction in $CeFe_4P_{12}$, was postulated by Meisner et al. (1985) to arise from hybridization of itinerant f electrons with the conduction electrons, the so-called c-f hybridization. In the jargon of solid-state physics, semiconductors with the band gap formed by this process are frequently referred to as hybridization gap semiconductors or, occasionally, as Kondo insulators or Kondo semiconductors. Readers interested in the concept of hybridization gap semiconductors might find it useful to read a review by P. S. Riseborough in *Advances in Physics* **49**, 257 (2000).

As the filler ion in the center of the skutterudite void is rather far apart from Sb atoms and from the transition metal atoms of the framework, one would normally expect a rather weak hybridization of orbitals. However, the fact that the coordination number of the filler is 12 pnicogens and eight transition metals makes it very plausible that the hybridization in filled skutterudites is actually quite strong, Martins et al. (1994). The idea was supported by the band structure calculations of Nordström and Singh (1996) and of Khenata et al. (2007), who showed that Ce $4f$ states hybridize with $3d$, $4d$, and $5d$ states of Fe, Ru, and Os, respectively, and with phosphorus p-states, opening the gap near the Fermi energy. To look more closely into the evolution of the semiconducting state of $CeRu_4P_{12}$ and to document the presumed metal–insulator (M–I) transition that had to materialize at some particular value of x in solid solutions of $La_{1-x}Ce_xRu_4P_{12}$, Shirotani et al. (1999) carried out detailed measurements of the electrical resistivity and magnetic susceptibility across a full range of Ce compositions from $x = 0$ to $x = 1$. The crossover from the metallic to the semiconducting domain of conduction was observed near the Ce content $x = 0.6$. Interestingly, solid solutions with $x = 0.5$, 0.6, and 0.7 possessed a small magnetic moment, reflecting the presence of $4f$ electrons associated with the Ce^{3+} state. The magnetic moment, however, rapidly disappeared as the content of Ce exceeded 70%. Surprisingly, measurements performed a few years later by Sugawara et al. (2006) on solid solutions of $La_{1-x}Ce_xFe_4P_{12}$, with Fe instead of Ru, revealed that the band gap is totally suppressed by merely 1 at% of La on the site of Ce. The feeble nature of the semiconducting state of $CeFe_4P_{12}$ compared to that of $CeRu_4P_{12}$ is the only notable difference between the two skutterudites; otherwise, their properties are very similar. Considering $CeOs_4P_{12}$, its band gap is even smaller, reflecting the general trend of the band gaps being inversely proportional to the lattice constant, Sugawara et al. (2005b). The physical properties are, however, very similar to other two Ce-filled phosphide skutterudites. The band gaps of various semiconducting Ce-filled skutterudites are collected in Table 3.8.

Rather sparse data are available regarding Ce-filled arsenides, and definitive assessments are further complicated by a strong dependence of the physical parameters on whether one measures polycrystalline or single-crystal samples. In the case of $CeFe_4As_{12}$, the early study by Grandjean et al. (1984), using polycrystalline samples, indicated a purely semiconducting behavior at all temperatures, from which a small band gap of 10 meV was extracted. More recent measurements by Maple et al. (2008) on a single crystal showed a metallic dependence down to near 150 K, at which point the resistivity started to increase as the temperature decreased, suggesting that the band gap was likely even smaller. A more glaring disparity in the transport behavior of polycrystalline and single-crystal samples was observed in $CeRu_4As_{12}$. Here, a polycrystal measured by Sekine et al. (2007a) behaved as a semiconductor with a small band gap of 4.3 meV, while a single crystal measured by Maple et al. (2008) displayed metallic conduction and revealed a spectrum of NFL characteristics in its properties, summarized in Figure 3.28. Particularly notable is the correlated electron behavior revealed by the robust $T^{1.4}$ temperature dependence of the electrical resistivity, which extends over two decades of temperature from 65 mK to 3.5 K. The other NFL aspects are the divergent nature of the specific heat and the unusual temperature dependence of the magnetic susceptibility between 1.9 K and 10 K that can be fitted by either a power law $\chi(T) \propto T^{-m}$ with $m = 0.49$ or a logarithmic $\chi(T) \propto -lnT$ dependence.

Regarding $CeOs_4As_{12}$, here the discord between the data obtained on polycrystals and single crystals is not as dramatic, and both forms of the skutterudite show a clear semiconducting trend,

TABLE 3.8
Band Gap Values (in eV) of Ce-Filled Skutterudites Determined by Various Experimental Techniques

Skutterudite	Resistivity	Optical	Inelast. nucl. scatt.	NMR	Reference
$CeFe_4P_{12}$	0.13	0.23	–	0.1	1–5
$CeRu_4P_{12}$	0.086	0.2	–	0.078	4,6,7
$CeOs_4P_{12}$	0.034	0.3	–	–	6,7
$CeFe_4As_{12}$	0.01	–	–	–	1,8
$CeRu_4As_{12}$	0.0043, 0[a]	0.1	–	–	7–9
$CeOs_4As_{12}$	0.0047, 0.0056[a]	0.05	–	–	7,10
$CeFe_4Sb_{12}$	0	0.003	0.04	–	1,7,11–13
$CeRu_4Sb_{12}$	0	0.047	0.03	0.03	7,14–18
$CeOs_4Sb_{12}$	0.0009	0.03	0.05	0.025	7,19–24

Source: The Table is constructed from the data of H. Sato et al., *Handbook of Magnetic Materials*, Vol. 18, Elsevier (2009). With permission from Elsevier.

[a] Single crystalline specimen.

1. Grandjean et al. (1984), 2. Meisner et al. (1985), 3. Sato et al. (2000), 4. Magishi et al. (2006), 5. Matsunami et al. (2008a), 6. Shirotani et al. (1999), 7. Matsunami et al. (2008b), 8. Maple et al. (2008), 9. Sekine et al. (2007a), 10. Sekine et al. (2008a), 11. Magishi et al. (2007b), 12. Mori et al. (2007), 13. Viennois et al. (2007), 14. Takeda & Ishikawa (2000a), 15. Takeda & Ishikawa (2001), 16. Dordevic et al. (2001), 17. Adroja et al. (2003), 18. Yogi et al. (2008), 19. Bauer et al. (2001c), 20. Matsunami et al. (2003), 21. Adroja et al. (2007), 22. Matsunami et al. (2009), 23. Nakanishi et al. (2007), 24. Yogi et al. (2005).

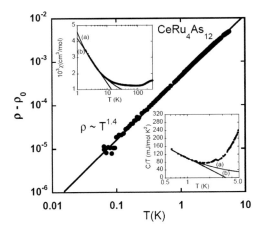

FIGURE 3.28 NFL aspects of $CeRu_4As_{12}$. The main panel depicts the highly unusual $T^{1.4}$ power law temperature dependence of resistivity extending over nearly two decades of temperature. The upper inset shows the temperature dependence of the magnetic susceptibility that at low temperatures can be fitted by either a power law function (curve a) or by a logarithmic function (curve b). The lower inset shows an upturn in the low-temperature specific heat C(T)/T that conforms to either a weak power law dependence (curve a) or a logarithmic dependence (curve b). Reproduced from a paper by M. B. Maple et al., *Journal of the Physical Society of Japan* **77**, Suppl. A, 7 (2008). With permission from the Physical Society of Japan.

albeit with different values of the band gap, 4.7 meV *vs.* 5.6 meV for a polycrystal, Sekine et al. (2008b), and single crystal, Maple et al. (2008), respectively. All three Ce-filled arsenides possess small and weakly temperature-dependent magnetic susceptibilities with room temperature values of $(1.3–1.4) \times 10^{-3}$ emu/mole.

Ce-filled antimonides, particularly those with the $[Fe_4Sb_{12}]$ framework, are among the most thoroughly explored skutterudites, primarily because their various substitutional forms and solid solutions have been considered to be prospective compositions for efficient thermoelectric materials for power generation applications, e.g., Sales et al. (1996), Uher (2001). A detailed discussion of their transport properties can be found in the study by Uher (2021). Here, I only mention studies related to their magnetic properties.

The first thing to note about Ce-filled antimonides is that there is no clear deviation from the lanthanide contraction, and thus, there is no reason to believe that the valence state of Ce is anything else but Ce^{3+}, at least at high temperatures. However, as the temperature decreases, there are indications, based on the evaluation of the temperature dependence of the magnetic susceptibility and the carrier concentration, Chen et al. (1997), that the average Ce valence deviates from 3+ toward 4+, particularly at liquid helium temperatures. To maintain the trivalent state and yet be a semiconductor implies that hybridization between the conduction electrons and the localized $4f$ electrons must be effective in CeT_4Sb_{12}. While low-temperature properties of Ce-filled phosphide skutterudites are dominated by strong c-f hybridization that leads to their Kondo semiconducting behavior, in antimonides, the c-f hybridization is weaker but still important in $CeRu_4Sb_{12}$ and $CeOs_4Sb_{12}$.

I should stress again that the physical properties, and magnetism in particular, are sensitive to the amount of the filler in the skutterudite voids. We shall see it in a number of situations and one of them is $CeFe_4Sb_{12}$.

From the perspective of magnetism, $CeFe_4Sb_{12}$ draws attention because both Ce and Fe can contribute to the magnetic state of the structure, in spite of Mössbauer measurements suggesting that the $[Fe_4Sb_{12}]$ polyanion has no moment, Kitagawa et al. (1998), Leithe-Jasper et al. (1999), and Long et al. (1999), as discussed in section 3.3.1. This is obvious from measurements of the magnetic susceptibility where, above 200 K, the Curie–Weiss behavior yields an effective magnetic moment $\mu_{eff} = 3.36 – 4.15 \, \mu_B$, depending on the level of impurities, the void occupancy, and the temperature range over which the Curie–Weiss behavior is assumed to be valid in different reports, Meisner and Morelli (1995), Gajewski et al. (1998), Ravot et al. (2001), and Viennois et al. (2004), (2007). In all cases, the effective moment is considerably larger than the free-ion value of Ce^{3+} of $2.54 \, \mu_B$, implying a significant magnetic contribution from the $[Fe_4Sb_{12}]$ polyanion. However, rather than being of localized ionic origin, Ravot et al. (2001) opined that the contribution of Fe has its origin in the itinerant form of magnetism. This is based primarily on measurements of magnetic susceptibility of $LaFe_4Sb_{12}$, where the Curie–Weiss law, holding well between 50 K and 300 K, results in $\mu_{eff} = 2.23 \, \mu_B$. Because La is non-magnetic, the entire moment must be associated with Fe. However, the ionic model predicts merely $1.73 \, \mu_B$, i.e., much less than the measured value.

Magishi et al. (2007b) explored low-temperature magnetic properties of $CeFe_4Sb_{12}$ by measurements of the electrical resistivity and ^{121,123}Sb-NQR. Their study was one of few paying attention to vacancies in the voids, i.e., to less than 100% void occupancy. Their polycrystalline samples had an occupancy of 87% as determined by electron probe microanalysis. Overall, the electrical resistivity was metallic with a gentle slope from room temperature down to about 100 K, followed by a much steeper decrease of resistivity at lower temperatures. Below 1 K, the resistivity became constant with no signs of superconductivity or any magnetic transition down to 0.3 K. Although the nuclear spin-lattice relaxation rate $1/T_1$ of $CeFe_4Sb_{12}$ and $LaFe_4Sb_{12}$ was identical at 200 K, an exponentially decreasing rate in $CeFe_4Sb_{12}$ led to its more than an order of magnitude smaller value at low temperatures, revealing a gap of $\Delta/k_B = 190$ K. Below 30 K, the relaxation rate became proportional to temperature, satisfying the Korringa relation $T_1T = $ const., expected for a metallic state. This suggested that some finite DOS exists within the gap. The gap value of 190 K was considerably smaller than the CEF energy splitting between the Γ_5^- ground state and the Γ_{67}^- first excited state, adding to a general assessment that the CEF splitting in the case of CeT_4Sb_{12} does not influence the low-temperature properties. Rather, the dominant role in the low-temperature properties of Ce-filled skutterudites is played by the hybridization gap.

In one of the rare reports where all skutterudites forms with a given filler ion were explored, Matsunami et al. (2008b) measured optical conductivity of CeT_4X_{12}, (T = Fe, Ru, and Os; X = P, As, and Sb) to systematically evaluate the formation of a gap in the DOS near the Fermi energy. The optical conductivity data as a function of photon energy are displayed in Figure 3.29. The first thing to note is a close similarity of the spectra for skutterudites having the same pnicogen atom, and a considerably weaker influence of the transition metal T. One also notes a series of peaks in a low-energy region of the spectra that arise due to infrared optical phonons. In the upper energy range of 1–3 eV, a pronounced peak reflects an interband transition from the top of the valence band formed mostly by d-states of transition metals and p-states of pnicogens to the $4f$-state of Ce just above the Fermi energy. Its considerable oscillator strength in the visible range of spectrum attests to the fact that all skutterudites have surprisingly high reflectivity, more like what one would expect of a metallic surface. Focusing on the band gap, the onset of optical conductivity in Ce-filled phosphide skutterudites sets in near 0.2–0.3 eV, as marked by the arrows. One also notes that as the temperature increases, the spectral weight below the onset point increases, reflecting some filling of states in the gap. In fact, in the case of $CeRu_4P_{12}$, and even more so in $CeOs_4P_{12}$, at high temperatures the onset of conduction is completely blurred. Ce-filled phosphides are clearly semiconductors. The gap values in the range of 0.2–0.3 eV are larger than the gaps measured by dc electrical resistivity. Because the optical dipole-allowed transitions take place between states having the same k-vector, the onsets of conduction in Figure 3.29(a) presumably mark the direct semiconducting gaps, while the smaller gaps from the resistivity indicate indirect gap values. In Ce-filled arsenides, Figure 3.29(b), the onset of conduction sets in at a much lower energy than in phosphides, implying significantly smaller direct gaps. Nevertheless, the optical gaps are, again, larger than the gaps determined from the electrical resistivity. However, prior to the onset, the spectral weight in the Ce-filled arsenides is significant, suggesting a small density of carriers even at 8 K. In Ce-filled antimonides, Figure 3.29(c), the spectra are entirely different, particularly those for $CeRu_4Sb_{12}$ and $CeOs_4Sb_{12}$, where a distinct mid-infrared peak develops that was interpreted as a measure of the hybridization gap. However, the pseudo-gap in the optical conductivity calculated based on the actual band structure of $CeRu_4Sb_{12}$ by Mutou and Saso (2004) was explained as arising not due to c-f hybridization, but rather as a gap between the pnicogen p-state-derived valence band and the conduction band formed predominantly by $4f$-states of Ce. Thus, the authors viewed $CeRu_4Sb_{12}$ as a band semiconductor rather than a heavy fermion metal. I just want to point out that less than complete filling of voids, so notorious in Ce-filled skutterudites, might influence the overall assessment.

Resistivity measurements of $CeRu_4Sb_{12}$ carried out by Takeda and Ishikawa (2000b) on polycrystalline samples and by Bauer et al. (2001c) on a single crystal show a metallic structure that exhibits a NFL behavior at low temperatures. This is concluded based on the observed $\rho(T) \sim T^{1.4}$ temperature dependence at low temperatures using a single crystal, and the exponents in the range of 1.6–1.7 observed with polycrystalline samples. Both sets of measurements thus clearly differ from the quadratic temperature dependence expected for a Fermi liquid conductor. The exponents of 1.6–1.7 are very close to 5/3, an exponent supposed to characterize FM critical fluctuations based on the SCR theory of Moriya and Takimoto (1995). $CeRu_4Sb_{12}$ was thus presumed to be close to a FM instability. Remarkably, applying a magnetic field or substituting a small amount of La for Ce, Takeda and Ishikawa (2001), tended to destroy the NFL character of $CeRu_4Sb_{12}$ at low temperatures and the skutterudite reverted to an ordinary regime of the Fermi-liquid behavior. Only one near-spherical Fermi surface sheet was detected in de Haas-van Alphen, Sugawara et al. (2002a), and Shubnikov-de Haas, Abe et al. (2002), measurements, yielding the cyclotron effective mass in the range (4.8–5.8) m_e, depending on the crystal orientation. Considering that the electronic specific heat has a magnitude $\gamma_e \sim 100$ mJK^{-2}mol^{-1}, the cyclotron mass was grossly inadequate and begged

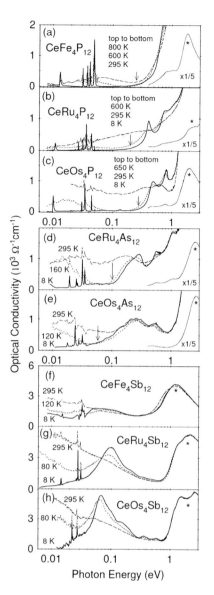

FIGURE 3.29 Optical conductivity spectra of (a) Ce-filled phosphide skutterudites, (b) Ce-filled arsenide skutterudites, and (c) Ce-filled antimonides obtained from reflectivity measurements on polished surfaces at various temperatures. Arrows indicate the onset of optical conductivity. Adapted from M. Matsunami et al., *Journal of the Physical Society of Japan* **77**, Suppl. A, 315 (2008). With permission from the Physical Society of Japan.

a question whether another Fermi sheet existed with a mass on the order of 50 m_e to back up the electronic specific heat.

An opening of the hybridization gap may affect both charge and spin aspects of the charge carriers. Optical studies by Dilley et al. (2000) and Dordevic et al. (2001) established the gap in the charge degree of freedom of $CeRu_4Sb_{12}$ of 47.1 meV, with the effective carrier mass m* ~ 80 m_e, the value more consistent with the high electronic specific heat on the order of 100 mJK^{-2}mol^{-1}. A spin gap magnitude of about 30 meV was determined by Adroja et al. (2003) based on a peak position in the INS probing the magnetic response of $CeRu_4Sb_{12}$ (obtained by subtracting the inelastic scattering of nonmagnetic $LaRu_4Sb_{12}$) at low temperatures. The spin gap persisted with little change up to

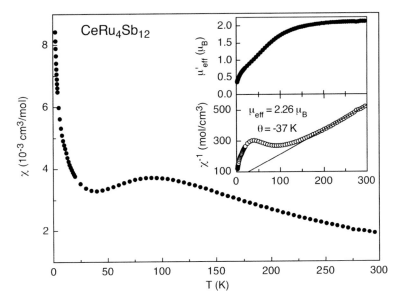

FIGURE 3.30 Temperature dependence of magnetic susceptibility of $CeRu_4Sb_{12}$. The upper inset shows the extracted effective magnetic moment in units of Bohr magnetons. The lower inset displays the inverse susceptibility as a function of temperature with a straight line indicating the Curie–Weiss law $\chi(T) = C/(T - \theta)$. Redrawn from E. D. Bauer et al., *Journal of Physics: Condensed Matter* **13**, 5193 (2001). With permission from the Institute of Physics.

100 K, at which temperature the gap started to close and was suppressed altogether by 150 K. The temperature of 100 K coincides with the temperature where the magnetic susceptibility exhibits a maximum, see Figure 3.30.

According to the single impurity Anderson model, Bickers et al. (1985), the Kondo temperature T_K is taken as three times the temperature where the susceptibility peaks, in this case yielding $T_K = 300$ K. It is interesting to compare the size of the spin and charge gaps in $CeRu_4Sb_{12}$, $\Delta_s/\Delta_c = 30/47.1 = 0.64$, consistent with the infinite-dimensional Anderson lattice model, Kumigashira et al. (2001). The already noted [121,123]Sb-nuclear quadrupole resonance measurements by Magishi et al. (2012) on $LaFe_4Sb_{12}$ also included samples of $CeFe_4Sb_{12}$ prepared under ambient pressure as well as under high pressure. In this case, with the increasing Ce filling achieved in high-pressure samples, the authors noted a significantly increased c-f hybridization.

Turning attention to $CeOs_4Sb_{12}$, contrary to calculations of Harima and Takegahara (2003) that predicted a metallic state, transport measurements by Bauer et al. (2001c) and subsequently by Sugawara et al. (2005b) invariably indicated that the metal-like temperature dependence of resistivity near room temperature crosses over to a semiconducting behavior at temperatures below about 50 K. From the exponential temperature dependence between 25 K and 50 K, Bauer et al. (2001c) determined a very narrow band gap of 0.9 meV. Establishing the Pauli nature of magnetic susceptibility in $CeOs_4Sb_{12}$, they associated a sharp upturn in $\chi(T)$ below 50 K with the Kondo effect, Kondo (1976), having the characteristic Kondo temperature determined from

$$T_K = \frac{N_A \mu_{eff}}{3k_B \chi(1.8\,K)}, \tag{3.95}$$

where N_A is the Avogadro number, μ_{eff} is the Ce^{3+} free-ion value of 2.54 μ_B, and $\chi(1.8$ K) is the susceptibility determined at the lowest temperature of measurements. The Kondo temperature turned out to be $T_K \sim 90$ K for $CeOs_4Sb_{12}$ and $T_K \sim 101$ K for $CeFe_4Sb_{12}$. Both values are in accordance with temperatures where the electrical resistivity experiences a distinct drop prior to acquiring a

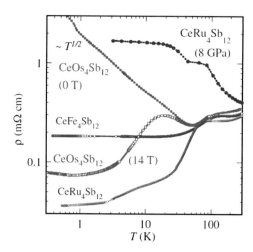

FIGURE 3.31 Temperature dependence of electrical resistivity of Ce-filled antimonide skutterudites. Note a small but distinct bend in the resistivity of CeOs$_4$Sb$_{12}$ at about 0.8 K marked by a red arrow. The effect of the applied magnetic field of 14 T on the resistivity of CeOs$_4$Sb$_{12}$ and of the external pressure of 8 GPa on the resistivity of CeRu$_4$Sb$_{12}$ is also presented. The erroneously stated temperature dependence of resistivity ~ $T^{1/2}$ of CeOs$_4$Sb$_{12}$ was corrected to the proper $T^{-1/2}$ dependence. Reproduced from H. Sato et al., *Journal of the Physical Society of Japan* **77**, Suppl. A, 1 (2008). With permission from the Physical Society of Japan.

negative temperature coefficient below 50 K. The drop reflects the decreasing rate of scattering as the Ce ions become effectively screened by conduction electrons. The semiconducting ground state was also supported by optical measurements of Matsunami et al. (2003) that indicated a gap opening at low temperatures. Sugawara et al. (2005b), on the other hand, having available also data on the temperature dependence of the Hall effect, described the unusual $\rho(T) \propto T^{-1/2}$ variation of low-temperature resistivity persisting over two decades of temperature, see Figure 3.31, as arising from the temperature-dependent carrier concentration (gradual reduction in the DOS at the Fermi level) combined with the charge carriers being scattered by spin fluctuations. Interestingly, below 1 K, the resistivity of CeOs$_4$Sb$_{12}$ developed a small but distinct knee marked by a red arrow in Figure 3.31. At a comparable temperature, Bauer et al. (2001c) previously noted a clear and sharp peak in their C/T vs. T^2 dependence of the specific heat. Having a small entropy content of merely 2% of $R\ln2$ and no other supporting data, they considered the peak to arise from a transition of some impurity phase containing Ce. This feature, however, proved to be essential and became a hot topic in subsequent studies.

Namiki et al. (2003b) reexamined the low-temperature specific heat behavior in CeOs$_4$Sb$_{12}$ by extending measurements to lower temperatures using a dilution refrigerator equipped with a SC magnet. Determining and subtracting the phonon contribution from measurements on LaOs$_4$Sb$_{12}$, a non-magnetic structure with very similar lattice properties, quite a spectacular peak was revealed in C/T near 0.9 K. The peak shifted to higher temperatures in an increasing field and diminished its magnitude. Rather than an impurity effect, the authors concluded that the peak reflects an intrinsic phase transition of electronic origin. They tentatively associated the ordering with the onset of a charge density wave (CDW) or spin density wave (SDW) but noted that the positive slope observed for dT/dH is highly anomalous for both the CDW and SDW.

A strong effect of magnetic field on the ordered state below 0.9 K observed by Namiki et al. (2003b) *via* the specific heat has proved to be even more dramatic in sub-Kelvin resistivity measurements in a magnetic field performed by Sugawara et al. (2005b). As shown in Figure 3.32, the semiconducting behavior turned progressively more metallic with the increasing magnetic field, and the small knee near 0.9 K in zero field traced a curve indicated by arrows in applied fields.

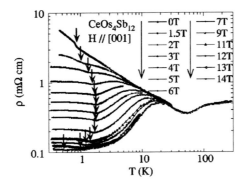

FIGURE 3.32 Temperature dependence of the electrical resistivity of a single crystal of $CeOs_4Sb_{12}$ measured at selected magnetic fields oriented parallel to the [001] axis of the crystal. Reproduced from H. Sugawara et al., *Physical Review B* **71**, 125127 (2005). With permission from the American Physical Society.

Bauer et al. (2001c), based on the analysis of their magnetic susceptibility of $CeOs_4Sb_{12}$, proposed the Γ_7 doublet (Γ_5^-) as the ground state with the Γ_8 quartet (Γ_{67}^-) positioned some 327 K above the ground state. However, inelastic neutron measurements by Yang et al. (2005) found no inelastic intensity at the expected energy of 28.2 meV, corresponding to the energy splitting of 327 K, thus questioning the CEF assignment. The absence of CEF excitations at the expected energy range was confirmed in subsequent INS measurements by Adroja et al. (2007). Instead, the authors identified two spin gap signatures in their data, one at 50 meV and the other at about 27 meV. They interpreted them as arising from direct ($\delta\mathbf{k} = 0$) and indirect ($\delta\mathbf{k} \neq 0$) transitions, respectively, as indicated later in Figure 3.36. In further neutron diffraction studies, Iwasa et al. (2008) extended measurements down to 0.1 K and observed AFM reflections characterized by a reduced wavevector $\mathbf{q} = (1,0,0)$. This is nicely illustrated by subtracting the scan taken at 1.4 K near (1,0,0) from the one at 0.1 K and plotting the difference, as shown in Figure 3.33. Interestingly, the AFM reflection peaks persisted, albeit with much reduced intensity, to temperatures near 1.6 K, considerably higher than the

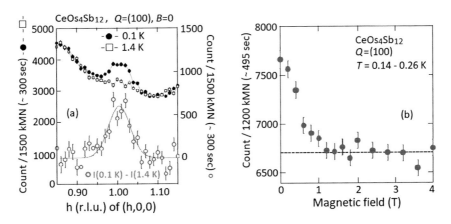

FIGURE 3.33 (a) Scan profiles of neutron diffraction obtained on $CeOs_4Sb_{12}$ through the AFM reciprocal lattice vector $q = (1, 0, 0)$ taken at 0.1 K (solid circles) and 1.4 K (open squares) at zero magnetic field. Open circles represent the subtraction of the 1.4 K data (above the ordering temperature) from the 0.1 K data (below the ordering temperature), depicting a well-defined peak at $(1, 0, 0)$. (b) Magnetic field dependence of the peak counts at $q = (1, 0, 0)$ taken near 0.2 K. The dashed line indicates the background level. Redrawn from K. Iwasa et al., *Journal of the Physical Society of Japan* **77**, Suppl. A, 318 (2008). With permission from the Physical Society of Japan.

ordering temperature of 0.9 K. The authors envisioned that the AFM ordering at $q = (1, 0, 0)$ arises from antiparallel arrangement of magnetic moments of Ce ions at the corner and at body-centered positions and evaluated the magnitude of the ordered moment at 0.1 K as being (0.07 ± 0.02) μ_B/Ce. Such a tiny magnetic moment is most likely associated with the SDW-like AFM instability. Taking scans in both zero and applied fields, the authors observed the suppression of the ordered phase by the magnetic field. In fact, at their lowest temperature of 0.1 K, the magnetic field of 1 T obliterated AFM reflections, replacing them with field-induced FM correlations. If one compares results of neutron scattering measurements, namely the AFM ordered phase below 0.9 K, and the induced FM state at fields above 1 T, with the transport behavior of $CeOs_4Sb_{12}$ at comparable temperatures, one notes a close correlation with the semiconducting, respectively, metallic state of the structure.

Elastic properties, too, are expected to depend strongly on the CEF energy scheme, yet the ultrasonic measurements by Nakanishi et al. (2007) indicated that CEF has little effect on the low-temperature elastic properties and demonstrated a dramatic discord between the experimental data and the behavior predicted based on the CEF energy splitting. Rather, the measurements of $CeOs_4Sb_{12}$ revealed a marked softening in the elastic coefficients C_{11} and C_{44} when the temperature approached and passed through the ordering temperature near 0.9 K, where anomalies in the electrical resistivity and the specific heat had been documented, as discussed above. As before, the applied magnetic field of a few Tesla completely suppressed the softening observed in zero magnetic field. The authors also noted a distinctly different behavior of $CeOs_4Sb_{12}$ with respect to its sister skutterudite $CeRu_4Sb_{12}$, the latter showing no anomalies and its elastic constants kept on stiffening as the temperature decreased regardless of the presence or absence of the magnetic field.

A dramatic effect of the magnetic field on the low-temperature properties of $CeOs_4Sb_{12}$ was also vividly revealed by the temperature dependence of the nuclear spin-lattice relaxation rate $1/T_1$ measured by Yogi et al. (2009) in their ^{121}Sb-NMR studies that extended down to 0.2 K in fields as high as 15 T. As indicated by Figure 3.34, at temperatures above about 100 K, the relaxation rate divided by absolute temperature, $1/T_1T$, is an exponential function of temperature with a gap size of about $\Delta \approx 320$ K and essentially independent of the applied magnetic field. This implies that the magnetic field does not alter the c-f hybridization gap. At temperatures below 20 K and in zero magnetic field, the temperature dependence of $1/T_1T$ followed the $T^{-1/2}$ power law for nearly two decades of temperature and was interpreted as a signature of AFM spin fluctuations in the authors' previous

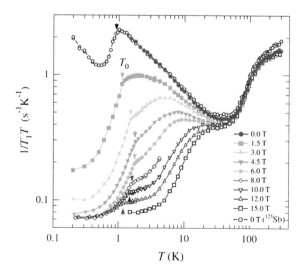

FIGURE 3.34 Temperature dependence of $1/T_1T$ for ^{121}Sb-NMR at various magnetic fields. Open circles indicate $1/T_1T$ multiplied by $(^{121}\gamma_n/^{123}\gamma_n)^2$ for ^{123}Sb-NQR. Redrawn from M. Yogi et al., *Journal of the Physical Society of Japan* **78**, 053703 (2009). With permission from the Physical Society of Japan.

^{123}Sb-NQR measurements that terminated just short of 1 K, Yogi et al. (2008). By rescaling these ^{123}Sb-NQR studies by $(^{121}\gamma_n/^{123}\gamma_n)^2$, where γ_n is the nuclear gyromagnetic ratio, they can be placed on the graph for the ^{121}Sb-NMR data in Figure 3.34 and are shown by open circles that perfectly overlap the zero field $1/T_1T$ data at temperatures below about 200 K. This documents the magnetic relaxation process that dominates in CeOs$_4$Sb$_{12}$. At an ordering temperature of 0.9 K, a sharp decrease in $1/T_1T$ was encountered, and below about 0.5 K, the relaxation rate started to increase again, but only in zero magnetic field, showing some residual low-lying excitations. The authors associated these excitations with a multipolar degree of freedom.

In the presence of a magnetic field, no further upturn in $1/T_1T$ was observed and, in fact, T_1T became constant, suggesting that the magnetic field suppressed any such excitations. Similarly as in the specific heat and resistivity, the increasing magnetic field shifted the maxima in $1/T_1T$ to higher temperatures. As the AFM spin fluctuations were continuously being suppressed, the temperature T_0, where the knee in the dependence of $1/T_1T$ vs. T is observed as the field increases, is indicated by small arrows in Figure 3.34. Plotting the magnetic fields that generated these temperatures T_0, a B-T phase diagram was constructed, as shown in Figure 3.35. Based on the measurements, Yogi et al. (2009) proposed the existence of two distinct ordered states: a zero-field AFM state and another ordered state at fields above 1 T, which forms as a result of freezing of multipole degree of freedom.

The electronic structure near the Fermi energy, particularly the evolution of the band gap or rather pseudo-gap as there seems to be a finite DOS within the gap, was probed by photoemission studies by Matsunami et al. (2009). A schematic how the hybridization evolves from a dispersive conduction band and the two lowest discrete CEF-split states of the Ce^{3+} ion, Γ_5^- and Γ_{67}^-, is illustrated in Figure 3.36. The main conduction band with the a_u symmetry hybridizes with the Γ_5^- states of Ce^{3+}, forming a gap in the DOS. High-resolution laser-photoemission spectra recorded at 4.2 K also revealed a strongly enhanced spectral DOS (after dividing by the Fermi–Dirac function) just above the Fermi level, forming as a consequence of fractional hybridization (fractional only because hybridization cannot take place between states of opposite parity at the zone center) between the a_g conduction band and the Γ_{67}^- states. The authors surmised that the high DOS associated with this fractionally hybridized $a_g - \Gamma_{67}^-$ state is the source of the heavy fermion nature of the charge carriers.

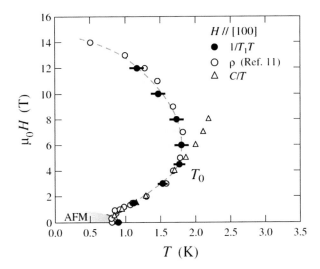

FIGURE 3.35 B–T phase diagram of CeOs$_4$Sb$_{12}$ for B // [001] constructed based on the ^{121}Sb-NMR and ^{123}Sb-NQR (full circles) measurements, electrical resistivity (open circles) measurements of Sugawara et al. (2005b), and specific heat (open triangles) measurements by Namiki et al. (2003b). A dark gray region is the AFM phase below 0.9 K. Redrawn from M. Yogi et al., *Journal of the Physical Society of Japan* **78**, 053703 (2009). With permission from the Physical Society of Japan.

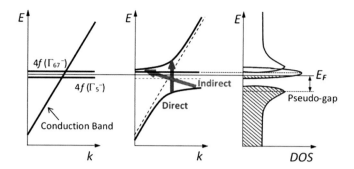

FIGURE 3.36 Schematic evolution of hybridization from (a) a dispersive conduction band and two lowest CEF-split discrete states of a Ce^{3+} ion. (b) The main conduction band with a_u symmetry hybridizes with only Γ_5^- states, giving rise to a hybridization gap around the Fermi energy. Γ_{67}^- states remain dispersionless because of only fractional hybridization with the conduction band of a_g symmetry. Direct and indirect optical transitions are marked by blue and red arrows, respectively. (c) The resulting band structure with a pseudo-gap and a sharp narrow band just above the Fermi energy believed to impart a heavy fermion character to the charge carriers. Redrawn and modified from M. Matsunami et al., *Physical Review Letters* **102**, 036403 (2009). With permission from the American Physical Society.

Thus, hybridization in this case plays a dual role, opening a gap in the DOS slightly below the Fermi energy, resulting in a semiconductor-like transport, and forming a sharply enhanced DOS just above the Fermi energy, imparting the heavy fermion character to the charge carriers.

As I already pointed out, a hybridization gap or pseudo-gap in the DOS may affect both spin and charge aspects of carriers. The spin gap can be extracted from inelastic neutron measurements as the energy where the magnetic scattering intensity peaks. In Kondo semiconductors, such as Ce-filled skutterudites, it has been established that the spin gap Δ_s is directly proportional to the Kondo temperature T_K, which, in turn, is obtained as three times the temperature T_{max} of a broad maximum in the temperature-dependent magnetic susceptibility, $T_K = 3T_{max}$. The charge gap Δ_c, on the other hand, follows either from photoemission or from optical spectroscopy. However, here one must be careful and consider the fact that photoemission and optical measurements do not necessarily yield the same gap value. The reason is that the optical measurements probe the electronic excitations from the occupied DOS to unoccupied DOS, while photoemission measures only a gap in the occupied range of the DOS, i.e., a gap that is potentially much smaller or perhaps even zero, such as in the case that the Fermi level is pinned at the top of the valence band. Taking this into account, Rayjada et al. (2010), based on their extensive core level and valence band photoemission spectroscopy measurements of $CeOs_4Sb_{12}$, were able to relate the charge gap to the Kondo temperature as $\Delta_c \sim 2k_B T_K$, establishing the coupled energy scaling of the spin and charge gaps in Kondo semiconductors. The decreasing DOS at the Fermi energy extracted from their photoemission measurements as the pseudo-gap opened was compared with the diminishing effective magnetic moment from measurements of susceptibility at the same temperatures. As indicated by Figure 3.37, excellent correlation between the two has resulted. In Table 3.9 are listed the Kondo temperatures and the spin and charge gaps based on photoemission and INS on the one hand and optical spectroscopy on the other hand.

While the phase transition from AFM to a high-field ordered phase was well documented by transport, photoemission, and neutron scattering studies, its thermodynamic grounding was lacking and had to wait until Tayama et al. (2015) used high-resolution dilatometry and magnetization measurements. The zero-magnetic field data are summarized in Figure 3.38, which combines a temperature-dependent plot of the linear thermal expansion coefficient α, specific heat, and dc magnetization. It is impressive how revealing are the rarely employed thermal expansion measurements, which

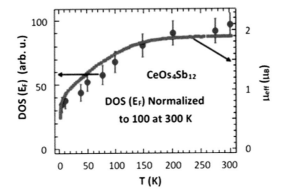

FIGURE 3.37 Correlation between the decreasing DOS at the Fermi energy and the diminishing effective magnetic moment as the hybridization gap opens near the Fermi energy. Adapted from P. A. Rayjada et al., *Journal of Physics: Condensed Matter* **22**, 095502 (2010). With permission from the IOP Publishing.

TABLE 3.9

Kondo Temperature T_K, Spin Δ_s, and Charge Δ_c Gaps of Some Ce-Filled Skutterudites. Data Collected from the Literature

Skutterudite	T_K (K)	Δ_s (meV)	Δ_c (meV)	Reference
$CeFe_4P_{12}$	1400	–	220–240	1
$CeRu_4Sb_{12}$	300	30	45–50	2–6
$CeOs_4Sb_{12}$	300	27	50–60	1,3,7

1. Rayjada et al. (2010), 2. Takeda & Ishikawa (2000b), 3. Adroja et al. (2007), 4. Dordevic et al. (1999), 5. Kanai et al. (2002), 6. Adroja et al. (2003), 7. Matsunami et al. (2003).

clearly depict not only a sharp peak at temperature $T_A = 0.87$ K, but also a dip near $T_N = 1.05$ K, both temperatures marked by dotted lines in the figure. Breaks in the temperature dependence of magnetization occur at the same T_A and T_N, while the specific heat shows a single broad peak unable to resolve the two temperatures. The authors associated the temperatures T_A and T_N with two successive phase transitions, namely, an entry into the AFM phase at the Néel temperature T_N and exiting it into a new phase at a lower temperature T_A.

Monitoring kinks and inflection points in the longitudinal magnetostriction at temperatures between 0.3 K and 3.0 K, believed to be associated with the phase transitions, and checking for any signs of hysteresis (none seen), the authors constructed a phase diagram shown in Figure 3.39. Instead of two ordered phases as depicted in Figure 3.35, now there are three ordered phases: the zero-field ground state marked as phase A, a narrow range of the AFM phase between 0.87 K and 1.05 K, and phase B representing high-field ground state. The authors assumed that the B phase is nonmagnetic. The B–P phase boundary is similar to the one in Figure 3.35.

Recently, Ho et al. (2016) extended magnetometry and high-frequency conductivity measurements to fields as high as 60 T to study the evolution of the field-induced transition and to obtain de Haas-van Alphen (dHvA) and Shubnikov-de Haas oscillations. They were able to show that the field-induced phase is characterized by a single nearly spherical Fermi surface section, and the phase boundary acquires an unusual curvature, likely due to quantum fluctuations associated with the AFM phase.

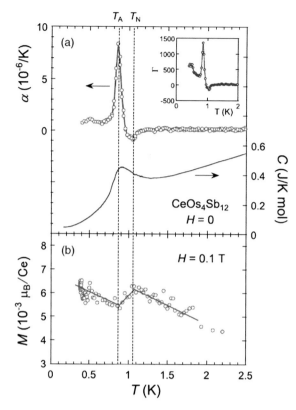

FIGURE 3.38 Coefficient of linear thermal expansion (top panel) in zero magnetic field. The dotted lines indicate two successive phase transitions at T_A and T_B. Specific heat measured by Namiki et al. (2003b) is shown in the middle panel. Temperature dependence of magnetization measured in field of 0.1 T is shown in the bottom panel. The solid line is a guide to the eye. Modified and redrawn from T. Tayama et al., *Journal of the Physical Society of Japan* **84**, 104701 (2015). With permission from the Physical Society of Japan.

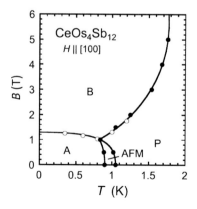

FIGURE 3.39 B–T phase diagram of $CeOs_4Sb_{12}$ for $B // [100]$. Open circles represent data obtained from measurements of thermal expansion and magnetization in magnetic field. Closed circles stand for the coefficient of linear expansion and magnetization as a function of temperature. P, AFM, A, and B indicate paramagnetic, AFM, A, and B phases, respectively. The lines are guides to the eye. All phase boundaries are of second order. Reproduced from T. Tayama et al., *Journal of the Physical Society of Japan* **84**, 104701 (2015). With permission from the Physical Society of Japan.

3.3.4.3 Pr-Filled Skutterudites

Of all the fillers in the skutterudite lattice, praseodymium has generated the greatest excitement and reports on properties of PrT_4X_{12} are the most voluminous. The reasons for such attention were spectacular physical properties, particularly the heavy fermion nature combined with an exotic form of superconductivity in $PrOs_4Sb_{12}$. Because an entire section 2.4 in this book is dedicated to SC properties of $PrOs_4Sb_{12}$, I will restrict my discussion here to only its magnetic properties. However, because the electronic structure determines both SC and magnetic properties of a material, there will be inevitably some repetition with respect to section 2.4. As much as $PrOs_4Sb_{12}$ drives the discussion, other Pr-filled skuterudites, too, show unusual properties and deserve at least a brief exposure.

Unlike Ce, which often reveals its intermittent valence between the magnetic Ce^{3+} and non-magnetic Ce^{4+}, praseodymium has a stable trivalent ion Pr^{3+}, i.e., its electronic configuration is $[Xe]4f^2$, losing its two $6s$ electrons and one $4f$ electron upon entering the skutterudite void. In comparison to the Ce^{3+} ion, where the CEF splitting between the ground state and the first excited state is large and thus has little effect on low temperature properties, the energy separation between the various CEF-split ground states and the first excited states in Pr-filled skutterudites is much smaller and, in many aspects, is the dominant factor governing low-temperature properties. It should also be noted that the $4f$-electrons of Pr^{3+} have a rather well localized character.

3.3.4.3.1 Pr-Filled Phosphides

Properties of $PrFe_4P_{12}$ are dominated by a spectacular behavior of its resistivity, magnetic susceptibility, and specific heat that display a sharp anomaly near $T_A = 6.5$ K, revealed first in measurements by Torikachvili et al. (1987) and confirmed by numerous subsequent investigations, e.g., Sato et al. (2000) and Tayama et al. (2004). The logarithmic rise in the resistivity between about 30 K and 150 K observed by Sato et al. (2000) and shown in Figure 3.40(a) reflects the apparent Kondo-like behavior. A plateau sets in at lower temperatures and is interrupted by a sudden rapid increase starting at 6.5 K and leading to a pronounced peak near 5 K. At still lower temperatures, the resistivity rapidly decreases. Equally impressive are anomalies in the susceptibility and the specific heat, the latter shown in Figures 3.40(b). The onset of the peak was originally associated with an AFM transition, but subsequent neutron diffraction studies by Keller et al. (2001) showed no magnetic

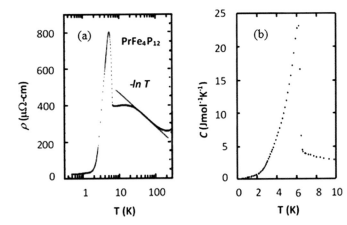

FIGURE 3.40 (a) Temperature dependence of the electrical resistivity of $PrFe_4P_{12}$ single crystal. The line indicates a logarithmic dependence of the resistivity on temperature between 30 K and 100 K. Modified and redrawn from H. Sato et al., *Physical Review B* 62, 15125 (2000). With permission from the American Physical Society. (b) Temperature dependence of specific heat near the phase transition at 6.5 K. Modified and redrawn from M. S. Torikachvili et al., *Physical Review B* **36**, 8660 (1987). With permission from the American Physical Society.

reflections, denying that the phase transition at T_A has any magnetic origin. This was also confirmed later by nuclear specific heat measurements by Aoki et al. (2002a), by X-ray diffraction by Iwasa et al. (2003), and additional neutron scattering studies by Hao et al. (2003). These latter neutron diffraction measurements, together with elastic measurements by Nakanishi et al. (2001), demonstrated that the phase transition at $T_A = 6.5$ K in zero field is of second order and is accompanied by a presumable antiferro-quadrupolar (AFQ) ordering. In publications, this non-magnetic phase is often referred to as the A-phase. One of the interesting features of the AFQ ordering is a crystal structure modulation with the wave vector $q = (1,0,0)$ seen in X-ray diffraction measurements by Iwasa et al. (2002) and attributed to the development of staggered local electronic states of the Pr^{3+} ions. This is the same wave vector believed to be responsible for nearly perfect nesting of the Fermi surface proposed theoretically by Harima and Takegahara (2002). Upon applying a magnetic field, the AFQ ordered state is suppressed and the system develops a heavy fermion character with the resistivity following the T^2 dependence but with a large coefficient $A = 2.5$ $\mu\Omega$cmK^{-2}, in accordance with a large electronic specific heat $\gamma \sim 1.4$ J mol^{-1}K^{-2} at 6 T, Aoki et al. (2005b). The AFQ ordering is totally destroyed ($T_A = 0$) by a magnetic field of 4–7 T, depending on the field orientation with respect to the crystallographic axes of the sample. The suppression of the ordered state is vividly demonstrated by a field variation of the resistivity shown in Figure 3.41. The heavy cyclotron mass $m^* = 81$ m$_e$ observed in the ω-branch of dHvA oscillations measured by Sugawara et al. (2002b) further corroborated the formation of a heavy fermion state. It should be noted, however, that the anomalies, including the mass enhancement, appear quite anisotropic, with the greatest effect when the magnetic field is applied parallel to [111], Namiki et al. (2003a). At higher magnetic fields oriented parallel to the [111] direction, Tayama et al. (2004) detected the presence of a new high-field ordered phase. This phase, often designated as phase B, forms only within a very narrow range of angles around [111] and is absent for any other field orientation. Although there were signs that the AFQ ordering fails in explaining some features, such as persistent isotropy of magnetic susceptibility in the ordered phase, its status as a substantially correct interpretation of the ordered A-phase remained unchallenged until measurements of the angular dependence of NMR by Kikuchi et al. (2007b) indicated no change in the local symmetry of Pr^{3+} ions as the $PrFe_4P_{12}$ crystal was cooled through the transition point. This led theorists to come up with several model scenarios based on the orbital degrees of freedom, generally referred to as scalar-type ordering, Kiss and Kuramoto (2006).

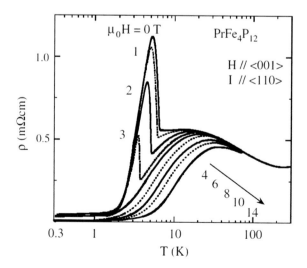

FIGURE 3.41 Temperature dependence of the electrical resistivity in magnetic fields marked in the figure in units of Tesla. Redrawn from Y. Aoki et al., *Journal of the Physical Society of Japan* **74**, 209 (2005). With permission from the Physical Society of Japan.

While the scalar order does not require lowering of the cubic site symmetry, in accordance with the experiment, it brings into considerations higher order multipoles, such as hexadecapoles, invoked by Kuramoto and Kiss (2008). The discussion of scalar-type ordering also depends critically on the CEF energy level scheme. This has never been nailed down exactly in the early studies, although it was clear that, in comparison to Ce^{3+} ions, the energy spacing of the ground state and the first excited state of the CEF-split $J = L - S = 5 - 1 = 4$ multiplet of a Pr^{3+} ion is much smaller. The issue was considered seriously by the same authors, who evaluated the ground state of $PrFe_4P_{12}$ as the Γ_1 singlet and the first excited state as the $\Gamma_4^{(1)}$ triplet with a possible participation of the Γ_{23} doublet, all six levels being almost degenerate at high temperatures. Taking a hint from neutron scattering experiments, where Iwasa et al. (2003) and Hao et al. (2005) noted two inelastic transitions in the ordered A-phase, Kuramoto and Kiss (2008) postulated that below T_A, one of the Pr sublattices features a singlet ground state while the other selects a doublet. However, high-resolution thermal expansion and magnetostriction measurements by Tayama et al. (2009) performed with a resolution a few orders of magnitude higher than what is possible with X-ray studies indicated the identical Γ_1 ground state for all Pr^{3+} ions, both in the A-phase and above T_A. The totally symmetric order parameter, i.e., the Γ_1 symmetry, was also implied for the A-phase based on ^{31}P NMR measurements by Kikuchi et al. (2007b), thus excluding any quadrupolar order.

Further complexity of the low-temperature state of $PrFe_4P_{12}$ was revealed by high-pressure studies that uncovered a gradual suppression of the A-phase and its demise at the M–I transition at pressures above 2.4 GPa detected in resistivity measurements by Hidaka et al. (2005), in symmetry lowering from cubic to orthorhombic at the M–I transition seen in X-ray diffraction by Kawana et al. (2006), and in three-orders of magnitude decrease in the density of holes at the M-I transition and suppression of susceptibility measured in Hall and magnetization investigations by Hidaka et al. (2006). One should point out that a transition from a metal to an insulator under pressure is rather atypical as the pressure usually induces a transition from the insulating state to the metallic state. The effect of pressure is perhaps most vivid in the behavior of the resistivity, shown in Figure 3.42. Two things are to be noted in Figure 3.42(a): the transition temperature T_A decreases with the applied pressure and, at 2.4 GPa, a distinct M–I transition sets in near 1 K. Pressures in excess of 2.7 GPa totally

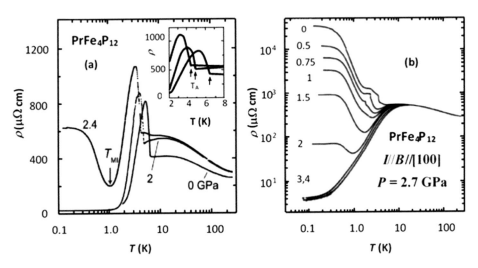

FIGURE 3.42 (a) Temperature dependence of the electrical resistivity under pressure. T_{MI} indicates the M–I transition first appearing at pressure of 2.4 GPa. Arrows in the inset display the temperature shift of the transition temperature T_A. (b) Effect of magnetic field on the electrical resistivity subjected to a fixed pressure of 2.7 GPa. The numbers indicate the magnetic field in tesla. Redrawn from H. Hidaka et al., *Physical Review B* 71, 073102 (2005). With permission from the American Physical Society.

suppress the A-phase (not shown) and the temperature of the M-I transition rises rapidly, moving close to 10 K at 3.6 GPa. As indicated in Figure 3.42(b), magnetic field very effectively destroys the insulating behavior and the resistivity converges to 4–5 $\mu\Omega$cm in fields higher than 3 T and attains a quadratic temperature dependence below about 1 K, again with a large value of the coefficient A. At 2.7 GPa and zero magnetic field, the energy gap was estimated as $E_g/k_B = 2.8$ K (0.24 meV), and it decreased rapidly with the magnetic field, vanishing at fields ~2.5 T. The pressure *vs.* temperature phase diagram based on the above studies is presented in Figure 3.43. Although an early hint that the M-I transition is triggered by an AFM ordering of f-electrons came originally from ^{31}P NMR studies by Hidaka et al. (2006), the definite proof of AFM in the insulating state of PrFe$_4$P$_{12}$ at pressures above 2.4 GPa had to wait until neutron diffraction measurements under high pressure were carried out by Osakabe et al. (2010). The AFM order with the propagation vector $q = (1, 0, 0)$ is identical to the wave vector in the A-phase and is closely related to the nesting ability of the Fermi surface. Applying magnetic field under pressure, the authors observed a gradual suppression of the AFM state and evolution of the FM state by a process of spin flopping. At the highest pressure of 3.8 GPa, the FM state took completely over in magnetic fields above 3 T.

Such a rich variety of low-temperature phases in PrFe$_4$P$_{12}$ stimulated much theoretical effort to provide a microscopic view of the structure. Among these are attempts by Hoshino et al. (2011) to explain the scalar order of the A-phase relying on a staggered Kondo-CEF singlet claimed to explain many experimental results, and general considerations regarding the role of competing p-f and d-f hybridizations in the stability of non-magnetic and magnetic orderings by Shiina (2012).

Turning attention to PrRu$_4$P$_{12}$, the first detailed transport measurements by Sekine et al. (1997) revealed a second-order M–I) transition near $T_{MI} \sim 60$ K in the electrical resistivity. The energy gap estimated from the exponentially rising resistivity was $E_g/k_B \approx 37$ K (3.19 meV). Optical measurements by Nanba et al. (1999) suggested a wider gap of about 100 cm^{-1} (12.4 meV). The gap seemed to be strengthened by the application of external pressure of a few GPa, showing a rate of increase of dT_{MI}/dP = 0.6 K/GPa. The Curie–Weiss susceptibility between 150 and 300 K yielded an effective magnetic moment of 3.84 μ_B/Pr^{3+}, a bit larger than the theoretical Hund's rule value of 3.58 μ_B/Pr^{3+}, the discrepancy explained as due to a positive FM exchange polarization of conduction electrons. The Curie–Weiss temperature was reported as $\theta_{CW} = -7$ K. Being an uncompensated metal (the [Ru$_4$P$_{12}$] polyanion is missing one electron to be compensated), to turn PrRu$_4$P$_{12}$ into an insulator necessitated the formation of more than a single unit cell below T_{MI}, and a question was what mechanism could accomplish the task. Because no anomaly in the magnetic susceptibility accompanied

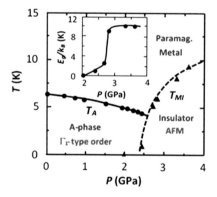

FIGURE 3.43 Pressure *vs.* temperature phase diagram for PrFe$_4$P$_{12}$. T_A is the transition temperature for entry into the non-magnetic A-phase, and T_{MI} is the M–I transition temperature. The inset shows the pressure dependence of the band gap in units of Kelvin. Redrawn from H. Hidaka et al., *Physical Review B* **71**, 073102 (2005). With permission from the American Physical Society.

the resistive transition, and the magnetic field had no effect on the λ-type anomaly in the specific heat at T_{MI}, Sekine et al. (2000a), magnetism was not a driving force. Likewise, XANES measurements by Lee et al. (1999) documented an unaltered trivalent state of Pr at temperatures above and below T_{MI}, implying that charge instability is an unlikely culprit either. Consequently, the focus was directed toward the behavior of the crystal lattice as a factor behind the M–I transition. Although the original measurements by Sekine et al. (1997) did not reveal any change in the powder diffraction pattern of $PrRu_4P_{12}$ upon cooling through T_{MI}, there were signs of considerable softening and broadening of a 380 cm^{-1} phonon mode below T_{MI} assigned to the vibrations involving phosphorus atoms in the Raman measurements by Sekine et al. (1999), suggestive of the transition being related to lattice vibrations. There were also superlattice diffraction spots observed below T_{MI} in electron diffraction measurements by Lee et al. (2001) that could be interpreted as due to some structural rearrangement. More definitive proof that the crystal lattice plays an important role in the M–I transition came from highly sensitive thermal expansion measurements by Matsuhira et al. (2000), where an anomalous jump was detected at T_{MI}.

An early theoretical assessment of the M–I transition by Harima et al. (2003) put forward a perfect 3-D nesting of the Fermi surface as the driving force behind the transition. However, because the T_{MI} of $PrRu_4P_{12}$ is too high to obtain dHvA oscillations, it was not possible to verify directly that the Fermi surface is, indeed, amenable to perfect nesting, even though band structure calculations by Harima and Takegahara (2002) suggested that $PrRu_4P_{12}$ should have a similar Fermi surface to $LaRu_4P_{12}$, where such nesting with the wave vector $q = (1, 0, 0)$ was observed. But even if so, there was a problem with the fact that, unlike $PrRu_4P_{12}$, its cousin $LaRu_4P_{12}$ did not show any M-I transition. Because the obvious difference between the two skutterudites is the presence of 4f-electrons in the former and the lack of them in the latter, it became clear that 4f-electrons of Pr^{3+} ions must play a major role in the transition.

Pressure was found a very useful parameter with which to explore the stability of $PrRu_4P_{12}$ and, specifically, its M–I transition. Shirotani et al. (2002) carried out two revealing experiments: room-temperature pressure measurements to 50 GPa using a diamond anvil cell that documented the structural stability of $PrRu_4P_{12}$ with a bulk modulus of 207 ± 12 GPa, indicating a quite hard material and resistivity measurements under a pressure of up to 8 GPa in a cubic anvil apparatus extending down to 4.2 K that showed a dramatic decrease of the resistivity in the insulating state. In accordance with the initial study of Sekine et al. (1997), the T_{MI} increased in pressures up to 4.5 GPa, but then a surprising turnaround set in and T_{MI} decreased with increasing pressure. While it is easy to understand the increasing T_{MI} with pressure as due to increased interaction of Pr ions that come closer together as the lattice constant shrinks under pressure, the subsequent reversed trend at still higher pressures is puzzling. Miyake et al. (2004) have shown that a gradual crossover from an insulating state to a more metallic temperature dependence of resistivity with increasing pressure is completed above 11 GPa. Moreover, at pressures above 12 GPa, $PrRu_4P_{12}$ became a superconductor with a transition temperature T_s = 1.8 K, joining the sister compounds $PrRu_4As_{12}$ and $PrRu_4Sb_{12}$, which are SC at ambient pressure with the transition temperatures of 2.4 K and 1.04 K respectively, for details see Chapter 2. Infrared reflectance spectroscopy under high pressure carried out by Okamura et al. (2011) confirmed that the M-I transition in $PrRu_4P_{12}$ is suppressed, albeit at a somewhat higher pressure of 14 GPa.

As in the case of $PrFe_4P_{12}$, where more and more sophisticated experiments continuously drove reevaluation of the microscopic nature of the non-magnetic A-phase, here, too, we see the same situation, except an issue is the microscopic nature of the M–I transition in $PrRu_4P_{12}$.

A new twist into the problem was added by Lee et al. (2004) who used an improved resolution of X-ray diffraction measurements performed with synchrotron radiation, which amplified the previously hinted superlattice reflections observed below the M–I transition. The superlattice reflection spots were identified as resulting from a surprising transition of the usual body-centered $Im\bar{3}$ space group of filled skutterudites into a simple cubic structure with the space group $Pm\bar{3}$, the process that doubles the unit cell and shifts the positions of Ru atoms. This creates different local environments for a filler atom Pr(1) situated at the unit cell origin (corner) and for a filler atom Pr(2) at the body center position of the unit cell, making them inequivalent. From the microscopic perspective, the doubling of the unit cell and the ensuing dramatic reduction in the carrier concentration could be viewed as nesting of the Fermi surface with the wave vector $q = (1, 0, 0)$. However, because $LaRu_4P_{12}$ with no f-electrons and a very similar Fermi surface does not undergo the M–I transition, the presence of the two $4f$-electrons of the Pr^{3+} ion and their interaction with conduction electrons, derived mostly from the pnicogen atoms, is expected to play an important role in the M-I transition. Moreover, the temperature evolution of the CEF-split energy levels of the $4f$-electrons is another pivotal factor to take into account. To help to visualize the transition from the $Im\overline{3}$ to the $Pm\overline{3}$ space group, I remind the reader about the crystal structure of filled skutterudites. In reference to Figure 1.1, each filler atom is surrounded by an icosahedron with the vertices occupied by 12 pnicogen atoms X (phosphorus in our case) and, at a slightly greater distance, a cube formed by eight transition metal atoms T (here ruthenium). In the transition to the lower $Pm\overline{3}$ space group, Ru atoms forming a cube around Pr(1) are displaced a bit outward parallel to the Pr(1)-Ru bond, while Ru atoms belonging to a cube surrounding the Pr(2) filler shift a bit inward, giving rise to cubes of unequal volume. The cubes are still connected at the corners in the (1, 1, 1) direction but now alternate in size, as depicted in Figure 3.44. The alternate volumes of Ru cubes imply different electron densities (larger density in a smaller cube), resulting in a modulation of electron density called a CDW. This charge density state is unusual and different from the conventional CDW, where the modulation is controlled by strong electron–phonon interaction. Here, in $PrRu_4P_{12}$, the modulation is due to the formation of two inequivalent Pr^{3+} ion sublattices, one associated with the Pr(1) site and the other one with the Pr(2) site. In their band structure calculations, Curnoe et al. (2004) have shown that the atomic displacements described above lead to a gap opening everywhere on the Fermi surface. By the way, since 2005, the M–I transition has often been referred to in publications as the charge order (CO) transition, and the insulating phase as the charge ordered phase. With the presence of two different praseodymium sites, a question arose what are the ground states of now two different Pr(1) and Pr(2) ions. This has been answered by detailed INS experiments by Iwasa et al. (2005b) that probed excitations between the four CEF-split energy levels: Γ_1 singlet, nonmagnetic Γ_{23} doublet, and two magnetic triplets $\Gamma_4^{(1)}$ and $\Gamma_4^{(2)}$. From the obtained energies, the field dependence of magnetization was fitted to determine the appropriate level scheme, the results shown in Figure 3.45. The key findings are different ground states of the two Pr^{3+} ions below about 40 K, with the corner Pr(1) ion maintaining its high temperature Γ_1 singlet, while the body center Pr(2) ion

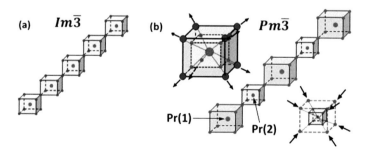

FIGURE 3.44   (a) Cubes of Ru atoms connected at the corners in the (1, 1, 1) direction of the $Im\overline{3}$ structure of $PrRu_4P_{12}$ skutterudite above the M–I transition. All cubes are identical because all Pr ions are equivalent.(b) Ru cubes still connected at the corners in the (1, 1, 1) direction but now with altering larger and smaller sizes due to shifts of Ru atoms belonging to two distinct Pr positions, Pr(1) at the unit cell origin (larger cube), and Pr(2) at the body center of the unit cell (smaller cube). The resulting structure forms as $PrRu_4P_{12}$ undergoes a M–I transition and has a reduced $Pm\overline{3}$ space group. Adapted from C. H. Lee et al., *Physical Review B* **70**, 153105 (2004). With permission from the American Physical Society.

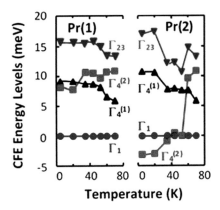

FIGURE 3.45   Temperature dependence of the CEF energy level scheme for the two inequivalent Pr(1) and Pr(2) ions in $PrRu_4P_{12}$ determined from INS. Note the different ground states below about 40 K. Drawn from the data in K. Iwasa et al., *Physical Review B* **72**, 024414 (2005). With permission from the American Physical Society.

attaining the triplet $\Gamma_4^{(2)}$ as its ground state. The CEF level scheme, with one-half of Pr^{3+} ions having a magnetic triplet $\Gamma_4^{(2)}$ as the ground state, was supported experimentally by low-temperature magnetization studies by Sakakibara et al. (2005), which reported the saturation magnetization of only 0.9 μ_B/Pr^{3+}, about one-half of 2 μ_B/Pr^{3+}, the value expected had all Pr^{3+} ions possessed the magnetic triplet ground state. One also notes in Figure 3.45 large shifts in the energy levels as the temperature increases toward the M–I transition, believed to arise from strong hybridization of 4f-electrons with the conduction electrons derived from pnicogen p-states. This makes an intuitive sense because the Pr ions are surrounded by icosahedrons of pnicogen atoms. In fact, Shiina and Shiba (2010) argued that the CO (insulating phase) arises exclusively from c-f mixing and the degree of mixing depends on the CEF energy levels. However, there were alternate approaches attempting to explain the M–I transition in $PrRu_4P_{12}$, and one of the more interesting microscopic viewpoints was provided by Takimoto (2006), who invoked an antiferro-hexadecapole order that does not break the local symmetry. His theory was apparently able to account also for the CEF level shifts. Having a magnetic triplet ground state at the Pr(2) site raised an issue why no magnetic order had developed even when the temperature was lowered down to 20 mK, Saha et al. (2005). This was answered partly by Aoki et al. (2011a) who showed that at sub-Kelvin temperatures there is considerable on-site hyperfine coupling between the $\Gamma_4^{(2)}$ triplet and the nuclear spin $I = 5/2$ of ^{141}Pr (100% abundance), resulting in what the authors called 4f-electron-nuclear hyperfine-coupled multiplets. Such multiplets were manifested by a peak structure in the specific heat of $PrRu_4P_{12}$ seen at 0.3 K in zero magnetic field that shifted to higher temperatures upon application of the field. The issue was also addressed theoretically by Shiina and Shiba (2010) who estimated that the orbital dependence of hybridization strongly favors (by an order of magnitude) charge interactions over magnetic interactions. Subsequently, Shiina (2011) also pointed out, based on a simple model Hamiltonian, that the transport properties are influenced by large f-orbital fluctuations arising from the singlet-triplet level crossing near 40 K associated with the Pr(2) site, as seen in Figure 3.45. In general, multipolar formalism, such as used by Takimoto (2006), has become important and essential to capture exotic electric and magnetic distributions encountered with ions of rare earth elements that possess f-electrons. In the hands of theorists, it has become a powerful approach to describe configurations of electric and magnetic moments in complex solids, such as strongly correlated electron systems, including certain filled skutterudites. The reader interested in the theory of electric and magnetic multipoles in condensed matter physics is referred to a recent review article by Suzuki et al. (2018).

The M–I transition in $PrRu_4P_{12}$ was also studied *via* substitutional effects at the Pr site using La, Ce, and Nd and by diluting the content of Ru by Rh. Typical substitutional levels explored were up to 20%, and the techniques used to detect changes in T_{MI} were the electrical resistivity, magnetic susceptibility, X-ray diffraction to monitor the onset of superlattice reflections, and, occasionally, INS to identify the antiferro-type ordered multipoles. The results are summarized in Figure 3.46. As follows from the trend shown, the transition temperature is suppressed by all types of substitutions, but to a different degree depending on where the substitution takes place and what kind of an atom is used. The weakest effect on T_{MI} seems to have a substitution of Nd for Pr, Iwasa et al. (2015), where the interaction between the Pr^{3+} ions and magnetic Nd^{3+} ions tends to stabilize the magnetic triplet ground state $\Gamma_4^{(2)}$ of Pr ions, and the antiferro-type electric multipole ordering of Pr 4*f*-electrons is least disturbed. In comparison, placing a non-magnetic La^{3+} ion at the Pr^{3+} site, as done by Sekine et al. (2000d), is merely a site-dilution effect that increases the spacing between the magnetic Pr(2) ions having the $\Gamma_4^{(2)}$ ground state. An interesting situation develops when Ce substitutes for Pr, as in the study by Sekine et al. (2011b). Just like La and Nd, Ce too suppresses T_{MI}, but the effect is much stronger. Moreover, the electrical resistivity of $Pr_{1-x}Ce_xRu_4P_{12}$ decreases with the increasing content of Ce and, for $x = 0.10$, a dramatic drop is observed near 7 K, highly suggestive of a phase transition, Figure 3.47. The anomalous low-temperature behavior of the resistivity was accompanied by a crossover in the magnetic susceptibility from a divergent ($x = 0, 0.05$) to a saturation nature ($x = 0.10$ and 0.15) and the development of a peak in magnetic fields. The authors interpreted the position of the peak as the Néel temperature in the newly developing AFM phase. A closer look into what is actually happening when Ce substitutes for Pr was taken by Saito et al. (2014) using X-ray and INS measurements, which revealed a rapidly decreasing intensity of superlattice reflections below 10 K and their total demise near 7 K. Surprisingly, powder neutron diffraction measurements near the previously assumed AFM ordering at 7 K showed no extra (magnetic) Bragg reflections, denying the existence of the AFM phase. Rather, the

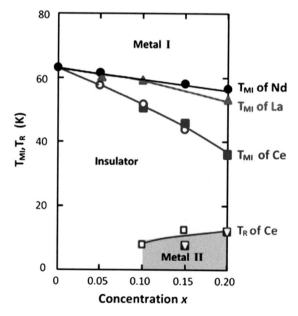

FIGURE 3.46 Phase diagram of $Pr_{1-x}R_xRu_4P_{12}$ for R = La, Ce, and Nd. Solid blue squares indicate T_{MI} obtained from X-ray measurements by Saito et al. (2014), and open blue circles are T_{MI} values from resistivity measurements by Sekine et al. (2011b). Open brown squares represent reentry temperatures T_R from susceptibility measurements by Sekine et al. (2011b), and brown squares with a yellow insert stand for T_R determined from susceptibility measurements by Saito et al. (2014). Adapted from K. Iwasa et al., *Physics Procedia* **75**, 179 (2015). With permission from Elsevier.

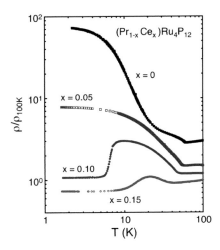

FIGURE 3.47 Temperature dependence of normalized resistivity, $\rho(T)/\rho(100\ K)$, for $Pr_{1-x}Ce_xRu_4P_{12}$ with Ce concentrations $x = 0$, 0.05, 0.10, and 0.15. Please note that the resistivity curves for $x \leq 0.10$ are shifted vertically for clarity. The curve for $x = 0$ is data of Sekine et al. (1997). Note a rapid drop in the resistivity of the $x = 0.10$ sample near 7 K, and a similar one, although not as sharper, for $x = 0.15$ near 14 K. These were originally interpreted as transitions to a new AFM phase, but later associated with the reentrant metallic transition. Redrawn from C. Sekine et al., *Journal of the Physical Society of Japan* **80**, SA024 (2011). With permission from the Physical Society of Japan.

compendium of results, including the analysis of the CEF excitations in $Pr_{1-x}Ce_xRu_4P_{12}$, was consistent with a reentrant metallic state, whereby below 10 K all Pr^{3+} ions possessed a non-magnetic Γ_1 ground state, just as was the case at temperatures above T_{MI}. Interestingly, a few percent of Rh substituting for Ru had a similar effect on the M-I transition and T_{MI} decreased, Sekine et al. (2006), Laulhé et al. (2010). What was more remarkable, however, was that the small amount of Rh strongly suppressed superlattice reflections below 14 K and eventually annihilated them, driving the system into a similar reentrant metallic state, Saito et al. (2014), as observed when Ce substituted for Pr.

Compared to a rich variety of phase transitions in $PrFe_4P_{12}$ and $PrRu_4P_{12}$, the last member of Pr-filled phosphide skutterudites, $PrOs_4P_{12}$, is an uneventful metal with no sign of any phase transition down to 50 mK. Magnetic susceptibility measurements by Sekine et al. (1997) and subsequent measurements by Yuhasz et al. (2007) indicated a Curie–Weiss behavior with the effective moment $\mu_{eff} = 3.63\ \mu_B$ and $\theta_{CW} = -17$ K. Specific heat measurements by Matsuhira et al. (2005b) revealed a Schottky-type anomaly near 13 K, from which the energy separation between the non-magnetic Γ_1 ground state and the first excited $\Gamma_4^{(2)}$ state was estimated as 3.78 meV (44 K). The extracted Sommerfeld constant $\gamma \approx 56.5$ mJmol^{-1}K^{-2} and the cyclotron effective mass obtained from dHvA measurements by Sugawara et al. (2009) yielded $m^* = 18\ m_e$, a modest enhancement due to 4f-electrons.

3.3.4.3.2 Pr-Filled Arsenides

Although the synthesis of filled arsenide skutterudites was demonstrated in the original studies by Braun and Jeitschko (1980a) and some basic data were obtained on polycrystalline forms of the structure, detailed studies of physical properties, including magnetism, commenced only after Henkie et al. (2008) developed high-pressure flux synthesis of single crystals based on a mineralization process in a molten Cd:As flux.

The three praseodymium-filled arsenides, $PrFe_4As_{12}$, $PrRu_4As_{12}$, and $PrOs_4As_{12}$, unlike other Pr-filled skutterudites, behave as the more conventional solids typified by FM ordering, BCS-type superconductivity, but also having somewhat heavy electrons. Their key magnetic properties are noted in the following few paragraphs.

The foundation work regarding magnetic properties of **PrFe$_4$As$_{12}$** is comprehensive measurements of Sayles et al. (2008) on single crystals that included magnetization, dc and ac susceptibility, electrical resistivity, ultrasonic attenuation, and specific heat carried out in various magnetic fields as well as under pressure. Electrical resistivity is entirely metallic and shows a clear break at $T_C = 18$ K, followed by a rapid, an order of magnitude decreasing resistivity, until below 2 K, it tends to saturate. The anomaly at 18 K reveals a transition to a FM state and the rapidly diminishing resistivity reflects a dramatically reduced charge carrier scattering associated with the ordered state. FM ordering was documented by the development of a spontaneous magnetic moment below 18 K and a steeply rising magnetic susceptibility at T_C. From the Curie–Weiss behavior between 100 K and 300 K, the effective magnetic moment $\mu_{eff} = 3.98\ \mu_B$/f.u. and $\theta_{CW} = 4.1$ K were obtained. Because the measured effective moment was larger than the Pr^{3+} free-ion moment of $\mu_{eff}^{Pr} = 3.58\ \mu_B$, this suggested that the polyanion [Fe$_4$As$_{12}$] carries a magnetic moment. Its value can be estimated from the measured effective moment μ_{eff} and the Pr^{3+} free-ion value μ_{eff}^{Pr} *via* equation

$$\mu_{eff} = \sqrt{\left(\mu_{eff}^{Pr}\right)^2 + \left(\mu_{eff}^{[Fe_4As_{12}]}\right)^2},\qquad(3.96)$$

and turned out to be $\mu_{eff}^{[Fe_4As_{12}]} = 1.74\ \mu_B$/f.u., remarkably close to a spin-only value of Fe^{3+} of $1.73\ \mu_B$. Magnetization measurements at 2 K indicated a change in the easy axis of magnetization from the [100] to the [111] direction at 0.8 T, and the saturation magnetization from a 2 K isotherm yielded $M_s = 2.3\ \mu_B$/f.u. in the field of 5.5 T. This value is not too far from $2.0\ \mu_B$/f.u., the value expected for the magnetic triplet $\Gamma_4^{(2)}$ ground state. The hysteresis loop was quite narrow, showing the coercive field of only 1.1 mT. Evaluation of the specific heat, characterized by a sharp discontinuity at T_C, yielded a rather substantial electronic specific heat $\gamma = 340$ mJmol^{-1}K^{-2} and the Debye temperature $\theta_D = 356$ K. A subsequent polarized neutron study by Wisniewski et al. (2011) documented the presence of two magnetic contributions, one due to Pr^{3+} ions and the other one due to Fe^{3+} ions. They were revealed as two magnetic sublattices, where on one sublattice the magnetic moments of praseodymium are fully aligned with the applied field (even at fields as low as 0.3 T), while the magnetic moments of iron on the other sublattice nearly compensate each other. Although the overall magnetic moment of all contributing Fe^{3+} ions also follows the orientation of the applied field, the individual iron moments point in various directions, depending on the crystal orientation. Situations depicted in Figure 3.48 refer to a magnetic field applied along the [001], $\left[01\bar{1}\right]$, and $\left[52\bar{8}\right]$ crystallographic directions.

As shown by Shirotani et al. (1997), PrRu$_4$As$_{12}$ is an ordinary paramagnetic metal that undergoes a SC transition at 2.4 K, documented also by a large diamagnetic response in its magnetic

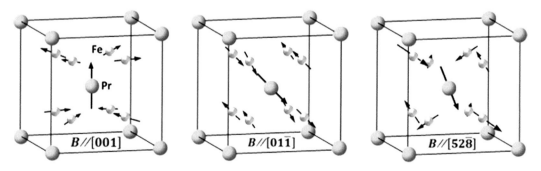

FIGURE 3.48 Magnetic structure of PrFe$_4$As$_{12}$ at 2 K and in magnetic field of 0.3 T applied along [001], $\left[01\bar{1}\right]$, and $\left[52\bar{8}\right]$ crystallographic directions. Redrawn from P. Wisniewski et al., *Journal of the Physical Society of Japan* **80**, SA012 (2011). With permission from the Japanese Physical Society.

susceptibility. In fact, most of the subsequent studies on $PrRu_4As_{12}$ were directed toward the characterization of its SC state using measurements of magnetic susceptibility, specific heat, magnetization, and [75]As-NQR, including the effect of an applied magnetic field, Namiki et al. (2007), Shimizu et al. (2007) and (2008), Maple et al. (2008), and Sayles et al. (2010). From the perspective of magnetism, the relevant issues are the Curie–Weiss behavior and the CEF energy level scheme.

Measurements of magnetic susceptibility by Namiki et al. (2007) made on a polycrystalline sample yielded the effective magnetic moment of $\mu_{eff} = 3.30\ \mu_B/Pr^{3+}$, a bit smaller than the Pr^{3+} free-ion value of 3.58 μ_B, and the Curie–Weiss temperature $\theta_{CW} = -11$ K. Below 10 K, susceptibility tended to saturate. From this and the fact that the skutterudite became a superconductor, it was surmised that the ground state is the non-magnetic singlet Γ_1 and the first excited state is the triplet $\Gamma_4^{(1)}$. The Γ_1 ground state is consistent with the specific heat measurements, from which it also followed that the electronic entropy at 8 K was only 20% of $R\ln2$, denying the existence of a degenerate ground state. As the best fit to their magnetic susceptibility data, Namiki et al. (2007) reported the energy spacing between Γ_1 and $\Gamma_4^{(1)}$ as about 2.58 meV (30 K). From fits of C/T vs. T^2 above the SC transition temperature, the electronic specific heat $\gamma \approx 95$ mJmol^{-1}K^{-2} was obtained. A similar set of measurements, but using a single crystal specimen with residual resistivity ratio (RRR) = 94, was performed by Sayles et al. (2010) and yielded $\mu_{eff} = 3.52\ \mu_B/Pr^{3+}$, in excellent agreement with the free-ion value, and $\theta_{CW} = -6.2$ K. A notable saturation trend in the magnetic susceptibility below 75 K was attributed to the effect of CEF splitting that was assessed as the Γ_1 ground state with $\Gamma_4^{(1)}$ some 8.2 meV (95 K) above, followed by Γ_{23} at 14 meV (162 K) and $\Gamma_4^{(2)}$ at 43.3 meV (502 K). The extracted electronic specific heat measured on the single crystal was $\gamma \approx 70$ mJmol^{-1}K^{-2}, some 26% lower value than that obtained for the polycrystalline sample that may have contained impurities.

Measurements of the physical properties of $PrOs_4As_{12}$ were carried out on single crystals by Yuhasz et al. (2006), and the authors identified multiple low-temperature phases. An increasing magnetic susceptibility as the temperature decreased was interrupted by a prominent peak at $T_1 = 2.3$ K, Figure 3.49a, that shifted to lower temperatures in an applied magnetic field. This strongly suggested the presence of an AFM transition. From two substantially linear regions on the plot of $\chi(T)^{-1}$ vs. T, $50K \leq T \leq 300$ K and 3.5 K $\leq T \leq 20$ K, indicating the Curie–Weiss behavior, the respective effective magnetic moments $\mu_{eff} = 3.81\ \mu_B/f.u.$ and $\mu_{eff} = 2.77\ \mu_B/f.u.$ and the corresponding Curie–Weiss

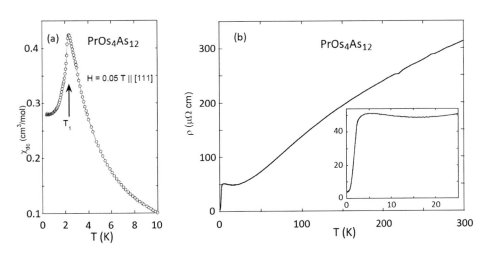

FIGURE 3.49 (a) Low-temperature dc susceptibility of the $PrOs_4As_{12}$ single crystal depicting a dramatic peak at temperature $T_1 = 2.2$ K. (b) Temperature dependence of electrical resistivity at low temperatures showing a Kondo behavior followed by a rapid decrease as the sample enters the ordered phase. Redrawn and adapted from W. M. Yuhasz et al., *Physical Review B* **73**, 144409 (2006). With permission from the American Physical Society.

temperatures $\theta_{CW} = -25.8$ K and $\theta_{CW} = 0.44$ K were obtained. In the upper temperature range, the effective moment was larger than the Pr^{3+} free-ion value and could possibly imply polarization of conduction electrons resulting in a constant Pauli paramagnetic contribution of about 5.5×10^{-4} cm³/mol. The authors mentioned that μ_{eff} and θ_{CW} for the lower T-range remained unchanged when this small Pauli contribution was included. From the value of $\mu_{eff} = 2.77$ μ_B/f.u. in the low-temperature range, it was surmised that the ground state of $PrOs_4As_{12}$ is a magnetic triplet $\Gamma_4^{(2)}$ (magnetic triplet Γ_5 in the cubic O_h symmetry), which has a theoretical effective moment of 2.83 μ_B/f.u. This assignment of the ground state was also supported by a saturation magnetization value $M_s \approx 2.7$ μ_B/f.u. attained at 2 K in the field of 5.5 T, which is fairly close to the theoretical value of 2.0 μ_B/f.u. expected for the $\Gamma_4^{(2)}$ ground state. Ultrasonic probing of elastic anomalies at temperatures near the AFM transition by Yanagisawa et al. (2008a) was also consistent with the $\Gamma_4^{(2)}$ ground state. A definitive confirmation of the CEF level assignment was based on measurements by Chi et al. (2008), where the magnetic triplet ground state $\Gamma_4^{(2)}$ with a closely lying first excited state Γ_1 singlet at 0.4 meV (4.6 K), was followed by a large separation to the magnetic triplet $\Gamma_4^{(1)}$ at 13 meV (151 K) and, finally, the non-magnetic doublet Γ_{23} at 23 meV (267 K). Magnetization measured in a field parallel to the [111] direction indicated no hysteresis. Apart from an inflection point, Yuhasz et al. (2006) noted two kinks on a field-dependent magnetization isotherm at 0.425 K, which were identified as fields H_1 and H_2. Both shifted to lower fields on higher temperature isotherms. Specific heat measurements on a collection of small single crystals displayed a well-developed anomaly, shown in Figure 3.50a, where two temperatures could be identified: temperature $T_2 = 2.3$ K associated with a shoulder and a prominent peak at $T_1 = 2.2$ K. Interestingly, as mentioned in the paper, the specific heat on a similar batch of single crystals was also measured independently at the Lawrence Livermore National Laboratory and the peak at T_1 appeared to be split into two peaks at T_1 and T_1' separated by 0.02 K, suggesting a possibility of yet additional transition or perhaps indicating a compositional variation among the single crystal batches. A careful analysis of the specific heat data assuming four contributing terms, namely, the nuclear Schottky, electronic, magnetic, and lattice terms, were weighted differently in different temperature regions and, above 10 K, returned the electronic specific heat $\gamma \approx 200$ mJmol^{-1}K^{-2} and the Debye temperature $\theta_D = 260$ K. In the ordered phase, at $T < 0.6$ T_1, assuming the nuclear Schottky term proportional to T^{-2} and writing the magnetic term as BT^{n+2}, the best fit was obtained for $n = 3.2$, close to the expected value of $n = 3$ for AFM spin waves. In this case,

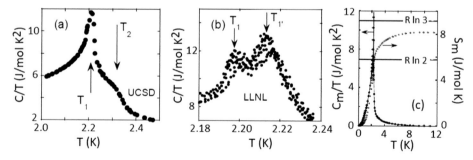

FIGURE 3.50 (a) Specific heat C divided by temperature T plotted as a function of temperature for a collection of single crystals of $PrOs_4As_{12}$ measured at the University of California San Diego. Temperature T_1 marks a phase transition to the AFM phase, while a slightly higher temperature T_2 indicates a transition to a phase where antiferro-quadrupolar interactions are believed to dominate. (b) A similar batch of single crystals of $PrOs_4As_{12}$ measured at the Lawrence Livermore National Laboratory. Here, the peak at T_1 appears to be split into two peaks separated by 0.02 K, possibly suggesting yet another phase boundary or perhaps a compositional variation between the single crystal batches. (c) Magnetic specific heat C_m divided by T vs. absolute temperature (left scale), and magnetic entropy S_m as a function of temperature (right scale). Redrawn and adapted from W. M. Yuhasz et al., *Physical Review B* **73**, 144409 (2006). With permission from the American Physical Society.

the electronic specific heat coefficient turned out to be quite large reaching $\gamma \approx 1$ Jmol^{-1}K^{-2}. The temperature-dependent coefficient γ has often been observed in heavy fermion compounds, Maple et al. (2004). By integrating C_m/T with temperature, after extracting the magnetic contribution to the specific heat C_m, the magnetic entropy S_m was obtained, as shown in Figure 3.50b. The entropy released is 90% of $R\ln3$ and is essentially constant above 10 K. A small difference compared to a triply degenerate ground state value of $R\ln3$ was explained as due to the hybridization of the localized 4f-electrons with the conduction electrons. The electrical resistivity of PrOs$_4$As$_{12}$ is strongly metallic but reaches a minimum near 16.5 K and then slightly rises until about 5 K. This documents the Kondo-like behavior. Below 5 K, the resistivity drops rapidly as the ordered state is reached, Figure 3.49b. From resistivity measurements in a magnetic field and assuming a single impurity ion with the spin of 1, the Kondo temperature T_K on the order of unity was estimated by Maple et al. (2006). Pressures up to 23 kbar (2.3 GPa) had little effect on the resistivity between 1K and 300 K. Although PrOs$_4$As$_{12}$ has a similar Fermi surface to its non-magnetic LaOs$_4$As$_{12}$ cousin, its effective masses are about a factor of 2 larger, regardless of the orientation of the magnetic field in which dHvA oscillations were observed by Singleton et al. (2008). This confirms the heavy fermion nature of carriers suggested by a large electronic specific heat coefficient in the AFM phase of the structure.

Taking into account all experimental studies related to PrOs$_4$As$_{12}$, a magnetic field $vs.$ temperature phase diagram was constructed by Yanagisawa et al. (2008a), which describes its rich phase structure at low temperatures, Figure 3.51. Thus, PrOs$_4$As$_{12}$ has an AFM ground state at temperatures below $T_1 = 2.2$ K. However, at a temperature merely a couple tens of a degree above T_1, this skutterudite forms another ordered phase that extends to between 3 and 4T, depending on the magnetic field direction. The phase is believed to be dominated by antiferro-quadrupolar (AFQ) interactions. Interestingly, ultrasonic measurements by Yanagisawa et al. (2008b) suggested that a phase boundary exists roughly in the middle of this AFQ phase, forming phases II and II', associated with orbital degrees of freedom, perhaps of the quadrupolar or octupolar interaction nature. In other words, a multipolar order is likely an important feature of the low-temperature physics of PrOs$_4$As$_{12}$.

3.3.4.3.3   Pr-Filled Antimonides

Of all filled skutterudites, Pr-filled antimonides have generated the greatest interest. The enthusiasm was driven primarily by the discovery of superconductivity with heavy fermion features in PrOs$_4$Sb$_{12}$. This exotic form of a skutterudite has rivaled the eminent position of heavy fermion structures, such as U-based compounds and other highly correlated electron systems. A full discussion of its SC properties and related issues is presented in section 2.4. Here, I will discuss only its magnetic properties. The other two Pr-filled antimonides, PrFe$_4$Sb$_{12}$ and PrRu$_4$Sb$_{12}$, are interesting in their own right, and their key magnetic properties are presented here also, in fact, starting with PrFe$_4$Sb$_1$.

The case of PrFe$_4$Sb$_{12}$ is a textbook example of what can happen when high-quality measurements are carried out on a structure that is highly defected and nowhere near the nominal composition. Specifically, the defects in this case arise from a severely incomplete filling of the structural voids. Although in the original report by Danebrock et al. (1996) the void occupancy was not stated, it must have been similar to a merely 73% filling reported in subsequent measurements by Bauer et al. (2002a) because both teams worked with polycrystalline samples prepared at atmospheric pressure. Remarkably, single crystals grown by a self-flux technique by Butch et al. (2005) have not resulted in a major improvement of the filling as some 13% of voids remained unfilled. All of the above studies provided interesting results, noting a formation of a long-range magnetic order near 4.5 K, and the fits to the magnetic susceptibility returning the CEF energy scheme with a trivalent magnetic $\Gamma_4^{(2)}$state (actually a magnetic triplet Γ_5 because they worked in the cubic O$_h$ point group) as the ground state. Only when fully filled PrFe$_4$Sb$_{12}$ was synthesized under a high pressure of 4 GPa by Tanaka et al. (2007) was it realized how sensitive the physical properties of skutterudites are on

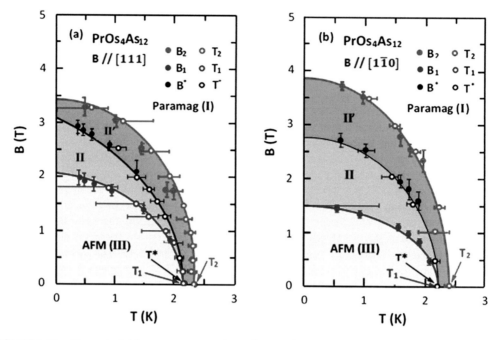

FIGURE 3.51 Magnetic field vs. temperature phase diagram for $PrOs_4As_{12}$ determined from measurements of the elastic constants measured as a function of temperature for magnetic fields oriented (a) along the [111] crystallographic direction and (b) along the $[1\bar{1}0]$ direction. T_2 and B_2 indicate the temperature and magnetic field where the first set of anomalies (phase transitions) occurs in C_{L111} as the sample is cooled and the field is gradually reduced from high fields. Temperature T_1 and field B_1 mark the entry into the AFM state at different fields and temperatures. Temperatures T^* and fields B^* indicate the entry into an additional phase, which splits the ordered phase lying between the AFM and paramagnetic phases in two, labeled as phases II and II'. The II – II' phase boundary is present for both B // [111] and B // $[1\bar{1}0]$ field orientations and may be associated with orbital degrees of freedom. Solid lines that separate various phases are guides for the eyes. Redrawn and adapted from T. Yanagisawa et al., *Journal of the Physical Society of Japan* **77**, Suppl. A, 225 (2008a). With permission from the Physical Society of Japan.

the filling fraction, and how unrepresentative the initial measurements were with respect to a fully stoichiometric $PrFe_4Sb_{12}$ skutterudite structure.

First, contrary to measurements on incompletely filled $Pr_xFe_4Sb_{12}$, no phase transition was observed down to 0.15 K in the specific heat. With no hints of a magnetic order, it could not have been the magnetic $\Gamma_4^{(2)}$ triplet as the ground state but, rather, a non-magnetic Γ_1 singlet. Moreover, from the released entropy of about $R\ln4$ at 25 K, it was likely that the triplet $\Gamma_4^{(2)}$ is the first excited state positioned some 1.9 meV (22 K) above. Another notable feature of fully filled $PrFe_4Sb_{12}$ was the absence of a spontaneous magnetic moment. Interestingly, the residual resistivity of a fully filled $PrFe_4Sb_{12}$ of 24 $\mu\Omega$cm was several times smaller than the residual resistivity of 127 $\mu\Omega$cm reported for an incompletely filled single crysta. Where the filling seemed to have little difference was the extended range of the Curie–Weiss behavior from 300 K down to about 25 K, yielding comparably large values (accounting for incomplete filling) of the effective magnetic moment in the range from $\mu_{eff} = 4.19$ μ_B/f.u to 4.6 μ_B/f.u., the values significantly exceeding the Pr^{3+} free-ion value of 3.58 μ_B. This implied a considerable magnetic moment of about 2.7 μ_B associated with the $[Fe_4Sb_{12}]$ polyanion, *via* the d^5 configuration of Fe^{3+}. The respective Curie–Weiss temperatures θ_{CW} fell into the range from −22 K to +4 K. Thus, unlike phosphite-based skutterudites where the magnetic moment is more-or-less tied to the rare-earth filler, in antimonide skutterudites both the rare earth and the $[Fe_4Sb_{12}]$ polyanion contribute to the overall magnetic moment. Magnetization measurements of stoichiometric $PrFe_4Sb_{12}$ were extended to very high pulsed fields of up to 52 T by Yamada et al. (2007) and revealed a shoulder-like feature around 20 T, speculated to arise from a CEF level crossing. At the highest field of 52 T, the magnetization attained 3.1 μ_B/f.u. After subtracting the magnetization of $LaFe_4Sb_{12}$, the resulting magnetization per Pr ion turned out to be 2.4 μ_B. A puzzling question regarding why the CEF energy levels are so sensitive to the filling fraction has not yet been definitely answered, although as the candidates are different environments of Pr^{3+} ions that have empty voids as their neighbors, a reduced Fermi energy of less than fully filled structures that shifts to a region of higher DOS, and possibly also altered interaction between 4f-electrons of Pr and 3d-electrons of Fe.

$PrRu_4Sb_{12}$ is a metallic compound that undergoes a SC transition at $T_c = 1.04$ K, as measured on a flux-grown polycrystalline sample with RRR = 25 by Takeda and Ishikawa (2000a). Single crystals with a RRR as high as 100 show substantially the same T_s values, Abe et al. (2002). The paramagnetic character of magnetic susceptibility typified by the Curie–Weiss behavior above 50 K yields the effective magnetic moment $\mu_{eff} = 3.58$ μ_B/Pr^{3+} in perfect agreement with the Pr^{3+} free-ion value and $\theta_{CW} = -11$ K. Below T_s, a full diamagnetic response documented the bulk nature of superconductivity. The transition to a SC state is also reflected in a distinct anomaly in the specific heat, and the extracted Sommerfeld coefficient of $\gamma = 59$ mJmol^{-1}K^{-2} indicated a slight mass renormalization, in accordance with the value of the cyclotron mass, $m_c = (1.5–1.8)$ m_e, measured by Matsuda et al. (2002) on all three observed dHvA frequency branches. The Debye temperature obtained from the fit was $\theta_D = 232$ K. With no magnetic transition at low temperatures and consistent with a small magnetization of 0.27 μ_B/Pr^{3+} at 2.2 K in the field of 4.7 T, $PrRu_4Sb_{12}$ is often classified as a van Vleck paramagnet. Being a superconductor, the ground state of $PrRu_4Sb_{12}$ must be non-magnetic and presumably Γ_1 singlet. The assignment was confirmed by Abe et al. (2002) by fitting the temperature dependence of resistivity with the CEF level scheme, which returned Γ_1 (0 K) ground state, the first excited state a magnetic triplet $\Gamma_4^{(2)}$ at about 6 meV (70 K) above the ground state, followed by a magnetic triplet $\Gamma_4^{(1)}$ at 7.8–12 meV (90–140 K), and a doublet Γ_{23} at 10.3 meV (120 K). More definitive determination of the ground state using INS, Adroja et al. (2005), and ultrasonic measurements, Kumagai et al. (2003), verified the order of the CEF energy levels.

As shown by Bauer et al. (2002c), $PrOs_4Sb_{12}$ becomes a superconductor below $T_c = 1.85$ K. The highly exotic SC state, coupled with a heavy fermion nature of its charge carriers, generated tremendous excitement and stimulated vigorous research activity. The essential issues are discussed in detail in section 2.4. The implied non-magnetic ground state, attested to also by the temperature dependence of the magnetic susceptibility (finite value as $T \rightarrow 0$), presented two choices: either a Γ_1 singlet or a non-Kramers Γ_{23} doublet. Initially, attempting to draw on an analogy with heavy fermion systems, the Γ_{23} doublet was favored as it ensured the necessary conditions for quadrupolar Kondo fluctuations believed to impart the heavy fermion nature to the charge carriers. However, INS experiments by Goremychkin et al. (2004) and Kuwahara et al. (2004) showed convincingly that the ground state of $PrOs_4Sb_{12}$ is the Γ_1 singlet. The first excited state, merely 0.7 meV (8.2 K) above the ground state, is a magnetic triplet $\Gamma_4^{(2)}$. These two states dominate the low-temperature physics. Higher CEF levels, $\Gamma_4^{(1)}$ at 11.62 meV (135 K) and Γ_{23} at 17.69 meV (205 K), lie far above and are irrelevant as far as low-temperature properties are concerned. Comparing the CEF level scheme of $PrRu_4Sb_{12}$ with that of $PrOs_4Sb_{12}$, we see the same sequence of the energy levels. The key difference between the two is a closeness of the ground state and the first excited state (only 8.2 K) in $PrOs_4Sb_{12}$, while in $PrRu_4Sb_{12}$ the spacing is an order of magnitude larger (70 K). This makes all the difference because the thermal energy k_BT is able to initiate excitations in $PrOs_4Sb_{12}$ already at liquid helium temperatures, while such excitations in $PrRu_4Sb_{12}$ are inaccessible until at much high temperatures. From a well-developed Curie–Weiss behavior above 50 K, the effective magnetic moment $\mu_{eff} = 2.97$ μ_B/Pr^{3+} was obtained, somewhat smaller than the Pr^{3+} free-ion value of 3.58 μ_B. The Curie–Weiss temperature was quoted as $\theta_{CW} = -16$ K. As discussed in Chapter 2, applied pressure reduces the SC transition temperature of $PrRu_4Sb_{12}$, Maple et al. (2002). Beyond that, the pressure also affects the boundary of the antiferro-quadrupolar (AFQ) phase due to its effect on the two lowest CEF-split energy levels, the ground state Γ_1 and the first excited state $\Gamma_4^{(2)}$. Specifically, the energy separation between the two decreases with increasing pressure, and at 1.45 GPa, the splitting is reduced by 16 %, according to Tayama et al. (2007).

3.3.4.4 Nd-Filled Skutterudites

Neodymium, with its electronic configuration $[\mathrm{Xe}]4f^46s^2$, enters the skutterudite void as a trivalent ion Nd^{3+} ($4f^3$) by losing its two $6s$-electrons and one $4f$-electron. The resulting ten-fold degenerate $J = 9/2$ Hund's rule multiplet is split by CEF into a Γ_5 doublet (Γ_6 doublet in O_h symmetry) and two quartets $\Gamma_{67}^{(1)}$ and $\Gamma_{67}^{(2)}$ ($\Gamma_8^{(1)}$ and $\Gamma_8^{(2)}$ quartets in O_h). On account of well-localized $4f$-electrons of Nd, all $\mathrm{NdT_4X_{12}}$ (T = Fe, Ru, and Os; X = P, As, and Sb) skutterudites show rather similar low-temperature properties characterized by the tendency to order ferromagnetically. Most of the interest has focused on properties of $\mathrm{NdFe_4P_{12}}$ and $\mathrm{NdOs_4Sb_{12}}$, with a considerably smaller number of publications dedicated to other Nd-filled skutterudites.

3.3.4.4.1 Nd-Filled Phosphides

All three Nd-filled phosphide skutterudites were first synthesized by Jeitschko and Braun (1977), and explorations of their physical properties commenced with measurements by Torikachvili et al. (1987) of $\mathrm{NdFe_4P_{12}}$ single crystals grown from molten Sn flux. Electrical resistivity revealed metallic temperature dependence with a broad minimum near 30 K followed by a sharp drop near 2 K, signaling the development of a magnetically ordered state. Magnetic susceptibility in a field of 1 T yielded a near-linear dependence of χ^{-1} vs. T with a slight downturn below about 150 K. The Curie–Weiss behavior at higher temperatures returned the effective magnetic moment of 3.53 μ_B, roughly equal to the Nd^{3+} free ion value of 3.62 μ_B, which would support the notion that the 4f-electrons of $\mathrm{NdFe_4P_{12}}$ are well localized. However, the slope below 150 K resulted in a smaller effective moment of 2.46 μ_B, the difference explained as the effect of CEF. AC magnetic susceptibility indicated a pronounced peak near 2 K and magnetization at 1.4 K and fields up to 5 T saturated with a value of 1.72 μ_B/Nd^{3+}, a considerably lower value than expected based on the Hund's rule 3.27 μ_B/Nd^{3+}. Hysteresis in the magnetization with a small coercive field suggested FM ordering. A well-developed λ-like anomaly in the specific heat peaking near 2 K correlated well with the sharp drop in the resistivity and the large peak in the ac susceptibility and provided the key experimental evidence for the bulk magnetic ordering taking place below about 2 K. The specific heat at temperatures below 2 K followed an approximately T^3-dependence, in conflict with the $T^{3/2}$-power law, one would expect for FM spin waves discussed in section 3.1.4.2. The extracted magnetic entropy of $\mathrm{NdFe_4P_{12}}$ reached $R\ln 4$ near 7.4 K, suggesting that the ground state is one of the Γ_{67} quartets. In a magnetic field, the sharply decreasing resistivity near 2 K became progressively more broadened with a decreased magnitude, indicating negative magnetoresistance associated with the suppression of spin-flip scattering. Essentially identical resistivity behavior was observed by Sugawara et al. (2000) and by Sato et al. (2000), except that the ordering temperature was pinpointed more precisely at $T_C = 1.9$ K. While a $-\ln T$ dependence of resistivity was notable in the range of 3K–15 K, the temperature dependence below T_C was described as being proportional to T^4, with twice as large exponent as expected for ordinary magnon scattering. Approximately doubled exponents in the power law of the specific heat and electrical resistivity below T_C are, of course, inconsistent with the usual magnon dispersion $\nabla\omega \propto q^2$ and would rather imply $\nabla\omega \propto q$. Longitudinal magnetoresistance of a single crystal of $\mathrm{NdFe_4P_{12}}$ with $B//\langle110\rangle$ and the field strength up to 5 T was measured by Sato et al. (2003a) and indicated negative values, which the authors were able to scale with the bulk magnetization as

$$\rho(B) = A_m \left[1 - \left(\frac{M(B)}{M(0)} \right)^2 \right], \tag{3.97}$$

where a single value of A_m reproduced all experimental curves over the entire range of magnetic fields.

The crucial identification of the magnetic order in $NdFe_4P_{12}$ came from neutron diffraction measurements by Keller et al. (2001), which confirmed the FM ordering at $T_C \approx 1.9$ K and determined the ordered magnetic moment of the Nd^{3+} ion as 1.61 μ_B at 1.5 K, in good agreement with the value 1.72 μ_B, obtained by Torikachvili et al. (1987) from the saturation magnetization measured at 1.4 K and 5 T. The CEF energy level scheme, the issue critical for the evaluation of magnetic properties of rare earth-filled skutterudites, was investigated by Nakanishi et al. (2004) *via* measurements of the elastic constants. Here, an ultrasonic pulse distorts the immediate environment of a rare-earth ion, and the generated strain couples to the quadrupolar moment of the 4f-electrons, perturbing their CEF potential. Provided the CEF ground state is orbitally degenerate with respect to the quadrupolar moment, softening in the elastic constants is expected. In addition, one also needs to take into account inter-ionic quadrupolar interaction among 4f-electrons located at different sites. Because the strain sensitively modulates the CEF levels, measurements of the elastic constants and fitting them with appropriate theoretical expressions are considered as one of the most reliable probes of the CEF energy level scheme. As documented in Figure 3.52, an obvious softening is observed in an elastic constant C_{11} as a sharp drop near T_C on an otherwise increasing (hardening) trend as the temperature decreases. In zero applied magnetic field, the drop is needle sharp, but becomes shallow and shifts to higher temperatures as the field increases. In the field of 3 T, it is completely wiped out.

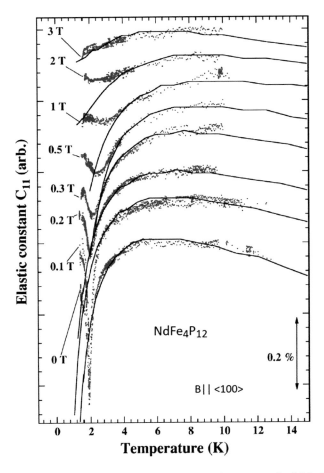

FIGURE 3.52 Temperature dependence of C_{11} measured at various magnetic fields $B//\langle 100 \rangle$ indicated by numbers in Tesla. The gray dots are experimental data points, and the solid lines are theoretical fits calculated using Equations 3.97 and 3.98. Redrawn from Y. Nakanishi et al., *Physical Review B* **69**, 064409 (2004). With permission from the American Physical Society.

The data points are indicated by small gray dots, and solid lines represent theoretical fits based on a formula for the temperature dependence of the elastic constant developed by Nakamura et al. (1994),

$$C_\Gamma\left(T\right) = C_\Gamma^0\left(T\right) - N g_\Gamma^2 \frac{\chi_\Gamma\left(T\right)}{1 - g_\Gamma' \chi_\Gamma\left(T\right)}. \tag{3.98}$$

Here, N is the number of Nd ions in the unit cell, $C_\Gamma^0\left(T\right)$ is the lattice contribution to the elastic constant originating from anharmonic effects having no quadrupolar-strain coupling g_Γ, g_Γ' is the quadrupolar interaction between Nd ions, and χ_Γ is the quadrupolar susceptibility for the $4f$ electronic state in the cubic CEF potential, given by

$$\chi_\Gamma\left(T\right) = \sum_{ik} \frac{exp\left(-E_{ik}^{(0)}/k_B T\right)}{Z} \left(\frac{1}{k_B T} \left| \langle ik | O_\Gamma | ik \rangle \right|^2 - 2 \sum_{jl} \frac{\left| \langle ik | O_\Gamma | jl \rangle \right|^2}{E_i - E_j} \right). \tag{3.99}$$

In Equation 3.99, $|ik\rangle$ stands for the k-th eigenfunction of the i-th CEF level, and O_Γ is the quadrupolar moment. Because no softening is expected in the case of the Γ_5 doublet being the ground state, the fitting was done with the $\Gamma_{67}^{(1)}$ quartet as the ground state ($\Gamma_{67}^{(2)}$ quartet yielded similarly good fits but $\Gamma_{67}^{(1)}$ was favored by magnetization measurements), consistent with the magnetic character of the ground state. The best fits shown in Figure 3.52 correspond to the $\Gamma_{67}^{(1)}$ quartet as the ground state, the doublet Γ_5 19.1 meV (222 K) above the ground state, and the $\Gamma_{67}^{(2)}$ quartet further 8.7 meV (101 K) above.

NdRu$_4$P$_{12}$, in the form of a single-phase polycrystalline sample, was reported by Sekine et al. (1998) as a FM metal with the Curie temperature $T_C = 1.5$ K. ^{101}Ru-NQR measurements by Masaki et al. (2008a) detected a well-developed signal, the intensity of which rapidly diminished as the temperature decreased below 4.2 K and disappeared altogether below 1.6 K when NdRu$_4$P$_{12}$ became a ferromagnet. Continuing with their assessment of NdRu$_4$P$_{12}$, Masaki et al. (2008b) added measurements of ^{31}P-NMR and ac magnetic susceptibility. Because ^{101}Ru-NQR and ^{31}P-NMR probe different sites, an element-specific microscopic viewpoint was gained. Furthermore, a combination of NMR with various magnetic fields and zero-field NQR allowed systematic investigations of the field dependence of spin correlations. The measurements confirmed the development of a FM ordering below 1.6 K and brought to light an interesting feature, namely, a coexistence of the FM spin correlation with $q = 0$ and AFM fluctuations with $q \neq 0$. Such dual spin fluctuations were explained by taking into account contributions of localized $4f$-electrons and conduction electrons that have the nesting property of the Fermi surface.

Regarding NdOs$_4$P$_{12}$, there is only one report, by Magishi et al. (2010), describing transport and magnetic properties of a single crystal sample. Similar to the other two Nd-filled phosphide skutterudites, NdOs$_4$P$_{12}$ develops an ordered FM state, in this case below $T_C = 1.15$ K. The metallic electrical resistivity displays a shallow minimum near 20 K, followed by $-\ln T$ dependence and a sharp drop starting at T_C. A very similar trend in the magnetic susceptibility with NdFe$_4$P$_{12}$ and NdRu$_4$P$_{12}$ shows a rapid rise below about 10 K that is suppressed by an external magnetic field of 5 T. The Curie–Weiss behavior above 50 K yields the effective magnetic moment $\mu_{eff} = 3.03$ μ_B/Nd^{3+} and a negative Curie–Weiss temperature $\theta_{CW} = -23$ K. Below a mild hump near 50 K, the Curie–Weiss law returns a smaller effective moment of 2.34 μ_B/Nd^{3+} and a positive $\theta_{CW} = 1.0$ K. From the presence of both negative and positive Curie–Weiss temperatures, the authors surmised that FM and AFM exchange interactions are intrinsic to the structure. ^{31}P-NMR spectra revealed an anisotropic Knight shift that increased rapidly below 10 K. The magnetic field had little effect on the Knight shift above 10 K, but diminished it at lower temperatures. The nuclear spin-lattice relaxation rate $1/T_1$ divided by absolute temperature, $1/T_1 T$, (the parameter usually plotted as a function of temperature), showed a sharp peak below 2 K that was gradually suppressed and shifted to somewhat higher temperature upon application of the magnetic field. Overall, the behavior was very similar to NdRu$_4$P$_{12}$, which prompted the authors to conclude that the spin fluctuations with $q = 0$ and $q \neq 0$ coexist in $NdOs_4P_{12}$ as well and are related to the Fermi surface nesting. However, unlike in the case of $NdFe_4P_{12}$ and $NdRu_4P_{12}$, dHvA measurements by Sugawara et al. (2009) have shown no nesting property in the case of $NdOs_4P_{12}$ and thus altogether different explanations will have to be found.

3.3.4.4.2 Nd-Filled Arsenides

As mentioned in the introduction to Nd-filled skutterudites and as is the general state of affairs with arsenide skutterudites, very few studies have been made on this family of skutterudites, primarily due to difficulties with their synthesis. Except for the lattice parameters determined by Jeitschko and Braun (1977), the first study of physical properties of **$NdFe_4As_{12}$** had to wait until Higashinaka et al. (2013) measured the electrical resistivity, specific heat, and magnetic susceptibility for small single crystals prepared by the As flux method under 4 GPa pressure. In accordance with all other Nd-filled skutterudites, $NdFe_4As_{12}$ is a ferromagnet, in this case at a higher $T_C = 14.6$ K. The ordering is reflected in a rapidly developing spontaneous magnetization, an anomaly in the specific heat, and as a sharply decreasing resistivity below T_C. The Curie–Weiss behavior above 150 K yielded the effective magnetic moment $\mu_{eff}^{HT} = 4.43$ μ_B/f.u. and the Curie–Weiss temperature $\theta_{CW}^{HT} = -35$ K. The effective moment was larger than the Nd^{3+} free-ion value of 3.67 μ_B, likely due to a contribution

from 3d electrons of Fe, estimated from $\mu_{eff}^{HT} = \left[\left(\mu_{eff}^{Nd} \right)^2 + \left(\mu_{eff}^{Fe} \right)^2 \right]^{1/2}$ to be $\mu_{eff}^{Fe} = 2.48$ μ_B (assuming

the same ordering temperature for both Nd and Fe moments). The negative value of θ_{CW} suggested the presence of AFM fluctuations around 15 K, in spite of the ordering being FM. However, from a substantially linear dependence of $\chi^{-1}(T)$ vs. T at lower temperatures, between 20 K and 37 K, the effective moment turned out to be $\mu_{eff}^{LT} = 3.15$ μ_B/f.u. with a positive Curie–Weiss temperature $\theta_{CW}^{LT} = 13.6$ K, fairly close to T_C. The actual value of $T_C = 14.6$ K was obtained from the Arrott plot (M^2 vs B/M, seeking an isotherm that passes through the origin). Hysteresis in the magnetization with a coercive field of about 0.001 T measured at 2 K documented the FM ordering. Magnetization was very close to saturation (~ 2.3 μ_B/f.u.) in the field of 7 T at 2 K. Above T_C, the temperature dependence of resistivity of $NdFe_4As_{12}$ followed closely that of $LaFe_4As_{12}$, the fact used to extract the resistivity contribution and its temperature dependence due to 4f-electrons, $\rho_{4f} = \rho_{Nd} - \rho_{La}^*$, where $\rho_{La}^* = a\rho_{La}$, with a scaling factor a taking care of errors in the geometrical sizes of very small single crystals.

Perhaps the most interesting aspect of the measurements was the temperature dependence of the specific heat showing a relatively small anomaly at T_C, but a huge, Schottky-like peak slightly above 5K, i.e., some 9 K below T_C and thus deep inside the FM state, Figure 3.53. External magnetic field had two effects on the specific heat: it strongly suppressed and eventually obliterated the anomaly at T_C, and it diminished the Schottky-like peak, making it broader and shifting it to higher temperatures. Assuming that the specific heat of $NdFe_4As_{12}$ consists of the electronic C_{el}, lattice C_{ph}, and magnetic C_{mag} contributions, the total specific heat is $C_{tot} = C_{el} + C_{ph} + C_{mag}$. Here, the phonon term includes a Debye contribution due to 16 atoms of $[Fe_4As_{12}]$ and an Einstein contribution due to a single filler atom. From the accompanying measurements on $LaFe_4As_{12}$ above T_C, the following parameters were extracted: $\gamma = 134$ mJmol^{-1}K^{-2}, $\theta_D = 373$ K, and $\theta_E = 125$ K. $NdFe_4As_{12}$ is expected to have the same electronic specific heat, a similar Debye temperature, and not too different Einstein temperature. In principle, below T_C, one should not forget to take into account a contribution of 3d-electrons of Fe toward C_{mag}, as the Fe moments order at T_C just as do the moments of Nd 4f-electrons. However, in reality, the partial entropy of the 3d magnetic contribution in $LaFe_4As_{12}$, obtained by point-by-point integration of $C_{mag,La} = C_{La} - (C_{el} + C_{ph})$ below 15 K, is no more than 5% of the entropy of $NdFe_4As_{12}$ at T_C, and neglecting it thus does not materially change the outcome. Although the magnetic entropy at T_C reached between $R\ln 4$ and $R\ln 6$, suggesting that the ground state is a quartet rather than a doublet, no direct information on the CEF energy level splitting in $NdFe_4As_{12}$ exists. Consequently, Higashinaka et al. (2013) assumed that the Schottky peak in C_{mag} arises due to

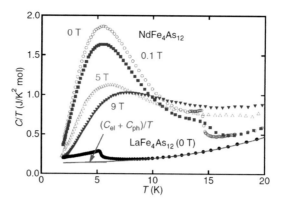

FIGURE 3.53 Temperature dependence of the specific heat of $NdFe_4As_{12}$ plotted as C/T vs. T for different values of the magnetic field. For reference, the figure also includes the behavior of $LaFe_4As_{12}$ with no 4f-electrons. Note a relatively small anomaly at T_C that is progressively wiped out by the magnetic field, and a huge Schottky-like peak that gets reduced and shifts to higher temperatures in magnetic field. Reproduced from R. Higashinaka et al., *Journal of the Physical Society of Japan* **82**, 114710 (2013). With permission from the Physical Society of Japan.

low-lying excitations between 4f-electron states of Nd^{3+} that are split *via* the Zeeman effect in the spontaneous internal field (Weiss molecular field) well within the FM state. In their rough estimates, they actually considered a quasi-degenerate sextet consisting of a quartet and a doublet, implying that the separation between the two is very small, on the scale of $k_B T_C$. This is doubtful, given that the energy splitting between the ground state and the first excited state in $NdOs_4As_{12}$ is some 200 K, Cichorek et al. (2014), and the breaks in the $\chi^{-1}(T)$ vs. T plots of the two Nd-filled arsenide skutterudites occur at comparable temperatures around 70–80 K.

As shown by Rudenko et al. (2016), $NdRu_4As_{12}$ orders ferromagnetically at a much lower temperature T_C = 2.3 K. Measurements were made on quite small (~ 0.3 mm size) dendritic crystals prepared by a self-flux method. From the Curie–Weiss behavior above 60 K, $\mu_{eff}^{HT} = 3.58$ μ_B/Nd^{3+}, very close to the theoretical Nd^{3+} free-ion value of 3.62 μ_B, documenting no magnetic moment associated with 4d-electrons of Ru. The Curie–Weiss temperature in this high-T range turned out to be negative at θ_{CW} = −37.5 K. From a straight-line section of $\chi^{-1}(T)$ vs. T below 10K, the corresponding low-temperature values were 2.21 μ_B/Nd^{3+} and a positive θ_{CW} = 2.3 K, in agreement with T_C. Ferromagnetism was evidenced by a rapidly rising magnetization below 2.5 K and a sharp peak at 2.3 K in the ac susceptibility. Specific heat developed a very sharp peak near 2.3 K, and its overall temperature dependence was modeled in a similar way as in the case of $NdFe_4As_{12}$, yielding γ = 240 mJmol^{-1}K^{-2}, θ_D = 370 K, and θ_E = 67 K. Similarly, the magnetic specific heat C_{mag} followed from $C_{mag} = C - \gamma T - C_{ph}$ (no complication with any magnetic contribution from the [Ru_4As_{12}] polyanion), and the magnetic entropy S_{mag} was obtained from C_{mag}/T vs. T by integration. As indicated in the inset of Figure 3.54, the magnetic entropy attained a value of $R\ln4$ already very near 2.3 K, implying the ground state being $\Gamma_{67}^{(2)}$, in accordance with a downward turn in $\chi^{-1}(T)$ vs. T near 60 K and the low-T value of the effective magnetic moment of 2.21 μ_B/Nd^{3+}. Although not resolved on the specific heat plot in Figure 3.54, at a finer temperature scale a clear hump was observed near 1.4 K = 0.6 T_C, similar to the one shown in Figure 3.55 for $NdOs_4As_{12}$ in zero field, and interpreted as due to low-lying Schottky anomaly arising from roughly Δ/k_B ~ 3 K splitting of the quartet ground state due to the Zeeman effect in the Weiss molecular field.

The last member of Nd-filled arsenide skutterudites, $NdOs_4As_{12}$, was studies by Cichorek et al. (2014) using single crystals grown from molten Cd/As flux. As all other Nd-filled skutterudites, this one, too, ordered ferromagnetically, but at a much lower temperature T_C = 1.1 K. Structural

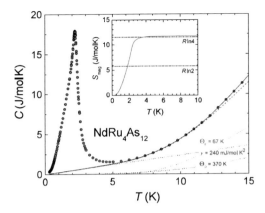

FIGURE 3.54 Specific heat of $NdRu_4As_{12}$ at low temperatures, showing also temperature-dependent contributions of the Debye and Einstein terms and the electronic specific heat γ. Due to the rapidly diminishing electronic and phonon specific heat below 5 K, the sharp peak is essentially all due to the magnetic contribution to the specific heat C_{mag}. The inset depicts the temperature dependence of magnetic entropy that levels near T_C to a value $R\ln4$. Reproduced from A. Rudenko et al., *Solid State Communications* **242**, 21 (2016). With permission from Elsevier.

refinement revealed full Nd occupancy of voids. Although the isotropic atomic displacement parameter U_{iso} of Nd at 300 K was significantly larger than that of either Os or As, it decreased rapidly with decreasing temperature, and at 11 K became only marginally larger than the atoms constituting the $[Os_4As_{12}]$ polyanion. A surprise was an increasing U_{iso} of Os as the temperature decreased (by some 25% at 11 K compared to 300 K), suggesting some form of disorder affecting Os cubes, discussed later. As with other two Nd-filled arsenide skutterudites, a plot of $\chi^{-1}(T)$ *vs.* T showed a downturn near 60 K, separating the high- and low-temperature Curie–Weiss ranges. In the higher range, the effective magnetic moments $\mu_{eff}^{HT} = 3.66\ \mu_B/Nd^{3+}$ was obtained, very close to the Nd^{3+} free-ion value of 3.62 μ_B. Because the polyanion $[Os_4As_{12}]$ does not carry a magnetic moment, the value of μ_{eff}^{HT} indirectly confirmed the full occupancy of voids. The Curie–Weiss constant was, however, negative at $\theta_{CW} = -15.2$ K. At lower temperatures, the corresponding values turned out to be 2.50 μ_B and the positive $\theta_{CW} \approx 1.2$ K, essentially the same as T_C. The inflection point of a rapidly rising magnetization below 2K was taken as a measure of T_C and agreed nicely with a peak in the ac susceptibility. The high quality of crystals (RRR $\geq$ 300) allowed a clear resolution of a $-\ln T$ dependence between about 10 K and down to 2 K, with a rapid drop as the magnetic ordering developed below 1.1 K. The logarithmic temperature dependence was entirely suppressed by the magnetic field of 2 T. Similar to $NdRu_4As_{12}$, the specific heat of $NdOs_4As_{12}$ showed a very sharp peak near its ordering temperature that, again, was dominated by the magnetic contribution C_{mag}, and a Schottky-like hump at 0.93 K = 0.85 T_C in zero magnetic field, see Figure 3.55. Performing similar least-squares fits to the specific heat data as discussed in the case of $NdRu_4As_{12}$ yielded $\gamma = 259$ mJmol^{-1}K^{-2}, $\theta_D = 350$ K, and $\theta_E = 61.2$ K. While the ordering temperature did not change in small magnetic fields, the Schottky anomaly shifted with the magnetic field to higher temperatures, coinciding with T_C at 0.05 T and moving to the right of T_C in increasing fields. Please note that all curves measured in magnetic fields in Figure 3.55 are displaced by the same $\Delta T = 0.25$ K for clarity. Just as in the case of $NdRu_4As_{12}$, the Schottky anomaly in $NdOs_4As_{12}$ in zero field could, too, be plausibly interpreted as arising from the presence of low-lying states split by about $\Delta/k_B \approx 2$ K *via* the Zeeman effect in the developed Weiss molecular field. However, above T_C, the Weiss molecular field no longer exists, and thus the authors sought an alternative explanation for the Schottky anomaly, and settled on an already hinted at structural disorder concerning Os atoms at the second nearest distance from the Nd fillers. As we shall see in a section discussing Nd-filled antimonides and, specifically $NdOs_4Sb_{12}$, the idea came

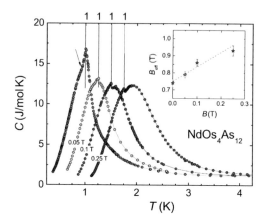

FIGURE 3.55 Specific heat of $NdOs_4As_{12}$ in zero and in small applied field as a function of temperature. Please note that all specific heat curves in the magnetic field are displaced by the same $\Delta T = 0.25$ K for clarity. In fact, small fields used had no effect on the ordering temperature T_C. However, even such small fields clearly shifted the Schottky peak, shown at $B = 0$ by an arrow, and passing through the T_C and to the right of T_C as the magnetic field increased. The inset shows the field dependence of the effective field B_{eff}, defined as $B_{eff} = B + B_{mol}$, where B is the applied field and B_{mol} is the Weiss molecular field, estimated at $0.85T_C$, where the hump in zero field specific heat is noted by an arrow, as 0.74 T. Thus, B_{eff} marks the field where the maximum in the Schottky anomaly is observed. Reproduced from T. Cichorek et al., *Physical Review B* **90**, 195123 (2014). With permission from the American Physical Society.

from a rather large distortion observed by Keiber et al. (2012) in their extended X-ray absorption fine structure (EXAFS) measurements. Interestingly, while the nearest neighbor Sb atoms were not affected, the second nearest neighbors to Nd fillers, i.e., Os atoms forming a cube around each Nd, were seriously affected and the cubes they form distorted into rhombuses, as shown in Figure 3.56. Such distortions lead to alternating long and short Nd-Os bonding, similar to what we have seen in the case of two non-equivalent Pr ions in $PrRu_4P_{12}$ shown in Figure 3.44, where the distortion led to a lowering of the symmetry from $Im\overline{3}$ to $Pm\overline{3}$. Because the number of affected Os cubes is on the order of the Avogardo number, such distortions may represent a significant fraction of the specific heat.

3.3.4.4.3   Nd-Filled Antimonides

First measurements of $NdFe_4Sb_{12}$ were performed by Danebrock et al. (1996) as part of a survey of magnetic properties of polycrystalline samples of RFe_4Sb_{12} (R = Ca, Sr, Ba, La-Nd, Sm, and Eu). Structural refinement of $NdFe_4Sb_{12}$ indicated only about 83–84% void occupancy, which obviously affected the obtained magnetic parameters. The FM transition, based on the $\chi^{-1}(T)$ vs. T plot, placed the Curie temperature of $NdFe_4Sb_{12}$ at $T_C \approx 13$ K. As with all skutterudites having $[Fe_4Sb_{12}]$ polyanions, there is a magnetic contribution associated with Fe, in addition to a magnetic moment of the rare-earth filler ion. From a straight line of $\chi^{-1}(T)$ vs. T above 100 K, an effective magnetic moment $\mu_{eff} = 4.5 \pm 0.2 \ \mu_B$/f.u. was extracted, of which 3.4 μ_B was estimated as due to the Nd^{3+} ion. The value of 3.4 μ_B, obtained by subtracting the susceptibility of $LaFe_4Sb_{12}$ with its non-magnetic La and assuming that the polyanion $[Fe_4Sb_{12}]$ contributes identical moment in both $NdFe_4Sb_{12}$ and $LaFe_4Sb_{12}$, is a bit smaller than the expected Nd^{3+} free-ion value of 3.62 μ_B. Magnetization loops indicated a nearly full saturation at 5 T with a moment of 2.3 μ_B, some 30% below the theoretical saturation value of 3.27 μ_B/Nd^{3+}. A contributing factor to both discrepancies (smaller experimental values) was likely only a partial occupancy of voids by Nd. Nevertheless, a small coercive field of about 0.004 T measured at 5 K clearly documented the FM ordering. Magnetism of polycrystalline

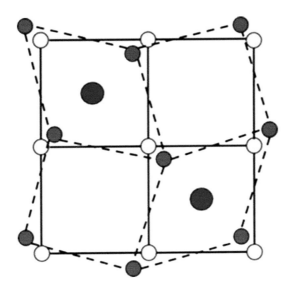

FIGURE 3.56 Distortion of the skutterudite unit cell (viewed along the 001 axis) that keeps the Os-Os distance constant but alters the originally fixed Nd-Os distance into larger and shorter set of distances as the cubes distort into rhombuses. Large atoms in the center of cubes are Nd and the atoms at the vertices of squares (rhombuses) are Os. Reproduced from T. Keiber et al., *Physical Review B* **86**, 174106 (2012). With permission from the American Physical Society.

samples of $NdFe_4Sb_{12}$, again with a deficiency of Nd of some 28%, was re-examined a few years later by Bauer et al. (2002b), who included also measurements of the specific heat and neutron diffraction. The Curie temperature, determined from the Arrott plots, turned out to be somewhat higher at 16.5 K. Refinement of the neutron diffraction pattern revealed an ordered FM moment of 2.04 μ_B originating from the Nd site and a collinear moment of Fe of 0.27 μ_B.

Properties of single crystals of $NdFe_4Sb_{12}$, grown from Sb self-flux, were studied by Ikeno et al. (2008). Even this single crystalline form of the skutterudite did not achieve full Nd filling, judged by the wavelength-dispersive electron-probe microanalysis (EPMA), and the Nd content varied between 77% and 90%. Having available several crystals with different contents of Nd, the dependence of the FM ordering temperature T_C on the Nd occupancy could be examined and is illustrated in Figure 3.57. Clearly, the occupancy of voids by Nd has a strong influence on T_C. At full occupancy, the data suggest the Curie temperature of about 20 K. Interestingly, the Nd occupancy had no obvious effect on the effective magnetic moment and the Curie–Weiss temperature θ_{CW}, as the data showed no particular trend and varied little.

Magnetic properties of $NdRu_4Sb_{12}$ were explored as part of a survey study of RRu_4Sb_{12} (R = La, Ce, Pr, Nd, and Eu) by Takeda and Ishikawa (2000a) and by Abe et al. (2002). In both cases, a Sb self-flux method was used to prepare samples in the form of brittle lumps and single crystals (RRR ~ 100), respectively, and both studies reported a magnetic ordering taking place at T_C = 1.3–1.4 K. The magnetic transition was revealed by anomalies in the *dc* and *ac* magnetic susceptibilities and in specific heat. From the Curie–Weiss behavior above 50 K, Takeda and Ishikawa (2000a) obtained the high-temperature effective magnetic moment μ_{eff}^{HT} =3.45 μ_B/Nd^{3+}, close to the Nd^{3+} free-ion value of 3.62 μ_B, and a negative θ_{CW} = −28 K. Below a downturn in the $\chi^{-1}(T)$ *vs.* T plot around 50 K (a typical common feature of all Nd-filled skutterudites), another straight-line section of the inverse susceptibility yielded μ_{eff}^{LT} = 2.31 μ_B/Nd^{3+} and a positive θ_{CW} = 1.29 K, essentially the same as the ordering temperature T_C.

The specific heat developed a pronounced peak at 1.3 K, substantiating the bulk nature of the magnetic transition, and the extracted magnetic entropy was close to $R\ln4$, implying one of the two

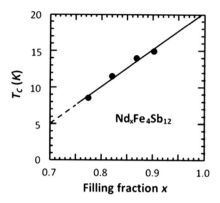

FIGURE 3.57 Dependence of the FM ordering temperature T_C on the filling fraction of voids by Nd. Constructed from the data of Ikeno et al., *Journal of the Physical Society of Japan* **77**, 309 (2008). With permission from the Physical Society of Japan.

quartets Γ_{67} as being the ground state. However, Takeda and Ishikawa mis-diagnosed the magnetic ordering as being AFM rather than FM, as clearly documented by a maximum in their *ac* susceptibility at 1.3 K and the positive low-temperature value of $\theta_{CW} = 1.29$ K. This was rectified by Abe et al. (2002), who obtained a comparable low-temperature value of $\mu_{eff}^{LT} = 2.36$ μ_B/Nd^{3+} and $\theta_{CW} \approx$ 1.3 K. With the additional input concerning the effect of a magnetic field on the resistivity, they described the ground state of NdRu$_4$Sb$_{12}$ as being of FM nature.

Early reports on NdOs$_4$Sb$_{12}$ focused on structural parameters, Jeitschko and Braun (1977) and Evers et al. (1995), the latter authors reporting 100% occupancy of voids by Nd. Sato et al. (2003b), in a paper describing Pr-filled skutterudites, included a Table in which NdOs$_4$Sb$_{12}$ was entered, with no comments, as a ferromagnet with the ordering temperature $T_C = 0.8$ K. The first detailed studies of magnetism in NdOs$_4$Sb$_{12}$ by Ho et al. (2005) used molten flux-grown single crystals and revealed the structure as being a ferromagnet with $T_C = 0.9$ K that also features exceptionally large electronic specific heat coefficient $\gamma \approx 520$ mJmol^{-1}K^{-2}, corresponding to an effective mass $m^* \approx 98$ m$_e$. It was this extraordinarily large carrier mass that implied the skutterudite is a heavy fermion system, which generated much excitement and motivated numerous further studies. Before I discuss them, I list the essential magnetic parameters obtained by Ho et al. (2005).

As in all other Nd-filled skutterudites, a downturn in the inverse susceptibility as a function of temperature near 50 K delineates the high- and low-temperature regions where the Curie–Weiss behavior is observed, and these yield the respective effective magnetic moments of $\mu_{eff}^{HT} = 3.84$ μ_B/Nd^{3+}, close to the free-ion value of 3.62 μ_B and $\mu_{eff}^{LT} = 2.35$ μ_B/Nd^{3+}. The corresponding Curie–Weiss temperatures were negative $\theta_{CW} = -43$ K and positive $\theta_{CW} \sim 1$K, indicating a FM order and an excellent agreement with the ordering temperature T_C. Based on the CEF splitting the ground-state multiplet of Nd into a doublet and two quartets, the authors determined the CEF contribution to the magnetic susceptibility $\chi_{CEF}(T)$ by fitting $\chi^{-1} = \chi_{CEF}^{-1} - \Lambda$ (Λ is a Weiss molecular field constant to account for the presence of a magnetic order in NdOs$_4$Sb$_{12}$) to the experimental data and imposed a constraint on the ground state by the value of $\mu_{eff}^{LT} = 2.35$ μ_B/Nd^{3+}. Without going into details, good fits were obtained for both a doublet Γ_5 as the ground state as well as a $\Gamma_{67}^{(2)}$ quartet as the ground state, corresponding, respectively, to the energy level scheme sequences of Γ_5 (0 K), $\Gamma_{67}^{(1)}$ (180 K), $\Gamma_{67}^{(2)}$ (420 K); and $\Gamma_{67}^{(2)}$ (0 K), $\Gamma_{67}^{(1)}$ (220 K), and Γ_5 (600 K), with $\Lambda = 1.39$ Nd-molcm^{-3} in both schemes. At 0.4 K, well below the Curie temperature, the magnetization saturated in the fields above 2 T at a value $M_s = 1.73$ μ_B/Nd^{3+}. Parallel isotherms of M^2 *vs.* B/M in the vicinity of the ordering temperature pinpointed T_C at 0.93 K.

The overall metallic temperature dependence of resistivity developed a slight negative curvature near 130 K and a well-defined shoulder near 1.2 K, where a rapidly decreasing $\rho(T)$ indicated diminished spin disorder as the sample passed through T_C. Magnetic field had a considerable effect on the resistivity by increasing the residual resistivity, altering the exponent of the temperature dependence from $n = 3$ to $n = 4$ (none reflecting either a typical Fermi liquid behavior with $n = 2$, or a distinctly NFL variation with $n < 2$), and shifting the shoulder in $\rho(T)$ to higher temperatures and eventually suppressing it altogether. The negative curvature near 130 K provided a hint of charge carrier scattering by CEF-split levels. A contribution of CEF states to the resistivity was evaluated by taking it as $\Delta\rho(T) = \rho(T) - \rho_{ph}(T) - \rho_{imp}$, where the resistivity due to phonons was estimated from the resistivity of $SmOs_4Sb_{12}$, known to have a nearly linear T-dependence above 100 K. The usual trick of subtracting the resistivity of isothermal non-magnetic $LaOs_4Sb_{12}$ would not work in this case as $LaOs_4Sb_{12}$ shows a particularly well developed negative curvature in its resistivity and this would skew the results. The two-level CEF scheme, used in fitting the magnetic susceptibility, proved satisfactory also in describing the incremental resistivity $\Delta\rho(T)$ due to CEF. However, a small aspherical Coulomb scattering term, proposed originally by Fulde and Peschel (1972), had to be added to prevent physically unreasonable negative impurity contribution to the resistivity. Ho et al. (2005) also explored the effect of pressure on the resistivity of $NdOs_4Sb_{12}$ and observed the resistivity dropping by 40% at low temperatures under a pressure of 28.4 kbar, and the ordering temperature increased to near 1.7 K under the same pressure. Both effects were more pronounced in comparison to changes the magnetic field was able to make.

The key and very surprising finding arose from measurements of the specific heat shown in Figure 3.58. A comparison of the specific heat of $NdOs_4Sb_{12}$ with that of non-magnetic $LaOs_4Sb_{12}$ is presented in Figure 3.58a, while in Figure 3.58b are shown the same $NdOs_4Sb_{12}$ data fit with $C_{el} + C_{ph}$, the latter including both the Debye and Einstein contributions modeled by

$$C_{Debye}(T) = (17 - r)\frac{12\pi^4}{5}R\left(\frac{T}{\theta_D}\right)^3, \tag{3.100}$$

and

$$C_{Ein}(T) = r\left[3R\frac{\left(\theta_E/T\right)^2 e^{\left(\theta_E/T\right)}}{\left[e^{\left(\theta_E/T\right)} - 1\right]^2}\right]. \tag{3.101}$$

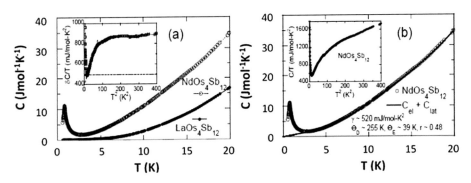

FIGURE 3.58 (a) Specific heat of $NdOs_4Sb_{12}$ and $LaOs_4Sb_{12}$ in zero magnetic field. The inset shows $\delta C/T$ vs. T^2 where $\delta C \equiv C(NdOs_4Sb_{12}) - C(LaOs_4Sb_{12})$. The two dashed lines extrapolated to T = 0 K indicate the lower (436 mJmol^{-1}K^{-2}) and upper (530 mJmol^{-1}K^{-2}) limits on the electronic specific heat γ. (b) The same $NdOs_4Sb_{12}$ data (open circles) fit with $C_{el} + C_{ph}$ (solid line) where C_{ph} includes both the Debye and Einstein specific heats given in Equations 3.99 and 3.100. The fit returned $\gamma = 520$ mJmol^{-1}K^{-2}, $\theta_D = 255$ K, $\theta_E = 39$ K, and $r = 0.48$. Reproduced from P.-C. Ho et al., *Physical Review B* **72**, 094410 (2005). With permission from the American Physical Society.

Here, R is the universal gas constant and $r \leq 1$ represents a fraction of the filler acting as Einstein oscillators. If all Nd fillers acted that way, $r = 1$, and their contribution to the Debye specific heat would not materialize and vibrations of just 16 atoms of $[Os_4Sb_{12}]$ would generate the Debye contribution. Designating by δC the difference between the specific heats of $NdOs_4Sb_{12}$ and $LaOs_4Sb_{12}$, a plot of $\delta C /T$ vs. T^2 is shown in the inset of Figure 3.58a. From the two dashed lines shown, the first estimate of the electronic specific heat γ falls between values of about 436 mJmol^{-1}K^{-2} and 530 mJmol^{-1}K^{-2}. From the fit of the experimental data with $C_{el} + C_{ph}$, where the latter includes both the Debye and Einstein contributions, the electronic specific heat comes out as $\gamma \approx 520$ mJmol^{-1}K^{-2}, well within the range of the values obtained by a simple subtraction of the specific heats of $NdOs_4Sb_{12}$ and $LaOs_4Sb_{12}$. Such a large Sommerfeld coefficient γ implies exceptionally large effective mass of the charge carriers, $m^* \approx 98\ m_e$, making $NdOs_4Sb_{12}$ among the three most notable heavy fermion skutterudites, the other two being $PrOs_4Sb_{12}$ with $\gamma \approx 600$ mJmol^{-1}K^{-2} and $SmOs_4Sb_{12}$ with $\gamma \approx 880$ mJmol^{-1}K^{-2}. The obtained Debye and the Einstein temperatures were $\theta_D = 255$ K and $\theta_E = 39$ K, with the parameter $r = 0.48$. Both θ_D and θ_E look reasonable, the first being close to the Debye temperature of $LaOs_4Sb_{12}$, while the other being comparable to the value of 45 K, extracted from the X-ray data *via* the thermal displacement parameter

$$U = \frac{\hbar^2}{2m_{Nd}k_B\theta_E} \coth\left(\frac{\theta_E}{2T}\right),$$
(3.102)

where m_{Nd} is the atomic mass of Nd. By integrating $\Delta C/T$ vs. T, where ΔC is the magnetic part of the specific heat taken as $\Delta C = C - C_{el} - C_{ph}$, the magnetic entropy S_{mag} at T_C reached 74% of its full value and leveled off above about 2 K with a value of 1.14R ($\approx R\ln3$). With the S_{mag} falling between $R\ln2$ and $R\ln4$, one cannot conclusively decide, based on the specific heat data, between the doublet Γ_5 and the quartet $\Gamma_{67}^{(2)}$ as the ground state.

Given that Sm has a mixed valence in $SmOs_4Sb_{12}$, see section 3.3.4.5.3, concerns were raised regarding the actual valence state of a neighboring rare-earth element Nd. Soft X-ray spectroscopy by Imada et al. (2007) confirmed that Nd is, indeed, trivalent in the bulk, but some Nd^{2+} ($4f^4$) state might possibly be present at the surface, as is often the case with other rare earths.

With no success to determine the CEF level scheme based on specific heat measurements, Kuwahara et al. (2008) made INS measurements attempting to settle the issue of the ground state in $NdOs_4Sb_{12}$. Energy spectra at the fixed final energy of 13.5 meV and at 41.0 meV, obtained after subtracting the phonon background estimated from the spectra of $LaOs_4Sb_{12}$ under the identical experimental conditions, revealed two inelastic peaks at energy transfers of 23 meV and 30.2 meV, identifying the first and the second excited CEF states. Applying the formalism developed by Lea et al. (1962) to describe the CEF energy levels, the authors considered altogether four CEF schemes that reproduced the inelastic neutron spectra. Using the respective parameters designating the four schemes, the calculated magnetic susceptibility could not reproduce the experimental data (no downturn in $\chi^{-1}(T)$ vs. T near 50 K) for two of the schemes. The remaining two CEF schemes both placed the $\Gamma_{67}^{(2)}$ quartet (actually the $\Gamma_8^{(2)}$ quartet as the authors used the O_h point group symmetry) as the ground state, but with different first excited states. Thus, one sequence of levels consisted of $\Gamma_{67}^{(2)}(0$ K$) - \Gamma_5(267$ K$) - \Gamma_{67}^{(1)}(350$ K$)$, while the other was $\Gamma_{67}^{(2)}(0$ K$) - \Gamma_{67}^{(1)}(267$ K$) - \Gamma_5(350$ K$)$. In both cases, a fairly large gap separated the first excited state from the ground state, meaning that the CEF should play only a very minor role in low-temperature properties, unlike a small gap (~ 8 K) in $PrOs_4Sb_{12}$, responsible for the fascinating heavy fermion SC state.

A contrast between a heavy fermion ferromagnet $NdOs_4Sb_{12}$ and an exotic heavy fermion superconductor $PrOs_4Sb_{12}$ offered an exciting opportunity to explore a crossover between the two skutterudites by forming solid solutions $Pr_{1-x}Nd_xOs_4Sb_{12}$. Ho et al. (2011) found out that solid solutions form across a full range of x values, and the authors were able to prepare single crystals using the

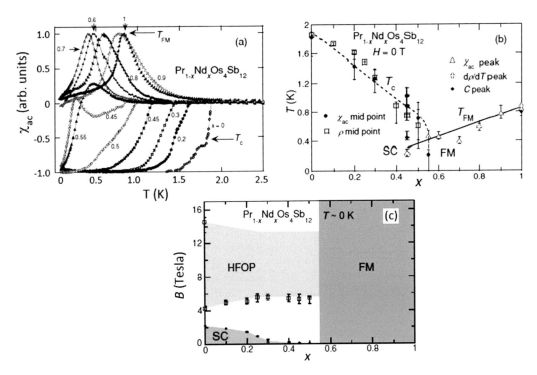

FIGURE 3.59 (a) Temperature dependence of ac susceptibility for various contents of Nd in $Pr_{1-x}Nd_xOs_4Sb_{12}$ solid solutions. (b) Superconducting (T_c) and Curie (T_C) temperatures as a function of the content x obtained from various experiments. (c) Phase diagram B-x at $T = 0$ K for $Pr_{1-x}Nd_xOs_4Sb_{12}$ solid solutions. Adapted from P.-C. Ho et al., *Physical Review B* **83**, 024511 (2011). With permission from the American Physical Society.

molten flux method. The evolving crossover from a SC state of $PrOs_4Sb_{12}$ with $T_c = 1.85$ K to a mean field $NdOs_4Sb_{12}$ ferromagnet with $T_C = 0.8$ K is perhaps most vividly demonstrated by the behavior of the low-temperature susceptibility shown in Figure 3.59. Here, the SC transition temperature, marked by a sharp (negative) drop in $\chi_{ac}(T)$, shifts progressively to lower temperatures for concentrations up to $x \approx 0.55$. Starting with $x = 0.60$, the behavior changes dramatically with $\chi_{ac}(T)$ turning positive and developing a peak that shifts to higher temperatures as x increases. The shift terminates at the peak at 0.8 K, corresponding to $NdOs_4Sb_{12}$. Thus, the crossover from the superconducing regime of Pr-rich solid solutions to the magnetically ordered structures of Nd-rich solid solutions takes place between $x = 0.55$ and 0.60. Interestingly, for $x = 0.45$, the data suggest a tantalizing possibility that the spins of Nd order within a SC phase, as shown in Figure 3.59(b), where the linear dependences of both T_c and T_C as a function of x are plotted. Magnetic susceptibility measurements also indicated that the correlations between the Nd moments in the SC domain change from FM for $0.3 \leq x \leq 0.6$ to AFM for $0 \leq x \leq 0.3$. The resulting B-x phase diagram of $Pr_{1-x}Nd_xOs_4Sb_{12}$ is depicted in Figure 3.59(c) and shows compositional and field ranges of the SC phase, the field-ordered phase (HFOP, also known as FIOP), and the FM phase. Details concerning the HFOP domain are given in Chapter 2.

It has been claimed, but rarely supported by solid experimental evidence, that some filler ions are not located in the center of the cage formed by the twelve pnicogen atoms and undergo off-center motion. Because the antimonide cage is the largest cage in skutterudites, it is certainly true that it offers large-amplitude excursions of the filler species, but it does not imply that the ions are located off the center or that they undergo off-center displacements, as some researchers have suggested based on their particular experiments, Goto et al. (2005), Yasumoto et al. (2008) and Yanagisawa et al. (2008c, 2009). Specifically, a large off-center displacement of 0.4 Å was claimed for Nd rattling in $NdOs_4Sb_{12}$ and one-half the above value for Pr in $PrOs_4Sb_{12}$. An important step toward resolving the issue of off-center motion of Nd and Pr ions in the icosahedral cage of Sb atoms was made by Keiber et al. (2012), who used EXAFS, a unique element-specific technique able to discern local distortions and the dynamics of various structural constituents. The experiment proved finally that, at least in the case of Nd and Pr fillers, their "rattling" motion relative to the Sb cage is on-center. However, a very unexpected finding concerned the second-neighbor distances, i.e., Nd-Os and Pr-Os, which were highly altered even at low temperatures. In the case of Pr fillers, we have already seen one type of dominant distortion setting in during the M–I transition in $PrRu_4P_{12}$, see section 3.3.4.3.1, leading to lowering of the symmetry from $Im\bar{3}$ to $Pm\bar{3}$ as smaller and larger Ru cubes altered along the [111] direction. Such distortion might very well be present also in the case of the Os cubes. Because the variation in the Nd-Os distance (the authors use an unfortunate word splitting) is about twice as large (~ 0.2 Å) as in the case of the Pr-Os distance, the distortion due to lowering of the symmetry from $Im\bar{3}$ to $Pm\bar{3}$ could account for one half of the effect, with the other half coming from another form of distortion, such as illustrated in Figure 3.56. Here the cubes of Os distort into rhombuses in such a way that the Os-Os distance is left intact while the fixed Nd-Os distance turns into a set of larger and smaller distances. Such a structural distortion might very well contribute to the development of a heavy fermion state in $NdOs_4Sb_{12}$. I should mention that in a prior EXAFS paper by Nitta et al. (2008), the authors did not consider interactions beyond the first nearest-neighbor distances from the filler, and their estimates for the Einstein temperatures of Nd and Pr came up way large in over 125 K, well beyond the range of 50–60 K based on the consensus of X-ray and other measurements. The discrepancy arose from using the reduced mass of a Nd-Sb (Pr-Sb) pair that is about one-half of the actual "rattler's" mass, given that the atomic masses of Sb (123), Nd (144), and Pr (140) are not much different.

Probing the Fermi surface cross-section of $NdOs_4Sb_{12}$ as a function of the orientation of the applied field, recent dHvA and Shubnikov-de Haas measurements by Ho et al. (2016) performed on single crystals down to 0.5 K and fields up to 60 T revealed a notable similarity in the spectra

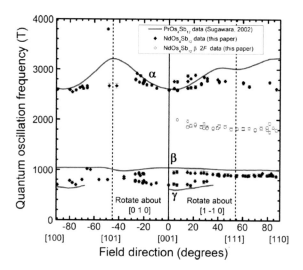

FIGURE 3.60 Quantum oscillation frequencies for different field orientations in $NdOs_4Sb_{12}$. Points are well-resolved frequencies observed in the data; the curves are analogous to the data for $PrOs_4Sb_{12}$ obtained by Sugawara et al. (2002c). The Fermi surface sections are labeled according to the scheme used in the $PrOs_4Sb_{12}$ paper. Reproduced from P.-C. Ho et al., *Physical Review B* **94**, 205140 (2016). With permission from the American Physical Society.

TABLE 3.10

Quantum Oscillation Frequencies and Effective Masses for Various Applied Field Orientations

Field Orientation	FS Section	NdOs$_4$Sb$_{12}$ F(T)	NdOs$_4$Sb$_{12}$ m*/m$_e$	PrOs$_4$Sb$_{12}$ F(T)	PrOs$_4$Sb$_{12}$ m*/m$_e$
B∥ [001]	α	2560 ± 20	3.1 ± 0.2	2610	4.1
B∥ [001]	β	950 ± 20	1.8 ± 0.3	1070	2.5
B∥ [001]	γ	690 ± 30	3.6 ± 0.4	710	7.6
B∥ [011]	β	870 ± 30	1.5 ± 0.4	875	3.9

Source: Adapted from P.-C. Ho et al, *Physical Review B* **94**, 205140 (2016). With permission from the American Physical Society.

Note: The data for NdOs$_4$Sb$_{12}$ are from Ho et al. (2016) and the entries for PrOs$_4$Sb$_{12}$ are from Sugawara et al. (2002c).

of NdOs$_4$Sb$_{12}$ and PrOs$_4$Sb$_{12}$, the latter measured previously by Sugawara et al. (2002c). Spectra of both filled skutterudites are shown in Figure 3.60. In spite of this similarity, the effective masses of NdOs$_4$Sb$_{12}$, obtained from the temperature dependence of the oscillation amplitude, turned up consistently smaller than those in PrOs$_4$Sb$_{12}$, see Table 3.10. Because the measured electronic specific heat coefficient (Sommerfeld coefficient) of 520 mJmol^{-1}K^{-2} and 650 mJmol^{-1}K^{-2} for NdOs$_4$Sb$_{12}$ and PrOs$_4$Sb$_{12}$, respectively, do not differ by more than 20%, the significantly larger difference between the respective effective masses suggests that some other effect also contributes to the large Sommerfeld coefficient of NdOs$_4$Sb$_{12}$, perhaps some local phonon mode.

3.3.4.5 Sm-Filled Skutterudites

As part of a large family of rare-earth-filled skutterudites, SmT$_4$X$_{12}$, T = Fe, Ru, and Os and X = P and Sb, Sm-filled skutterudites were first synthesized by Braun and Jeitschko (1977). Although initially these compounds did not command much interest, measurements during the mid-2000s revealed a wealth of fascinating physical properties unique to essentially each Sm-filled skutterudite structure. Even though these skutterudites lacked any signs of superconductivity, they started to draw interest for their unusual transport and magnetic behavior to a degree that the number of scientific reports devoted to their properties has been exceeded only by reports on Pr-filled skutterudites, the charm and attraction of which were their exotic SC properties. With the exception of SmOs$_4$Sb$_{12}$, where Sm might have a mixed valence, all other Sm-filled skutterudites are believed to possess trivalent Sm ions. With the electronic configuration [Xe]4f^5, L = 5, S = 5/2, and J = L-S = 5/2, the ground multiplet of Sm^{3+} is ^{6}H$_{5/2}$, which gets split by the CEF into a doublet Γ_5 [Γ_7] and a quartet Γ_{67} [Γ_8], using the T_h [O_h] point group symmetry.

3.3.4.5.1 Sm-Filled Phosphides

The first attempt to get some insight into the physical properties of SmFe$_4$P$_{12}$ was made by Jeitschko et al. (2000) who reported on the electrical resistivity and magnetic susceptibility and described the structure as a van Vleck paramagnet. A closer examination of transport, magnetic, and thermodynamic properties by Takeda and Ishikawa (2003) on flux-grown polycrystalline samples revealed a Kondo effect in the temperature dependence of resistivity, FM transition at T_C = 1.6 K, and a large electronic specific heat coefficient γ = 370 mJmol^{-1}K^{-2}. Measurements of specific heat carried out at the same time by Giri et al. (2003) marked the transition temperature at 1.5 K. Overall, the results pointed toward a heavy fermion Kondo lattice with a FM ground state, rather than a van Vleck paramagnet. The notion that the Kondo effect is important in transport and magnetic properties of

$SmFe_4P_{12}$ came from a typical shoulder-like feature in the otherwise metallic temperature dependence of the electrical resistivity that set in below about 90 K and persisted down to 30 K, at which point the resistivity fell sharply, Takeda and Ishikawa (2003). By subtracting the phonon contribution from the overall electrical resistivity, a negative temperature coefficient resulted above 50 K, documenting the Kondo lattice. In subsequent measurements by Kikuchi et al. (2006), the magnetic contribution actually showed a peak near 40 K followed at higher temperatures by the typical $-\ln T$ dependence, a clear sign of the influence of the Kondo effect. Temperature dependence of the electrical resistivity of all three Sm-filled phosphide skutterudites as well as detailed studies of the specific heat, including the effect of a magnetic field, was shown by Matsuhira et al. (2005a). In each case, the corresponding non-magnetic La-filled phosphide was measured for comparison to reveal the effect of 4f-electrons. Specifically, the authors isolated the magnetic contribution to the specific heat C_m using the following procedure: (i) They accounted for the lattice contribution to the specific heat of La-filled phosphides $C_{La,L}(T)$ by subtracting from the measured $C_{La}(T)$ the electronic contribution $\gamma_{La}T$, (ii) they assumed that the Sm-filled structures have the same electronic specific heat coefficient γ as the corresponding La-filled compounds, and (iii) they made the correction to the Debye temperature θ_D by assuming $\theta_D \sim (m)^{-1/2}$, where m is a mean atomic mass. The temperature scale was thus normalized to $T_S = \left(m_{La} \big/ m_{Sm} \right)^{1/2} T$, where m_{La} and m_{Sm} are masses of La and Sm compounds, respectively. (iv) Finally, they obtained $C_m(T)$ from $C_m(T) = C_{Sm}(T) - C_L(T) - \gamma_{La}T$. Plots of the magnetic specific heat, together with the magnetic entropy, are shown in Figure 3.61. Because Sm represents only 1 out of 17 atoms of the structure, the magnetic specific heat contribution at 70 K is comparable to the experimental uncertainty ($\sim 3\%$) and the error bars become very large. A remarkable feature of $SmFe_4P_{12}$ is a very small magnetic entropy released below $T_C = 1.6$ K, amounting to no more than 16% of $R\ln2$ (8% of $R\ln4$), suggesting that the FM transition is a very weak one. From such a very low value, it would be difficult to argue that the ground state is a Γ_5 doublet. However, from the fits to the magnetic specific heat $C_m(T)$ by a Schottky term calculated assuming the Γ_5 ground state doublet (thin solid line in Figure 3.61(a)) and the Γ_{67} ground state quartet (dash-dotted line), respectively, a better fit was obtained assuming the former as the ground state. The authors also estimated the energy spacing between the ground state and the excited state as about 6.7 meV (70 K). Interestingly, as first noted by Takeda and Ishikawa (2003) and subsequently confirmed by others, the specific heat below T_C does not obey the expected $T^{3/2}$ dependence predicted for FM spin waves but varies as T^3, similar to the case of a clearly FM $NdFe_4P_{12}$ skutterudite discussed in section 8.3.4.4. The FM nature of the phase transition at T_C was well documented by the steeply rising magnetic susceptibility below 3 K and sharply rising magnetization below 1.6 K with its ultimate saturation at very low temperatures as reported by Takeda and Ishikawa (2003). Additional support for the FM transition at $T_C = 1.6$ K came from muon spin rotation (μSR) measurements carried out at zero field by Hachitani et al. (2006a), who detected the presence of a static internal field below T_C. Moreover, their ^{31}P-NMR measurements yielded the temperature independent spin relaxation rate $1/T_1$ above 30 K, while at lower temperatures, between 7–30 K, the Korringa relation ($1/T_1 \propto T$) was followed, the behavior typical of a Kondo system. In their subsequent studies of Sm-filled phosphides that included all three transition metal elements Fe, Ru, and Os, Hachitani et al. (2007) actually determined the internal field in $SmFe_4P_{12}$ at 0.31 K of 0.065 T and smaller values of 0.0075 T for $SmRu_4P_{12}$ and 0.025 T for $SmOs_4P_{12}$ at 1.9 K. The smaller internal fields measured in $SmRu_4P_{12}$ and $SmOs_4P_{12}$ compared to $SmFe_4P_{12}$ are a consequence of antiferromagnetically ordered structures (discussed in subsequent paragraphs) that partially cancel the transferred hyperfine fields from Sm moments. dHvA measurements on a single crystal of $SmFe_4P_{12}$ with RRR = 54 by Kikuchi et al. (2005) made at 30 mK and fields up to 17 T picked up only fundamental dHvA frequency branch γ with a weak angular dependence over the entire angular range, implying a nearly spherical Fermi surface. From the temperature dependence of the dHvA oscillation amplitude, a rather anisotropic cyclotron effective mass of $m_c^* = 4.3\ m_e$ for $B\|[100]$ and $m_c^* = 9.2\ m_e$ for

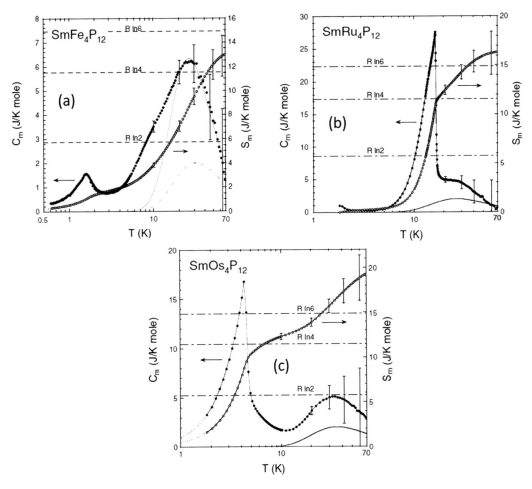

FIGURE 3.61 Magnetic contribution to the specific heat C_m and the magnetic entropy S_m for (a) $SmFe_4P_{12}$, (b) $SmRu_4P_{12}$, and (c) $SmOs_4P_{12}$. The thin solid line in (a) shows the Schottky peak calculated with the Γ_5 ground state, and the dash-dotted line shows the Schottky peak calculated with the Γ_{67} ground state. Clearly, the Γ_5 ground state is a better fit in the case of $SmFe_4P_{12}$. In $SmRu_4P_{12}$ and $SmOs_4P_{12}$, the thin solid line depicts the Schottky peak calculated using the Γ_{67} ground state. Redrawn from K. Matsuhira et al., *Journal of the Physical Society of Japan* **74**, 1030 (2005). With permission from the Physical Society of Japan.

$B\|[110]$ was obtained. Although the effective mass was significantly enhanced, the value was not large enough to support the large electronic specific heat coefficient of 370 mJmol^{-1}K^{-2} reported by Takeda and Ishikawa (2003). Assuming the spherical Fermi surface and using the measured cyclotron effective masses, only about 15 mJmol^{-1}K^{-2} would be consistent with the dHvA measurements. Such a large discrepancy in the Sommerfeld coefficient γ might arise from the fact that other sheets of the Fermi surface, specifically α and β frequency branches that could have been associated with heavier masses, were not detected in the dHvA measurements. Although it was always tacitly assumed that Sm is trivalent in $SmFe_4P_{12}$, and Takeda and Ishikawa (2003), indeed, documented that the magnetic susceptibility cannot be fitted with divalent Sm, it was important to see solid spectroscopic evidence for the Sm^{3+} state in $SmFe_4P_{12}$. The proof was provided by X-ray absorption spectrum measurements by Mizumaki et al. (2007). Pushing the limits of available magnetic fields, Takeda et al. (2008) performed magnetization measurements on $La_{1-x}Sm_xFe_4P_{12}$ at pulsed fields to 42 T and observed a metamagnetic crossover at 22 T at temperatures below 4 K and no saturation of magnetization even at 42 T, where the magnetization reached only 0.4 μ_B, much less than the saturation magnetization of 0.71 μ_B expected for Sm^{3+}. Interestingly, the metamagnetism persisted even in paramagnetic samples with $x = 0.7$ and 0.8. The authors conjectured that the metamagnetism arises because of the DOS that has a sharp peak structure near the Fermi energy. If, upon an application of magnetic field, the Fermi level is swept through the peak in the DOS, metamagnetism might occur. Recent combined magnetization and INS measurements performed on a single crystal sample of $SmFe_4P_{12}$ by Konno et al. (2015) revealed a distinct hump in a plot of $1/\chi(T)$ vs. T near 12 K, which was reproduced by taking the Γ_5 doublet as the ground state and Γ_{67} quartet as the excited state 6 meV (70 K) above the ground state, Figure 3.62. While no simple Curie–Weiss behavior was observed over a broad temperature range, the authors were able to fit the data with a modified Curie–Weiss law of the form

$$\chi(T) = \frac{N_A \mu_{eff}^2}{3k_B(T - \theta_{CW})} + \chi_0, \tag{3.103}$$

where N_A is the Avogardo number, θ_{CW} is the Curie–Weiss temperature, and χ_0 is a temperature-independent van Vleck term due to coupling with the $J = 7/2$ multiplet, which the authors calculated to be $20N_A\mu_B^2/7\Delta k_B$, where Δ is the energy splitting between the $J = 5/2$ and $J = 7/2$ multiplets. This is the first time we come across a contribution from the higher Sm^{3+} multiplet, which theoretically is some 130 meV (~ 1500 K) spaced from the ground state $J = 5/2$ multiplet and surely cannot influence low-temperature properties! Be as it may, with thus modified Curie–Weiss law, the authors then obtained $\theta_{CW} = 0.1$ K, $\mu_{eff} = 0.79$ μ_B (less than 0.85 μ_B expected for the Sm^{3+} free ion), and $\Delta = 49$ meV (569 K). Although the extracted multiplet splitting Δ is much smaller than the theoretical value, the INS measurements detected a small excitation below about 100 meV with high incident neutron energy of 300 meV (thermal neutrons are exceptionally strongly absorbed by Sm), in my opinion, it is still too high energy splitting to make a meaningful effect on low-temperature properties.

$SmRu_4P_{12}$, just like $PrRu_4P_{12}$ discussed previously, undergoes a M–I transition, in this case at $T_{MI} = 16.5$ K, documented first by Sekine et al. (2000e). However, unlike $PrRu_4P_{12}$, where magnetic field had no role in the mechanism of the transition, in $SmRu_4P_{12}$ the magnetic field plays a pivotal role. Perhaps the most vivid evidence for the magnetic nature of the transition is provided by specific heat measurements by Matsuhira et al. (2002b), shown in Figure 3.63(a). In zero magnetic field, a large and sharp anomaly at 16.5 K marks the M–I transition. In an applied field of 9 T, the sharp peak shifts slightly to a higher temperature and a new more rounded peak, designated as T^*, appears close to 14 K. The magnetic entropy, shown in Figure 3.61(b) is approximately $R\ln4$ at

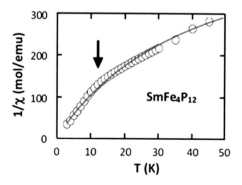

FIGURE 3.62 Temperature dependence of inverse susceptibility of $SmFe_4P_{12}$ single crystal. The line indicates a fit to Equation 3.103. Redrawn using the data of S. Konno et al., *Journal of Physics: Conference Series* **592**, 012029 (2015). With permission from the IOP Publishing.

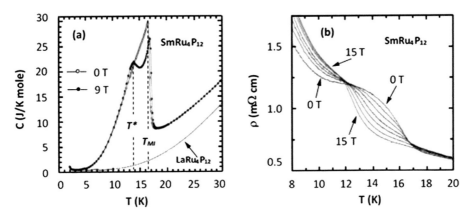

FIGURE 3.63   (a) Specific heat of a polycrystalline sample of $SmRu_4P_{12}$ measured at zero field and at 9 T. A sharp peak designates the temperature of the M–I transition T_{MI}. Note the round peak developing in magnetic field at temperature T^* close to the transition point. For comparison, specific heat of $LaRu_4P_{12}$ at zero field is also shown. (b) Temperature dependence of the electrical resistivity measured at various magnetic fields near the M–I transition. Note a negative magnetoresistance between about 12 K and 16.5 K and the development of a kink near 12 K. Redrawn from K. Matsuhira et al., *Journal of the Physical Society of Japan* **71**, 237 (2002). With permission of the Physical Society of Japan.

T_{MI} , implying that the ground state is more likely a quartet Γ_{67} (quartet Γ_8 in O_h) than a doublet Γ_5 (doublet Γ_7). Resistivity is metallic down to about 50 K, where a minimum is observed followed by a rapidly rising resistivity with a notable hump developing at 16.5 K and ceasing to exist near 12 K. Resistivity $\rho(T)$ continues to increase as temperature decreases. Applied field tends to diminish the hump, decreases the resistivity between 12 K and 16.5 K, and slightly shifts T_{MI} to higher temperatures, and a clear kink develops at 12 K, Figure 3.63(b). Magnetization is small but clearly reveals a sharp drop at T_{MI} and a more rounded peak at T^*. Obviously, the magnetic field brings into view a field-induced phase that develops below T_{MI} and extends for a few degrees to T^*. Below T^*, the system becomes an ordinary antiferromagnet. A tentative phase diagram, based on measurements of Matsuhira et al. (2002b), is shown in Figure 3.64 and consists of three phases: Phase I at $T > T_{MI}$ is a paramagnetic metal albeit likely influenced by the Kondo effect responsible for the resistivity upturn

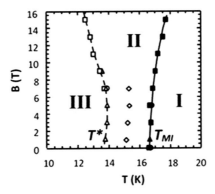

FIGURE 3.64   *B-T* phase diagram of $SmRu_4P_{12}$ constructed from the specific heat anomalies (open and solid circles), electrical resistivity (open and solid squares), and magnetization (open and solid triangles). The solid symbols designate the M–I transition and open symbols stand for T*. Diamonds show the peak position in *M(T)*. Redrawn from the data of K. Matsuhira et al., *Journal of the Physical Society of Japan* **71**, 237 (2002). With permission from the Physical Society of Japan.

below 50 K; Phase II for temperatures $T^* < T < T_{MI}$ is a mysterious magnetic field-induced phase with negative magnetoresistance; and Phase III at $T < T^*$ is an insulator with antiferromagnetically arranged magnetic moments. The fact that T_{MI} increased with magnetic field while T^* decreased, the behavior of $\rho(T)$ in a magnetic field and the Γ_{67} ground state that harbors both orbital and magnetic degrees of freedom were reminiscent of the behavior in CeB_6, where successive transition of orbital and magnetic ordering were well documented and explained *via* AFQ ordering by Effantin et al. (1985). As used many times to explain unusual behavior in skutterudites, similarities with other strongly correlated electron systems, in this case CeB_6, were enough to claim the presence of AFQ ordering in Phase II of $SmRu_4P_{12}$. While some subsequent studies supported the AFQ ordering in Phase II, e.g., Sekine et al. (2003), softening of the elastic constants observed in Phase II in ultrasonic measurements by Yoshizawa et al. (2004, 2005, 2006) suggested the breakdown of the time reversal symmetry (TRS) even in zero magnetic field, a situation incompatible with the AFQ ordering. Rather, group-theoretical arguments favored octupolar ordering, supported by predictions by Hotta (2005) that the body-centered cubic structure of skutterudites enhances octupolar fluctuations. Firm evidence of the presence of a static internal magnetic field and, hence, the broken TRS, came from μSR measurements in zero field carried out by Hachitani et al. (2006b, 2006c) and Itoh et al. (2007), the first group of authors actually determining the magnitude of the static internal field as 0.0065 T (65 Oe). The presence of the spontaneous internal field below T_{MI}, i.e., dipoles or octupoles but not quadrupoles, was also confirmed independently in NMR measurements by Mito et al. (2006). Moreover, because the internal field developed right below T_{MI} and not just below T^*, the magnetic transition was clearly at the heart of the M–I transition. As the mysterious Phase II of $SmRu_4P_{12}$ drew more and more attention, many different experimental techniques were applied to shed light on the underlying physical mechanism. The effect of pressure on $\rho(T)$ was explored by Miyake et al. (2006) and indicated a total suppression of the rising resistivity below 50 K as the pressure exceeded 7.5 GPa. While T_{MI} shifted to higher temperatures with the applied pressure, the metallic temperature dependence between T^* and T_{MI} was observed already at a pressure of 3.5 GPa. Below T^*, the resistivity maintained its insulating character. The effect of pressure on elastic properties of polycrystalline $SmRu_4P_{12}$ in zero and applied magnetic field was studied by Sun et al. (2007), who detected softening in elastic constants above T_{MI} that was amplified by the applied pressure. Moreover, the team observed a large increase in the Grüneisen parameter under pressure, reaching a magnitude of 12 at 0.95 GPa, compared to its value of unity at ambient pressure. A very large value of the Grüneisen parameter at T^* of 28 was measured previously at ambient pressure by Yoshizawa et al. (2006). XAS carried out by Tsutsui et al. (2006) using the Sm L_3 absorption edge confirmed the purely trivalent state of Sm down to the lowest temperature of the experiment of 14 K.

The now prevailing viewpoint that the octupolar ordering is responsible for Phase II was further supported by specific heat measurements on a single crystal of $SmRu_4P_{12}$ extended down to 0.2 K by Aoki et al. (2007), who determined a small, but non-zero, ordered Sm dipole moment of 0.29 μ_B from a hyperfine-enhanced Sm nuclear contribution. Such strongly suppressed dipolar contribution was argued as evidence for the primary octupolar order parameter. However, because the dipole and octupole belong to the same representation Γ_{4u} in O_h and T_h symmetries, they mix with each other and the ordering below T_{MI} could also be considered as the antiferro-octupolar one. Single crystals were used also in magnetization and specific heat measurements in fields up to 8 T by Kikuchi et al. (2007a). While a very small anisotropy was detected in magnetization and T_{MI} was entirely isotropic, the temperature T^* showed clear evidence of anisotropy, $T^*_{\langle 111 \rangle} < T^*_{\langle 110 \rangle} < T^*_{\langle 100 \rangle}$. The authors also reported a marginally higher $T_{MI} \sim 17.5$ K for their single crystals as opposed to 16.5 K typical of polycrystalline samples. Angle-dependent ultrasonic measurements under extreme conditions of ultra-low temperatures, high pressures, and high magnetic fields carried out by Yoshizawa et al. (2008) revealed a weak two-fold angular dependence in Phase II. Remarkably, below T^*, i.e., in Phase III, a four-fold angular dependence (in addition to a weak two-fold dependence) set in. However, as the temperature decreased further, both two-fold and four-fold dependences weakened and gradually disappeared. The four-fold angular dependence in Phase III could be due either to the presence of domains or to a quadrupolar contribution and its coupling to elastic strains. The authors noted that the Phase III–Phase II boundary is a type of crossover in zero applied field, which develops into a full-fledged phase transition with the assistance of an external magnetic field. Again, as an underlying mechanism of the M–I transition was favored octupolar ordering.

As we have seen in the case of $PrFe_4P_{12}$, substitutional studies were very useful in providing information regarding the strength of interactions taking place in skutterudites. Hence, it is no surprise that a similar approach was tried by Sekine et al. (2007b, 2012) in two studies with $SmRu_4P_{12}$. In the first one with $Sm(Ru_{1-x}Os_x)_4P_{12}$, a substitution was made at the site of Ru with up to $x = 0.75$ content of Os. While the minimum in $\rho(T)$ near $T_{min} \approx 50$ K persisted for all values of x, the rising resistivity at $T < T_{min}$ was substantially curtailed (but not eliminated) and a clear $-\ln T$ temperature dependence developed above $x = 0.25$, manifesting the Kondo system. The cusp in the resistivity characterizing T_{MI} was washed out above $x = 0.25$, while the anomaly at T^* survived at even $x = 0.5$. In other words, the magnetic phase associated with the M–I transition was more sensitive to substitution than the AFM phase below T^*. In the second paper, the dilution effect was explored by substituting La or Y at the site of Sm, i.e., forming $(Sm_{1-x}R_x)Ru_4P_{12}$ filled skutterudites with R = La, Y. In both cases, concentrations up to 30% were prepared by the high-temperature/high-pressure synthesis. Although the presence of La and Y had an opposite effect on the lattice parameter (La decreasing, while Y increasing the lattice constant), both elements dramatically suppressed the rising $\rho(T)$ (La more strongly than Y), and the M–I transition was no longer apparent on the temperature dependence of $\rho(T)$ with $x = 0.3$, Figures 3.65(a) and 3.65(b). Moreover, the double peak, so typical of the specific heat of $SmRu_4P_{12}$ under applied magnetic field, as shown in Figure 3.63(a), was completely annihilated and an identical $C(T)$ dependence was observed in $(Sm_{0.9}La_{0.1})Ru_4P_{12}$ at both zero and at 7 T. Both substitutional reports clearly documented the fragility of Phase II to substitution, i.e., to the disruption of periodicity of the lattice, irrespective of whether it took place at the site of the transition metal or at the void site. Such a fragile nature of Phase II was in accordance with multipole interactions.

So what kinds of octupole moments have been considered? There are three possible types of octupoles under the O_h symmetry: T_{xyz}, $T\alpha$, and $T\beta$. The isotropic T_{xyz} octupole, possibly hinted at by the T_{MI} invariance with respect to the field direction, was ruled out of considerations as a primary order parameter on account of ^{101}Ru-NQR measurements by Masaki et al. (2007) that indicated

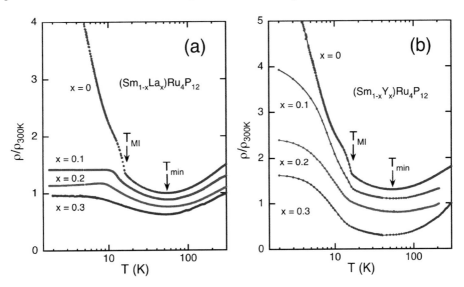

FIGURE 3.65   Temperature dependence of the normalized resistivity $\rho(T)/\rho(300\ K)$ for (a) $(Sm_{1-x}La_x)Ru_4P_{12}$, ($x = 0$, 0.1, 0.2, and 0.3) and (b) $(Sm_{1-x}Y_x)Ru_4P_{12}$, ($x = 0$, 0.1, 0.2, and 0.3). Reproduced from C. Sekine et al., *Journal of Physics: Conference Series* **391**, 012061 (2012). With permission from IOP Publishing.

lowering of the local symmetry at Ru sites below T_{MI}. Based on [149]Sm nuclear resonant forward scattering by Tsutsui et al. (2006) and specific heat measurements by Aoki et al. (2007), it was argued that the primary order parameter that supports non-zero Sm moment of 0.29 μ_B is a $T\alpha$ octupole of the Γ_{4u} irreducible representation. Because [101]Ru-NQR measurements by Masaki et al. (2007) indicated line splitting below T_{MI} with the local symmetry axis along the [111] direction, T_{111}^{α} was chosen. The antiferro-octupolar ordered state is depicted in Figure 3.66, where the heavy red arrows designate the induced Sm dipole moments. Model calculations by Hotta (2007) suggested that the Γ_{4u} multipole is, indeed, the major candidate for the primary order parameter when the CEF splitting is not too large. However, arguing that the $T\alpha$ octupole ordering would bring simultaneous ordering of dipole moments and thus no successive transitions would be possible, Yoshizawa et al. (2005) explained the temperature-dependent behavior of elastic constants with the aid of $T\beta$ octupoles. The presence of dipole moments below T_{MI} was attributed to the mixing of $T\alpha$ and $T\beta$, allowed under the actual T_h symmetry. A few years later, Yoshizawa et al. (2013) extended ultrasonic measurements of the elastic constants to very high magnetic fields up to 55 T generated by pulsed magnets. Although these measurements were prone to excessive Joule and eddy current heating, the field dependence of the transverse elastic constant $C_T = (C_{11} - C_{12} + C_{44})/4$ revealed two anomalies: one due to a magnetic transition at T^* and the other a new one associated with a phase boundary within Phase III.

This new phase was designated as Phase IV and extended along Phase II and above a dome-like Phase III, Figure 3.67.

Although most of the studies agreed on the presence of octupole ordering below T_{MI}, they all were indirect measures of the magnetic order, and it was essential that the magnetic structure is determined by a direct probe. But how to go about it when Sm is such a strong absorber of neutrons, the most discriminating probe of the magnetic order? Actually, the culprit is the naturally occurring Sm, which contains strongly absorbing isotopes. By enriching Sm to 98.69% with the [154]Sm isotope, and using this material to grow a single crystal, the problem was solved by Lee et al. (2012), who made pivotal neutron diffraction measurements. Neutron diffraction revealed superlattice reflections at $Q = (h, k, l)$ with $h + k + l =$ odd integer below T_{MI}, as shown in Figure 3.68(a). Moreover, because the reflections were not accessible by X-rays, their origin was long-range AFM ordering with the propagation vector $q = (1, 0, 0)$. As indicated by the data in Figure 3.68(b), the onset of reflections

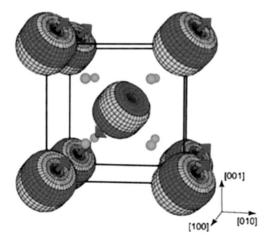

FIGURE 3.66   Antiferro-octupolar T_{111}^{α} ordered state of SmRu$_4$P$_{12}$ below T_{MI}. The heavy red arrows indicate the induced Sm dipole moments. The apple-like shape represents the induced charge-density distortion of 4f electrons. The color on the surface denotes the sign of the non-dipolar magnetization density at $T = 0$ appearing around each Sm ion due to the primary T_{111}^{α} octupoles. Ru atoms are shown as small spheres. Reproduced from Y. Aoki et al., *Journal of the Physical Society of Japan* **76**, 113703 (2007). With permission from the Physical Society of Japan.

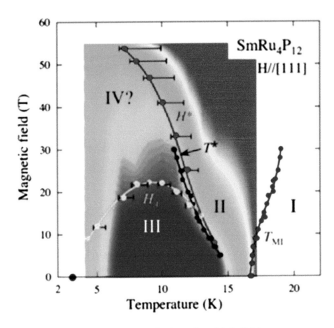

FIGURE 3.67 Magnetic phase diagram of $SmRu_4P_{12}$ based on high field measurements of elastic properties, depicting a new Phase IV developing at high fields and showing a dome-like shape of the phase boundary of Phase III. Temperatures T^* (solid black circles) and T_{MI} (solid blue circles) indicate phase boundaries from the work of Sekine et al. (2003) and Yoshizawa et al. (2006). Reproduced from M. Yoshizawa et al., *Journal of the Physical Society of Japan* **82**, 033602 (2013). With permission from the Physical Society of Japan.

was at 16.5 K, and they grew rapidly as the temperature decreased. The result provided crucial evidence that the M–I transition relates to AFM ordering, in other words, $T_{MI} \equiv T_N$, where T_N is the Néel temperature. Interestingly, no anomaly was detected at $T^* \sim 14$ K. However, because no magnetic field was applied to the crystal during the measurements, the absence of any feature near T^* was not a surprise. In fact, it was in accordance with a single peak at T_{MI} in the specific heat measurements carried out in zero field in Figure 3.63(a). The magnetic moment of a Sm ion extracted from neutron diffraction was 0.30 ± 0.02 μ_B, in excellent agreement with the value of 0.29 μ_B reported by Aoki et al. (2007), but only about one half of the Sm^{3+} free ion.

Efforts were also made on purely theoretical grounds to explain the unconventional nature of the M–I transition in $SmRu_4P_{12}$. Kiss and Kuramoto (2009) proposed a CEF-based model of localized $4f^5$ electrons that form a pseudo-sextet from the presumed Γ_{67} quartet ground state and a closeby Γ_5 doublet excited state and included intersite interactions between multipoles within the mean-field theory. This led to a multipole order (hexadecapole) with irreducible representation Γ_{5u}, which could explain some, but not all, features of the M–I transition. A more successful approach was developed in a series of papers by Shiina (2013, 2014, 2016). Based on a previous approach consisting of a detailed analysis of hybridization of a p-band formed by orbitals of the icosahedral cage of pnicogen atoms having a nested Fermi surface with the CEF states of f-electrons, Shiina and Shiba (2010) and Shiina (2011) succeeded in explaining an unconventional CO phase in $PrRu_4P_{12}$ observed below its T_{MI}. This, coupled with the fact that the conduction bands of $PrRu_4P_{12}$ and $SmRu_4P_{12}$ are formed primarily by molecular p-orbitals of the same pnicogen icosahedral cage, served as a motivation to attempt to tackle the magnetic nature of the M–I transition in $SmRu_4P_{12}$. Critical in such theory was a recognition that hybridization can only happen when the p-band has the symmetry compatible with the CEF states. Specifically, the hybridization of the p-band is allowed only with the Γ_5 doublet (Γ_7 doublet in O_h) state. The crucial effective p-f interaction arising from hybridization is of the form

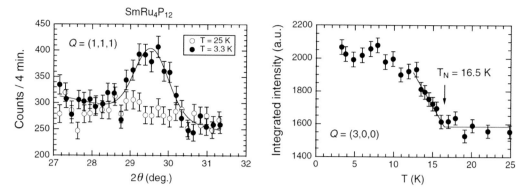

FIGURE 3.68   (a) Neutron scattering patterns of SmRu$_4$P$_{12}$ at $T = 25$ K (open circles) and $T = 3.3$ K (filled circles) at $Q = (1, 1, 1)$. The solid line is a Gaussian fit. (b) Temperature dependence of magnetic Bragg peak intensity at $Q = (3, 0, 0)$. The Néel temperature T_N at 16.5 K is identical with the M–I transition T_{MI}. The solid line depicts calculated values. Reproduced from C.-H. Lee et al., *Journal of the Physical Society of Japan* **81**, 063702 (2012). With permission from the Physical Society of Japan.

$$H_{pf} = J_s \sum_i \overline{\sigma}_i . \overline{s}_i + J_c \sum_i \delta \overline{n}_i . \overline{\phi}_i, \tag{3.104}$$

where the two terms represent magnetic and charge parts of the interaction with J_s and J_c being the magnetic and charge coupling constants, respectively, expected to satisfy $J_s = 2J_c > 0$. Here, $\overline{\sigma} = \sum_{\xi\eta} c_\xi^+ \overline{\rho} c_\eta$ is the spin density of conduction electrons given by the creation and annihilation operators and the Pauli matrix $\overline{\rho}$. Assuming $\overline{s} | \Gamma_{67} \rangle = 0$, $\overline{s}$ is the $s = \frac{1}{2}$ pseudo-spin operator in the Γ_5 doublet. $\delta \overline{n}_i = \overline{n}_i - 1$ with $\overline{n}_i$ the number of conduction electrons at the i-th site, and $\overline{\phi}$ represents the f-state occupancy, $\overline{\phi} = 1$ for the occupied Γ_5 state, and $\overline{\phi} = 0$ for the occupied Γ_{67} state. The outcome of hybridization is a magnetic field-induced charge ordering just below $T_{MI} \equiv T_N$ that gives rise to an unconventional hybrid phase comprising an AFM phase (formed from the Γ_5 doublet) with the magnetic moment parallel to the applied magnetic field and the CO phase. At temperatures below T^*, the model returns a conventional AFM phase with the moments lined up perpendicular to the applied field. It is the interaction between the AFM and CO states within the hybrid phase (Phase II) and the general instability of the hybrid phase as the temperature falls that results in remarkable thermodynamic and transport properties observed in SmRu$_4$P$_{12}$ at low temperatures. In the following paper, Shiina (2014), it was shown that the coexisting AFM and CO states develop only when the splitting between the quartet Γ_{67} and the doublet Γ_5 is small and not when either Γ_{67} or Γ_5 are isolated ground states. Shiina also raised a question whether the $R\ln4$ entropy value measured at 16.5 K was a sufficient reason to claim Γ_{67} as the ground state when the Γ_5 ground state, on account of its exclusive hybridization with the p-band, would be a more likely choice. Regardless, Shiina has shown that the hybrid AFM + CO phase requires the presence of finite magnetic field. However, in case Γ_{67} is the ground state, the hybrid phase exists even in zero field.

   In spite of searching for some evidence of atomic displacements associated with the M–I transition and its unconventional nature, none was found until high-resolution synchrotron X-ray diffraction was used by Matsumura et al. (2014, 2016). In their first report, non-resonant Thomson scattering with the wavevector $q = (1,0,0)$ induced by the magnetic field in Phase II was attributed to atomic displacements arising from charge ordering in the p-band, precisely a scenario predicted to happen by Shiina. Moreover, simultaneously with the appearance of the CO state, the AFM moment on Sm ions was enhanced in the direction of the applied field, suggesting that a staggered ordering of the $\Gamma_5 - \Gamma_{67}$ CEF split states was developing, a situation we have already encountered in discussions of magnetic properties of PrRu$_4$P$_{12}$. The staggered ordering of CEF states implied that Sm ions at the corner and the center of the bcc lattice were no longer identical as they "attract" different numbers of conduction electrons due to charge ordering. In turn, these p-electrons couple with the CEF split states of Sm^{3+}. When the density of electrons around Sm^{3+} ions increases, the Γ_5 doublet is favored as the ground state. In their second paper, Matsumura et al. (2016) were actually able to measure the atomic displacements associated with the lattice distortion in the ordered phase. This was accomplished by (i) determining displacements of $q = (1, 0, 0)$ in Phase II and (ii) by detecting uniform lattice distortion by precise measurements of the lattice parameter. The first experiment revealed displacements of Ru and P atoms in Phase II assuming the same $Pm\overline{3}$ space group as used in the case of PrRu$_4$P$_{12}$. In the second experiment, from the splitting of fundamental Bragg reflections, the authors identified a rhombohedral distortion induced along the [111] axis that coincided with the direction of the AFM moments. Below T^*, the AFM moments aligned perpendicular to the applied field, again in excellent agreement with predictions by Shiina. Overall, the staggered arrangement of $\Gamma_5 - \Gamma_{67}$ states, alternating larger and smaller cubes of Ru, including shifts of P atoms, is similar to that depicted in Figure 3.44 for PrRu$_4$P$_{12}$.

In comparison to SmFe$_4$P$_{12}$ and SmRu$_4$P$_{12}$, the third member of Sm-filled phosphide skutterudites, SmOs$_4$P$_{12}$, having the largest lattice parameter, appears as a rather ordinary AFM compound with the Néel temperature $T_N = 4.6$ K. The structure is apparently not easy to synthesize, and all studies so far have been performed for high-pressure/high-temperature prepared polycrystalline specimens. Weak, metallic temperature dependence of the electrical resistivity observed by Giri et al. (2003) turned into a broad minimum near 30 K, suggesting that the Kondo effect is relevant in this skutterudite. Near 4.6 K, the resistivity rapidly decreased, reflecting the development of the ordered AFM state that was confirmed by a distinct cusp in the otherwise rising magnetic susceptibility. A sharp peak at 4.6 K in the specific heat provided additional evidence of the phase transition. The extracted magnetic entropy at $T_N = 4.6$ K in zero field, see Figure 3.61(c), reached 87% of $R\ln4$, implying that the ground state was a Γ_{67} quartet. The peak in the specific heat shifted to lower temperatures with the increasing magnetic field, as expected for AFM, and a simple magnetic phase diagram, based on measurements by Matsuhira et al. (2005a), is depicted in Figure 3.69. The development of the AFM order was also documented by a dramatically increased broadening of the ^{31}P-NMR line width in measurements by Hachitani et al. (2006d, 2007). Analyzing the Knight shift, the hyperfine coupling constant of ~ 0.09 T/μ_B was determined. The essentially temperature

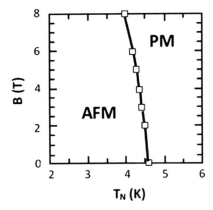

FIGURE 3.69   Magnetic phase diagram of SmOs$_4$P$_{12}$ based on specific heat measurements. Reproduced from K. Matsuhira et al., *Journal of the Physical Society of Japan* **74**, 1030 (2005a). With permission from the Physical Society of Japan.

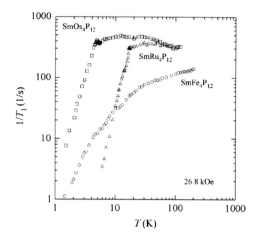

FIGURE 3.70 Temperature dependence of the spin-lattice relaxation rate $1/T_1$ of SmT_4P_{12} (T = Fe, Ru, and Os) measured in a magnetic field of 2.68 T. Redrawn from K. Hachitani et al., *Journal of Physics and Chemistry of Solids* **68**, 2080 (2007). With permission from Elsevier.

independent spin-lattice relaxation rate $1/T_1$ above T_N turned into a rapidly decreasing one below T_N, following approximately the T^3 power law dependence. The spin-lattice relaxation rate of all three Sm-filled phosphide skutterudites is shown in Figure 3.70. μSR measurements in zero field by the same authors revealed the presence of a static internal field of 0.025 T (250 Oe) related to the AFM ordering below T_N. The effect of external pressure of up to 1.4 GPa on the electrical resistivity and magnetization measured by Kawamura et al. (2012) indicated the rate of increase in the Néel temperature T_N of about 0.4 K/GPa. Elastic properties of polycrystalline $SmOs_4P_{12}$ investigated by Nakanishi et al. (2013) in fields up to 12 T indicated a monotonically increasing longitudinal C_L and transverse C_T modes as the temperature decreased. The trend was interrupted below 20 K for C_L and below 40 K for C_T by a small softening, the rate of which dramatically increased below T_N = 4.6 K, representing a 4% and 7% drop in C_L and C_T, respectively. In the case of C_L, the magnetic field shifted the transition to a slightly lower temperature, while the transverse mode was essentially immune to the effect of magnetic field. Clearly, in order to make elastic measurements more meaningful, single crystals are absolutely essential.

3.3.4.5.2 Sm-Filled Arsenides

Difficulties with the synthesis of arsenide skutterudites in general, combined with a narrow range where $SmFe_4As_{12}$ forms, have contributed to a dearth of studies on this filled skutterudite, and only two substantive studies by Kikuchi et al. (2008a) and by Namiki et al. (2010) have been reported. In both cases, polycrystalline $SmFe_4As_{12}$ was made by high-pressure synthesis near 4 GPa, and the samples turned out to be of surprisingly high quality, as judged by their RRR = 20 and 40, respectively. Likewise, for comparison purposes, $LaFe_4As_{12}$ possessing no 4f-electrons was prepared under similar conditions and used as a reference material. In the earlier, more survey-type study by Kikuchi et al. (2008b), $SmFe_4As_{12}$ was established as a FM skutterudite with the Curie temperature T_C = 39 K. The transition was revealed as a distinct kink on the otherwise metallic resistivity and confirmed by a rapidly rising magnetic susceptibility that saturated at liquid helium temperatures, with the magnetization showing a clear hysteresis at 2 K. At a magnetic field of 7 T, the magnetization reached 0.62 μ_B/f.u., somewhat smaller than the Sm^{3+} free ion value of 0.71 μ_B. The transition was also notable as a well-developed anomaly in the specific heat at 39 K, yielding the Sommerfeld constant $\gamma \sim 170$ mJmol^{-1}K^{-2}. More detailed measurements by Namiki et al. (2010) confirmed all key parameters, and the higher quality sample clearly resolved the $-\ln T$ temperature dependence in the 4f-electron contribution to the resistivity, $\rho_m(T) = \rho_{SmFe4As12}(T) - \rho_{LaFe4As12}(T)$, above T_C extending

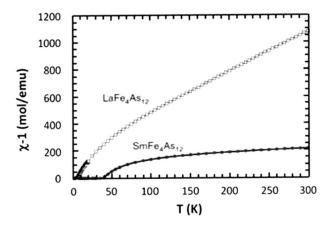

FIGURE 3.71 Temperature dependence of inverse susceptibility for $SmFe_4As_{12}$ (solid blue circles) and $LaFe_4As_{12}$ (open squares). Redrawn from T. Namiki et al., *Journal of the Physical Society of Japan* **79**, 074714 (2010). With permission from the Physical Society of Japan.

to at least 150 K, documenting the presence of the Kondo effect in the paramagnetic phase. The saturated magnetization of about 0.53 μ_B/f.u was smaller than that in the first report. The small value of M_s, coupled with the fact that the inverse susceptibility in Figure 3.71 did not follow the usual Curie–Weiss behavior but, rather, could be well fitted by an expression for a ferrimagnetic substance from Chikazumi (1997), where C, χ_0, ξ, and θ are the parameters representing various magnetic interactions

$$\left(M/H\right)^{-1}(T) = \frac{T}{C} + \frac{1}{\chi_0} - \frac{\xi}{T-\theta}, \tag{3.105}$$

and distribution of the magnetic ions on the two sublattices, prompted the authors to claim that $SmFe_4As_{12}$ is a ferrimagnetic structure. Of course, to provide a solid proof of the ferrimagnetic state in $SmFe_4As_{12}$, a sharp probe of the magnetic structure, namely, neutron scattering, would be required. However, given that the naturally occurring Sm strongly absorbs thermal neutrons, the synthesis would have to be done with a sample highly enriched with the [154]Sm isotope, as done in the studies of Lee et al. (2012) with $SmRu_4P_{12}$.

There are no reports on magnetic or transport properties of $SmRu_4As_{12}$ and $SmOs_4As_{12}$.

3.3.4.5.3 Sm-Filled Antimonides

Studying magnetic susceptibilities and magnetization of several rare earth-filled iron antimonide skutterudites, Danebrock et al. (1996) identified $SmFe_4Sb_{12}$ as a FM structure with the Curie temperature $T_C = 45$ K. The room-temperature susceptibility amounted to an effective magnetic moment $\mu_{eff} = 1.6 \pm 0.2 \mu_B$/f.u., and the saturation magnetization at 5 K in a field of 5.5 T yielded $M_s = 0.7 \mu_B$/f.u., in good agreement with the value of 0.71 μ_B for the Sm^{3+} free ion. A well-developed hysteresis at 5 K indicated a coercive field of 0.24 T. The sample was polycrystalline prepared by reacting powders in sealed silica tubes at 640 °C for 24 hours and quenching in water. Although the authors did not specify the filling fraction of Sm, the above synthesis process is known to result in a large number of vacancies in the void sites. Judging from the filling fraction of Nd of only 83% the authors reported for similarly prepared $NdFe_4Sb_{12}$, the $SmFe_4Sb_{12}$ sample likely had similar deficiency of Sm. To achieve near full occupancy of voids ($x \sim 0.95 \pm 0.05$), Ueda et al. (2008) prepared their polycrystalline $SmFe_4Sb_{12}$ using a high-pressure synthesis under 3.5 GPa. All three key macroscopic parameters, electrical resistivity, magnetic susceptibility, and specific heat, evidenced distinct anomalies in their temperature dependence at $T_C \sim 43$ K, two degrees lower than in the measurements of Danebrock et al. (1996). The anomalies reflected the onset of long-range FM order, further evidenced by a hysteresis in M vs. B dependence. The kink in $\rho(T)$ and a subsequent rapidly decreasing resistivity indicated much reduced charge carrier scattering as the spins spontaneously ordered below T_C. From the specific heat measurements that displayed a distinct peak at 43 K, the authors extracted the Sommerfeld coefficient $\gamma = 72$ mJmol^{-1}K^{-2}. The value of the saturation magnetization $M_s \sim 0.53$ μ_B/f.u. was viewed as favoring a quartet Γ_{67} for the ground state. But, this should be taken with caution because of the unknown contribution of Fe $3d$ electrons. Comparing the two sets of measurements and taking into account that the first set of data was collected on samples likely having a serious deficiency of Sm in the voids, the magnetic properties are remarkably consistent, and SmFe$_4$Sb$_{12}$ seems to be robust with respect to the void vacancies, in clear contrast to Pr$_x$Fe$_4$Sb$_{12}$, where we have seen the nature of magnetism to change with the filler content x. I should also mention ^{149}Sm and ^{57}Fe nuclear resonant inelastic scattering measurements by Tsutsui et al. (2012) on SmFe$_4$X$_{12}$, for all three pnicogen elements P, As, and Sb, that revealed a strong dependence of hybridization between Sm and Fe on the pnicogen element X.

Surprisingly, beyond the lattice parameter of SmRu$_4$Sb$_{12}$ provided by Evers et al. (1995), and a report by Sekine et al. (2009c) describing the high-pressure synthesis of SmRu$_4$Sb$_{12}$ at 2 GPa and temperatures 700–760°C, there are no reports in the literature devoted to properties of this particular filled skutterudite.

Two reports, authored by Sanada et al. (2005) and Yuhasz et al. (2005), which started to draw attention to SmOs$_4$Sb$_{12}$ as an interesting and unconventional electronic structure and initiated a subsequent feverish research activity, described the behavior of electrical resistivity, magnetic properties, and specific heat of single crystals. The reports substantially agreed on SmOs$_4$Sb$_{12}$ being a weak FM skutterudite with the Curie temperature $T_C \approx 2.6$ K that shows a heavy fermion nature, as judged by an extraordinarily large value of the Sommerfeld constant γ of 820 mJmol^{-1}K^{-2} and 880 mJmol^{-1}K^{-2} extracted from the two respective specific heat measurements. For comparison, the only other Sm-filled potentially heavy fermion skutterudite is SmFe$_4$P$_{12}$ with $\gamma = 370$ mJmol^{-1}K^{-2}. Using the formula

$$\gamma = \frac{\pi^2 \left(Z \middle/ \Omega \right) k_B^2 m^*}{\nabla^2 k_F^2}, \tag{3.106}$$

with Z the number of charge carriers per unit cell ($Z = 2$), Ω the unit cell volume, and assuming a spherical Fermi surface, i.e., taking $k_F = \left(3\pi^2 Z \middle/ \Omega \right)^{1/3}$, the effective mass of charge carriers becomes $m^* \approx 170$ m$_e$, indeed, exceptionally heavy charge carriers. The heavy fermion nature also followed from a large coefficient of the quadratic term in the electrical resistivity, $A \approx 0.88$ $\mu\Omega$cmK^{-2} and 0.39 $\mu\Omega$cmK^{-2} in the two respective reports. However, the ratio A/γ^2, the so-called Kadowaki–Woods relation, Kadowaki and Woods (1986), used to characterize heavy fermion systems, was an order of magnitude smaller than $A/\gamma^2 = 1 \times 10^{-7}$ Ωm(mol-K/J)^2, the value relevant to most of the heavy fermion systems. This discrepancy was rationalized by Tsuji et al. (2005) as a consequence of high-orbital degeneracy of Sm^{3+} ions, and Kontani (2004) subsequently generalized the Kadowaki–Woods relation to take this into account, confirming the status of SmOs$_4$Sb$_{12}$ as a heavy fermion skutterudite. A plot of $\chi^{-1}(T)$ vs. T shown in Figure 3.72 indicates no obvious temperature region where a clear linear dependence, reflecting the usual Curie–Weiss behavior, could be identified. Sm is unusual in the sense that the Sm^{3+} ion has a relatively small energy separation between its different total angular momentum states. While in most of the rare-earth ions it suffices to consider the Hund's rule ground state, in Sm-filled skutterudites and, specifically, in SmOs$_4$Sb$_{12}$, the first excited total angular momentum multiplet $J = 7/2$ is often needed to be taken into account in addition to the $J = 5/2$ ground state. With this *proviso*, Yuhasz et al. (2005) fitted the susceptibility with an expression

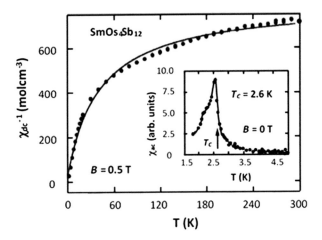

FIGURE 3.72 Inverse susceptibility as a function of temperature measured in field of 0.5 T for $SmOs_4Sb_{12}$ single crystal. The solid line is a fit of the data using Equation 3.107. The inset shows ac susceptibility with a sharp peak near $T_C = 2.6$ K. The T_C was defined as a mid-point of the transition on the paramagnetic side. Redrawn from W. M. Yuhasz et al., *Physical Review B* **71**, 104402 (2005). With permission from the American Physical Society.

$$\chi(T) = \frac{N_A}{k_B}\left[\frac{\mu_{eff}^2}{3(T - \theta_{CW})} + \frac{\mu_B^2}{\delta}\right], \tag{3.107}$$

where the first term is the usual Curie–Weiss law and the second term is the temperature independent van Vleck term with $\delta = 7\Delta/20$, where Δ is the energy splitting (in units of temperature) between the $J = 5/2$ and $J = 7/2$ multiplets. Equation 3.107 is the same expression as Equation 3.103 except here it is written out fully in terms of δ. The fit of $\chi^{-1}(T)$ *vs.* T using Equation 3.107 is indicated by a solid line in Figure 3.73, and it returned $\theta_{CW} = -0.99$ K, $\mu_{eff} = 0.63$ μ_B/f.u., and $\delta = 300$ K. The last parameter resulted in $\Delta = 20\delta/7 = 850$ K, a considerably smaller value that the theoretical spacing (in units of temperature) of ~ 1500 K, van Vleck (1932).

A real revelation that ordering of magnetic moments takes place at low temperatures came from ac magnetic susceptibility measured by Yuhasz et al. (2005) that showed a sharp peak at $T_C = 2.6$ K. A clear hysteresis in the magnetization, seen in both reports, with the coercive field of about 2.5×10^{-3} T, documented the ordering as being of the FM nature. Where the reports departed was in the CEF ground state of $SmOs_4Sb_{12}$, with Sanada et al. (2005) favoring a Γ_{67} quartet (Γ_8 quartet in O_h symmetry) as the ground state, while Yuhasz et al. (2005) arguing for Γ_5 doublet (Γ_7 doublet) as the ground state. The splitting between the two lowest CEF states was stated as 1.64 meV (19 K) and 3.27 meV (38 K), respectively. A remarkable and unusual feature was the total insensitivity of the Sommerfeld constant γ, and only marginal sensitivity of the quadratic coefficient A in $\rho(T)$ to the applied magnetic field, implying that the heavy quasiparticles have an unconventional non-magnetic origin. From the facts that the magnetic ordering left no mark whatsoever on the $\rho(T)$ behavior, the saturation magnetization was very small, and only a meager anomaly was seen on the plot of C/T *vs.* T, both teams of researchers noted that the FM state must be very weak. Sanada et al. (2005) proposed itinerant heavy quasiparticles as the origin of such weak ferromagnetism, while Yuhasz et al. (2005) speculated whether the ordering might be due to some impurity phase present in their crystals.

The above two reports established the baseline properties of $SmOs_4Sb_{12}$, including its puzzling heavy fermion nature immune to the effect of magnetic field. Subsequent numerous studies have

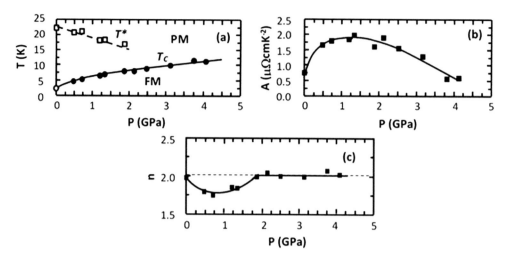

FIGURE 3.73 (a) Pressure-temperature phase diagram of $SmOs_4Sb_{12}$. Solid circles indicate T_C obtained from measurements of the resistivity and an open circle stands for T_C from NQR measurements. (b) Pressure dependence of the quadratic term in the resistivity A. (c) Pressure dependence of the exponent n in the temperature-dependent resistivity $\rho(T) = \rho_0 + AT^n$. Adapted and redrawn from H. Kotegawa et al., *Physical Review Letters* **99**, 156408 (2007). With permission from the American Physical Society.

focused on clarifying the FM order and finding an appropriate model to describe the unusual heavy fermion state of quasiparticles. The usual macroscopic parameters, such as the magnetization and specific heat, were addressed in several studies, including measurements at very high magnetic fields and samples subjected to external pressure. As an example, magnetization measurements by Yamada et al. (2007) were extended to 45 T with the field oriented either along the [100] or [110] directions, hoping to check on any anisotropy and inquiring about the magnitude of the magnetic moment in such high applied fields. Detecting no measureable anisotropy was inconsistent with the Γ_{67} state being the favored ground state of the system. The magnetization remained very low at 0.4 μ_B/Sm^{3+} even in such enormous fields and barely reached one half of the Sm^{3+} free ion value of 0.71 μ_B/Sm^{3+}. Concerns regarding the non-intrinsic origin of ferromagnetism in $SmOs_4Sb_{12}$ were dispelled by NQR studies and by measurements of resistivity and magnetization under pressure in a series of studies by Kotegawa et al. (2005, 2007, and 2008). Specifically, [123]Sb-NQR spectra showed a strong broadening at 1.4 K compared to narrow lines at 4.2 K and 10 K, indicating the presence of an internal magnetic field (~ 0.02 T) along the [100] direction. This together with a clear peak in the spin-lattice relaxation rate $1/T_1$ near T_C supported the bulk form of ferromagnetism. Moreover, the applied pressure of up to 4.5 GPa seemed to decrease the Kondo temperature and stabilize the magnetic ordering, a trend opposite to the behavior observed in Ce-filled skutterudites. The pressure-temperature phase diagram and the pressure dependence of the coefficient of the quadratic term in the resistivity A as well as the exponent n are shown in Figures 3.73(a)–(c), respectively. In their subsequent [121]Sb- and [123]Sb-NQR measurements under pressure, Kotegawa et al. (2007, 2008) concluded that their temperature and pressure dependence of the relaxation rate $1/T_1$ demonstrates the presence of the magnetic Kondo effect, which attains a coherent state below the temperature T^*, determined by the coupling between the c-f hybridization and the charge fluctuations in the system. They proposed that this might be a plausible explanation for the robust heavy fermion behavior observed in $SmOs_4Sb_{12}$. A peculiar feature observed in these measurements was a decrease in magnetization with increasing pressure above its peak value at 0.44 GPa, in spite of the transition temperature T_C showing a dramatic rise with applied pressure. Weak temperature dependence of magnetization and its small value was explained by Aoki, Y. et al. (2006) as due to screening of Sm magnetic moments in the heavy fermion state, and a suggestion was made that "rattling" of Sm ions in large size voids of the $SmOs_4Sb_{12}$ skutterudite might be connected with the non-magnetic type of the mass enhancement. In their subsequent work, making use of zero-field μSR measurements, Aoki et al. (2009) demonstrated the intrinsic bulk nature of ferromagnetism in $SmOs_4Sb_{12}$ below 2.6 K by detecting a small spontaneous magnetic moment developing along the [001] crystal orientation with a tiny magnitude of 0.07 μ_B/Sm.

As we have seen already, because the macroscopic strain of symmetry Γ can couple directly to the quadrupolar operator O_Γ, the quadrupolar susceptibility is a faithful description of the elastic constants of compounds containing rare-earth elements with their substantially localized $4f$ electrons. Typically, elastic constants tend to increase monotonically as the temperature decreases. However, any significant departure from this trend is a sign of something extraordinary happening with the state of electrons. The case in point is ultrasonic measurements of elastic constants of $SmOs_4Sb_{12}$ by Nakanishi et al. (2006), showing a rapid drop in the C_{11} constant in zero magnetic field close to T_C. Because neither of the two ground state scenarios (Γ_5 or Γ_{67}) could fit adequately the temperature dependence of the elastic constants, the controversial issue of the ground state remained unresolved and had to wait for future studies. Instead, the authors injected a new idea into the problem by suggesting that valence instability of Sm at low temperatures might have something to do with the temperature dependence of the elastic constants. So far, Sm has always been considered to enter the voids as a trivalent magnetic Sm^{3+} ion. The presumed valence instability implied that some Sm atoms might take on a divalent non-magnetic form Sm^{2+} with the $4f^6$ electron configuration. This was an impulse to initiate serious studies of the valence state of Sm in $SmOs_4Sb_{12}$, the task taken on by Yamasaki et al. (2007) and by Mizumaki et al. (2007), who used hard X-ray ($h\nu = 7932$ eV) photoemission spectroscopy (XPS) and XAS, respectively. The first group made use of the fact that the mean free path of photo-generated electrons increases with the increasing energy of photons, and thus bulk properties, rather than just the surface, were probed. The Sm $3d$ core-level photoemission spectrum (open circles) obtained at 18 K is compared to the calculated spectrum (solid line) in Figure 3.74. The larger peak corresponds to Sm^{3+}, and a smaller peak at a lower energy is due to the presence of Sm^{2+}. The average valence extracted from measurements at 100 K and at 18 K was 2.763 and 2.726, respectively, indicating also that the Sm valence is temperature-dependent. More detailed measurements of the temperature dependence of the photoemission spectrum below 300 K were carried out by Mizumaki et al. (2007). They revealed two temperatures: $T_A \sim 150$ K, above which the average Sm valence is constant at 2.83, and $T_B \sim 20$ K, below which it is constant at 2.76. Between T_A and T_B, the average valence decreases as depicted in Figure 3.75. The authors have speculated that the presence of Sm^{2+}, together with the orbital degrees of freedom, is crucial in order to understand

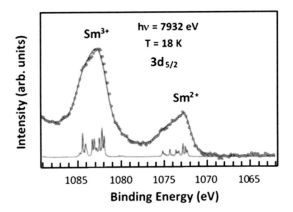

FIGURE 3.74 Experimental Sm $3d$ core-level photoemission spectrum (open circles) obtained at 18 K with a photon incident energy of 7932 eV and calculated spectrum (solid line). Adapted and redrawn from A. Yamasaki et al., *Physical Review Letters* **98**, 156402 (2007). With permission from the American Physical Society.

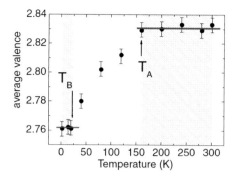

FIGURE 3.75   Temperature dependence of the average valence of Sm in $SmOs_4Sb_{12}$. Temperatures $T_A \approx 150$ K and $T_B \approx 20$ K indicate the onset and sunset, respectively, of the temperature-dependent average valence. The shaded region indicates temperatures where the average valence of Sm is constant. Reproduced from M. Mizumaki et al., *Journal of the Physical Society of Japan* **76**, 053706 (2007). With permission from the Physical Society of Japan.

the large electronic specific heat and its insensitivity to the external magnetic field. If true, this would mean that the heavy fermion state in $SmOs_4Sb_{12}$ has the origin in a non-magnetic valence, a highly unusual situation compared to a typical magnetically-linked mass enhancement. It is intriguing that, although $SmOs_4Sb_{12}$ orders magnetically at low temperatures, it is the non-magnetic Sm^{2+} ion that is more stable than the magnetic Sm^{3+} ion. Although there is no doubt, as follows from the following discussion, that the Sm average valence is less than 3 and that it is temperature-dependent between about 20 K and 150 K, I caution against blindly adopting the numerical values obtained by Mizumaki et al. (2007). My hesitation stems from more recent theoretical evaluation of Sm XAS and XPS spectra by Nanba et al. (2013), where the authors pointed out that basing the average Sm valence on the intensity ratio I^{2+}/I^{3+} of the Sm $2p$ XAS spectra without considering hybridization between the Sm $4f$ and Sb $5p$ orbitals in the final state underestimates the average valence by about a factor of 0.05.

The reduced valence of Sm, as the temperature decreases, has consequences regarding the structural parameters of $SmOs_4Sb_{12}$, and such correlations were eagerly sought in a number of studies. While no notable changes were detected in the otherwise slowly decreasing lattice parameter a investigated by Tsubota et al. (2008) using high-resolution synchrotron radiation powder diffraction measurements, see Figure 3.76(a), the authors observed clear changes in the Sb-Sm bond length (distance r_{Sm-Sb}) *via* altered positional parameters y and z, Figure 3.76(b). Comparing the trend in Figures 3.76(a) and 3.76(b), it is obvious that the temperature dependences of a and r_{Sm-Sb} are entirely different. The temperature T_A marks a point where the bond length starts to decrease rapidly as the temperature falls. Because $SmOs_4Sb_{12}$ increases its fraction of Sm^{2+} ions below T_A, and a divalent Sm ion has a larger ionic radius than the trivalent Sm ion, an increase in the bond length would be expected as the temperature decreases below T_A. The trend in Figure 3.76(b) is, however, opposite, and this implies that Sb atoms are moving toward their nearest Sm atoms. Judging by the last two low-temperature points, the movement of Sb atoms toward Sm ceases at the temperature T_B, where the Sm valence attains a constant value. Similar studies by Tsutsui et al. (2009) agreed on the insensitivity of the unit cell (i.e., the lattice parameter a) to valence changes of Sm and revealed a notable reduction in the ratio of volumes of the Sb-based icosahedral cage to the unit cell, expressed *via* the positional parameters as $(y^2 + z^2)^{3/2}$. The shrinking of the cage as the temperature decreases below T_A, of course, enhances hybridization between $4f$-electrons and the conduction electrons (mostly $5p$-states of Sb), and the authors speculated that this might be a factor behind the development of the heavy fermion state in $SmOs_4Sb_{12}$. From their diffraction measurements, they also established the atomic displacement parameter (ADP) of Sm and, using Equation 3.102, evaluated the Einstein temperature θ_E as 40.1 K.

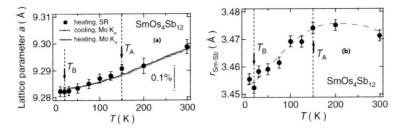

FIGURE 3.76 (a) Temperature dependence of the lattice parameter a of $SmOs_4Sb_{12}$. Solid circles represent the synchrotron data while lines are data from Mo K_α radiation, blue line for cooling and red line for heating. The error bars on solid circles represent 3σ standard deviations obtained in the Rietveld analysis. Errors in Mo K_α data are much smaller than the width of the lines. (b) Temperature dependence of the Sm-Sb bond length $r_{Sm\text{-}Sb}$ in $SmOs_4Sb_{12}$. The icosahedral cage start to shrink at $T_A \approx 150$ K and the trend continues to a lower temperature $T_B \approx 20$ K. Reproduced from M. Tsubota et al., *Journal of the Physical Society of Japan* **77**, 073601 (2008). With permission from the Physical Society of Japan.

Such localized low-energy optical modes became of much interest ever since first observed by Keppens et al. (1998), as their presence is considered vital in reducing the thermal conductivity of filled skutterudites, making them useful as thermoelectric materials. Here, the interest stemmed from an observation by Ogita et al. (2006) that this localized mode shifts from $\theta_E \approx 63$ K in non-magnetic $LaOs_4Sb_{12}$ down to $\theta_E \approx 40$ K in $SmOs_4Sb_{12}$. The observed shift of the localized optical mode was a motivation for Matsuhira et al. (2007) to have a closer look at what effect such a shift might have on the specific heat. The analysis indicated that a peculiar shoulder observed on C/T vs. T plot near 10 K is precisely due to this shift toward lower temperatures and has nothing to do with any Schottky anomaly arising from CEF energy splitting.

Coupling between the localized optical phonon mode and the electronic system was viewed as yet another non-magnetic mechanism potentially able to explain the immunity of the Sommerfeld constant γ and the quadratic coefficient of resistivity A to the external magnetic field. The theoretical work along these lines was done by Hattori et al. (2005), Mitsumoto and Ono (2005), Yotsuhashi et al. (2005), and Hotta (2006). The idea that "rattling" mode of $SmOs_4Sb_{12}$ is an important component in the formation of the heavy fermion state and its field insensitivity has appeared time and again. In one of the more recent variants by Aoki et al. (2011b), it is formulated as a correlation between the developing mass enhancement, rattling phonon excitations, and charge fluctuations, all appearing in the same temperature range between 10 K and 100 K. The $4f$-electrons get modified by the rattling Sm ions and/or charge fluctuations and absorb such non-magnetic degrees of freedom. As the temperature decreases below 100 K, the mass enhancement strengthens due to c-f hybridization, and at temperatures below T_K (~ 20 K), the fully developed heavy fermion state becomes field insensitive. However, whatever role a potential coupling of the localized low-energy optical phonon mode and conduction electrons may play in a large mass enhancement in $SmOs_4Sb_{12}$, one should keep in mind the fact that "rattling", i.e., large atomic displacement of the filler ions compared to displacements of the other species constituting a filled skutterudite, is observed in many filled skutterudite structures, yet a large mass enhancement is found rather rarely.

The temperature-dependent valence of Sm drew considerable attention, and attempts were made to find its microscopic origin. One of the more interesting models, which tied the reduced valence of Sm to an anharmonic motion of Sm ions in the voids, was developed by Tanikawa et al. (2009). In this, the so called two-level Kondo model, it was assumed that the filler ion tunnels in a two-well potential and, as the temperature decreases through the Kondo temperature (~ 20 K), it attracts the conduction electrons progressively more strongly, *de facto* reducing the valence of positive Sm ions. The basic physics of this interaction was laid down by Kondo himself in 1983 and further elaborated on by Vladar and Zawadowski (1983).

In Figure 3.75, we have seen the surprising temperature dependence of the Sm valence in $SmOs_4Sb_{12}$. An intriguing question is whether the Sm valence might also be pressure-dependent. This has been addressed by Tsutsui et al. (2013) who carried out Sm $L3$-edge XAS under an applied pressure of up to 3.5 GPa. Figure 3.77(a) documents the presence of non-magnetic divalent Sm even at a pressure of 3.5 GPa. Taking the spectra at different pressures and observing shifts in the temperature T_A, the pressure dependence of T_A is depicted in Figure 3.77(b). The initially decreasing T_A reaches its minimum value near 1 GPa and then increases to 250 K at 3.5 GPa. This implies not only a strong effect of pressure on the valence state of Sm, but also that the characteristic pressure for the valence state of Sm is about 1 GPa, the value that coincides with the pressure where the resistivity shows its largest deviation from the Fermi-liquid behavior (judged by the exponent n deviating most from the value of 2) and near where the quadratic coefficient of resistivity A attains its maximum value, see Figure 3.73(b). Unfortunately, the lowest temperature the experiment could reach did not extend below 15 K, and it was not possible to track the pressure dependence of the temperature T_B, which, on account of its proximity to the Kondo temperature, might have otherwise provided a better measure of correlation between the valence state of Sm and parameters, such as the Kondo effect.

Unlike in the $[Fe_4Sb_{12}]$ anion framework, where the magnetic contribution of Fe has been amply documented, it is usually assumed that Ru- and Os-containing frameworks contribute no magnetic moment. But is it really the case? Except in band structure calculations, the exact role played by Ru and Os in the electronic structure has never really been investigated. One rare occasion where it has been attempted were measurements of the XMCD at the Os L-edge under extreme conditions of low temperature (2.2 K), high magnetic field of 10 T, and pressures up to 3.5 GPa carried out by Kawamura et al. (2009). XMCD is a technique that makes use of a difference spectrum of two X-ray absorption spectra collected in an applied magnetic field, one using a left circularly polarized light and the other a right circularly polarized light. The difference spectrum reveals element-specific information on the orbital and spin magnetic moments. The measurements provided a microscopic picture of Os $5d$ electronic states and indicated a small but clear dichroic signal at a 0.1% level at ambient pressure. The very small orbital ($-0.001\ \mu_B$) and spin magnetic moments ($+0.012\ \mu_B$) of Os $5d$-electrons were antiferromagnetically coupled with respect to the Sm magnetic moment. Under an applied pressure, the dichroic signal peaked near 0.6 GPa and then disappeared altogether above 3.5 GPa. The behavior mimicked the trend in the magnetization, pointing toward an unconventional FM state at high pressure. The authors raised a possibility that Os $5d$-electrons might play some role in the exotic electronic properties of $SmOs_4Sb_{12}$, but given such a tiny AFM-coupled moment, the influence is unlikely to be of consequence.

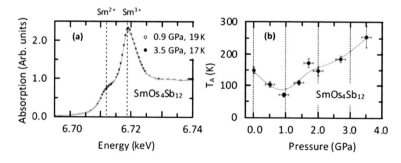

FIGURE 3.77   (a) Sm L_3 X-ray absorption spectra of $SmOs_4Sb_{12}$ taken at low temperatures and pressures of 0.9 GPa (open circles) and 3.5 GPa (solid circles). The presence of divalent Sm even at 3.5 GPa is clearly documented. (b) Pressure dependence of temperature T_A in $SmOs_4Sb_{12}$. The curve is a guide to the eye. Reproduced from S. Tsutsui et al., *Journal of the Physical Society of Japan* **82**, 023707 (2013). With permission from the Physical Society of Japan.

Returning to a controversial issue of what constitutes the ground state in $SmOs_4Sb_{12}$, the latest report by Mombetsu et al. (2016) describes ultrasonic measurements of elastic constants in pulsed field of up to 58 T. The premise of the experiment is to make use of ultrasound to obtain elastic constant, from which quadrupolar susceptibilities can be evaluated. Because the latter strongly depends on the CEF energy level splitting, such interdependence can be profitably used to determine the ground state by differentiating between the Γ_5 doublet supporting magnetic dipoles and the Γ_{67} quartet that contains magnetic dipoles, electric quadrupoles, and magnetic octupoles. A relative change in the elastic constant, $\Delta C_{44}/C_{44}$, as a function of magnetic field measured at temperatures 1.5 K, 4.2 K, and 8.7 K is illustrated in Figure 3.78(a), and in Figures 3.78(b) and 3.78(c) are displayed calculated values assuming that the ground state is the Γ_5 doublet and the Γ_{67} quartet, respectively. Clearly, the Γ_{67} quartet ground state captures the minimum and its temperature dependence, followed by a reduction in absolute value (i.e., the rising trend), while the Γ_5 doublet does not. The qualitative agreement is especially good at fields above 20 T where the 4f-electrons should be localized, as assumed in calculations shown in Figure 3.78(c). At low temperatures and fields well below 20 T, Kondo-like screening due to c-f hybridization should dominate.

In a recent theoretical study trying to shed light on the origin of a field-insensitive mass enhancement in $SmOs_4Sb_{12}$, Shiina (2018) used an effective two-orbital Anderson model for the respective $4f^5$ and $4f^6$ states of Sm^{3+} and Sm^{2+} and has shown that the Fermi-liquid state with an enhanced Sommerfeld coefficient γ and negligibly small Wilson ratio forms in the vicinity of a quantum critical point (QCP) and satisfactorily accounts for the unconventional heavy-Fermion state. To make computations based on the numerical renormalization method tractable, the model makes a simplification by assuming that the $4f^5$ and $4f^6$ electron configurations are a single Γ_{67} multiplet (Γ_8 designation of the equivalent O_h symmetry is used in the paper), occupied by one electron in the case of Sm^{3+} and two electrons in the case of Sm^{2+}. This effectively reduces the problem to $f^2 - f^1$ valence fluctuations, neglecting, i.e., effectively freezing, participation of all other electrons. What effect such a drastic simplification makes on the outcome of computations remains a question.

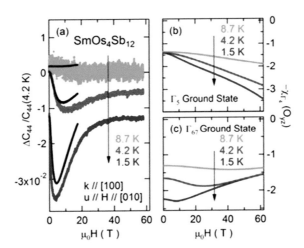

FIGURE 3.78 (a) Relative change in the elastic constant $\Delta C44/C44$ plotted as a function of applied magnetic field at three different low temperatures. Pulsed field data for k // [100] and u // H // [010] are presented in color while static field data (obtained using a SC solenoid) are shown as black curves. (b) Magnetic field dependence of the quadrupolar susceptibility $\chi_{\Gamma4}(O_{yz})$, calculated for the Γ_5 doublet ground state with the excited state Γ_{67} quartet 38 K above the ground state. (c) The same as in (b), except the ground state is the Γ_{67} quartet and the excited state Γ_5 doublet is about 20 K above the ground state. Reproduced from S. Mombetsu et al., *Journal of the Physical Society of Japan* **85**, 043704 (2016). With permission from the Physical Society of Japan.

The lastest report addressing the issue of a very large Sommerfeld coefficient γ in $SmOs_4Sb_{12}$ and its insensitivity to applied magnetic field describes ^{149}Sm synchrotron radiation-based Mössbauer spectroscopy measurements by Tsutsui et al. (2019). The work aims to substantiate the already mentioned XAS measurements by directly observing Sm fluctuating valence. From the changes in the isomer shift with temperature that mimicked temperature variation of the Sm valence in XAS measurements, the authors established that the valence of Sm at temperatures below 40 K fluctuates on the scale of 10 MHz. Such low-temperature Sm valence fluctuations enhance the entropy, which, in turn, increase the Sommerfeld coefficient γ, and are a viable non-magnetic mechanism giving rise to the heavy fermion state in $SmOs_4Sb_{12}$.

3.3.4.6 Eu-Filled Skutterudites

Unlike most of the other rare-earth elements, Eu with its electronic configuration [Xe] $4f^7 6s^2$ enters the skutterudite void as a divalent Eu^{2+} or mixed valence ion rather than a trivalent Eu^{3+}. The respective electronic configurations are then [Xe]$4f^7$ with the term symbol $^8S_{7/2}$, i.e., with a half-filled f-shell, and [Xe]$4f^6$ with the term symbol $^7F_0$, for the non-magnetic Eu^{3+} ion. The first hint for the divalent form of Eu came from a larger than expected cell volume of Eu-filled skutterudites synthesized by Jeitschko and Braun (1977). Mössbauer measurements were the early probe of the microscopic origin of magnetism and played an important role in clarifying its nature. Except for $EuRu_4As_{12}$, where no sign of any phase transition was detected down to 2 K, all other Eu-filled skutterudites are FM structures, some of them with surprisingly high Curie temperatures. There are not many studies on Eu-filled phosphide and arsenide skutterudites, primarily for the lack of intriguing physical properties and also on account of their very limited practical potential, namely, as thermoelectric materials. The situation is very different in the case of antimonides, where filling by Eu has resulted in more prospective thermoelectric structures and this is reflected in a considerably larger volume of the literature.

3.3.4.6.1 Eu-Filled Phosphides

All Eu-filled phosphide skutterudites are rather ordinary FM metals with ordering taking place on the Eu sublattice, but with significantly different ordering temperatures. Early studies by Gérard et al. (1983) and Grandjean et al. (1984) identified $EuFe_4P_{12}$ as a FM skutterudite with the high Curie temperature $T_C = 100 \pm 3$ K and the effective magnetic moment $\mu_{eff} = 6.2$ μ_B/f.u. Because the ^{57}Fe Mössbauer measurements detected no magnetic moment associated with Fe, the measured moment is entirely due to Eu. Given the fact that the measured moment is smaller than the theoretical value of 7.94 μ_B/Eu^{2+}, and taking into account that the Eu^{3+} ion is non-magnetic, the presence of about 22% of Eu^{3+} ions for the intermediate valence of +2.22 is implied. The Mössbauer experiment with ^{151}Eu resulted in a single absorption line down to 100 K, which split into several lines at lower temperatures, documenting that the magnetic state is a result of the ordering on the Eu lattice. A large and temperature independent isomer shift of – 6 mm/s and a giant hyperfine field of −67 T implied a large s-electron charge density at the nucleus as well as large unpaired spin density at the nucleus. First-principles full-potential linearized augmented plane wave (FP-LAPW) calculations with the PBE-GGA approximation implemented in the WIEN2K code, performed by Shankar and Thapa (2013), gave a magnetic moment of 5.62 μ_B using the local spin-density approximation (LSDA) and much larger 8.18 μ_B using the LSDA+U method, where U indicates Coulomb repulsion. The calculated DOS resulted in a small but non-zero DOS, i.e., a pseudogap, just above the Fermi level in both spin channels. As in all filled skutterudites, this pseudogap is crossed by a nearly linearly dispersing band discussed in more details in Chapter 1.

^{151}Eu Mössbauer measurements were also instrumental in establishing $EuRu_4P_{12}$ as a FM structure, albeit at a much lower Curie temperature $T_C = 18$K, Grandjean et al. (1983). Subsequent ^{151}Eu Mössbauer studies by Indoh et al. (2002), shown in Figure 3.20, revealed three subspectra, Eu^{2+}(A), Eu^{2+}(B), and Eu^{3+} with isomer shifts of −13.4 mm/s, −9.9 mm/s, and +0.3 mm/s, respectively. The ratio of the respective ions (from the relative area of the absorption peak) was Eu^{2+}(A): Eu^{2+}(B):

Eu^{3+} = 20%: 56%: 24%, of which only Eu^{2+}(B) ions were supposedly participating in magnetic ordering with the other two groups viewed as due to impurity phases present in the sample. Remarkably, however, for the sample claimed to contain 44% of impurity phases, the Curie temperature of 17.8 K did not differ much from the T_C of 18 K measured by Grandjean et al. (1983).

First direct measurements of magnetization, magnetic susceptibility, and electrical resistivity on small Sn-flux-grown single crystals of $EuRu_4P_{12}$ were performed by Sekine et al. (2000b). The overall metallic temperature dependence of resistivity developed a knee near 18 K, with $d\rho(T)/dT$ providing a more sharply defined transition at T_C = 17.8 K. Magnetic ordering was verified to be of the FM nature by the magnetization measurements and magnetic susceptibility. From the Curie–Weiss behavior of $\chi^{-1}(T)$ *vs.* T above 50 K, somewhat anisotropic effective moments of μ_{eff} = 7.75 μ_B/Eu ‖ [100]; 7.81 μ_B/Eu ‖ [110]; and 7.69 μ_B/Eu ‖ [111] with θ_{CW} = 20 K ‖ [100]; 21 K ‖ [110]; and 21 K ‖ [111] were measured. The measured moments were slightly smaller than the theoretical value of 7.94 μ_B/Eu, and the Curie–Weiss temperature θ_{CW} was nearly equal to T_C. The rapidly rising magnetization at 2 K tended to saturate at 0.3 T with the moment of 6 μ_B/Eu (theoretically should have been 7 μ_B/Eu), showing only a small difference between the three field orientations. Hysteresis was observed in all three cases, further documenting the FM state. Taking into account the slightly smaller effective moments and saturation magnetization than the purely Eu^{2+} state should have generated, it was concluded that Eu in the structure is in a valence fluctuating state composed of 90% Eu^{2+} and 10% of Eu^{3+} ions. The magnetic transition near 18 K was also indirectly reflected in the behavior of the nuclear spin-lattice relaxation rate $1/T_1$ measured by Magishi et al. (2007a) that, above 50 K, was temperature independent and two orders of magnitude larger than in $LaRu_4P_{12}$, but rapidly diminished below T_C as spin fluctuations were suppressed when magnetic ordering set in. The calculated value of the effective magnetic moment of 6.56 μ_B/Eu obtained by Shankar et al. (2013) is significantly smaller than any of the experimental values above. By the way, in their Table 2, this value is erroneously associated with $EuFe_4P_{12}$, instead of $EuRu_4P_{12}$.

$EuOs_4P_{12}$ was prepared by a high-pressure technique under 4–5 GPa, and basic transport and magnetic measurements performed by Kihou et al. (2004) revealed a metallic structure undergoing a FM ordering near 15 K, i.e., about 3 K lower than the T_C of $EuRu_4P_{12}$. From the Curie–Weiss behavior above 20 K, the effective magnetic moment μ_{eff} = 7.64 μ_B/Eu was obtained, fairly close to but smaller than the theoretical moment of 7.94 μ_B/Eu^{2+}, suggesting, again, that a small fraction of Eu ions, perhaps around 10%, is likely to be trivalent. The same density functional theory (DFT)-based computations, Shankar et al. (2015), as done for the other two Eu-filled phosphide skutterudites, returned the magnetic moment of a Eu ion of 6.61 μ_B, considerably below the experimental value of 7.64 μ_B.

3.3.4.6.2   *Eu-Filled Arsenides*

As with all arsenide skutterudites, detailed studies were delayed by the difficulties with the synthesis. Nevertheless, both the Cd/As flux method and the high-pressure synthesis were eventually able to provide suitable samples for measurements. In the first study dedicated to the basic transport and magnetic properties of EuT_4As_{12} (T = Fe, Ru, and Os), Sekine et al. (2009b) identified $EuFe_4As_{12}$ and $EuOs_4As_{12}$ as FM metals, while $EuRu_4As_{12}$ maintained its metallic paramagnetic state down to 2 K, the lowest temperature of the experiment. Temperature dependence of magnetic susceptibility of $EuFe_4As_{12}$ and $EuOs_4As_{12}$ is shown in Figures 3.79(a) and 3.79(b). Sharp increases at 152 K and 25 K, respectively, indicate the FM ordering.

Table 3.11 summarizes the key magnetic parameters of all Eu-filled skutterudites reported. Magnetization of the three Eu-filled arsenide skutterudites is depicted in Figure 3.80(a) and shows that the saturation value M_s measured at 2 K in a field of 1 T is 4.5 μ_B/Eu for $EuFe_4As_{12}$ and 5.2 μ_B/Eu for $EuOs_4As_{12}$, both significantly below 7 μ_B/Eu expected theoretically for the half-filled 4f shell of Eu^{2+}. Magnetization of $EuRu_4As_{12}$ kept on increasing even at 1 T as the paramagnetic spins were gradually lined up by the external field. A significantly smaller value of μ_{eff} of $EuFe_4As_{12}$ than

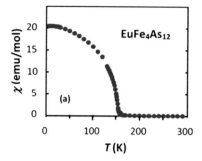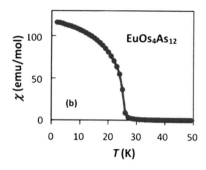

FIGURE 3.79 Susceptibility of (a) $EuFe_4As_{12}$ and (b) $EuOs_4As_{12}$ as a function of temperature. The rapid rise indicates the Curie temperature T_C. $EuRu_4As_{12}$ remains paramagnetic down to 2 K. Adapted and redrawn from C. Sekine et al., *Journal of the Physical Society of Japan* **78**, 093707 (2009). With permission from the Physical Society of Japan.

TABLE 3.11

Magnetic Properties of Eu-Filled Skutterudites Listing the Curie Temperature T_C, Effective Magnetic Moment μ_{eff}, the Curie–Weiss Temperature θ_{CW}, and the Respective Reference. The Data Collected from the Literature

Compound	T_C (K)	μ_{eff} (μ_B)	θ_{CW} (K)	Reference
$EuFe_4P_{12}$	99	6.2	-	Grandjean et al. (1984)
$EuRu_4P_{12}$	18	7.8	21	Sekine et al. (2000b)
$EuOs_4P_{12}$	15	7.64	-	Kihou et al. (2004)
$EuFe_4As_{12}$	152	6.93	46	Sekine et al. (2009b)
$EuRu_4As_{12}$	paramag.	8.31	−7.4	Sekine et al. (2009b)
$EuOs_4As_{12}$	25	7.29	9.7	Sekine et al. (2009b)
$EuFe_4Sb_{12}$	82	8.4	−13	Danebrock et al. (1996)
	84	7.28	19	Bauer et al. (2001a)
	89	8.4	−18	Bauer et al. (2004)
$EuRu_4Sb_{12}$	3.3	7.2	6.1	Takeda and Ishikawa (2000a)
	4	8.0	3	Bauer et al. (2004)
$EuOs_4Sb_{12}$	9	7.3	8	Bauer et al. (2004)

its theoretical Hund's rule free-ion value of 7.94 μ_B/Eu^{2+}, its non-linear $\chi^{-1}(T)$ *vs. T* behavior (not shown here), and considerably reduced saturation magnetization were interpreted by the authors as a form of canted ferromagnetism or perhaps even a ferrimagnetic ordering below T_C. In the case of $EuOs_4As_{12}$, the discrepancy with the theoretical estimates of μ_{eff} and M_s was less drastic, nevertheless significant, and a question begs an answer whether these experimental underestimates might have something to do with a significant fraction of Eu ions being trivalent to perhaps an even greater degree than in the case of Eu-filled phosphides. There is also an issue how large magnetic moment is supported by Fe in the $[Fe_4As_{12}]$ polyanion, and whether such moment might oppose the dominant magnetic moment of the Eu^{2+} ions. While the temperature dependence of resistivity of $EuRu_4As_{12}$ was perfectly smooth and monotonic with no knees or kinks to hint at ordering, the Curie temperature left a distinct mark on the resistivity of $EuFe_4As_{12}$ and $EuOs_4As_{12}$ as a break in the slope at T_c of the former and a small kink in the case of the latter, Figure 3.80(b).

As we have already seen, pressure is a useful variable with which to observe changes in the electronic structure as the bond distances are altered. $EuFe_4As_{12}$, with its by far the highest Curie temperature of 152 K, drew interest, and thus it is no surprise that it was chosen as the focus of pressure studies of magnetism. Using an opposed anvil pressure cell capable of generating pressures up to 4.1 GPa, Kawamura et al. (2015) measured magnetization of $EuFe_4As_{12}$ with an eye on the variation of the Curie temperature and the magnitude of magnetization. The results are shown in Figure 3.81 and indicate that the Curie temperature increases with pressure at a rate $dT_c/dP \approx 5.5$ K/GPa, while the magnetization decreases with pressure, especially in low magnetic fields. There are several possible explanations for such a trend, and they include the presence of trivalent Eu, the content of which might be even temperature-dependent, and perhaps even more relevant issue of the actual magnetic contribution of Fe and the nature of its coupling (ferrimagnetic?) to the dominant moment of Eu^{2+} ions. Here, more than anywhere else, detailed X-ray absorption studies to determine

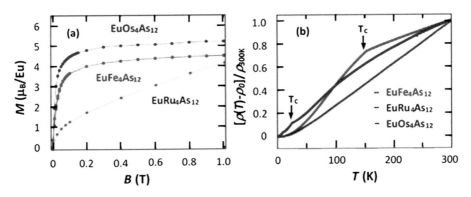

FIGURE 3.80   (a) Magnetization of all three Eu-filled arsenide skutterudites as a function of magnetic field. (b) Temperature-dependent part of the electrical resistivity normalized to the room temperature value. Magnetic ordering temperatures of $EuFe_4As_{12}$ and $EuOs_4As_{12}$ are indicated by arrows. Adapted and redrawn from C. Sekine et al., *Journal of the Physical Society of Japan* **78**, 093707 (2009). With permission from the Physical Society of Japan.

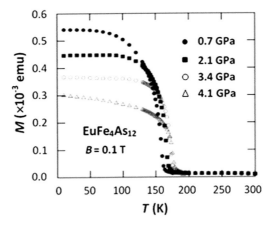

FIGURE 3.81   Temperature dependence of magnetization of $EuFe_4As_{12}$ measured in the field of 0.1 T under various external pressures. Redrawn from Y. Kawamura et al., *Journal of Physics: Conference Series* **592**, 012033 (2015). With permission from the IOP Publishing.

the valence state of Eu, and precise [57]Fe Mössbauer studies to identify the magnetic contribution due to Fe would be highly desirable in order to shed light on the magnetic state of the Eu-filled arsenide skutterudites.

3.3.4.6.3 Eu-Filled Antimonides

The idea of Slack (1995), quickly verified by Morelli and Meisner (1995), that fillers in the skut-terudite voids (cages) may dramatically reduce the otherwise too high lattice thermal conductivity of skutterudites and thus make them appealing for thermoelectric applications ignited a vigorous search for the fillers that would maintain the excellent electronic properties of skutterudites yet lead to a maximum suppression of their thermal conductivity, the concept known as a phonon-glass electron-crystal. However, it was soon realized that the $[Co_4Sb_{12}]$ framework can accept only a very limited number of trivalent rare-earth fillers because the structure gets rapidly saturated with nega-tive charges of the donated electrons. Divalent fillers or fillers with a mixed 2+ to 3+ valence, such as Eu, could be accommodated more readily and with significantly higher fractions of up to about 54% in $[Co_4Sb_{12}]$, Jeitschko et al. (2000), and near full filling in $[Fe_4Sb_{12}]$, Danebrock et al. (1996). This provided a strong impetus for numerous subsequent explorations of Eu-filled antimonide skut-terudites. I should also mention that because the electronic configuration of Eu^{2+} ions has the angular momentum $L = 0$, CEF is expected to have little effect on the physical properties.

In their broad survey of filled antimonide skutterudites, Danebrock et al. (1996) established $EuFe_4Sb_{12}$ as a FM metal with the Curie temperature $T_C = 82$ K. An approximately linear form of $\chi^{-1}(T)$ *vs.* T well above 150 K yielded the effective moment $\mu_{eff} = 8.4$ μ_B/f.u., of which 2.6 μ_B was due to a contribution of Fe^{3+} (higher than the theoretical spin-only value of 1.73 μ_B expected for the Fe^{3+} free ion), the value likely enhanced due to spin-orbit coupling. From $\mu_{eff}^{meas} = \sqrt{\left(\mu_{eff}^{Eu}\right)^2 + \left(\mu_{eff}^{Fe}\right)^2}$,

the value of 6.8 μ_B was assigned to Eu. The effective moment of 6.8 μ_B for Eu was less than the Eu^{2+} free-ion value of 7.94 μ_B. The magnetic moment of 4.9 μ_B/f.u. measured in the field of 1 T at 5 K was also well below the theoretical value of 7 μ_B/Eu^{2+}. However, in the absence of structural refine-ment to provide the actual occupancy of the voids and the lack of X-ray absorption data to assess the valence of Eu, nothing specific could be said about why the experimental magnetic moments came short of the theoretical values. Although the $[Fe_4Sb_{12}]$ framework is far more hospitable to the rare-earth fillers than is $[Co_4Sb_{12}]$, it still remained a challenge to make $EuFe_4Sb_{12}$ stoichiometric, as follows from measurements of magnetic properties by Bauer et al. (2001a), where the filling frac-tion of Eu was not higher than 83%. The key magnetic parameters are presented in Table 3.11. The authors also explored the effect of pressure on their $Eu_{0.83}Fe_4Sb_{12}$ sample and observed a rising T_C at a rate of about 0.5 K/kbar (0.05 K/GPa). From the pressure response and taking the bulk modulus $B_0 = 1000$ kbar (100 GPa), a value of the Grüneisen parameter came to about 6, typical of rare-earth intermetallics.

As already noted, the filling fraction of Eu depends critically on how much Co and Fe is con-tained in the antimonide framework. Detailed assessment of the Eu occupancy was provided by studies of Grytsiv et al. (2002) on a series of $Eu_yFe_{4-x}Co_xSb_{12}$ solid solutions. The results indicated a gradual decrease of Eu in the voids from $y = 0.83$ in the $[Fe_4Sb_{12}]$ polyanion framework down to only $y = 0.44$ in $[Co_4Sb_{12}]$, i.e., even lower value than $y = 0.54$, originally estimated by Jeitschko et al. (2000). Concomitant with the decreasing Eu filling, the valence of Eu tended to increase. Starting with substantially divalent Eu at the maximum filling, the valence of Eu reached 2.6 when $y = 0.2$. Of course, such drastic changes in the Eu occupancy and its valence state had far-reaching conse-quences for the transport and magnetic properties. Thus, the Curie temperature gradually decreased from 84 K in $Eu_{0.83}Fe_4Sb_{12}$ down to 8 K in $Eu_{0.44}Co_4Sb_{12}$ and, upon further reducing the content of Eu to $Eu_{0.2}Co_4Sb_{12}$, the long-range magnetic order disappeared, as shown in Figure 3.82(a). The effective magnetic moment and the Curie–Weiss temperature are plotted as a function of Co content in Figure 3.82(b).

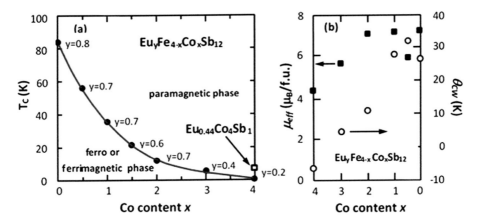

FIGURE 3.82 (a) Dependence of the magnetic ordering temperature T_C on the content of cobalt in polycrystalline $Eu_yFe_{4-x}Co_xSb_{12}$ (b) Dependence of the effective magnetic moment (solid squares and left-hand scale) and the Curie–Weiss temperature (open circles and right-hand scale) on the content of Co. Adapted and redrawn from A. Grytsiv et al., *Physical Review B* **66**, 094411 (2002). With permission from the American Physical Society.

All the studies above were performed for polycrystalline samples that might have been inhomogeneous and contained impurity phases at the grain boundaries. To secure more intrinsic results, Bauer et al. (2004) measured magnetic properties, electrical resistivity, specific heat, and Eu L_{III} edge X-ray absorption in single crystals of all three Eu-filled antimonides EuT_4Sb_{12} (T = Fe, Ru, and Os) prepared by a Sb flux growth method. Except for $EuFe_4Sb_{12}$, where the structural refinement indicated Eu occupancy of $y = 0.96$, all other crystals had 100% of voids filled by Eu. Clearly, the filling fraction in single crystals was superior to the occupancy achieved in polycrystalline samples. The key magnetic and thermodynamic parameters of the three Eu-filled antimonide skutterudite crystals are given in Table 3.12.

From the Eu L_3-edge absorption data shown in Figure 3.83, it follows that Eu in these single crystalline antimonide skutterudites is essentially divalent. The low experimental value of the saturation magnetization of $Eu_{0.96}Fe_4Sb_{12}$ (5.1 μ_B/Eu) in comparison to the expected saturation moment of Eu^{2+} of 7 μ_B was, again, interpreted as a possible ferrimagnetic coupling between the moments of Eu and Fe or some form of a canted FM order. As in a vast majority of skutterudites having the $[Fe_4Sb_{12}]$ polyanion as their fundamental structural unit, there is strong evidence for Fe contributing to the overall magnetic state. In this case, the susceptibility results are consistent with an effective moment originating from the polyanion of 3.8 μ_B, the value even higher than 2.6 μ_B originally estimated by Danebrock et al. (1996). While having considerably lower magnetic ordering temperatures

TABLE 3.12
Magnetic and Thermodynamic Parameters of EuT_4Sb_{12}
(T = Fe, Ru, and Os)

Skutterudite	T_C (K)	μ_{eff}^{Eu} (μ_B/Eu)	θ_{CW} (K)	μ_{sat} (μ_B/Eu)	γ (mJmol^{-1}K^{-2})	θ_D (K)	θ_E (K)
$Eu_{0.96}Fe_4Sb_{12}$	87	7.7	−18	5.1	85	348	84
$EuRu_4Sb_{12}$	4	8.0	3	7.3	73	262	78
$EuOs_4Sb_{12}$	9	7.3	8	6.0	135	304	74

Source: The entries are the data of E. Bauer et al., *Journal of Physics: Condensed Matter* **16**, 5095 (2004). With permission from IOP Publishing.

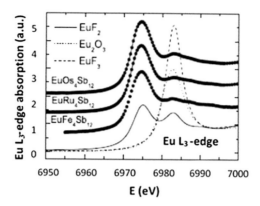

FIGURE 3.83   Eu L_3 absorption edge spectra for EuT$_4$Sb$_{12}$ (T = Fe, Ru, and Os), including three reference compounds EuF$_2$ (Eu^{2+}), EuF$_3$ (Eu^{3+}), and Eu$_2$O$_3$ (Eu^{3+}) at 300 K. The data of Eu-based skutterudites are shifted vertically for clarity. Divalent Eu dominates with a small contribution of trivalent Eu present. Reproduced from E. Bauer et al., *Journal of Physics: Condensed Matter* **16**, 5095 (2004). With permission from the IOP Publishing.

of 4 K, respectively 9 K, the effective magnetic moments and the saturation moments of EuRu$_4$Sb$_{12}$ and EuOs$_4$Sb$_{12}$ are much closer to the expected values for divalent Eu. Extensive Mössbauer studies of Eu-filled skutterudites, including nearly filled as well as incompletely filled Eu$_y$Fe$_4$Sb$_{12}$, are discussed in section 3.3.1.

The puzzling nature of magnetism in EuFe$_4$Sb$_{12}$, the high Curie temperature combined with a small value of the saturated magnetic moment, called for more studies to clarify the magnetic state in this skutterudite. Moreover, the widely different estimates of the magnetic moment contributed by the [Fe$_4$Sb$_{12}$] polyanion further clouded the picture. Employing XMCD spectroscopy, an element-specific technique briefly mentioned in section 3.3.4.5.3 when discussing properties of SmOs$_4$Sb$_{12}$, in conjunction with XAS, Krishnamurthy et al. (2007, 2009) described in detail their measurements for a polycrystalline sample of Eu$_{0.95}$Fe$_4$Sb$_{12}$ having comparable Eu filling as achieved in single crystals. In the first of the two papers, by detecting and analyzing XMCD signals at Eu M$_{4,5}$-edges (corresponding to electron excitations from the 3d shell to 4f states of Eu) and Fe L$_{2,3}$-edges, together with measuring XAS at the Eu L$_3$-edge to monitor the valence of Eu, the authors determined that their polycrystalline sample of composition Eu$_{0.95}$Fe$_4$Sb$_{12}$ contained about 15% of Eu^{3+} irrespective of temperature. Moreover, each Eu^{2+} ion carried a total magnetic moment of 7.2 ± 0.3 μ_B, composed of the small orbital contribution μ_{orb} = 0.13 ± 0.1 μ_B and the dominant spin contribution of μ_{spin} = 7.07 ± 0.3 μ_B. The dichroic signal, obtained near the Fe L$_3$-edge at 2.3 K and the field of 2 T and consisting of peaks (at 705.5 eV and 710 eV) and minima (at 707.5 eV and 709 eV), is shown in Figure 3.84. The dichroic response documents the ordered magnetic moment of Fe in Eu$_{0.95}$Fe$_4$Sb$_{12}$. Using the established magneto-optic sum rules, Thole et al. (1992) and Carra et al. (1993), the authors extracted the magnetic moment of Fe from the intensities of the XAS and XMCD spectra. The total magnetic moment associated with each Fe ion turned out to be −0.21 ± 0.03 μ_B, consisting of the orbital part of −0.07 ± 0.02 μ_B and the spin part of −0.14 ± 0.02 μ_B. The Fe contribution in the formula Eu$_{0.95}$Fe$_4$Sb$_{12}$ is thus −0.84 ± 0.12 μ_B. More important than the actual size of the Fe moment is its negative sign, which implies that Fe couples antiferromagnetically to the positive moment of Eu^{2+}, resulting in a ferrimagnetic structure. Taking into account the 87% fraction of Eu^{2+} ions and the y = 0.95 filling fraction, the net moment of Eu$_{0.95}$Fe$_4$Sb$_{12}$ is then about 5.1 μ_B, in excellent agreement with the experimental saturation moment in Table 3.12. In their second paper, Krishnamurthy et al. (2009) added temperature-dependent measurements of XMCD at the Eu M$_{4,5}$ edges and at Eu L$_{2,3}$-edges (electron transitions

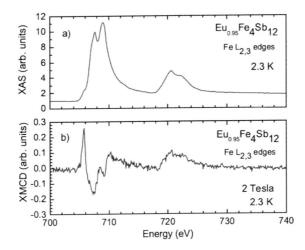

FIGURE 3.84 Fe $L_{2,3}$-edges of (a) XAS and (b) XMCD in an $Eu_{0.95}Fe_4Sb_{12}$ polycrystal documenting that Fe moments are ordered. Reproduced from V. V. Krishnamurthy et al., *Physical Review Letters* **98**, 126403 (2007). With permission from the American Physical Society.

from $2p$ to $5d$ states of Eu) and focused on the role of Eu $5d$ states and whether they are magnetically polarized *via* interatomic $5d$-$3d$ exchange interaction. The measurements evidenced the dominant role of Eu in the onset of the ferrimagnetic state at $T_C \approx 88$ K as well as an important role of band-like $5d$ states of Eu that seem to mediate the coupling of localized $4f$ moments of Eu with the near-ferromagnetically ordered conduction band of the $[Fe_4Sb_{12}]$ framework.

DFT-based calculations of electronic and magnetic properties of $EuFe_4Sb_{12}$ by Shankar et al. (2014), using their favorite FP-LAPW method within the framework of the LSDA, suggested the coexistence of localized $4f$-electrons of Eu and $3d$-electrons of Fe with high DOS near the Fermi level. However, the magnetic moments came, again, somewhat lower.

The magnetic and thermodynamic properties of $EuRu_4Sb_{12}$ and $EuOs_4Sb_{12}$ are presented in Table 3.12. In addition, magnetism of polycrystalline $EuRu_4Sb_{12}$ was studied by Takeda and Ishikawa (2000a) and indicated a FM transition T_C at about 3.3 K. The Curie–Weiss behavior above 10 K yielded an effective moment of 7.20 μ_B/Eu^{2+}, smaller but not too far from the theoretical Hund's rule value of 7.94 μ_B/Eu^{2+}. Because no information was given on the valence of Eu and no structural refinement was performed regarding the filling fraction of Eu, it is not clear why the experimental effective moment was somewhat lower. The Curie–Weiss temperature was quoted as 6.1 K, about twice the value in Table 3.12. The specific heat developed a pronounced peak at T_C and the extracted magnetic entropy turned out to be close to $R\ln8$, a bit smaller than expected for the $J = 7/2$ multiplet. The magnetization at 2.1 K, i.e., below T_C, saturated above 1 T and yielded a moment of 6.2 μ_B, below the expected 7.0 μ_B. This might have indicated the presence of about 11% of Eu^{3+} ions.

The substantially divalent nature of Eu, rather than the trivalent state of most rare-earth ions, was also reflected in distinctly different dHvA frequencies observed in measurements of Sugawara et al. (2008b) on single crystals of $EuRu_4Sb_{12}$. Two closely lying branches, labeled a and a' in Figure 3.85, were detected for both $B\|\langle100\rangle$ and $B\|\langle110\rangle$. For comparison purposes, dashed curves in Figure 3.85 indicate branches of $LaRu_4Sb_{12}$ filled with trivalent La. From their very weak dependence on the direction of the magnetic field, nearly spherical Fermi surfaces were conjectured for $EuRu_4Sb_{12}$. The cyclotron effective masses for the two branches and two field orientations were 2.1 (1.6) for the branch a and 2.0 (1.7) for the branch a'.

Existing magnetic parameters of $EuOs_4Sb_{12}$ are presented in Table 3.12.

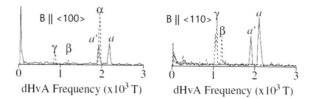

FIGURE 3.85 Fourier transforms of the dHvA spectra of $EuRu_4Sb_{12}$ measured in two different field orientations. The panel on the left depicts the results obtained with the field along the $\langle 100 \rangle$ direction and the panel on the right in the field along the $\langle 110 \rangle$ direction. Dashed curves indicate Fourier transforms of dHvA spectra of $LaRu_4Sb_{12}$ with a typical trivalent La ion. Reproduced from H. Sugawara et al., *Journal of the Physical Society of Japan* **77**, 297 (2008). With permission from the Physical Society of Japan.

3.3.4.7 Yb-Filled Skutterudites

Ytterbium is the second last element of the rare earth series and has electronic configuration $[Xe]4f^{14}6s^2$, i.e., a fully filled f-shell. The last and heaviest element Lu has also a fully filled f-shell, but possesses an extra $5d$ electron. I am discussing Yb prior to the group of six heavy rare earths, Gd, Tb, Dy, Ho, Er, and Tm, that lay in the periodic table before Yb, primarily because Yb-filled skutterudites were synthesized early without the need for high pressure on account of the substantially divalent (and thus larger) non-magnetic Yb^{2+} ion having the electronic configuration $[Xe]4f^{14}$ (closed shell). Had Yb been trivalent, i.e., with the configuration $[Xe]4f^{13}$, the lanthanide contraction would make this magnetic Yb^{3+} ion too small to form bonds with the $[T_4X_{12}]$ framework, and the synthesis under ambient pressure would fail. Already in the early studies of $Yb_yCo_4Sb_{12}$ and $YbFe_4Sb_{12}$ skutterudites, the Yb filler proved very effective (a small and heavy rattler) in lowering the lattice thermal conductivity, and this fact intensified the interest in Yb-filled skutterudites as efficient thermoelectric materials. Here, I present magnetic properties of Yb-filled skutterudites.

3.3.4.7.1 Yb-Filled Phosphides

The first study of magnetic properties of $YbFe_4P_{12}$ was performed by Shirotani et al. (2005) as part of their search of skutterudites with heavy lanthanide fillers. The polycrystalline samples were synthesized by a high-pressure high-temperature technique and the larger than expected lattice constant of $YbFe_4P_{12}$ provided a hint that the valence state of Yb is not trivalent but intermediate between Yb^{2+} and Yb^{3+}. This suspicion was confirmed by measurements of the magnetic susceptibility, where the Curie–Weiss behavior above 70 K yielded the effective magnetic moment of 3.58 μ_B/f.u., significantly below the Hund's rule Yb^{3+} free ion value of 4.54 μ_B, and $\theta_{CW} = -67$K. No sign of a magnetic transition was noted down to 2K, the lowest temperature of the experiment, even though the metallic resistivity developed a minimum near 45 K and started to rise at lower temperatures. Subsequent X-ray measurements by Shironati et al. (2006) at ambient pressure and at pressures up to 20 GPa revealed an interesting structural similarity between the $4f^{13}$ electronic configuration of Yb^{3+} and the $4f^1$ configuration of Ce^{3+}. These "one-hole", respectively "one-electron" state skutterudites have very similar lattice parameters (7.787 Å and 7.792 Å) and bulk moduli (167 GPa and 162 GPa, respectively). On the other hand, their transport and magnetic properties are entirely different. $YbFe_4P_{12}$ is a metallic paramagnet while $CeFe_4P_{12}$ is an insulator with a temperature independent magnetic susceptibility. Yamamoto et al. (2006, 2007) used ^{31}P-NMR and ^{121}Sb-NQR to study the nuclear spin-lattice relaxation rate $1/T_1$ to shed light on a supposed development of the heavy Fermi-liquid state in $YbFe_4P_{12}$ at temperatures below about 10 K, hinted at in an unpublished workshop report by Wakeshima et al. (2005). The report was to provide data on the low-temperature specific heat of $YbFe_4P_{12}$ characterized by a Schottky-type peak near 30 K (likely due to CEF splitting), a λ-type peak near 0.7 K (likely associated with the YbP impurity), and a divergent behavior of C/T below about 8 K yielding a large Sommerfeld coefficient $\gamma \approx 300$ mJmol^{-1}K^{-2}, suggesting a heavy fermion state. Yamamoto et al. concluded that, at low fields, $YbFe_4P_{12}$ is located very close to the non-Fermi-liquid state and the field of merely 0.2 T brings it to the QCP.

There are no reports on magnetic properties of $YbRu_4P_{12}$ and $YbOs_4P_{12}$.

3.3.4.7.2 Yb-Filled Arsenides

There are no reports on magnetic properties of Yb-filled arsenide skutterudites.

3.3.4.7.3 Yb-Filled Antimonides

As we have already seen, in many skutterudites with the $[Fe_4Sb_{12}]$ framework, even if a filler ion is large enough and the compound can be synthesized, rarely a full occupancy of the voids can be achieved using synthesis routes under ambient pressure. This is also the case of $YbFe_4Sb_{12}$, where the development of the high-pressure synthesis method at typically 4 GPa was much instrumental in achieving the near 100% occupancy of Yb. In turn, after a long and tortuous path, this was the key to uncovering the intrinsic magnetic properties of this filled skutterudite.

The early magnetic characterization of $YbFe_4Sb_{12}$ was performed by Dilley et al. (1998) who used polycrystalline samples synthesized by mixing stoichiometric quantities of elements and heating them to about 950°C, followed by annealing at 600°C for 20 h. This classical baking/annealing synthesis worked because Yb enters the voids as an intermediate ion with the dominant Yb^{2+} valence and, hence, a larger radius able to form bonds with the $[Fe_4Sb_{12}]$ framework. The resistivity developed a broad shoulder near 70 K, but otherwise was metallic at all temperatures. Magnetic susceptibility followed the Curie–Weiss behavior down to about 100 K, but a notable upturn developed in $\chi(T)$ below 50 K. An effective magnetic moment $\mu_{eff} = 3.09\ \mu_B$/f.u. with a Curie–Weiss temperature of 40 K was obtained above 100 K. The effective moment was distinctly smaller than the free ion moment of Yb^{3+} of 4.54 μ_B. However, because XAS was not performed and the exact magnetic contribution of the $[Fe_4Sb_{12}]$ polyanion was not known, no conclusion could be made concerning the fraction of the Yb^{2+} ions present in the sample. At low temperatures, the linear magnetic susceptibility saturated at a rather large value of $\chi_0 = 3.45 \times 10^{-2}$ cm^{3}mol^{-1}. Results of the specific heat measurement are shown in Figure 3.86. From a simple fit of the data in the range 5 K < T < 15 K to $C(T) = \gamma T + \beta T^3$, a slightly enhanced Sommerfeld coefficient $\gamma \approx 75$ mJmol^{-1}K^{-2} was obtained with the Debye temperature $\theta_D = \left(12\pi^4 N_i R / 5\beta\right)^{1/3} = 190$ K. Here, $N_i = 17$ is the number of ions in a formula unit and R is the universal gas constant. However, below 5 K, the electronic specific heat coefficient $\gamma(T) = \dfrac{C(T)}{T} - \beta T^2$ increased linearly with decreasing temperature, see the lower right inset in Figure 3.86, and easily reached 140 mJmol^{-1}K^{-2} before a sharp upturn ($\sim T^{-2}$) developed below 1 K, interpreted as the tail of a Schottky anomaly due to trace impurities. Such a large value of γ indicated a moderate enhancement of the effective mass, consistent with the Wilson–Sommerfeld ratio $\dfrac{\chi_0(0)}{\gamma(0)} \dfrac{\pi^2 k_B^2}{\mu_{eff}^2} = 2.62$. As possible explanations of such moderate heavy fermion behavior were offered the intermediate valence of Yb, and a Kondo lattice of screened Yb^{3+} moments.

Numerous other studies soon followed. Leithe-Jasper et al. (1999), based on their susceptibility measurements of single crystals with presumed full occupancy of voids, obtained an effective magnetic moment of 4.49 μ_B/f.u. and $\theta_{CW} = 13.8$ K, the differences with the data of Dilley et al. (1998) likely due to a different form of the samples (single crystal vs. polycrystals) resulting in the different filling fraction of Yb and different ratio of Yb^{2+}/Yb^{3+} ions. Moreover, they interpreted their [57]Fe Mössbauer measurements as if the polyanion $[Fe_4Sb_{12}]$ was contributing no magnetic moment, a highly controversial step given the fact that the effective magnetic moment of $LaFe_4Sb_{12}$ is about 3 μ_B/f.u. and must be carried by the polyanion as La is non-magnetic, Danebrock et al. (1996). Based on their XANES measurements, Leithe-Jasper et al. (1999) obtained the Yb intermediate valence of 2.68. A comparable intermediate valence of 2.62 was also found by Okane et al. (2003) in their

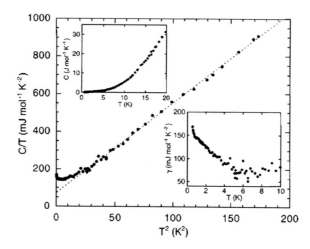

FIGURE 3.86 Specific heat of $YbFe_4Sb_{12}$ plotted as $C(T)/T$ vs. T^2. The upper left inset is a plot of $C(T)$ vs. T below 20 K, and the lower right inset depicts the electronic specific heat (Sommerfeld coefficient) γ extracted from $\gamma(T) = \dfrac{C(T)}{T} - \beta T^2$ at low temperatures. Reproduced from N. R. Dilley et al., *Physical Review B* **58**, 6287 (1998). With permission from the American Physical Society.

high-resolution ultra-violet photoemission spectroscopy spectra. A much smaller value of 2.16 was reported by Bérardan et al. (2003) for their arc-melted and annealed $Yb_{0.93}Fe_4Sb_{12}$. However, this intermediate valence of Yb increased linearly with the content of Ce added to Yb and, in their double-filled $Ce_{0.85}Yb_{0.05}Fe_4Sb_{12}$ sample, the Yb valence reached 2.71. Nevertheless, for $Yb_{0.93}Fe_4Sb_{12}$, the valence of 2.16 implied that most of the magnetic moment must come from the $[Fe_4Sb_{12}]$ polyanion as Yb^{2+} is nonmagnetic, and there could not be enough Yb^{3+} ions to support magnetism. Bauer et al. (2000) used reaction sintering to prepare and study Yb-filled antimonides with different transition metals T = Fe, Co, Rh, and Ir. Because the authors made detailed refinements of their structures, they established the occupancy of Yb in $Yb_yFe_4Sb_{12}$ as y = 0.83 and considerably smaller occupancies in the other three transition metal skutterudites. For their $Yb_{0.83}Fe_4Sb_{12}$ sample, the effective magnetic moment of 3.36 μ_B/f.u. and the Curie–Weiss temperature of 26.5 K were reported. Unfortunately, no information on the valence state of Yb was provided. A very different outcome was reported by Schnelle et al. (2005) in their measurements of strongly inter-grown crystals of $Yb_{0.95}Fe_4Sb_{12}$. Here, the XANES measurements were dominated by a peak near the energy of 8940 eV, assigned to Yb^{2+}, which was shifted 8 eV down from a peak of Yb_2O_3 that was used to mark exclusively an Yb^{3+} ion. Consequently, the authors concluded that Yb in their sample is essentially divalent. But then, because Yb^{2+} is non-magnetic, the effective magnetic moment of 1.50 μ_B obtained from the Curie–Weiss behavior had to come entirely from the itinerant $3d$ electrons of Fe in the $[Fe_4Sb_{12}]$ polyanion. All subsequent studies either assumed, tried to justify, or sometimes actually measured, Dedkov et al. (2007), the essentially divalent form of Yb in their samples.

An important step forward was made in measurements by Tamura et al. (2006, 2007). By comparing magnetic properties of two $Yb_yFe_4Sb_{12}$ structures with different filling fraction y (a single crystal with y = 0.94 and a polycrystal with y = 0.89), they obtained very different outcomes depicted in Figures 3.87(a) and 3.87(b). The polycrystalline structure with a high density of Yb vacancies clearly ordered ferromagnetically below about 13 K, as seen in a rapidly rising susceptibility, and a clear hysteresis loop taken at 2 K. Because the spontaneous magnetic moment is very small, the magnetic state was categorized as weakly FM and likely associated with itinerant Fe $3d$ electrons. In contrast, the single crystal with fewer Yb vacancies remained paramagnetic. Thus, the importance of the filling fraction of Yb as a crucial parameter deciding the magnetic state of $Yb_yFe_4Sb_{12}$ was laid down. Alleno et al. (2006), in their measurements of magnetic susceptibility on a polycrystal with

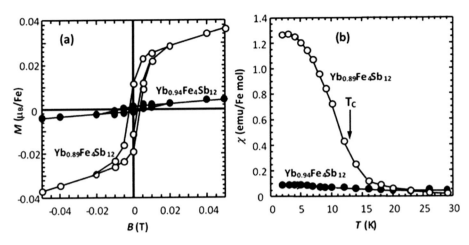

FIGURE 3.87 (a) Magnetic hysteresis loops for $Yb_{0.89}Fe_4Sb_{12}$ (open circles) and $Yb_{0.94}Fe_4Sb_{12}$ (solid circles). (b) Low-temperature magnetic susceptibility of $Yb_{0.89}Fe_4Sb_{12}$ (open circles) and of $Yb_{0.94}Fe_4Sb_{12}$ (solid circles) measured in the field of 0.01 T. Constructed from the data of I. Tamura et al., *Journal of the Physical Society of Japan* **75**, 014707 (2006).

a comparably few Yb vacancies, $y = 0.93$, came to the same conclusion that such high void filling leaves $YbyFe_4Sb_{12}$ paramagnetic down to at least 2 K. Measuring a series of single crystals with the Yb filling fraction in the range $0.875 \leq y \leq 0.91$, Ikeno et al. (2007) documented a strong (linear) dependence of the Curie temperature T_C and a remanent moment m_0 on y, as shown in Figure 3.88(a). Obviously, the FM order exists for filling fractions of Yb below about $y = 0.93$ but disappears when the density of Yb vacancies becomes small. Applied pressure increases the Curie temperature, as shown for one of the FM samples ($Yb_{0.882}Fe_4Sb_{12}$) in Figure 3.88(b). An independent confirmation of the development of a spontaneous magnetic moment in samples with $y < 0.93$ was provided by the divergence in the nuclear spin-lattice relaxation rate $1/T_1$ and the rapidly increasing linewidth in the ^{121}Sb-NQR spectra observed below $T_C = 18$ K in measurements on $Yb_{0.89}Fe_4Sb_{12}$ by Kawaguchi et al. (2006). Later ^{121,123}Sb-NQR measurements by Magishi et al. (2014) on fully-filled (high pressure-synthesized) $YbFe_4Sb_{12}$ showed no significant line broadening down to 1.5 K, documenting that no magnetic transition has taken place, in contrast with the measurements on $Yb_{0.89}Fe_4Sb_{12}$.

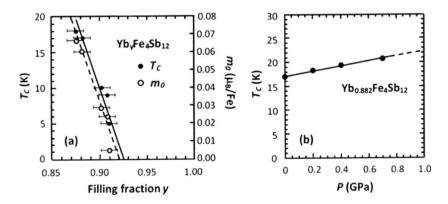

FIGURE 3.88 (a) Curie temperature T_C as a function of the filling fraction y of Yb. (b) Pressure dependence of the Curie temperature T_C for $Yb_{0.882}Fe_4Sb_{12}$. Drawn from the data of T. Ikeno et al., *Journal of the Physical Society of Japan* **76**, 024708 (2007).

Regardless of how hard researchers have tried, all samples, including single crystals prepared under ambient pressure, fell significantly short of full Yb occupancy. In cases where the report implied full occupancy by discussing $YbFe_4Sb_{12}$, the actual occupancy was not adequately documented. A major progress regarding higher levels of filling was achieved following the development of high-pressure synthesis. Samples prepared by this technique, both polycrystals and single crystals, attained a much higher level of filling of over 99%. Comparing magnetic properties of such high-pressure-synthesized Yb-filled skutterudites with structures prepared under ambient pressure, Saito et al. (2011) observed significant differences at low temperatures. Rather than a continuously rising magnetic susceptibility with decreasing temperature, typical of samples synthesized under ambient pressure, the high pressure-grown samples developed a notable peak near 50 K, shown in Figure 3.89. Moreover, from the analysis of the specific heat came a much smaller Sommerfeld coefficient $\gamma \approx 100$ mJmol^{-1}K^{-2}, significantly below values of 140 mJmol^{-1}K^{-2} reported by Dilley et al. (1998).

Anomalous softening observed by Möchel et al. (2011) near 50 K in the measurements of elastic constants of high pressure-synthesized $YbFe_4Sb_{12}$ occurred at the same temperature where the samples displayed a pronounced peak in the magnetic susceptibility. Noting a robust signal starting to develop at 50 K and growing as the temperature decreases in their measurements of electron paramagnetic resonance (EPR), see Figure 3.90, the authors suggested that the signal arises from a change of the Yb valence, whereby a growing fraction of Yb^{2+} ions converts to the 3+ valence. Obviously, an EPR signal cannot come from a closed $4f$-shell of Yb^{2+}.

It would have been nice to have on hand T-dependent measurements of XAS to accompany the measurements of the elastic constants so that an independent assessment of the Yb valence evolution could be made. This had to wait until Yamaoka et al. (2011) demonstrated a strong coupling between the Yb $4f$ valence instability and the weak Fe $3d$-driven ferromagnetism in $Yb_yFe_4Sb_{12}$. To do so, the authors applied resonant X-ray emission spectroscopy (RXES) and partial fluorescence yield X-ray absorption spectroscopy (PFY-XAS) to two Yb-filled skutterudite single crystals with different filling fractions of $y = 0.88$ and $y = 0.97$ (the latter prepared under a pressure of 4 GPa) and studied their electronic structure as a function of temperature and pressure. Magnetic susceptibility

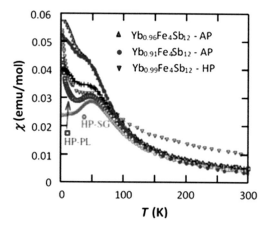

FIGURE 3.89 Susceptibility of several $Yb_yFe_4Sb_{12}$ samples with different filling fractions y. Samples labeled AP were synthesized under ambient pressure. Samples denoted as HP were prepared under high pressure. Upside red triangles designate the data of Schnelle et al. (2005) with the filling fraction $y = 0.96$; blue circles are data of Ikeno et al. (2007) with the filling fraction $y = 0.91$; and down red triangles are susceptibilities measured by Yamamoto et al. (2008) on nearly fully filled samples prepared under high pressure. HP-SG (red open circles) and HP-PL (blue open circles) designate a single crystal and polycrystal synthesized under high pressure. Adapted and redrawn from T. Saito et al., *Journal of the Physical Society of Ja*pan **80**, 063708 (2011). With permission from the Physical Society of Japan.

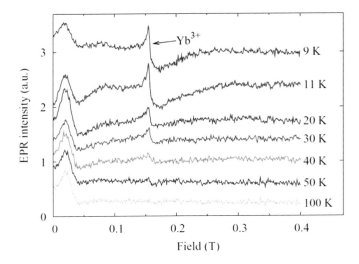

FIGURE 3.90 Temperature-dependent EPR signal of $YbFe_4Sb_{12}$ at a frequency of 9.5 GHz. The signal develops below 50 K and is assigned to Yb^{3+} because the closed $4f$-shell of Yb^{2+} cannot generate an EPR signal. Reproduced from A. Möchel et al., *Physical Review B* **84**, 184306 (2011). With permission from the American Physical Society.

of the two samples behaved similarly as in the measurements of Saito et al. (2011), i.e., the susceptibilities were identical above 150 K, but $Yb_{0.97}Fe_4Sb_{12}$ developed a peak near 50 K while the susceptibility of $Yb_{0.88}Fe_4Sb_{12}$ kept on increasing and attained an order of magnitude larger value at low temperatures. In fact, $Yb_{0.88}Fe_4Sb_{12}$ became a weak ferromagnet with $T_C = 17$ K. As was also noted in measurements of Tamura et al. (2006), see Table 3.13, the extracted paramagnetic effective moments, here 3.19 μ_B/f.u. for $Yb_{0.97}Fe_4Sb_{12}$ and 3.13 μ_B/f.u. for $Yb_{0.88}Fe_4Sb_{12}$, were remarkably similar given the fact that the Yb ions in one structure are divalent and the skutterudite remains paramagnetic, while the other skutterudite contains some trivalent Yb ions and becomes a ferromagnet. This is an indirect validation of the idea that the magnetism in $Yb_yFe_4Sb_{12}$ has its origin in the itinerant $3d$-electrons of Fe and not $4f$-electrons of Yb. After all, as the band structure calculations have amply shown, Schnelle et al. (2005), Sichelschmidt et al. (2006), and Dedkov et al. (2007), the $4f$-electrons of $YbFe_4Sb_{12}$ do not contribute to the DOS near the Fermi level as their contribution lies well below E_F. While the mean valence of Yb in the two skutterudites estimated from PFY-XAS measurements turned out to be quite similar and temperature independent above 60 K, namely, 2.11 ± 0.03 for $Yb_{0.97}Fe_4Sb_{12}$ and 2.13 ± 0.03 for $Yb_{0.88}Fe_4Sb_{12}$, the latter skutterudite had undergone a dramatic increase in its average valence below 50 K (the valence of $Yb_{0.97}Fe_4Sb_{12}$ remained temperature independent down to the lowest temperatures). The higher valence, i.e., an increased component of Yb^{3+} ions in $Yb_{0.88}Fe_4Sb_{12}$, was assumed to arise from the electron transfer from Yb sites neighboring cages where Yb was missing. Applied pressure slightly increased the valence state of Yb in both samples, Figure 3.91, in accordance with the notion that the trivalent Yb is a more favored state under pressure as it has a smaller radius. I remind the reader that the pressure tended to enhance the Curie temperature T_C of $Yb_yFe_4Sb_{12}$ skutterudites that had a significant number of Yb vacancies, Figure 3.88(b). In fact, the valence change at the pressure-induced transition is of similar order as the temperature-induced transition in $Yb_{0.88}Fe_4Sb_{12}$. The idea that the valence of Yb increases in the cages located in close proximity of cages where Yb is missing was supported by Mössbauer spectra collected by Tamura et al. (2012) on weakly FM $Yb_{0.88}Fe_4Sb_{12}$. Here, the quadrupole splitting deviated from the $T^{3/2}$ dependence obeyed by many metallic lattices. Moreover, the typical double-peak Mössbauer spectrum developed a notable asymmetry below 45 K because of the increased Yb valence and consequent charge transfer from such cages to the neighboring cages where Yb was absent. Thus, while the weak form of ferromagnetism in $Yb_yFe_4Sb_{12}$ with $y \leq 0.93$ is

TABLE 3.13
Lattice Constant, Effective Magnetic Moment, Curie–Weiss Temperature, and Valence of Yb of the Existing YbFe$_4$Sb$_{12}$ Skutterudites Collected from the Literature

Compound	Lat. const. (Å)	μ_{eff}/f.u. (μ_B)	θ_{CW} (K)	Valence of Yb	Reference
YbFe$_4$Sb$_{12}$	9.158	3.09	40	2–3	Dilley et al. (1998)
YbFe$_4$Sb$_{12}$	9.1571	4.49	13.8	2.68	Leithe-Jasper (1999)
Yb$_{0.83}$Fe$_4$Sb$_{12}$	9.150	3.36	26.5	–	Bauer et al. (2000)
Yb$_{0.9}$Fe$_4$Sb$_{12}$	9.154	–	–	2.16	Bérardan et al. (2003)
Yb$_{0.95}$Fe$_4$Sb$_{12}$	9.1587	3.0	49	≈ 2	Schnelle et al. (2005)
Yb$_{0.89}$Fe$_4$Sb$_{12}$	9.15	3.7	31.9	2, assumed	Tamura et al. (2006)
Yb$_{0.94}$Fe$_4$Sb$_{12}$	9.15	3.72	22.3	2,assumed	Tamura et al. (2006)
Yb$_{0.91}$Fe$_4$Sb$_{12}$	9.1524	2.92	57.7	2, assumed	Ikeno et al. (2007)
Yb$_{0.991}$Fe$_4$Sb$_{12}$	9.156	2.8	50	2, assumed	Saito et al. (2011)
Yb$_{0.95}$Fe$_4$Sb$_{12}$	9.158	4.6	–	2, but rises below 50 K for y < 0.93	Möchel et al. (2011)
Yb$_{0.97}$Fe$_4$Sb$_{12}$	–	3.19	–	2.11	Yamaoka et al. (2011)
Yb$_{0.88}$Fe$_4$Sb$_{12}$	–	3.13	–	2.13	Yamaoka et al. (2011)

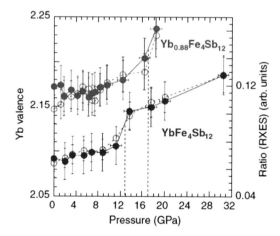

FIGURE 3.91   Pressure dependence of the valence of Yb (solid circles) in Yb$_{0.88}$Fe$_4$Sb$_{12}$ (red symbols) and YbFe$_4$Sb$_{12}$ (actual Yb content $y = 0.97$, blue symbols) based on the analysis of the PFY-XAS spectra. The intensity ratio of Yb^{3+}/Yb^{2+} (open circles and RHS scale) is based on the RXES spectra taken at a photon energy of 8939 eV. Reproduced from H. Yamaoka et al., *Physical Review Letters* **107**, 177203 (2011). With permission from the American Physical Society.

tied to the itinerant $3d$ electrons of Fe, Yb vacancies control whether the ordered magnetic state is actually realized.

There are no reports on magnetic properties of YbRu$_4$Sb$_{12}$. This is undoubtedly due to the reduced structural stability of YbRu$_4$Sb$_{12}$ compared to that of YbFe$_4$Sb$_{12}$ arising from an expanded cage of the [Ru_4Sb_{12}] framework and thus difficulties in establishing bonds between Yb and the neighboring Sb. DFT calculations by Chen et al. (2013) support the weak structural stability of $YbRu_4Sb_{12}$.

$YbOs_4Sb_{12}$ was first synthesized by Kaiser and Jeitschko (1999), but only structural parameters were provided. Galván (2011) reported on his theoretical calculations of the band structure, which indicated strong hybridization between f-electrons of Yb, d-electrons of Os and p-electrons of Sb, suggesting that the skutterudite might have a heavy fermion character. No magnetic instability was predicted. The only experimental report on $YbOs_4Sb_{12}$ is that by Kunitoshi et al. (2016), where transport, magnetic, and thermal properties were investigated on very small single crystals prepared under ambient pressure. An attempted high-pressure synthesis yielded far too small crystals ≤ 0.1 mm. Overall, the behavior is similar to that of $YbFe_4Sb_{12}$, including a broad shoulder on the otherwise metallic temperature dependence of resistivity, this time at a somewhat higher temperature of about 100 K, and a maximum in $\chi(T)$ around 50 K. Although the crystals were presented as $YbOs_4Sb_{12}$, no Rietveld refinement was performed to ascertain the actual occupancy of Yb in the voids, and no XAS was carried out to confirm the presumed divalent form of Yb. Regardless of such vital information, there were no signs of any phase transition down to 2 K. Setting aside a peak in the specific heat at 2.3 K, undoubtedly associated with a significant presence of AFM Yb_2O_3 impurity with the Néel temperature of 2.3 K, no unusual features were detected on the temperature dependence of the specific heat. From a straight line of C/T vs. T^2 between 2K and 3.5 K, a small Sommerfeld coefficient $\gamma \approx 37$ mJmol^{-1}K^{-2} and a Debye temperature of 226 K were obtained. Deviations from the straight line above 3.5 K were ascribed to an Einstein-like phonon mode with $\theta_E \approx 40$ K. Detailed XAS studies would be highly desirable to ascertain the valence state of Yb and any temperature dependence it might show.

3.3.4.8   Gd-Filled Skutterudites

Gadolinium is usually considered the starting member of the heavy and small rare-earth elements. However, while all elements following Gd cannot form stable bonds with the icosahedral cage of skutterudites under any ambient pressure synthesis conditions and require high-pressure synthesis at 4–5 GPa, Gd has occasionally been filled into the cage of phosphide skutterudites at ambient pressure. Moreover, because Gd-filled skutterudites have attracted considerable interest that resulted in several publications, I am treating it separately from the other heavy rare earth-filled skutterudites. Except for one report describing properties of $GdFe_4As_{12}$, all other publications focus on Gd-filled phosphides, as they have smaller cage sizes than arsenide and certainly smaller cages than antimonide skutterudites.

Gd enters the skutterudite voids as a trivalent ion Gd^{3+} representing a half-filled $4f^7$ shell. Using the Hund's rules, i.e., within the LS coupling scheme, maximizing the total spin angular momentum S we obtain $S = 7/2$. Because a single electron occupies every one of the seven $4f$ orbitals, the total orbital angular momentum L is zero. Then, the total angular momentum $J = L + S = 7/2$. The ground state multiplet (octet) is thus spherically symmetric, carries no quadrupoles, and CEF cannot split it.

$GdFe_4P_{12}$ was reported by Jeitschko et al. (2000) and subsequently by Kihou et al. (2004) as a FM metal with the Curie temperature $T_C = 22 \pm 3$ K, and 23 K, respectively. The effective magnetic moment determined from the Curie–Weiss behavior above 100 K turned out to be $\mu_{eff} = 7.6 \pm 0.1$ μ_B/f.u. and 8.19 μ_B/f.u., respectively. Both values are quite close to the Hund's rule ground state value for the Gd^{3+} free-ion moment $\mu_{eff} = g[J(J+1)]^{1/2} = 7.94$ μ_B. The occupancy of Gd, based on a structural refinement, was essentially 100%. $GdFe_4P_{12}$ is a very soft ferromagnet with a coercive field at 4 K of only 0.0014 T and a remanence of 0.18 ± 0.04 μ_B. The entirely metallic temperature dependence of resistivity measured by Jeitschko et al. (2000) by a two-probe technique (samples were simply too small) lacked any sign of the FM ordering. The four-probe resistivity measurements by Kihou et al. (2004) documented clearly the onset of ferromagnetism by a cusp at 23 K followed by a rapidly decreasing resistivity as the ordered moments of Gd presented less resistance than their disordered state above T_C. Further attestation of the onset of the FM order was provided by ^{31}P-NMR measurements by Magishi et al. (2008) that indicated a nearly constant relaxation rate $1/T_1$ above T_C (^{31}P nucleus interacting with fluctuating local moments at Gd^{3+} sites), which gave way to a rapidly decreasing $1/T_1$ below T_C as spin fluctuations got suppressed by the developed magnetic order. At low temperatures, below about 10 K, the relaxation rate varied linearly with T, indicating the Korringa-like behavior usually expressed as $T_1 T$ = const and signaling the relaxation dominated by the conduction electrons of the $[Fe_4P_{12}]$ polyanions.

Sekine et al. (2000c) used the high-pressure technique to synthesize $GdRu_4P_{12}$ and, unlike its cousin $GdFe_4P_{12}$, the skutterudite turned out to be an antiferromagnet with the Néel temperature $T_N = 22$ K. The linear $\chi^{-1}(T)$ vs. T behavior above 100 K resulted in the effective magnetic moment $\mu_{eff} = 8.04~\mu_B/Gd^{3+}$, in excellent agreement with a theoretical value of 7.94 μ_B. The Curie–Weiss temperature was positive 23 K, indicating FM correlations in the system even though the ordering was stated as AFM. Below 22 K, i.e., in the AFM regime, the magnetization increased linearly with the applied field, started to level off above 6 T, and reached saturation at 18 T with about 7 μ_B/Gd, in excellent agreement with the theoretical Gd^{3+} saturation moment of 7 μ_B. The onset of AFM ordering was reflected also in the otherwise metallic resistivity, which developed a broad minimum near 25 K and below 22 K increased rapidly and reached a peak near 12 K, followed by an equally rapid decrease. The sharp upturn in the resistivity at T_N leading to the peak was strongly suppressed by the magnetic field in measurements by Matsuhira et al. (2006). In fact, in the field of 7 T, the peak was no longer present, Figure 3.93. The field dependence of T_N was described by

$$T_N(B) = T_N(0)\left[1 - \left(\frac{B}{B_c}\right)^2\right]^{\delta},\tag{3.108}$$

where B_c is the critical field at zero temperature. Fits to the data, shown in the inset of Figure 3.92, returned $T_N(0) = 21.38$ K, $B_c = 6.48$ T, and $\delta = 0.58$. The onset of the AFM order left a mark also on the temperature dependence of the Seebeck coefficient that rose sharply at T_N and peaked near 12 K, mimicking the trend in $\rho(T)$. However, measurements of the magneto-Seebeck coefficient were not pursued. As shown by Matsunami et al. (2005), the emergence of the AFM ordering at T_N was accompanied by the development of a broad peak in the energy-dependent optical conductivity $\sigma(\omega)$ below 22 K. The anomaly was ascribed to electronic excitations across a pseudogap (32 meV at 8 K) near the Fermi level induced by AFM ordering. Specific heat measurements by Sekine et al. (2008a) depicted a sharp rise at 22 K associated with the onset of AFM ordering. Upon application of a magnetic field, the anomaly was gradually suppressed and shifted to lower temperatures and disappeared altogether

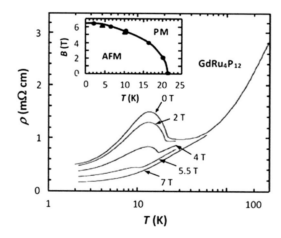

FIGURE 3.92 Electrical resistivity of $GdFe_4P_{12}$ in various magnetic fields. The inset shows the phase diagram based on electrical resistivity measurements. The line through the data is the best fit to Equation 3.108, yielding $T_N(0) = 21.38$ K and the critical field $B_c = 6.48$ T. Redrawn from K. Matsuhira et al., *Physica B* **378–380**, 235 (2006). With permission from Elsevier.

in fields above 7 T, resulting in an essentially identical phase diagram as the one shown in the inset of Figure 3.92. As done in all previous cases, the specific heat of non-magnetic $LaRu_4P_{12}$ served as a reference to determine the magnetic fraction of the specific heat of $GdFe_4P_{12}$, $C_m(T)$, from which, by integration, the temperature-dependent magnetic entropy $S_m(T)$ was obtained. At 22 K, the magnetic entropy reached about 17 Jmol^{-1}K^{-1}, in excellent agreement with the entropy $S_m = R\ln(2J+1) = R\ln8 = 17.3$ Jmol^{-1}K^{-1} expected based on the $J = 7/2$ ground state of Gd. Subsequent X-ray measurements by Sekine et al. (2009a), looking into whether the AFM transition was accompanied by some structural anomaly, turned out negative as no anomaly was detected on a smoothly temperature varying lattice constant. However, their precision thermal expansion measurements indicated the same trend as observed previously in measurements of the resistivity and specific heat, namely, the linear thermal expansion coefficient developed a sharp jump at T_N that was gradually suppressed, shifted to lower temperatures, and eventually completely obliterated when the magnetic field was applied. The evaluated Grüneisen parameter, $\Gamma_G = 3\alpha B_0 V_m/C_v$, where α is the linear thermal expansion coefficient, B_0 is the bulk modulus, V_m is the molar volume, and Cv is the specific heat, came to 4.4, and was actually even larger than the ones obtained for $PrRu_4P_{12}$ (2.8) and $SmRu_4P_{12}$ (3.3), the structures that underwent M–I transitions at 63 K and 16 K, respectively.

The large jump in the resistivity (representing some 50% increase) as the sample passed across the ordering temperature at T_N, Figure 3.92, implies a rather dramatic decrease in the carrier concentration, because the carrier scattering would be expected to weaken as the moments order below T_N. This was borne out by a detailed analysis of the Hall coefficient and magnetoresistance carried out by Watanabe et al. (2010) at low temperatures, which indicated that merely one-tenth of the charge carriers in the paramagnetic state supports conduction in the AFM regime.

$GdOs_4P_{12}$ was prepared by high-pressure synthesis by Kihou et al. (2004) as part of their survey of properties of heavy rare earth-filled skutterudites having the $[Os_4P_{12}]$ framework. Similar to $GdFe_4P_{12}$, but unlike $GdRu_4P_{12}$, this last member of the Gd-filled phosphide skutterudite family became a ferromagnet, albeit at a much lower temperature $T_C \approx 5$ K. Its effective magnetic moment of $\mu_{eff} = 8.54$ μ_B/Gd^{3+} was, again, in a reasonable agreement with a theoretical value of 7.94 μ_B for a trivalent Gd, and the Curie–Weiss temperature turned out to be positive at 2.9 K. Interestingly, because the nearest Gd-Gd distance is fairly long and systematically increases from Fe to Ru to Os (6.751 Å to 6.960 Å to 6.984 Å), the direct magnetic interaction among Gd ions cannot be very large. Moreover, the FM order in $GdFe_4P_{12}$ and $GdOs_4P_{12}$ is not shared with the antiferromagnetically ordered $GdRu_4P_{12}$. Thus, interactions of Gd with the transition metals must play a role in determining what ordered state the skutterudite will choose. Figure 3.93 offers a comparison of electrical

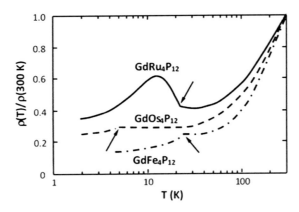

FIGURE 3.93 Temperature dependence of the resistivity $\rho(T)/\rho(300$ K) for the three Gd-filled phosphide skutterudites. Arrows indicate the Curie temperatures of $GdFe_4P_{12}$ and $GdOs_4P_{12}$ and the Néel temperature of $GdRu_4P_{12}$. Adapted from C. Sekine et al., *Journal of the Physical Society of Japan* **77**, Suppl. A, 135 (2008). With permission from the Physical Society of Japan.

resistivities of the three Gd-filled phosphite skutterudites, showing the respective ordering temperatures as anomalies on otherwise smooth curves.

Sekine et al. (2011a) succeeded in preparing $GdFe_4As_{12}$ by high-pressure synthesis, the only Gd-filled skutterudite with the cage made of other than phosphorus atoms. The structure is metallic and becomes a ferromagnet at $T_C = 56$ K, revealed by a sharply rising susceptibility and a kink in the temperature-dependent resistivity. The Curie–Weiss behavior above 150 K gives the effective magnetic moment $\mu_{eff} = 8.09\ \mu_B$/f.u., a slightly larger value than the theoretical Gd^{3+} free-ion $\mu_{eff} = 7.94\ \mu_B/Gd^{3+}$. Perhaps, there is a small contribution coming from the magnetic moment of Fe. Magnetization measurements at 2 K indicated a rapid rise in fields below 0.5 T that changed into a slow, nearly linear increase up to the highest field of 5 T, where the magnetization reached slightly above 6 μ_B/f.u. This is below the expected saturation magnetization of 7 μ_B/Gd^{3+}. The authors suggested that the smaller experimental magnetization coupled with a rather high Curie temperature might indicate a ferrimagnetic or canted FM ground state rather than an ordinary FM state. However, until reliable information on the magnetic contribution of Fe becomes available, the nature of the magnetic ground state will remain no more than a speculation.

In a theoretical paper by Niikura and Hotta (2012), the authors pointed out that the usual assumption concerning the validity of the Hund's rules, i.e., the *LS* coupling scheme, may not be adequate to describe Gd^{3+} ions because the fundamental premise that the magnitude of the Hund's rule interaction is very much larger than the spin-orbit coupling, $U \gg \lambda$, is not satisfied and a *j-j* coupling scheme should be used instead. If so, while the ground state in the *j-j* coupling scheme maintains $J = 7/2$, the octet consists of $j = 5/2$ sextet and one electron in a $j = 7/2$ octet. Here, j indicates the one-electron total angular momentum. As the single electron in the $j = 7/2$ octet possesses an orbital degree of freedom, the Gd^{3+} state is subject to CEF splitting within the *j-j* coupling scheme, and the authors predicted that a softening of the elastic constants should be detectable. So far, no reports on the effect have surfaced.

3.3.4.9 Heavy Lanthanide-Filled Skutterudites

By heavy lanthanides one understands rare-earth elements Tb, Dy, Ho, Er, Tm, and Lu that, because of their small ionic size, cannot be inserted into any skutterudite cage in syntheses carried out under ambient pressure. Yb is also a heavy rare earth, situated between Tm and Lu, but because of its substantially divalent nature, an ambient pressure synthesis can be used and, because of their importance, I have discussed Yb-filled skutterudites separately in section 3.3.4.7. Occasionally, one finds an element Yttrium also included among the rare earths. I discuss Y-filled skutterudites separately in the following section. From a scientific perspective, an enticing feature of skutterudites with heavy lanthanide fillers is a prospect for observing interesting physics associated with the orbital degree of freedom that benefits from highly degenerate 4f orbitals compared to light lanthanide fillers. Unfortunately, due to the need to use special high-pressure synthesis to prepare samples, the number of reports in the literature is rather limited. All heavy lanthanides in this section enter the skutterudite cage as trivalent ions.

The first Tb-filled skutterudite, $TbRu_4P_{12}$, was prepared by Sekine et al. (2000c) using high-pressure synthesis. In many aspects, its properties are very similar to those of $GdRu_4P_{12}$. Specifically, $TbRu_4P_{12}$ shows a metallic temperature dependence of resistivity and undergoes an AFM transition at $T_N = 20$ K. However, the magnetic susceptibility develops an additional anomaly at a lower temperature of about 10 K, designated as T_1. Magnetization below T_1 seems to proceed *via* a two-step metamagnetic transition, showing anomalies at fields $B_{c1} = 0.8$ T and $B_{c2} = 2.5$ T when measured at 4.2 K. Above T_N, the magnetization increases monotonically. In the field of 18 T and temperature of 4.2 K, the magnetization reached 8.1 μ_B/Tb^{3+}, a somewhat smaller value than the theoretical saturation moment of the Tb^{3+} ion of 9 μ_B. Mimicking the resistivity behavior of $GdRu_4P_{12}$, the resistivity of $TbRu_4P_{12}$ increases rapidly below T_N, reaches a peak near 17 K, and subsequently decreases. No anomaly was seen on the temperature-dependent resistivity that would correspond to a hump in the magnetic susceptibility at T_1.

The peculiar nature of the AFM state with its two defining temperatures T_N and T_1 drew much attention and was subsequently studied by a variety of experimental techniques. Neutron diffraction measurements by Kihou et al. (2005) revealed extra magnetic peaks at 2.3 K that were missing in the diffraction pattern taken at 30 K, documenting the AFM ordering. Plotting the intensity of one of the extra magnetic peaks as a function of temperature, Figure 3.94, depicted a weakening moment and its total demise at T_N. The measurements nicely correlated with the magnetic susceptibility shown in the inset, except that no signature of the second AFM transition at T_1 was detected in the powder neutron diffraction pattern. However, there is clearly a wide temperature gap in the diffraction data near T_1, and more points should have been collected near the T_1 to see if there is or is not a close correlation with the magnetic susceptibility. In any case, the authors surmised that the AFM order is a result of opposite orientation of Tb^{3+} spins at $(0, 0, 0)$ and at $(½, ½, ½)$ lattice sites. Specific heat measurements carried out in magnetic fields up to 8 T by Sekine et al. (2005) revealed shifts of temperatures T_N and T_1, as shown in Figure 3.95(a). Here, the specific heat is plotted as C/T vs. T,

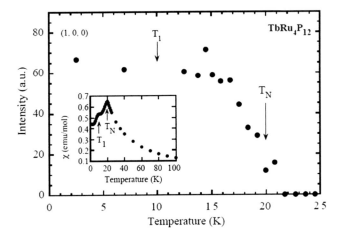

FIGURE 3.94 Temperature dependence of normalized magnetic intensity at (100) reflection of $TbRu_4P_{12}$. The inset shows the magnetic susceptibility of $TbRu_4P_{12}$ measured by Sekine et al. (2000c). Reproduced from K. Kihou et al., *Physica B* **359–361**, 859 (2005). With permission from Elsevier.

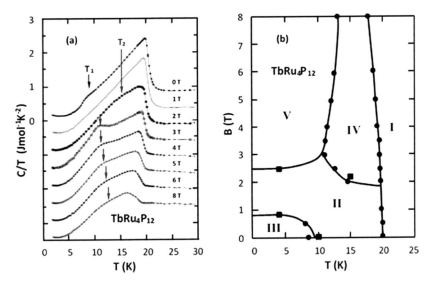

FIGURE 3.95 (a) Temperature dependence of the specific heat of $TbRu_4P_{12}$ plotted as C/T vs. T at various magnetic fields. Temperatures T_1 and T_2 are marked with a black arrow and blue arrows, respectively. (b) B-T phase diagram constructed based on anomalies in the specific heat (solid circles). Solid squares indicate anomalies in the magnetic susceptibility from measurements of Sekine et al. (2000c). Redrawn from C. Sekine et al., *Physica B* **359–361**, 856 (2005). With permission from Elsevier.

and the curves for various magnetic fields are displaced for clarity by the same amount. As the field increases, the jump in the specific heat at T_N becomes smaller, less sharp, and shifts slightly to lower temperatures. In contrast, the hump at T_1 is strongly affected and disappears in fields above 1 T. A new anomaly is observed on the 2 T curve at T_2 and shows peculiar field dependence: the anomaly first plunges down to 11 K in the field of 3 T and then slowly climbs back up as the field increases beyond 3 T. The anomaly is still notable in the field of 8 T where it has shifted to 13 K. Obviously, with at least four and perhaps even five different magnetic phases, the B-T phase diagram is necessarily complex, as shown in Figure 3.95(b).

Because the electrical resistivity and neutron diffraction were "blind" to the anomaly at T_1, it was of interest to look whether this anomaly affects other properties, beyond the susceptibility and specific heat. In previous sections, we have seen that, as a function of temperature and magnetic field, elastic constants are surprisingly very sensitive to the electronic structure. This has been confirmed by Fujino et al. (2008) also in the case of $TbRu_4P_{12}$, where both longitudinal and transverse elastic constants revealed a clear anomaly at T_1, beyond a huge softening taking place at T_N, with the field dependence perfectly matching the one seen in the behavior of the specific heat. However, because the samples were polycrystalline, the ground state of the Tb^{3+} ion could not be uniquely determined. In subsequent measurements of elastic constants and magnetic susceptibility under pressure, Nakanishi et al. (2009) noted a strong effect of pressure on the former while the susceptibility changed little. Although the temperatures T_N and T_1 hardly shifted under pressure, the steep dip (softening) of elastic constants observed near T_N at ambient pressure was strongly suppressed by applied pressure, and the hump at T_1 became undetectable at pressures above 0.5 GPa, Figure 3.96.

The internal magnetic field that must exist in the ordered phase of $TbRu_4P_{12}$ was studied by Fukazawa et al. (2009) using a combination of ^{31}P-NMR, ^{101}Ru-NQR, and μSR techniques. Interestingly, while the transition at T_N near 20 K was detected in all three experiments, the transition at T_1 near 10 K left an imprint only on the spectra of μSR. It is not clear why some otherwise sensitive techniques that probe both macroscopic and microscopic origin of magnetism are insensitive to the anomaly at T_1.

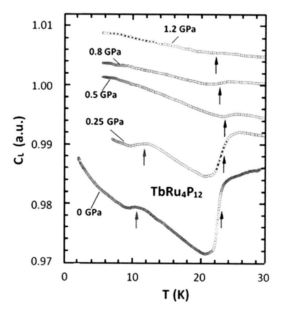

FIGURE 3.96 Temperature dependence of the longitudinal elastic constant $C_L(T)$ in zero magnetic field and at selected applied pressures. The blue arrows indicate the phase transformation point T_1, and the black arrows designate the ordering temperature T_N determined by $dC_L(T)/dT$ vs. T. Redrawn from Y. Nakanishi et al., *Physica B* **404**, 3271 (2009). With permission from Elsevier.

$TbFe_4P_{12}$ and $TbOs_4P_{12}$ have attracted much less interest, and their properties are described in only a handful of publications. Shirotani et al. (2006) reported $TbFe_4P_{12}$ as a metal that undergoes a FM transition near 10 K. The Curie–Weiss behavior above 15 K yielded an effective magnetic moment of 9.48 μ_B/f.u., close but smaller than the Hund's rule value of 9.72 μ_B/Tb^{3+}. It is not clear if the $[Fe_4P_{12}]$ polyanion contributes any magnetic moment.

Sekine et al. (2008a) described $TbOs_4P_{12}$ as a paramagnetic metal with no ordering down to 2 K. The effective magnetic moment and the Curie–Weiss temperature are included in Table 3.14, where the magnetic properties of all heavy lanthanide-filled skutterudites are listed. The resistivity of the three Tb-filled phosphide skutterudites is shown in Figure 3.97. Microscopic viewpoint of magnetism in $TbFe_4P_{12}$ was gained through ^{31}P-NMR measurements by Magishi et al. (2008), which revealed a large and substantially temperature independent relaxation rate $1/T_1$ above T_C, indicative of the ^{31}P nucleus interacting with fluctuating local moments at Tb sites. The fluctuations rapidly diminished as the sample cooled through its T_C and the ordering set in, documented by a sharply decreasing relaxation rate by three orders of magnitude. At temperatures below 2 K, the relaxation rate became linearly dependent on temperature (Korringa relation) and comparable to the relaxation rate in $LaFe_4P_{12}$, reflecting relaxation *via* conduction electrons originating from $[Fe_4P_{12}]$ polyanions.

Even fewer publications describe properties of the remaining heavy lanthanide-filled skutterudites. Resistivity of DyT_4P_{12} is displayed in Figure 3.98, and the magnetic data are listed in Table 3.14. Fe- and Os-versions are low-temperature ferromagnets, while $DyRu_4P_{12}$ orders antiferromagnetically at 15 K. $HoFe_4P_{12}$ is a ferromagnet with $T_C = 5$ K, and measurements of elastic constants by Yoshizawa et al. (2007) showed a distinct softening of the elastic constants at T_C. Both $ErFe_4P_{12}$ and $TmFe_4P_{12}$ remain paramagnetic down to at least 2 K, with the latter having an order of magnitude smaller susceptibility than all other heavy lanthanide skutterudites. For $LuFe_4P_{12}$, only the lattice constant (7.7771 Å) is given in the literature, Shirotani et al. (2003b).

TABLE 3.14

Lattice Constant and Magnetic Parameters of Heavy Lanthanide-Filled Skutterudites Collected from the Literature

Skutterudite	a (Å)	Ordering	μ_{eff} (μ_B/f.u.)	θ_{CW} (K)	Reference
$GdFe_4P_{12}$	7.7964	$T_C = 23$ K	7.90	17.5	Sekine et al. (2008a)
$GdRu_4P_{12}$	8.0375	$T_N = 22$ K	8.04	23	Sekine et al. (2000c)
$GdOs_4P_{12}$	8.0657	$T_C = 5$ K	8.54	2.9	Kihou et al. (2004)
$GdFe_4As_{12}$	8.3024	$T_C = 56$ K	8.09	23	Sekine et al. (2011a)
$TbFe_4P_{12}$	7.7926	$T_C = 10$ K	9.48	10	Kihou et al. (2004)
$TbRu_4P_{12}$	8.0338	$T_N = 20$ K	9.76	8	Sekine et al. (2000c)
$TbOs_4P_{12}$	8.0631	Paramagnet	10.86	3	Sekine et al. (2008a)
$TbFe_4As_{12}$	8.2961	$T_C = 38$ K	9.76	16	Sekine et al. (2011a)
$DyFe_4P_{12}$	7.7891	$T_C = 4$ K	10.70	0.6	Sekine et al. (2008a)
$DyRu_4P_{12}$	8.0294	$T_N = 15$ K	12.23	−0.1	Sekine et al. (2008a)
$DyOs_4P_{12}$	8.0601	$T_C = 2$ K	13.55	4.4	Sekine et al. (2008a)
$HoFe_4P_{12}$	7.7854	$T_C = 5$K	10.43	–	Shirotani et al. (2005)
$ErFe_4P_{12}$	7.7832	Paramagnet	9.59	–	Shirotani et al. (2005)
$TmFe_4P_{12}$	7.7802	Paramagnet	6.58	–	Shirotani et al. (2005)

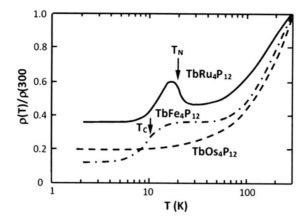

FIGURE 3.97 Temperature dependence of electrical resistivity normalized to its room-temperature value for $TbFe_4P_{12}$ (dash-dotted curve), $TbRu_4P_{12}$ (solid curve), and $TbOs_4P_{12}$ (dashed curve). T_C and T_N indicate the respective Curie and Néel temperatures. Traced from the data of C. Sekine et al., *Journal of the Physical Society of Japan* **77**, 135 (2008). With permission from the Physical Society of Japan.

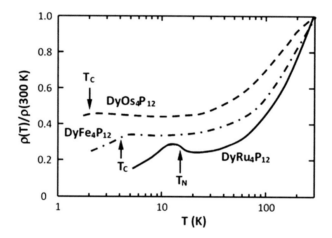

FIGURE 3.98 Temperature dependence of electrical resistivity normalized to its room temperature value for $DyFe_4P_{12}$ (dash-dotted curve), $DyRu_4P_{12}$ (solid curve), and $DyOs_4P_{12}$ (dashed curve). T_C and T_N indicate the respective Curie and Néel temperatures. Traced and modified from the data of C. Sekine et al., *Journal of the Physical Society of Japan* **77**, 135 (2008).

3.3.4.10 Yttrium-Filled Skutterudites

Yttrium is the first *d*-block element in the fifth period of the periodic table and has the electronic configuration $[Kr]4d^1 5s^2$. Yttrium is often counted among the lanthanide elements on account of its similar ionic radius and its reactivity that falls somewhere between that of Tb and Dy. It enters the skutterudite voids invariably as a trivalent Y^{3+} ion. All three Y-filled phosphide skutterudites are superconductors and are discussed in Chapter 2. Here, I only summarize some of their properties in Table 3.15.

3.3.4.11 Alkali Metal-Filled Skutterudites and Tl-Filled Skutterudites

In this section are discussed $[Fe_4Sb_{12}]$ skutterudite frameworks filled with monovalent fillers that include alkali metal ions Na^{1+} and K^{1+} and a heavier monovalent Tl^{1+} ion. The reason for mixing here a heavy Tl filler with the light alkali metal fillers is the fact that the monovalent nature of the filler

TABLE 3.15
Lattice Parameters, SC Transition Temperatures, Bulk Moduli, and Where Known the Electronic Specific Heat of Y-Filled Phosphide Skutterudites. Data Collected from the Literature

Compound	a (Å)	T_c (K)	B_o (GPa)	γ (mJmol^{-1}K^{-2})	Reference
YFe$_4$P$_{12}$	7.7896	7[a]	144[b]	27.2	Shirotani et al. (2003b)
YRu$_4$P$_{12}$	8.0298	8.5	183[b]	–	Shirotani et al. (2005)
YOs$_4$P$_{12}$	8.0615	3	189[b]	–	Kihou et al. (2004)

[a] The value quoted later as 5.6 K by Cheng et al. (2013).
[b] Bulk moduli obtained by Hayashi et al. (2010).

ions has the overriding influence on the properties of filled skutterudites and results in very similar magnetic and thermal characteristics.

First alkali metal-filled skutterudites, NaFe$_4$Sb$_{12}$ and KFe$_4$Sb$_{12}$ with the lattice constants 9.1767 Å and 9.1994 Å, respectively, were synthesized by Leithe-Japser et al. (2003) and found FM at a temperature $T_C \approx 85$ K. Although magnetic contributions originating from the [Fe$_4$Sb$_{12}$] framework had been argued for before as part of the overall magnetism of certain rare earth-filled iron antimonide skutterudites, there was always an issue how to aportion the magnetism arising from $4f$ electrons of rare earths and from $3d$ states of Fe. Here, the issue was clear as the source of magnetism is the [Fe$_4$Sb$_{12}$] polyanion. The Curie–Weiss behavior above 100 K resulted in an effective magnetic moment of 1.6–1.8 μ_B/Fe for both NaFe$_4$Sb$_{12}$ and KFe$_4$Sb$_{12}$. The Curie–Weiss temperatures were positive and very close to T_C. Interestingly, although the Curie temperature was rather high, the remanent magnetic moment in a well-developed hysteresis yielded only 0.28 μ_B/Fe at 1.8 K, possibly depressed by critical fluctuations, Figure 3.99. In a magnetic field of 14 T, where the fluctuations should be suppressed, the magnetic moment increased to 0.60 μ_B/Fe and, extrapolated to 1/B $\rightarrow$ 0, where it topped at 0.67 μ_B/Fe. Both real and imaginary parts of the ac susceptibility developed a large peak near T_C, and the bulk nature of magnetism was also documented by a smaller anomaly at T_C in the plot of the specific heat as a function of temperature. The Sommerfeld coefficient, determined from the specific heat below 14 K, was quoted as $\gamma = 145$ mJmol^{-1}K^{-2}, contributing 62 states eV^{-1}(f.u.)$^{-1}$ to the density of states at the Fermi level. Metallic resistivity with a quadratic $\rho(T) \sim AT^2$ temperature

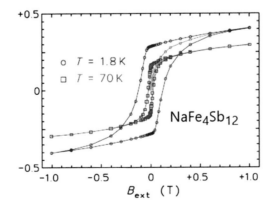

FIGURE 3.99 Isothermal hysteresis loops for NaFe$_4$Sb$_{12}$ taken at 1.8 K and at 70 K. Remanent field at 1.8 K is about 0.28 μ_B/Fe. Adapted from A. Leithe-Jasper et al., *Physical Review Letters* **91**, 037208 (2003). With permission from the American Physical Society.

dependence below 20 K yielded $A = 1.78 \times 10^{-9}$ ΩmK^{-2} for sodium and $A = 1.207 \times 10^{-9}$ ΩmK^{-2} for potassium-filled structures. Combining the coeffient A with the Sommerfeld coefficient γ, the ratio A/γ^2 fell near the Kadowaki-Woods value of 1.0×10^{-7} Ωm(mol-K/J)^2. The authors also made ^{23}Na-NMR and [121,123]Sb-NQR spectroscopy measurements, providing a microscopic view of magnetism. The nuclear spin-lattice relaxation rate $1/T_1$ developed a sharp dip at Tc, followed by the Korringa-like behavior ($1/T_1$ linearly dependent on T) at lower temperatures. Band structure DFT-based calculations using the LSDA, indeed, revealed a FM ground state with a total moment of 2.97 μ_B/cell. Assuming that only Fe carries a magnetic moment resulted in 0.74 μ_B/Fe, the value considerably larger than the experimental 0.28 μ_B/Fe. However, if one takes a position that the experimental value was depressed by critical fluctuations, which, in turn, were suppressed by a high field of 14 T at which the moment of 0.60 μ_B/Fe was measured, the disagreement is not that large. In a subsequent study, Leithe-Jasper et al. (2004) added investigations of the thermal expansion (the coefficient of linear thermal expansion $\alpha = 11 \times 10^{-6}$ K^{-1}), the pressure dependence of the unit cell to 9.5 GPa (at zero pressure the bulk modulus of NaFe$_4$Sb$_{12}$ was $B_0 = 85$ GPa). Bonding was assessed by the electron localization function (mean value of the charge transfer from the Na filler to the [Fe$_4$Sb$_{12}$] polyanion of 0.98 electrons). From [57]Fe and [121]Sb Mössbauer spectroscopy, a weak hyperfine magnetic field $\mu_0 H_{hf}$(Fe) = 1.64 T and a small hyperfine field $\mu_0 H_{hf}$(Sb) = 1.04 T were obtained. Yes, indeed, non-magnetic Sb appeared to be actually polarized! The authors also added DFT calculations of the electronic structure focusing on spin polarization defined as (DOS$_\downarrow$ – DOS$_\uparrow$)/(DOS$_\downarrow$ + DOS$_\uparrow$) and found NaFe$_4$Sb$_{12}$ an almost perfect half-metallic skutterudite with the spin polarization equal to 0.99. This offerred tantalizing possibilities of using the skutterudite also in spintronic applications. Of course, a prediction of such significance had to be verified experimentally, and it was done by point-contact Andreev reflection measurements at 2.8 K on polished polycrystalline samples of KFe$_4$Sb$_{12}$ and NaFe$_4$Sb$_{12}$ by Sheet et al. (2005). The authors recorded differential conductivity $G(V)$ as a function of applied voltage V using sharp SC Nb and Pb tips as contacts to the sample. The conductance spectra of NaFe$_4$Sb$_{12}$ using both Nb and Pb tips are shown in Figure 3.100(a). The transport spin polarization P_t as a function of the barrier parameter Z (the entity characterizing the strength of the potential barrier at the interface between

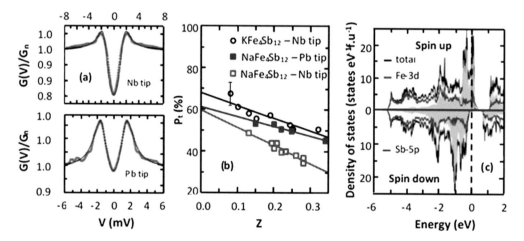

FIGURE 3.100 (a) Conductance spectra for NaFe$_4$Sb$_{12}$ normalized to the normal conductance G_n at high voltage. Solid lines are fits with the following parameters: top panel (Nb tip), the barrier parameter $Z = 0.15$, transport spin polarization $P_t = 53.5$ %, and SC energy gap of the Nb tip $\Delta = 1.4$ meV; bottom panel (Pb tip), $Z = 0.13$, $P_t = 48.2$ %, and $\Delta = 1.4$ meV. (b) Dependence of the transport spin polarization P_t on the value of Z for different samples and tip combinations. (c) Calculated total DOS for KFe$_4$Sb$_{12}$ (black curve), partial Fe-3d contribution (red curve), and partial Sb-5p contribution (blue curve). Adapted and redrawn from G. Sheet et al., *Physical Review B* **72**, 180407(R) (2005). With permission from the American Physical Society.

the sample and the SC contact) for KFe_4Sb_{12} and $NaFe_4Sb_{12}$ with different tips is displayed in Figure 3.100(b). Note a nice agreement between the transport spin polarization in $NaFe_4Sb_{12}$ taken with two different SC tips, giving confidence that a realistic value is being measured. As indicated in Figure 3.100(b), the experimental transport polarizations (67 % for KFe_4Sb_{12} and about 60% for $NaFe_4Sb_{12}$) are smaller than the theoretically predicted values of more than 99 %. The calculated total and partial DOSs for KFe_4Sb_{12}, based on which the theoretical spin polarizations are based, are shown in Figure 3.100(c). Similar densities of states were obtained also for $NaFe_4Sb_{12}$. As the figure indicates, the DOS at the Fermi level (zero eV) is essentially fully polarized with the spin up channel formed by mostly two bands of nearly pure Fe-3d character, while there is hardly any DOS at the Fermi level in the spin down channel. The reason why the transport spin polarization P_t is significantly lower than the calculated spin polarization has to do chiefly with the fact that strong spin fluctuations, typical in weak itinerant ferromagnets, tend to significantly reduce spin polarization. Moreover, the SC contacts were made on polycrystalline samples rather than on single crystals and, as such, may have had a more diffusive rather than ballistic nature, which too, reduces the transport spin polarization. In any case, it is interesting that in spite of the structure not being ideal, the spin polarization values are quite impressive.

Trying to shed more light on the magnetic ground state of $NaFe_4Sb_{12}$, Leithe-Jasper et al. (2014) carried out neutron diffraction and scattering measurements on powders of the structure. Although the magnetic scattering here is quite weak, the authors were able to document it by taking a difference plot of the data collected at 95 K (well above the ordering temperature of about 85 K) and at 2 K. Analyzing the data, they established an ordered magnetic moment associated with the Fe site of 0.6 μ_B/Fe, in good agreement with the magnetization measurements. Assuming that the orbital magnetic moment is fully quenched by the crystal electric field and counting the number of spins above the T_C (from the effective magnetic moment of 1.6 μ_B/Fe) with respect to the number of spins contributing to the saturated magnetic moment at low temperatures, the ratio of the two came to about 1.7. As already noted, deviations of this so-called Rhodes and Wohlfarth ratio from unity is a measure of the itinerant nature of magnetic moments. The ratio of 1.7 reflects some degree of delocalization of the Fe spins. On the other hand, diffuse magnetic scattering experiments with polarized neutrons revealed the presence of localized moments in the paramagnetic domain, the strength of which, however, rapidly decreased with the rising temperature.

Pressure measurements of the Curie temperature of $NaFe_4Sb_{12}$ carried out by Schnelle et al. (2008) yielded a large positive rate of about 3 K/GPa, accompanied by about 20% increase in the coercive field at a pressure of 1.42 GPa at 1.8 K. Overall, a compendium of measurements established $NaFe_4Sb_{12}$ and KFe_4Sb_{12} as itinerant weakly FM systems where spin fluctuations play an important role.

Thallium, as a filler in skutterudites, was initially explored in several papers by Sales' group in the context of lowering the lattice thermal conductivity of partially filled $CoSb_3$-based skutterudites, demonstrating a large atomic dispacement parameter (ADP) of Tl compared to the other constituent elements of the structure and identifying a low-frequency Einstein mode associated with "rattling" of the Tl ion in an oversized void of the skutterudite, Sales et al. (2000a, 2000b), Long et al. (2002), Hermann et al. (2003), and Sales (2003). The actual fully-filled $TlFe_4Sb_{12}$ skutterudite with the lattice constant $a = 9.1973$ Å was synthesized by Leithe-Jasper et al. (2008), and its properties were thoroughly explored by a combination of transport, magnetic, thermal, and Mössbauer measurements. With the lattice parameter very similar to that of $NaFe_4Sb_{12}$ and KFe_4Sb_{12}, there was no doubt that Tl is monovalent. Moreover, the structural refinement confirmed that essentially all voids are filled with Tl. Sharp increases in the magnetic susceptibility and the behavior of magnetization identified the Curie temperature $T_C \approx 80$ K. The effective magnetic moment obtained from the Curie–Weiss behavior amounted to 1.69 μ_B/Fe and $\theta_{CW} = 88.3$ K, the latter value a bit larger than T_C but comparable. It is obvious that the magnetic properties of $TlFe_4Sb_{12}$ are very similar to those of $NaFe_4Sb_{12}$ and KFe_4Sb_{12}, and this assessment is also quickly reached by glancing at the data in Table 3.16 and at Figure 3.101, where the paramagnetic electronic DOS are presented for all three

TABLE 3.16

Experimental Lattice Parameters, taken from Kaiser and Jeitschko (1999), Calculated Total and Partial DOS is from Takegahara et al. (2008), and the Electronic Specific Heat $_\Gamma$ Extrapolated from C/T to $T = 0$ by Takabatake et al. (2006)

Skutterudite	a (Å)	DOS (Total)	A-partial	T-partial	Sb-partial	γ (mJmol^{-1}K^{-2})
CaFe$_4$Sb$_{12}$	9.1620	32.20	0.42	17.35	5.29	118
SrFe$_4$Sb$_{12}$	9.1812	22.57	0.22	11.91	3.85	87
BaFe$_4$Sb$_{12}$	9.200	20.88	0.18	11.53	3.29	104
SrRu$_4$Sb$_{12}$	9.289	10.06	0.13	3.27	2.68	11
BaRu$_4$Sb$_{12}$	9.315	3.91	0.05	1.44	0.94	9
SrOs$_4$Sb$_{12}$	9.3313	18.70	0.37	6.21	4.49	44
BaOs$_4$Sb$_{12}$	9.3351	15.92	0.33	5.23	3.80	46

Note: The values of the total and partial DOS were originally given in units of states per Ry-formula unit and in Table 8.16 are entered in units of states per eV-formula unit. With permission from Elsevier.

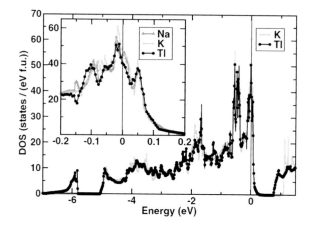

FIGURE 3.101 Total paramagnetic electronic DOS for TlFe$_4$Sb$_{12}$ (black symbols and line) and KFe$_4$Sb$_{12}$ (blue line). In the inset is shown an expanded region near the Fermi energy for all three skutterudites with a monovalent filler, with NaFe$_4$Sb$_{12}$ depicted in orange color. The electronic structure was calculated by a full-potential local-orbital (FPLO) scheme within the local (spin) density approximation with the exchange and correlation potentials of Perdew and Wang (1992). Reproduced from A. Leithe-Jasper et al., *Physical Review B* **77**, 064412 (2008). With permission from the American Physical Society.

skutterudites filled with the monovalent fillers. On account of the high DOS near the Fermi level dominated by Fe-3d states, the ground state of TlFe$_4$Sb$_{12}$ is also nearly half-metallic with a static spin polarization of 96%.

3.3.4.12   Skutterudites Filled with Alkaline-Earth Metals

First skutterudites with alkaline-earth fillers, BaFe$_4$Sb$_{12}$ and BaRu$_4$Sb$_{12}$, were synthesized by Stetson et al. (1991), and the structure was identified as cubic with the space group $Im\bar{3}$, i.e., a typical filled skutterudite. Subsequent structural studies by Evers et al. (1994) evidenced seven more skutterudites with Ca, Sr, and Ba fillers, all of them having the characteristic $Im\bar{3}$ space group. The first measurements of magnetic properties of AFe_4Sb_{12} with A = Ca, Sr, and Ba were conducted by Danebrock et al. (1996) who described them as paramagnetic structures obeying the Curie–Weiss law with similar effective magnetic moments of 3.7 μ_B/f.u., 3.8 μ_B/f.u., and 4.0 μ_B/f.u., respectively, ascribed to localized 3d spins of Fe. The Curie–Weiss temperatures turned out to be +3 K, −17 K, and −36 K, respectively. Based on the magnetic, transport, and thermal studies, Matsuoka et al. (2005) categorized AFe_4Sb_{12} (A = Ca, Sr, and Ba) as nearly FM metals with the moments arising from the itinerant magnetism of Fe 3d electrons rather than being of localized nature. An incontrovertible argument for the itinerant character of magnetic moments was an exceptionally small magnetization at 2 K, which in a field of 5 T attained only 0.08 μ_B/Fe. As shown also in subsequent papers by Matsuoka et al. (2006) and Takabatake et al. (2006), all three alkaline-earth-filled iron antimonides have a similar temperature dependence of resistivity with a notable knee near 70 K, depicted in Figure 3.102. This is ascribed to the presence of spin fluctuation scattering that exerts its strongest influence (maximum in dρ/dT) near T_{sf} = 50 K. Saturation of spin scattering above T_{sf} is believed to be the origin of the knee in the resistivity of AFe_4Sb_{12} (A = Ca, Sr, and Ba) near 70 K. From the Curie–Weiss behavior above 200 K, the effective magnetic moments of 1.52, 1.47, and 1.50, respectively, were obtained for the Ca-, Sr-, and Ba-filled [Fe_4Sb_{12}] frameworks. As one goes from Ca to Sr to Ba, the paramagnetic Curie–Weiss temperatures of 54 K, 53 K, and 31 K follow, indicating that the FM interactions are considerably stronger than in $LaFe_4Sb_{12}$ where θ_{CW} = 3 K. Consequently, alkaline-earth-filled AFe_4Sb_{12} skutterudites are rightfully classified as being the near FM systems. Interestingly, the knee-dominant temperature dependence of resistivity seems to be restricted to the [Fe_4Sb_{12}] and [Os_4Sb_{12}] frameworks only as the alkaline-earth-filled [Ru_4Sb_{12}] framework shows no such anomaly and decreases monotonically. However, as shown in Figure 3.103, the magnetic susceptibility of alkaline-earth-filled osmite antimonides is some 50 times smaller than the susceptibility of AFe_4Sb_{12}, and it is unlikely that vigorous spin fluctuations are the origin of the knee in AOs_4Sb_{12}. Rather, as suggested by Matsuoka et al. (2006), strong electron–phonon interaction might be a more relevant source. The [Ru_4Sb_{12}] framework filled with alkaline-earth ions is entirely diamagnetic at all temperatures. Room-temperature resistivities of AFe_4Sb_{12} and ARu_4Sb_{12} are about 350–450 $\mu\Omega$cm and about 600 $\mu\Omega$cm for AOs_4Sb_{12}. All polycrystalline structures have a RRR not higher than 3. Interestingly, a pronounced knee in $\rho(T)$ vs. T also develops in single crystals of $SrOs_4Sb_{12}$ and $BaOs_4Sb_{12}$ studied by Narazu et al. (2008), but at a somewhat higher temperature of around 100 K. These single crystals were apparently of high quality as judged by their lower

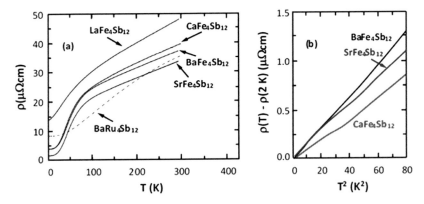

FIGURE 3.102 (a) Temperature dependence of the electrical resistivity of alkaline-earth-filled skutterudites. Note a sharp knee on the curves of $CaFe_4Sb_{12}$, $BaFe_4Sb_{12}$, and $SrFe_4Sb_{12}$ in distinct contrast to a smooth dependence of $BaRu_4Sb_{12}$. Resistivity of $LaFe_4Sb_{12}$ is also shown for comparison. (b) A near quadratic variation of the resistivity of AFe_4Sb_{12} (A = Ba, Sr, Ca) at low temperatures. Drawn from the data of E. Matsuoka et al., *Journal of the Physical Society of Japan* **74**, 1382 (2005).

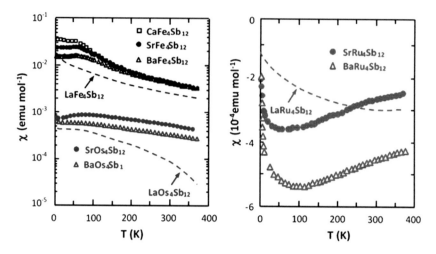

FIGURE 3.103 Magnetic susceptibility of alkaline-earth-filled antimonide skutterudites. The left panel shows the behavior of iron antimonides (black symbols) and osmium antimonides (blue symbols). The right panel shows negative susceptibilities of ruthenium antimonide skutterudites. Dashed curves in both panels stand for the magnetic susceptibility of the respective La-filled structures. Note that the susceptibility of AOs_4Sb_{12}, while positive, is some 50 times smaller than the susceptibility of AFe_4Sb_{12}. Adapted from T. Takabatake et al., *Physica B* **383**, 93 (2006). With permission from Elsevier.

values of room temperature resistivity of 130–280 $\mu\Omega$cm and RRR $\approx$ 20. Again, because of much smaller susceptibilities (4–6 $\times$ 10^{-4} emu mol^{-1}), spin fluctuations are likely irrelevant as far as the knee is concerned. From specific heat measurements below 6 K, Matsuoka et al. (2005) obtained the Sommerfeld coefficients γ of 118, 87, and 104 mJmol^{-1}K^{-2} for Ca, Sr, and Ba-filled [Fe_4Sb_{12}] frameworks, respectively. Here, γ is taken per mole of Fe. Combined with the quadratic coefficient of the electrical resistivity A, the ratio $A/\gamma^2 = 8.9 \times 10^{-8}$, 2.8×10^{-7}, and 2.5×10^{-7} Ωm(mol-K/J)^2 was obtained for the three alkaline-earth-filled iron antimonides, respectively. The values are not too different from the Kadowaki–Woods value of 1.0×10^{-7} Ωm(mol-K/J)^2. Later measurements by Schnelle et al. (2008) yielded considerably smaller values in the range of $(0.24–0.29) \times 10^{-8}$ Ωm(mol-K/J)^2.

Magnetization of alkaline-earth-filled skutterudites was studied by Yoshii et al. (2006) in pulsed fields as high as 50 T and indicated a development of an interesting metamagnetic enhancement at 1.3 K that depended on the field strength and the type of the filler. In $CaFe_4Sb_{12}$, it sets in at a critical field B_c (taken as the maximum value on dM/dB) near 13 T and decreased to 12 T and 11 T in $SrFe_4Sb_{12}$ and $BaFe_4Sb_{12}$, respectively. The metamagnetic anomaly became weaker as the temperature increased and disappeared altogether at $T_0 = 40$ K. Going from Ca to Sr to Ba fillers, the metamagnetic transition temperature decreased rapidly to 15 K (Sr) and down to 4.2 K (Ba).

Band structure calculations for alkaline-earth-filled antimonide skutterudites, AT_4Sb_{12}, performed by Takegahara et al. (2008) using a FLAPW method attempted to clarify the magnetic properties by evaluating the DOS near the Fermi energy. The largest DOS was found for T = Fe, followed by T = Os, and a considerably smaller DOS for T = Ru. In the [Fe_4Sb_{12}] framework, about half of all states were contributed by Fe d-states. Interestingly, the calculations predicted a stable FM ground state not just for $NaFe_4Sb_{12}$, but also for AFe_4Sb_{12}, where A stands for Ca, Sr, and Ba. Lattice parameters, DOS values, together with respective partial contributions, and the experimental electronic specific heat are collected in Table 3.16. Small DOSs at the Fermi level of ARu_4Sb_{12} are reflected in very small coefficients γ of these skutterudites. Optical reflectivity measurements by Sichelschmidt et al. (2006) suggested a formation of a nearly identical pseudogap structure (~ 12 meV) in the far-infrared spectral range of all alkaline-earth-filled iron antimonide skutterudites (including a divalent rare earth Yb) at temperatures below 90 K. The observation was in accordance with the authors' FPLO calculations carried out within the LSDA using the exchange and correlation potential of Perdew and Wang (1992). The onset of a pseudogap appears at a temperature not too far from where the knee is observed on $\rho(T)$ and likely imposes its presence on the transport behavior. Higher temperatures tended to smear out the gap. The fact that the LSDA calculated single particle band structure mapped the optical data so faithfully was an indirect confirmation that Coulomb correlations play only a minor role, if any, in alkaline-earth-filled skutterudites. This is an entirely different situation than in some of the 4-f heavy fermion skutterudites, which develop similar optical spectra. Subsequent optical conductivity measurements by Kimura et al. (2006, 2007) focused on a comparison of spectra of alkaline-earth-filled antimonide skutterudites having different transition metals T = Fe, Ru, and Os in the framework. For this purpose, the authors also modeled the respective optical conductivities using band structure calculations based on the FLAPW + the local orbital method implemented in the WIEN2K code, Blaha et al. (2001). Comparing experimental and theoretical spectra for various structures revealed a strong influence of different wavefunction forms of 3d (Fe), 4d (Ru) and 5d (Os) orbitals of the transition metals on the detailed features within the infra-red region of the spectra, which, in turn, were reflected in low-temperature nuances of the magnetic properties. In contrast, altering positional parameters seemed to have little impact on the optical conductivity spectra. While, overall, the computed spectra replicated the experimental optical spectra reasonably well, one should not look too closely at the infrared range of frequencies where very large discrepancies arose, as depicted in Figure 3.104.

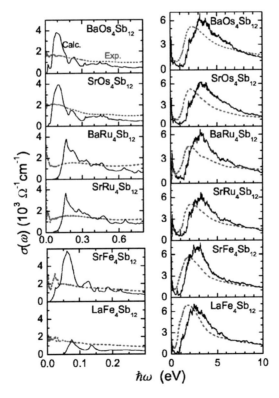

FIGURE 3.104 Experimental optical conductivity spectra (indicated by red dashed lines) and the computed spectra (shown by black solid curves) for various alkaline-earth-filled skutterudites and for LaFe$_4$Sb$_{12}$. The right column indicates an overall range of energies of measurements at 300 K, while the left column shows an expanded low-energy region measured at 7 K. Reproduced from S. Kimura et al., *Physical Review B* **75**, 245106 (2007). With permission from the American Physical Society.

Microscopic insight into the magnetic state of alkaline-earth-filled skutterudites and, specifically, the effect of FM correlations of itinerant $3d$ electrons of Fe was sought by Matsumura et al. (2005) by conducting ^{121,123}Sb-NQR measurements. Observing neither splitting nor broadening of NQR spectra attested to the absence of any FM ordering in both $SrFe_4Sb_{12}$ and $BaFe_4Sb_{12}$. Moreover, the temperature-dependent parameter $1/T_1T$, where T_1 is the nuclear spin-lattice relaxation time at the Sb site, developed a broad maximum near 60 K, i.e., close to the maximum in the magnetic susceptibility, Figure 3.105(a), followed by an unusual suppression at lower temperatures, mimicking the decreasing $\chi(T)$ below 70 K, before the presence of impurity phases in the sample reversed the trend. Decomposing $1/T_1T$ above 80 K into a constant term $1/T_1T_0 = const$ and a T-dependent spin term $(1/T_1T)_{spin}$ and showing that the spin term is linearly proportional to a spin-part of the magnetic susceptibility (via the Curie–Weiss behavior) confirmed a widely held belief that, while the alkaline-earth-filled $[Fe_4Sb_{12}]$ framework is close to a FM instability, it never attains an ordered FM state because the susceptibility is weakened by the formation of the pseudogap structure near the Fermi energy. Subsequent extraction of the Knight shift K at a site of Sb in AFe_4Sb_{12} (A = Sr and Ca) from ^{123}Sb-NQR measurements by Sakurai et al. (2008) allowed a determination of the hyperfine coupling constant $A_{hf} = -1.49$ T/μB-f.u. Moreover, the measurements also specified the actual intrinsic value of the susceptibility due to spins χ_{spin} via a general relation $K_{spin}(T) = \dfrac{A_{hf}}{N_A\mu_B}\chi_{spin}$, where N_A is the Avogadro number. The decomposition of $1/T_1T$ into spin-dependent $(1/T_1T)_{spin}$ and spin-independent $(1/T_1T_0)$ parts, shown in Figure 3.105(b), revealed that a marked decrease in the measured $1/T_1T$ (coincidental and proportional to the decreasing $\chi(T)$ below 70 K) is primarily due to spin-gap-like suppression of the spin-independent contribution. It is generally believed that such a strong suppression of χ_{spin} below 70 K is the reason why alkaline-earth-filled skutterudites never become ferromagnets.

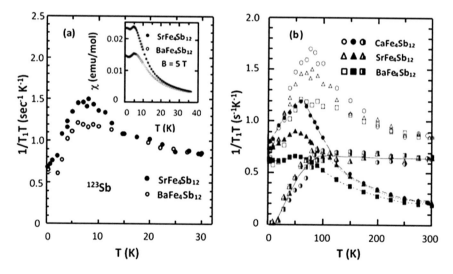

FIGURE 3.105 (a) Temperature dependence of $1/T_1T$ for $SrFe_4Sb_{12}$ and $BaFe_4Sb_{12}$ showing a broad maximum near 60 K, similar to the maximum in the magnetic susceptibility displayed in the inset. The solid and dotted lines in the inset are the least-squares fits to $\chi(T) = \chi_0 + C/(T - \theta)$. Redrawn from M. Matsumura et al., *Journal of the Physical Society of Japan* 74, 2205 (2005). With permission from the Physical Society of Japan. (b) Decomposition of $1/T_1T$ measured on AFe_4Sb_{12} (A = Ca, Sr, and Ba) into spin-dependent and spin-independent parts. Open symbols are experimental values of $1/T_1T$, solid symbols represent the spin-dependent part $(1/T_1T)_{spin}$, and half-solid symbols stand for the spin-independent part obtained from $1/T_1T - (1/T_1T)_{spin}$. Redrawn from A. Sakurai et al., *Journal of the Physical Society of Japan* **77**, 063701 (2008). With permission from the Physical Society of Japan.

Beyond antimonides, alkaline-earth-filled skutterudites were also prepared based on arsenides, making use of high-pressure synthesis at about 4 GPa and temperatures around 825°C. Specifically, $BaFe_4As_{12}$ and $BaRu_4As_{12}$ with lattice constants of 8.3975 Å and 8.5555 Å, respectively, were synthesized by Takeda et al. (2010) and Takeda et al. (2011), and SrT_4As_{12} (T = Fe, Ru, and Os) were prepared by Nishine et al. (2017) with lattice parameters of 8.351 Å (Fe), 8.521 Å (Ru), and 8.561 Å (Os). All structures showed metallic temperature dependence and, except for $SrRu_4As_{12}$, displayed a typical bending (knee) in their $\rho(T)$ vs. T dependence above about 100 K. Surprisingly, $SrOs_4As_{12}$ turned out be a new superconductor with the transition temperature near 4.8 K. Magnetic properties of $BaFe_4As_{12}$ were explored by Sekine et al. (2015), who found the magnetic susceptibility to develop a broad peak near 50 K, and the Curie–Weiss behavior above 150 K yielded the effective magnetic moment $\mu_{eff} = 2.96\ \mu_B$/f.u., i.e., 1.46 μ_B/Fe, very close to the effective magnetic moment of 1.50 μ_B/Fe measured for $BaFe_4Sb_{12}$. Unexpectedly, and in contradiction to the case of $BaFe_4Sb_{12}$, the Weiss temperature turned out to be negative −57 K, for reasons still unknown. Magnetization measured at 2 K revealed a substantially linear dependence on the magnetic field with a very small value of 0.02 μ_B/Fe at 7 T. An enormous discrepancy between this moment and the effective magnetic moment vividly documents the itinerant form of magnetism associated with Fe $3d$ electrons. Extrapolating low-temperature specific heat data $C(T)/T$ to $T = 0$, the electronic specific heat term $\gamma = 62$ mJmol^{-1}K^{-2} was obtained. Nishine et al. (2017) reported on magnetic measurements of $SrFe_4As_{12}$, which returned comparable magnetic parameters as for $BaFe_4As_{12}$, notably $\mu_{eff} = 1.36\ \mu_B$/Fe and the electronic specific heat coefficient $\gamma = 58$ mJmol^{-1}K^{-1}, except that the Weiss temperature was definitely positive at 36 K, contrary to the negative −57 K reported by Sekine et al. (2015) for $BaFe_4As_{12}$.

An interesting DFT-based theoretical study of the evolution of the magnetic state as Ru gradually replaces Fe in $BaFe_{4-x}Ru_xAs_{12}$ was performed by Shankar et al. (2017). Surprisingly, as one quarter of Fe atoms was replaced with Ru atoms, the near FM state of $BaFe_4As_{12}$ with the calculated effective magnetic moment of 1.88 μ_B/Fe became enhanced by a factor of more than three and was reported as strongly FM. The FM state with close to twice as high effective moment as $BaFe_4As_{12}$ persisted until 75% of Fe was replaced by Ru, only to plunge to zero as $BaRu_4As_{12}$ turned out to have a paramagnetic ground state. Unfortunately, no insight was offered why and how the presence of Ru might lead to such dramatic changes in magnetic properties. Clearly, this theoretical work should be corroborated by experimental investigations.

Microscopic viewpoint of magnetism in $SrFe_4As_{12}$ was provided by detailed ^{75}As NMR and NQR measurements by Ding et al. (2018). The lack of broadening or splitting of the NQR line at 56.3 MHz confirmed the absence of any structural or magnetic transitions in the experimental range of 4 K–300 K. The Knight shift K at the As nucleus, normally assessed by performing NMR measurements under high magnetic field, was inaccessible due to a complex nature of the NMR spectrum (complicated by a signal from an As impurity phase amounting to some 12%). Instead, the authors used a low-magnetic field NQR technique to determine the Knight shift that relied on monitoring the NQR resonance frequency $\nu_{NQR}(B)$ as a function of applied magnetic field B, written as

$$\nu_{NQR}(B) = \nu_{NQR}(0) \pm \frac{\gamma_N}{2\pi} A(\eta)(1+K)BF(\theta), \tag{3.109}$$

where $A(\eta)$ is a factor dependent on the asymmetry parameter of the electric field gradient at the As site η, $\gamma_N/2\pi = 7.2919$ MHz/T, and $F(\theta) = \dfrac{\cos\theta}{2}\left[3 - \left(4\tan^2\theta + 1\right)^{1/2}\right]$. Because $A(\eta)$ differs from unity, depending on the value of η, it had to be determined from a position of the lower edge in the spin-echo intensity spectra as a function of magnetic field, see Figure 3.106(a). With the value of $A(\eta) = 0.9794$ thus determined, the Knight shift at the ^{75}As nucleus at 4.3 K turned out to be $K = (-5.7 \pm 2.5)\%$ with essentially no anisotropy. The temperature dependence of the Knight shift obtained by repeating

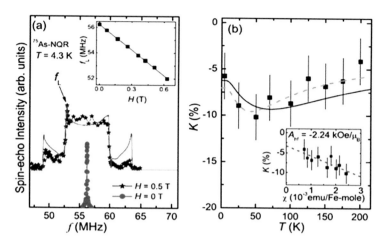

FIGURE 3.106 (a) ^{75}As-NQR spectra taken at 4.3 K in zero field (solid red circles) and in a field of 0.5 T (black stars). The blue curve is a simulated powder-pattern spectrum with $f_Q = 54.8$ MHz, $\eta = 0.4$, and $B = 0.5$ T. The arrow shows the position of the lower-frequency edge f_L, calculated from $\theta = \pi$ and also $\theta = 0$. The inset depicts the external field dependence of f_L. (b) Temperature dependence of ^{75}As Knight shift K. The solid curve is the calculated results, and the inset shows the temperature-dependent Knight shift $K(T)$ plotted *versus* magnetic susceptibility $\chi(T)$. The red dashed line is a linear fit. Reproduced from Q.-P. Ding et al., *Physical Review B* **98**, 155149 (2018). With permission from the American Physical Society.

the above procedure at each experimental temperature yielded a curve depicted in Figure 3.106(b), which resembles closely the temperature-dependent magnetic susceptibility, except that a minimum near 50 K is seen as a maximum on $\chi(T)$ vs. T, due to a negative hyperfine coupling constant $A_{hf} = (-2.24 \pm 0.6)$ T/μ_B. The authors were also able to model the temperature dependence of the nuclear spin-lattice relaxation rate divided by temperature, $1/T_1T$, by a simple band structure depicted in Figure 3.107(a), mimicking a pseudogap symmetrically located around the Fermi energy. Because

the spin part of the Knight shift is proportional to the DOS at the Fermi energy while $1/T_1T$ is propor-

tional to its square, one can write $T_1TK_s^2 = \dfrac{\hbar}{4\pi k_B}\left(\dfrac{\gamma_e}{\gamma_N}\right)^2 \equiv S = 8.97 \times 10^{-6}$ K.s for the ^{75}As nucleus.

Defining the Korringa ratio as $K(\alpha) = \dfrac{2S}{T_1TK_s^2}$ with a factor of 2 in the numerator to account for the number of nearest neighbor Fe ions, the value of $K(\alpha)$ determines whether AFM or FM correlations dominate in the structure. Specifically, $K(\alpha) > 1$ indicates the dominance of AFM spin correlations, while $K(\alpha) < 1$ favors FM correlations. As shown in Figure 3.107(b), the values of $K(\alpha)$ in SrFe$_4$As$_{12}$ are substantially less than unity at all temperatures; hence, the analysis evidences the dominance of FM fluctuations. A similar outcome with even smaller values of $K(\alpha) \sim 0.005$ was obtained by Sakurai et al. (2008) in the case of SrFe$_4$Sb$_{12}$ and CaFe$_4$Sb$_{12}$ skutterudites, suggesting that the type of a ligand does not matter while the overall behavior is determined by the number of valence electrons of a filler.

An additional insight into why monovalent and divalent filler species in the [Fe$_4$Sb$_{12}$] framework lead to different magnetic outcomes was provided by hard X-ray absorption measurements by Mounssef et al. (2019) that focused on the electronic and structural assessment of the FeSb$_6$ octahedral building blocks of the skutterudite structure. The pre-edge region of the Fe K-edge XAS spectra, in conjunction with theoretical calculations that provided specific contributions of orbital states of all constituents, revealed strong orbital mixing of Fe 3d and Sb 5p states and was interpreted as a confirmation of strong hybridization of Fe and Sb at the Fermi level. There was also an

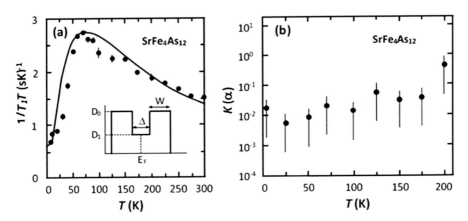

FIGURE 3.107 (a) Temperature dependence of $1/T_1T$ in SrFe$_4$As$_{12}$. The solid line is the calculated result based on a simple band structure shown in the inset with the parameters Δ = 88 K, W = 220 K, and $r \equiv D_1/D_0$ = 0.38. (b) Temperature dependence of the Korringa ratio $K(\alpha)$ in SrFe$_4$As$_{12}$. Adapted and redrawn from Q.-P. Ding et al., *Physical Review B* **98**, 155149 (2018). With permission from the American Physical Society.

indication that the pre-edge structure of spectra of the alkaline-earth-filled skutterudites is split into two components (the effect most notable in the case of CaFe$_4$Sb$_{12}$), something that was not found in the magnetically ordered alkali metal-filled skutterudites. The EXAFS analysis was used to probe coordination of Fe. Although no specific distortion of the Fe coordination was found, there emerged a significant Fe-Sb bond disorder, again most notable in CaFe$_4$Sb$_{12}$, but present also in SrFe$_4$Sb$_{12}$ and BaFe$_4$Sb$_{12}$, and considerably larger than in the structures filled with monovalent fillers Na and K. The authors interpreted this finding as suggesting that a large Fe-Sb bond disorder may be behind the absence of magnetically ordered states in alkaline-metal-filled [Fe$_4$Sb$_{12}$] frameworks.

DFT-based calculations were performed by Bhat and Gupta (2018), using the WIEN2K package and employing both GGA and Becke–Johnson, Becke and Johnson (2006), potentials, associated SrFe$_4$As$_{12}$ with a FM ground state, and the effective magnetic moment of 1.35 μ_B/Fe, nearly identical with the experimental value of 1.36 μ_B/Fe measured by Nishine et al. (2017). Using the mean-field approximation, the Curie temperature of SrFe$_4$As$_{12}$ was estimated from $T_C = 2\Delta E/3k_B$ as 735 K, where ΔE is the total energy difference between the FM and AFM states. We see here, again, a tendency for the theory to overestimate FM interactions, even though in reality, the structure never attains the FM state.

Two alkaline-earth-filled phosphides synthesized recently under high-pressure conditions by Deminami et al. (2017) and by Kawamura et al. (2018), BaOs$_4$P$_{12}$ with a = 8.124 Å and CaOs$_4$P$_{12}$ with a = 8.084 Å, are both superconductors with the respective bulk transition temperatures of 1.8 K and 2.5 K. Details concerning their SC properties are given in Chapter 2.

3.3.4.13 Skutterudites Filled with Actinide Elements

There are three actinide elements that were used to fill the skutterudite structure: thorium (Th), uranium (U), and neptunium (Np). Of these, Th- and U-filled skutterudites have been described in several publications, while there is a single report on the properties of Np-filled skutterudites.

3.3.4.13.1 Th-Filled Skutterudites

With the atomic configuration of [Rn] $5f^06d^27s^2$, the most natural oxidation state of Th is tetravalent Th^{4+} having neither $5f$ nor $6d$ electrons. One would expect tetravalent thorium to perfectly saturate the [T$_4$X$_{12}$]$^{4-}$ polyanion and, consequently, result in a distinctly semiconducting behavior. This viewpoint was supported by FLAPW calculations made within the LDA approximation by Takegahara and Harima (2003), predicting ThFe$_4$P$_{12}$ to have a rather substantial direct band gap of 0.45 eV at the

Γ point. A similar result, with an even slightly larger direct gap of 0.52 eV, was obtained by Khenata et al. (2007) in their calculations of the elastic constants and electronic and optical properties using the identical computation approach. An indirect band gap was predicted based on the calculations by Cheng et al. (2008). In contrast to all the above theoretical results, tight-binding calculations by Galván et al. (2003) predicted a metallic character of $ThFe_4P_{12}$. In reality, the resistivity measurements by Torikachvili et al. (1987) and infrared reflection spectroscopy by Dordevic et al. (1999) revealed metallic temperature dependence although with a rather poor RRR < 1.8. At a glance, the resistivity of $ThFe_4P_{12}$ (about 200 $\mu\Omega$cm at 300 K) obviously exceeds the value of 150 $\mu\Omega$cm, taken as the Yoffe–Regel limit for typical metallic carrier densities $\sim 10^{22}$ cm^{-3}. However, the plasma edge frequency $\omega_p^2 = \dfrac{4\pi n e^2}{m_e} \approx 3500\,\text{cm}^{-1}$ implies the carrier density $\sim 10^{20}$ cm^{-3}, and such two orders of magnitude lower carrier concentration, of course, shifts the Yoffe–Regel criterion to a much higher resistivity.

Apart from phosphides, thorium was also filled into the framework of [Fe_4As_{12}] by Wawryk et al. (2016), resulting in $ThFe_4As_{12}$. Single crystals with edge dimensions up to about 1 mm were prepared by mineralization in a molten Cd:As flux at high temperatures and pressures, as described by Henkie et al. (2008). The crystals had a lattice constant $a = 8.2970$ Å. Although a single crystalline material, the RRR, defined in this case as $\rho(300\ \text{K})/\rho(2\ \text{K}) \approx 2.7$, was only marginally better than in the case of polycrystalline samples of $ThFe_4P_{12}$. Nevertheless, all single crystals displayed metallic temperature dependence down to about 25 K, followed by a sample-dependent minimum and either flat temperature dependence or mildly rising resistivity at the lowest temperatures. Magnetic susceptibility followed a modified Curie–Weiss dependence of the form $\chi(T) = \dfrac{C}{(T - \theta_{CW})} + \chi_0$, where under χ_0 were swept temperature independent orbital and diamagnetic contributions of Th, Fe, and As. The resulting effective magnetic moment turned out to be $\mu_{\text{eff}} = 1.05\ \mu_B$/f.u., and the Curie–Weiss temperature θ_{CW} was negative at −83.7 K. Specific heat, shown in Figure 3.108, displayed no anomaly down to 0.4 K and was unaffected by the magnetic field even at 9 T. A simple fit of the form

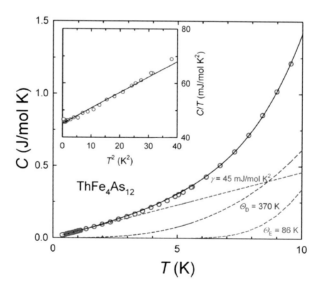

FIGURE 3.108 Temperature dependence of specific heat of $ThFe_4As_{12}$ below 10 K. In this temperature range the specific heat is well described by the expression $C(T) = \gamma T + C_{Debye} + C_{Einst}$ with the least-squares fit yielding the parameters $\gamma = 45$ mJmol^{-1}K^{-2}, $\theta_D = 370$ K, and $\theta_E = 86$ K. The inset shows the specific heat plotted as C/T vs. T^2. Reproduced from R. Wawryk et al., *Journal of Alloys and Compounds* **688**, 478 (2016). With permission from Elsevier.

$C(T) = \gamma T + C_{Debye} + C_{Einstein}$ with the parameters $\gamma = 45$ mJmol^{-1}K^{-2} ($\gamma_{Fe} = 11.2$ mJmol$_{Fe}^{-1}$K^{-2}), $\theta_D = 373$ K, and $\theta_E = 86$ K, replicated the experimental data below 10 K very well. Interestingly, the ratio A/γ_{Fe}^2 for ThFe$_4$As$_{12}$ falls within the range of values $(4.7 \pm 0.4) \times 10^{-7}$ and $(6.4 \pm 0.5) \times 10^{-7}$ in units of Ωcm(mol$_{Fe}$-K/J)^2 and is thus considerably larger than the Kadowaki–Woods value of 1×10^{-7} in the same units, commonly delineating a threshold for heavy fermion systems. This rather large value of A/γ_{Fe}^2 arises chiefly from a large quadratic resistivity term A, not from an ordinary value of $\gamma_{Fe} = 11.2$ mJmol$_{Fe}^{-1}$K^{-2}. Consequently, Wawryk et al. (2016) concluded that ThFe$_4$As$_{12}$ is not a heavy Fermion system, but rather an ordinary paramagnet that happens to have overlapping valence and conduction bands, i.e., being a semimetal with the charge carrier density on the order of 2×10^{20} cm^{-3}. Nevertheless, the large experimental Sommerfeld coefficient $\gamma = 45$ mJmol^{-1}K^{-2} remains unexplained.

3.3.4.13.2 U-Filled Skutterudites

Uranium has the electron configuration [Rn]$5f^36d^17s^2$ and enters compounds mostly as U^{4+} and sometimes as U^{6+} ions. Taking that in the skutterudite structure uranium is tetravalent, it then has two $5f$ electrons in the valence shell. Uranium-filled skutterudite, UFe$_4$P$_{12}$, was first reported on by Meisner et al. (1985) as a FM semiconductor with the ordering temperature $T_C = 3.15$ K. The sample studied was a single crystal grown from molten Sn flux and had the lattice constant $a = 7.7729$ Å, the second smallest among all filled skutterudites. Starting with a room temperature resistivity greater than 1000 $\mu\Omega$cm (the irregular sample shape did not allow determination of a precise value), the resistivity increased by 7 orders of magnitude as the temperature decreased down to 4.2 K. A knee developing below 85 K prevented to specify the actual activated behavior. The FM transition at T_C was detected via ac susceptibility, and hysteresis was observed below T_C. The Curie–Weiss behavior below about 80 K resulted in the effective magnetic moment $\mu_{eff} = 2.25$ μ_B/U atom. According to Hund's rule, the tetravalent uranium ion with $J = 4$ ground state multiplet should yield a free ion value of 3.58 μ_B/U atom. The authors suggested that the reduced value of the experimental effective moment is due either to the CEF effect or to strong hybridization (a small lattice parameter might favor the latter). Magnetization measured in fields up to 3 T at 1.9 K reached 1.9 μ_B/U, smaller than the expected 3.2 μ_B/U. Subsequent specific heat measurements by Torikachvili et al. (1986) revealed a sharp peak at 2.86 K, close to the $T_C = 3.15$ K, superimposed against a broad background that was assumed to be due to an electronic Schottky anomaly associated with CEF splitting of the $J = 4$ multiplet. The authors also presented their longitudinal magnetoresistance measurements that indicated pronounced negative magnetoresistance in the temperature range of 5 K–20 K, while the magnetoresistance below T_C trended initially positive, turned around, and in fields above 1 T became negative with an even more rapid rate than at $T > T_C$. Most likely, the negative longitudinal magnetoresistance arises because the external field forces progressively greater spin alignment and thus reduced scattering. Optical properties studied by Dordevic et al. (1999) confirmed the insulating transport character of UFe$_4$P$_{12}$ and revealed a high reflection in the mid-infrared range of the spectrum, suggesting strong interband transitions. The originally proposed hybridization of $5f$ electrons of U with the conduction electrons leading to a gap opening in the DOS near the Fermi energy, Meisner et al. (1985), found support in the enhanced value of the high frequency dielectric constant $\varepsilon_\infty = 17$, linked to hybridization by Öğüt and Rabe (1996). The effect of hydrostatic pressure on the Curie temperature of UFe$_4$P$_{12}$ was explored by Guertin et al. (1987) to pressures of 16 kbar (1.6 GPa) and indicated a rather large dT_C/d$P = 0.26$ K/kbar (2.6 K/GPa), nearly an order of magnitude larger rate than that observed for the isomorphic ferromagnet NdFe$_4$P$_{12}$ (dT_C/d$P = 0.03$ K/kbar). A definitive proof of FM ordering in UFe$_4$P$_{12}$ was gained by neutron-diffraction measurements on a single crystal carried out by Nakotte et al. (1999) at temperatures between 2.8 K and 15 K. Although no extra purely magnetic reflections were observed below T_C, additional magnetic contributions on top of nuclear reflections were detected, in accordance with the FM ground state. Temperature-dependent intensity of the (220) reflection taken upon heating (open symbols) and cooling (solid symbols)

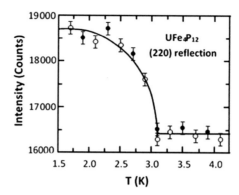

FIGURE 3.109 Temperature dependence of the intensity of the (220) reflection measured on a triple-axis spectrometer at the high-flux reactor. Open circles indicate measurements taken upon heating and solid circles upon cooling. The curve and line are guides to the eye. Redrawn from H. Nakotte et al., *Physica B* **259–261**, 280 (1999). With permission from Elsevier.

is shown in Figure 3.109, clearly documenting magnetic ordering setting in at about 3.1 K. In an attempt to explain the key magnetic features, namely, the ordering temperature T_C, the temperature dependence of magnetic susceptibility, a small saturation moment, and the excess magnetic specific heat, Matsuda et al. (2004) extended magnetic characterization of UFe_4P_{12} single crystals to 800 K and pulsed fields as high as 50 T. They also considered CEF schemes relevant to the $5f^2$ electronic structure of the U^{4+} ion. Unlike in previous resistivity measurements, the authors were able to extract the energy gap $E_g = 53$ meV from a fit of the form $\rho(T) \propto \exp\left(\frac{E_g}{2k_BT}\right)$. The Curie–Weiss behavior was observed only above 400 K and resulted in an effective moment of 3.0 μ_B/U, still smaller but now by not as much in comparison to the theoretical 3.58 μ_B/U and the $\theta_{CW} = -98$ K. Nevertheless, the observed full saturation value in fields above 10 T and temperature of 1.3 K remained small at 1.3 μ_B/U, compared to $5f^2$ Hund's rule value of 3.2 μ_B/U. Magnetization data also revealed a notable anisotropy for magnetic fields below 2 T, with the easy axis along the <100> direction. The specific heat, typified by a sharp peak near T_C and by an elevated magnetic specific heat, overall similar to the behavior observed by Torikachvili et al. (1986), is shown in Figure 3.110. The magnetic entropy was extracted by subtracting the specific heat of $LaFe_4P_{12}$ from the specific heat of UFe_4P_{12} and integrating C_{mag}/T over the temperature. The magnetic entropy at the T_C of $0.65R\ln2$ is rather small and only near 20 K approaches $R\ln3$. Thus, the magnetic entropy did not provide a reliable hint as to what CEF energy scheme is the most likely one.

Using the T_h site symmetry to reflect the environment of a filler in the skutterudite cage, the nine-fold degenerate multiplet associated with $J = 4$ of the U^{4+} ion having two $5f$ electrons is split into a singlet Γ_1, doublet Γ_{23} (Γ_3 in the O_h cubic CEF scheme), and two triplet states $\Gamma_4^{(1)}$ and $\Gamma_4^{(2)}$ (Γ_4 and Γ_5 in O_h symmetry). As pointed out in section 8.1.4.4, the distinction between the CEF Hamiltonians appropriate for the T_h and O_h symmetries is the last term in

$$\mathcal{H}_{CEF} = A_4\left(O_4^0 + 5O_4^4\right) + A_6\left(O_6^0 - 21O_6^4\right) + A_6'\left(O_6^2 - O_6^6\right) \qquad (3.110)$$

where A_4, A_6, and A_6' are the CEF parameters and O_m^n are the Stevens operators, Stevens (1952).

In the O_h cubic symmetry $A_6' = 0$, while it is finite in the T_h symmetry. In the latter case, this leads to mixing between the Γ_4 and Γ_5 states, resulting in $\Gamma_4^{(1)}$ and $\Gamma_4^{(2)}$ states. As in most of the situations, the physical properties depend crucially on what constitutes the ground state and the first excited state, higher excited levels typically having little effect, if any. Lacking a clear guidance from the

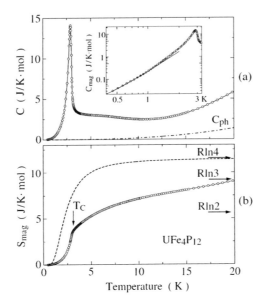

FIGURE 3.110 (a) Temperature dependence of specific heat of UFe_4P_{12} below 20 K. C_{ph} indicates the lattice specific heat estimated from measurements of $LaFe_4P_{12}$. (b) Magnetic entropy as a function of temperature below 20 K, obtained by subtracting the specific heat of $LaFe_4P_{12}$ from the specific heat of UFe_4P_{12} and thus obtained magnetic contribution to the specific heat C_{mag} is divided by temperature and integrated over temperature. The FM ordering temperature is indicated by T_C. The dashed curve indicates the theoretical entropy based on the CEF model A depicted in Figure 3.111. Adapted from T. D. Matsuda et al., *Journal of the Physical Society of Japan* **73**, 2533 (2004). With permission from the Physical Society of Japan.

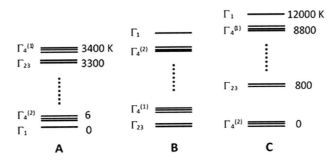

FIGURE 3.111 Possible CEF schemes applicable to UFe_4P_{12} as explored by Matsuda et al. (2004). Reproduced from T. D. Matsuda et al., *Journal of the Physical Society of Japan* **73**, 2533 (2004). With permission from the Physical Society of Japan.

trend in the magnetic entropy, Matsuda et al. (2004) thus explored three versions of the CEF scheme to fit all their magnetic measurements. They are shown in Figure 3.111. Of the three CEF schemes, the scheme B did not fit the data well. Scheme A with the singlet Γ_1 ground state spaced by only 6 K from the first excited state formed by the triplet $\Gamma_4^{(2)}$ provided a reasonable fit to the magnetic susceptibility and the field dependence of magnetization but, as indicated in Figure 3.110(b), grossly overestimated the magnetic entropy. Scheme C with the triplet ground state $\Gamma_4^{(2)}$, widely spaced (~800 K) from the first excited state Γ_{23} doublet, provided the best overall fit to the magnetic data, even though it somewhat overestimated the saturation magnetization as well as the temperature dependence of the magnetic moment obtained from neutron scattering measurements by Nakotte et al. (1999), both shown in Figures 3.112(a) and 3.112(b), respectively. Moreover, the triple ground state is also more consistent with the magnetic entropy gradually reaching $R\ln 3$ above 20 K.

Microscopic insight into magnetism in UFe_4P_{12} was provided by ^{31}P-NMR measurements by Tokunaga et al. (2005). Using a single crystal sample, the authors were able to trace the angular dependence of the Knight shift, while its temperature dependence was obtained on a powder sample by grounding several small crystals. The extracted isotropic K_{iso} and anisotropic K_{ani} Knight shifts have identical temperature dependence as the magnetic susceptibility, Figure 3.113(a), and from the

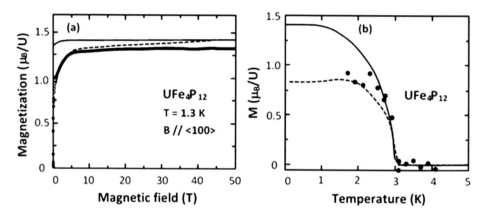

FIGURE 3.112 (a) High-field magnetization of UFe_4P_{12} for the B // <100> direction at 1.3 K. Heavy solid curve and open circles are experimental data. A thin solid curve and dashed curve represent CEF calculated magnetization assuming the level scheme C and A, respectively. (b) Solid circles are experimental temperature-dependent magnetic moment obtained from neuron scattering measurements by Nakotte et al. (1999). The solid curve and the dashed curve are calculated moments assuming CEF schemes C and A, respectively. Adapted and redrawn from T. D. Matsuda et al., *Journal of the Physical Society of Japan* **73**, 2533 (2004). With permission from the Physical Society of Japan.

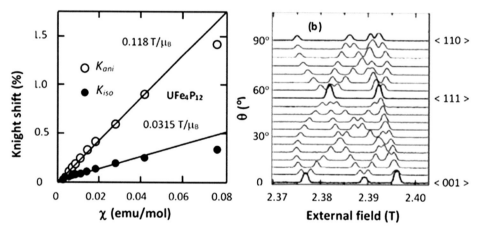

FIGURE 3.113 (a) Anisotropic and isotropic Knight shifts plotted against the magnetic susceptibility with temperature as an implicit parameter. Slopes of the straight lines give values of anisotropic and isotropic hyperfine coupling constants A_{ani} and A_{iso}. (b) Angular dependence of the ^{31}P-NMR spectrum of a single crystal of UFe_4P_{12} taken at 50 K. θ stands for the polar angle of the applied field relative to the [001] crystal axis as the field rotates from <001> to <110> through the <111> direction of the crystal. The line splitting is caused by the presence of anisotropic hyperfine field acting at the site of P. Adapted and redrawn from Y. Tokunaga et al., *Physical Review B* **71**, 045124 (2005). With permission from the American Physical Society.

slopes the isotropic and anisotropic hyperfine coupling constants $A_{iso} = 0.0315$ T/μ_B and $A_{ani} = 0.118$ T/μ_B were determined. For comparison, the above values of the hyperfine coupling constants are much larger than in the case of isostructural $PrFe_4P_{12}$ (0.0137 T/μ_B and 0.0139 T/μ_B, respectively). Such a marked difference was explained as a consequence of much stronger hybridization between phosphorus and uranium orbitals in UFe_4P_{12}. Note also that a much smaller isotropic Knight shift than the anisotropic one implies a rather small contribution of conduction electrons to the ^{31}P hyperfine field. Because the magnetic susceptibility $\chi(T)$ in the paramagnetic range is isotropic, this means that the large anisotropic Knight shift necessarily implies anisotropy in the hyperfine coupling itself. Although there are twelve crystallographically equivalent P sites in the cubic structure of skutterudites, the local symmetry at P sites is lower than cubic, leading to six inequivalent directions in an external magnetic field when using a single crystal. In a powder sample, because of averaging over all directions, the pattern corresponding to a single P site is obtained. The NMR line splitting observed in the angular dependence of the ^{31}P-NMR spectrum shown in Figure 3.113(b) arises because of the anisotropic hyperfine field acting at the P site. The line splits up to a maximum of six peaks that change positions and merge into four, three, and two peaks as the field coincides with the <110>, <001>, and <111> crystallographic directions. The line splitting is of purely magnetic origin as the ^{31}P nucleus ($I = \frac{1}{2}$) has no electric quadrupole moment.

Tokunaga et al. (2005) also measured the nuclear spin-lattice relaxation rate $1/T_1$, which was substantially temperature independent in the paramagnetic regime but developed a strong temperature dependence $\propto T^{7/2}$ in the FM range. As was previously suggested by Beeman and Pincus (1968), similar behavior in ferromagnets with anisotropic hyperfine interaction is indicative of nuclear spin flips, and thus spin-lattice relaxation, caused by three-magnon processes. The ^{31}P-NMR studies were fully consistent with a localized nature of $5f$ electrons of uranium.

3.3.4.13.3 Np-Filled Skutterudites

The electronic configuration of neptunium is alternately quoted as either $[Rn]5f^46d^17s^2$ or $[Rn]5f^36d^27s^2$, and it readily attains valence states from Np^{3+} to Np^{7+}. Assuming that in the skutterudite structure it is tetravalent, there are then three $5f$ electrons in the valence shell. This first transuranium filler was successfully entered into the $[Fe_4P_{12}]$ framework by Aoki, D. et al. (2006), forming a single crystalline $NpFe_4P_{12}$ skutterudite. Its lattice constant at $a = 7.7702$ Å is even smaller than that of UFe_4P_{12} and, in fact, is the smallest of all filled skutterudites. Magnetic and transport measurements identified $NpFe_4P_{12}$ as a ferromagnetically ordered structure below $T_C = 23$ K, a substantially higher ordering temperature than in the case of UFe_4P_{12}. While isotropic in the paramagnetic regime, the magnetization becomes distinctly anisotropic below T_C, with the easy axis identified as the <100> direction. With a magnetic field in this direction, the magnetic moment saturated at a value of 1.35 μ_B/Np already well below 1 T at 5 K. In the other two field orientations, B // <110> and B // <111>, the magnetization reached 1.0 μ_B/Np and 0.9 μ_B/Np, respectively, at 0.2 T and 5 K. No clear-cut Curie–Weiss behavior was observed below 300 K as the $\chi^{-1}(T)$ vs. T dependence continuously curved. Electrical resistivity, see Figure 3.114(a), trended metallic down to about 150 K, at which point the resistivity started to rise and developed a rather sharp peak near 30 K, suggesting that this temperature is characteristic of spin fluctuations. Below 30 K, the resistivity fell rapidly and became more-or-less constant below 4 K. Applied magnetic field of 5.5 T dramatically suppressed the rising resistivity (curtailing spin fluctuations) and the now broader peak shifted to near 50 K. Given the fact that the resistivity was rather high (~ 65 mΩcm at 300 K!), the authors surmised that $NpFe_4P_{12}$ is likely a semimetal with a low carrier concentration, and the Np $5f$-electrons are localized. Another notable feature was a rather high negative magnetoresistance (nearly 65% in a field of 15 T) observed in the temperature range between about 10 K and 50 K, indicating the development of FM spin-fluctuations around T_C. The FM ordering near 23 K left a clear mark on the temperature-dependent specific heat in the form of a cusp shown in Figure 3.114(b). Interestingly, rather than the expected $T^{3/2}$ power law describing

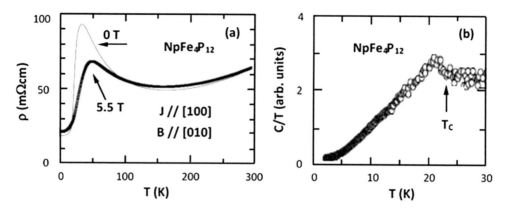

FIGURE 3.114   (a) Temperature dependence of electrical resistivity of $NpFe_4P_{12}$ in zero field and in a field of 5.5 T. (b) Specific heat plotted as C/T versus temperature for $NpFe_4P_{12}$. The λ-like anomaly near T_C is clearly visible. Adapted and redrawn from D. Aoki et al., *Journal of the Physical Society of Japan* **75**, 073703 (2006). With permission from the Physical Society of Japan.

the magnetic specific heat, the authors obtained good fits with equation $C/T = \gamma + BT^2$, which returned the electronic specific heat coefficient $\gamma = 10$ mJmol^{-1}K^{-2}.

The same research team that reported on ^{31}P-NMR measurements of UFe_4P_{12} made a similar study on a single crystal of $NpFe_4P_{12}$, Tokunaga et al. (2009), including the field-orientation dependence of ^{31}P-NMR line splitting and the temperature dependence of the nuclear spin-lattice relaxation rate $1/T_1$. The NMR line splitting in an applied magnetic field proceeded in essentially identical way to that in UFe_4P_{12}, with a maximum of up to six peaks and their dependence on the polar angle θ as depicted in Figure 3.113(b), reflecting on the anisotropy and the angular dependence of the hyperfine field at the P site. By subtracting the dipolar field generated by the Np $5f$ time-averaged spin moment $\bar{\mu}_j$ (= 0.044 μ_R, estimated from magnetization at 100 K) at site j and sensed at site i of the ^{31}P nucleus,

$$H_{dip}\left(i,j\right) = 3\frac{\bar{\mu}_j.\bar{r}_{ij}}{r_{ij}^5} - \frac{\bar{\mu}_j}{r_{ij}^3}, \tag{3.111}$$

from the NMR line splitting defined as $\Delta H \equiv (f_{res} - f_0)/\gamma$, where f_{res} is the center of gravity of the observed NMR peak, and from $f_0 = \gamma H_0$ with $\gamma = 17.237$ MHz/T for the ^{31}P nucleus, one obtains the so-called transferred hyperfine field $H_{tr} = \Delta H - H_{dip}$. From angular dependence and the magnitude of the transferred hyperfine field, the authors concluded that a fraction of Np $5f$ spin moment is transferred chiefly to $3p$ orbitals of phosphorus. Evaluating the isotropic and anisotropic parts of the Knight shift in a similar fashion as in the case of UFe_4P_{12}, except explicitly considering the three anisotropic contributions $K_{ani}^1, K_{ani}^2,$ and K_{ani}^3 and plotting them as a function of the magnetic susceptibility, Figure 3.115(a). The slopes resulted in the isotropic hyperfine coupling constant $A_{iso} = 0.0887$ T/μ_B and the anisotropic coupling constants $A_{ani}^{1,2,3} = (-0.051, -0.0491,$ and $+0.1001)$ T/μ_B. Note that the isotropic Knight shift K_{iso} and the anisotropic K_{ani}^3 component both increase with the decreasing temperature, while the anisotropic Knight shift components K_{ani}^1 and K_{ani}^2 decrease. Hence, A_{iso} and A_{ani}^3 are positive as opposed to negative coupling constants A_{ani}^1 and A_{ani}^2. The temperature dependence of the relaxation rate $1/T_1$, Figure 3.115(b), documented a strong suppression of FM fluctuations by the applied field. This finding offers an explanation for a large negative magnetoresistance seen in the transport measurements by Aoki, Y. et al. (2006) at temperatures between 20 K and 80 K. Specifically, by forcing a progressively greater alignment of local magnetic moments of Np, the increasing magnetic field dramatically weakens magnetic scattering, giving rise to negative magnetoresistance.

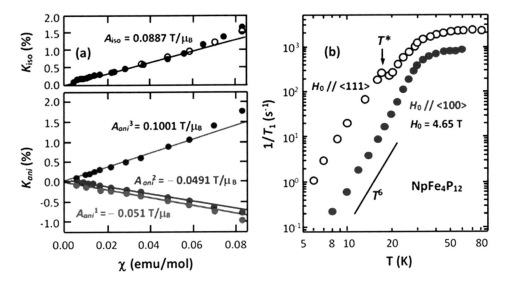

FIGURE 3.115 (a) Isotropic (top panel) and anisotropic (bottom panel) Knight shifts measured in $NpFe_4P_{12}$ and plotted as a function of the magnetic susceptibility. Slopes of the lines are isotropic and anisotropic hyperfine coupling constants with the values indicated in the figure. (b) Temperature dependence of the nuclear spin-lattice relaxation rate $1/T_1$ in a magnetic fields of 4.65 T applied parallel to <111> (black open circles) and parallel to <100> (solid blue circles). Note a rapidly falling relaxation rate below about 30 K associated with dramatically decreased magnetic scattering. Adapted and redrawn from Y. Tokunaga et al., *Physical Review B* **79**, 054420 (2009). With permission from the American Physical Society.

3.3.5 Magnetic Properties of Skutterudites with the $[Pt_4Ge_{12}]$ Framework

As first reported by Bauer et al. (2007) and almost simultaneously by Gumeniuk et al. (2008), it is possible to form a skutterudite cage entirely of Ge atoms, avoiding the volatile and often toxic pnicogen elements X = P, As, and Sb. It is achieved by forming a $[Pt_4Ge_{12}]$ framework, where the dominant 4p states of Ge combine with 5d states of Pt. Just as in the pnicogen-based skutterudites, where the role of a filler is to electronically stabilize the $[T_4X_{12}]$ polyanion, here, too, alkaline-earth metals or rare-earth species take on the task of stabilizing the $[Pt_4Ge_{12}]$ polyanion. Because the "radius" of the Ge-based cage is considerably smaller than that formed by Sb, the filler ions have much less space in which to move, are more strongly bonded to the Ge atoms, and consequently, many filler species in $[Pt_4Ge_{12}]$-based skutterudites show ADPs similar to other atoms of the structure. In other words, such fillers display no "rattling" behavior. Several of the synthesized APt_4Ge_{12} skutterudites were shown to be superconductors, and their SC properties are described in Chapter 2. Here, I focus mostly on $[Pt_4Ge_{12}]$-based skutterudites that have either magnetic ground state or do not become superconductors. Magnetic properties of APt_4Ge_{12} skutterudites are summarized in Table 3.17.

3.3.5.1 Ce-Filled $[Pt_4Ge_{12}]$ Framework

$CePt_4Ge_{12}$ was originally synthesized by Gumeniuk et al. (2008) with the lattice parameter $a = 8.6156$ Å and described as a non-magnetic skutterudite with a fluctuating valence. Subsequently, Toda et al. (2008) studied the structure down to 0.37 K also with no signs of magnetic ordering. Of the non-SC skutterudites with the $[Pt_4Ge_{12}]$ framework, $CePt_4Ge_{12}$ samples had the smallest RRR = 24. Although showing a metallic temperature dependence of resistivity, the $\rho(T)$ developed a strong "bowing" seen in many skutterudites. Subtracting the phonon part of resistivity of non-magnetic $LaPt_4Ge_{12}$,

TABLE 3.17

Lattice Constants and Some Key Magnetic Parameters of Filled Skutterudites Having the [Pt₄Ge₁₂] Framework

Compound	a (Å)	Ground State	$T_c/T_C/T_N$ (K)	μ_{eff} (μ_B/ion)	θ_{CW} (K)	γ (mJ/mole-K²)	Reference
LaPt₄Ge₁₂	8.6235	SC	8.27	–	–	56–76	1–3
CePt₄Ge₁₂	8.6156	metal	–	2.51–2.69	–39 to –84	86–105	1,4–6
PrPt₄Ge₁₂	8.6111	SC	7.91	3.58–3.67	+8.1 to –21	45–109	1,2,4–11
NdPt₄Ge₁₂	8.6074	AFM	0.67	3.72	–31		1,5,12
SmPt₄Ge₁₂	8.6069	metal	–	–	–	92	13
EuPt₄Ge₁₂	8.6363	AFM	1.78	7.35–7.9	–11 to –16	220	1,11,12,14
ThPt₄Ge₁₂	8.5926	SC	4.62			35–40	15,16
UPt₄Ge₁₂	8.5887	metal	–			156	16
SrPt₄Ge₁₂	8.6601	SC	5.10	–	–	41	17
BaPt₄Ge₁₂	8.6928	SC	5.35	–	–	42	17

1. Gumeniuk et al. (2008), 2. Sharath Chandra et al. (2016), 3. Pfau et al. (2016), 4. Huang et al. (2014), 5. Toda et al. (2008), 6. Gumeniuk et al. (2011), 7. Zhang et al. (2013), 8. Maisuradze et al. (2009), 9. Maisuradze et al. (2010), 10. Jeon et al. (2016), 11. Jeon et al. (2017), 12. Nicklas et al. (2011), 13. Gumeniuk et al. (2010), 14. Grytsiv et al. (2008), 15. Kaczorowski & Tran (2008), 16. Bauer et al. (2008), 17. Bauer et al. (2007).

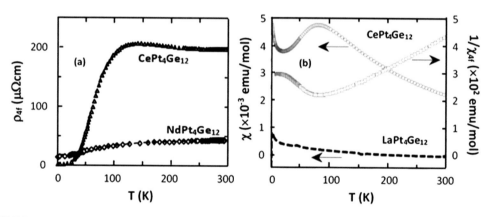

FIGURE 3.116 (a) Temperature dependence of the $4f$ contribution to the resistivity of CePt₄Ge₁₂ and NdPt₄Ge₁₂ obtained by subtracting the phonon part of the resistivity of LaPt₄Ge₁₂ from the measured resistivity of CePt₄Ge₁₂ and NdPt₄Ge₁₂. The dashed curve through the data of NdPt₄Ge₁₂ indicates a fit obtained assuming that the CEF ground state is a quartet $\Gamma_{67}^{(2)}$. (b) Magnetic susceptibility (left scale) and the inverse susceptibility of the $4f$ magnetic contribution of CePt₄Ge₁₂ obtained by subtracting the susceptibility of non-magnetic LaPt₄Ge₁₂. The broad maximum in the magnetic susceptibility near 80 K provided a firm hint that the structure might possess fluctuating valence state of Ce. Adapted from M. Toda et al., *Journal of the Physical Society of Japan* **77**, 124702 (2008). With permission of the Physical Society of Japan.

assuming its commonality with that of CePt₄Ge₁₂, a contribution due to the $4f$ electrons $\rho_{4f}(T)$ was obtained and is shown in Figure 3.116(a). It initially increases as the temperature decreases down to 150 K, develops a broad peak near 130 K, and rapidly decreases thereafter, dropping below the $\rho_{4f}(T)$ value of NdPt₄Ge₁₂ at temperatures lower than about 40 K. Invoking a similarity with the behavior of $CeFe_4Sb_{12}$ and $CeRu_4Sb_{12}$, Toda et al. (2008) postulated that the rapid decrease of $\rho_{4f}(T)$ observed in $CePt_4Ge_{12}$ might be a sign of a developing coherence in Kondo scattering.

Magnetic susceptibility of $CePt_4Ge_{12}$ is depicted in Figure 3.116(b) and displays a prominent peak near 80 K (a feature common to compounds with intermediate valence), a minimum at about 20 K, and an upturn at yet lower temperatures. From the Curie–Weiss behavior above 200 K, the effective magnetic moment of $\mu_{eff} = 2.58\ \mu_B$/f.u. and $\theta_{CW} = -62$ K were extracted. The effective moment is very close to the free-ion moment of Ce^{3+} of 2.54 μ_B.

Toda et al. (2008) also collected the ^{195}Pt-NMR spectra on $CePt_4Ge_{12}$. From the spectra, they assessed the ^{195}Pt Knight shift dominated by the 4f-based susceptibility via the transferred hyperfine interaction that has its origin in the hybridization of Ce 4f electrons with the conduction electrons. They also evaluated the nuclear spin-lattice relaxation rate $1/T_1$. Again, the Knight shift components parallel and perpendicular to the principal axis plotted against $\chi(T)$ yielded a linear relation with the slopes related to the isotropic and anisotropic hyperfine fields at the ^{195}Pt nucleus, $A_{iso} = 0.78$ T/μ_B and $A_{ani} = 0.03$ T/μ_B, documenting that the Knight shift in $CePt_4Ge_{12}$ is substantially isotropic. From the exponential temperature dependence of the nuclear spin-lattice relaxation rate $1/T_1$ above 100 K, the authors concluded the existence of an energy gap $\Delta/k_B = 150$ K. At temperatures below 20 K, the relaxation rate became proportional to the temperature. The microscopic view of $CePt_4Ge_{12}$ is thus consistent with the presence of a pseudogap arising due to hybridization between Ce 4f and conduction electrons (c-f hybridization).

The often-claimed fluctuating valence state of Ce in $CePt_4Ge_{12}$ was looked at more closely by Gumeniuk et al. (2011), who, in addition to transport and magnetic studies on samples with RRR ≈ 100, actually made X-ray absorption near edge structure (XANES) measurements to determine the valence state of Ce. Notable transport and magnetic results included a quadratically varying electrical resistivity between 0.35 K and 10 K, reflecting the Fermi liquid character of the transport, magnetization studied in fields as high as 60 T (pulsed magnets), where the magnetization reached only 0.37 μ_B, and a failure of the Coqblin-Schrieffer model of Rajan (1983) (appropriate for dilute rare-earth impurities, such as Ce^{3+}) to account for the behavior of the magnetic susceptibility. From measurements of the specific heat, the authors extracted a moderately enhanced Sommerfeld coefficient of the electronic specific heat $\gamma = 105$ mJmol^{-1}K^{-2}, the Debye temperature $\theta_D = 206$ K, and concluded that the specific heat data are consistent with a ground state doublet separated by a gap of $\Delta/k_B = 180$ K from a quartet constituting the first excited state. Although all transport and magnetic measurements faithfully reproduced previous studies, apparently validating the fluctuating valence of Ce that was believed to govern the low-temperature transport and magnetic properties of this skutterudite, XANES measurements invariably documented that the valence of cerium is none other than trivalent over a broad range of temperatures from 300 K down to 10 K. However, as it is expected that a transition from the Kondo lattice behavior to the fluctuating valence state takes place when the valence of Ce is larger than 3.10, it does not have to be a very large deviation from the trivalent Ce state for a system to acquire fluctuating valence characteristics. ^{195}Pt-NMR studies on $CePt_4Ge_{12}$ by Baenitz et al. (2010) affirmed the trivalent state of cerium at high temperatures, but a maximum on the perpendicular Knight shift $K\perp$ near 60 K, and its saturation below 10 K, was interpreted as arising from an unstable Ce valence that shifted from Ce^{3+} toward Ce^{4+}. The Knight shift result was consistent with the temperature dependence of the nuclear spin-lattice relaxation rate $1/T_1$ of the ^{195}Pt nucleus, which followed the Korringa relation ($1/T_1T$ = const) below 60 K. Such dependence of the relaxation rate was nearly identical to the behavior of $ThPt_4Ge_{12}$, where Th is clearly tetravalent, but at higher temperatures deviated notably and merged with the relaxation rate of $LaPt_4Ge_{12}$, where La is trivalent. Galéra et al. (2015) applied INS in an attempt to resolve whether $CePt_4Ge_{12}$ is a valence fluctuating system or rather a Kondo lattice structure. The INS spectra were dominated by phonon contributions, including clearly resolved rattling modes and evidenced a complete lack of crystal field excitations. This was interpreted as being inconsistent with the Kondo behavior and more in favor of the fluctuating valence state of Ce.

While $CePt_4Ge_{12}$ remains paramagnetic down to the lowest temperatures and displays typical Fermi liquid characteristics, replacing some Ge atoms with Sb opens the door to a plethora of interesting electronic structures. This is due to a progressively larger unit cell, and the fact that Sb brings with it an extra electron that tunes the electronic properties. Nicklas et al. (2012) has shown that there is a surprisingly wide range of stability of $CePt_4Ge_{12-x}Sb_x$ extending up to $x = 3$. In the process of increasing x, the electronic structure progressively evolves. At low Sb content $x \leq 0.5$, the structure maintains its Fermi liquid character. As the amount of substituted Sb increases, the electrical resistivity increases and acquires a logarithmic temperature dependence at the lowest temperatures, i.e., a NFL regime is reached. At concentrations $x \geq 1.5$, a peak appears on $\rho(T)$ that increases and shifts to higher temperatures as x increases, Figure 3.117(a). The resistivity attains its largest magnitude at a concentration $x = 1.9$, at which point a distinct peak anomaly develops at $T_N = 0.89$ K on the specific heat, as the structure undergoes AFM ordering. This proceeds in a delicately balanced fashion where on the one hand local $4f$ moments start to develop but are rather effectively screened by the Kondo effect. On the other hand, Kondo correlations are being progressively weakened by the expansion of the lattice and by the disorder introduced by Sb substituting for Ge. As the content of Sb increases beyond x = 1.9, the Néel temperature increases, and at its highest Sb concentration $x = 3$, shifts to $T_N = 1.46$ K, representing a robust AFM state, Figure 3.117(b). As the evolution from $CePt_4Ge_{12}$ to $CePt_4Ge_9Sb_3$ proceeds, the Sommerfeld coefficient γ undergoes an interesting change from a modestly enhanced value of 105 mJmol^{-1}K^{-2} to a very large $\gamma = 1.74$ Jmol^{-1}K^{-2} at $x = 1.5$, documenting a strongly correlated ground state and then dropping back for $x \geq 1.9$ on account of weakened Kondo correlations. The $CePt_4Ge_{12-x}Sb_x$ presents an interesting system that evolves from a borderline case of an intermediate valence paramagnet ($CePt_4Ge_{12}$) to a NFL structure ($x \sim 1.5$) and eventually to a robust antiferromagnet with localized 4f moments of Ce ($x > 1.9$). The above evolution of the system as the substituted content of Sb increases leaves a distinct mark on the behavior of the Seebeck coefficient, as evidenced by measurements of White et al. (2014) shown in Figure 3.118. In fact, such Seebeck coefficient measurements might be even more revealing of the state of a system than measurements of the specific heat and do not even require studies at sub-Kelvin temperatures.

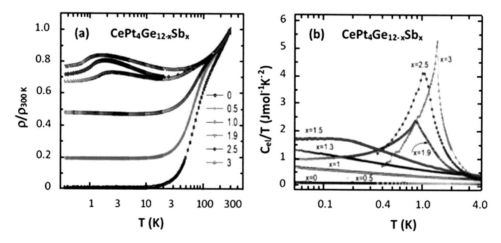

FIGURE 3.117 (a) Temperature dependence of the reduced resistivity of $CePt_4Ge_{12-x}Sb_x$ with different contents of Sb. The development of peak signals AFM ordering. (b) Electronic specific heat of $CePt_4Ge_{12-x}Sb_x$ plotted as C_{el}/T vs. T at low temperatures. Sharp peaks indicate AFM ordering in structures with $x \geq 1.9$. Adapted from M. Nicklas et al., *Physical Review Letters* **109**, 236405 (2012). With permission from the American Physical Society.

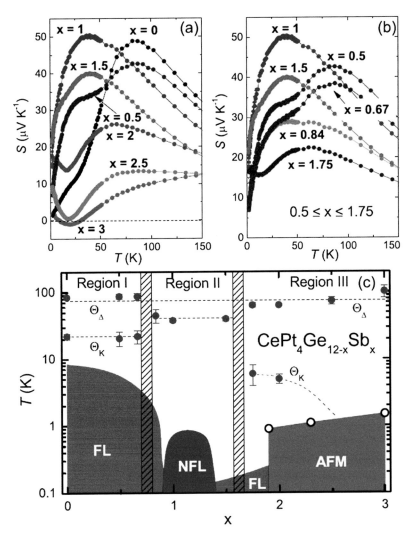

FIGURE 3.118 (a) Thermoelectric power S *vs* temperature T for $CePt_4Ge_{12-x}Sb_x$. A dashed line indicates $S(T) = 0 \ \mu V \ K^{-1}$. (b) S *vs.* T for $0.5 < x < 1.75$. (c) $T - x$ phase diagram for $CePt_4Ge_{12-x}Sb_x$. The Fermi-liquid, NFL, and AFM regions, explicitly labeled and colored red, blue, and green, respectively, were reported originally by Nicklas et al. (2012). The AFM phase boundary should go to zero temperature at a concentration in the range $1.5 < x < 1.9$. Striped bands separate regions corresponding to distinct $S(T)$ behavior. Blue circles represent the maxima of features observed in $S(T)$ at temperatures near 85 K and near 20 K, respectively. The crossover from two distinct features on $S(T)$ to a single feature in the range of Sb susbtitutions $0.84 \le x < 1.5$ coincides with the onset of NFL behavior driven by strong Kondo disorder. Dashed lines are guides to the eye. Reproduced from B. D. White et al., *Physical Review B* **90**, 235104 (2014). With permission from the American Physical Society.

3.3.5.2 Nd-Filled [Pt₄Ge₁₂] Framework

At the same time $CePt_4Ge_{12}$ was synthesized, Gumeniuk et al. (2008) also prepared $NdPt_4Ge_{12}$ and $EuPt_4Ge_{12}$. The lattice constant of $NdPt_4Ge_{12}$ was $a = 8.6074$ Å, and the structure was reported to undergo AFM ordering below 0.67 K. A more detailed study by Toda et al. (2008) revealed a metallic temperature dependence of resistivity and a quite high residual resistance ratio RRR = 100. At temperatures below 0.7 K, the resistivity dropped rapidly as the sample passed through the AFM ordering temperature. Accounting for a contribution of the 4*f* electrons, the authors used the same approach as in the case of $CePt_4Ge_{12}$, namely subtracted the resistivity due to lattice vibrations by taking it the same as in non-magnetic $LaPt_4Ge_{12}$. The resulting $\rho_{4f}(T)$, shown in Figure 3.116(a), is much smaller than in the case of $CePt_4Ge_{12}$ and gradually decreases with decreasing temperature, likely the consequence of decreasing magnetic scattering associated with the CEF excitations.

Magnetic susceptibility of $NdPt_4Ge_{12}$ is depicted in Figure 3.119(a) and shows a rapidly rising magnetization below about 50 K. The figure also depicts the temperature dependence of $\chi_{4f}^{-1}(T)$, a magnetic contribution due to the $4f$ electrons of Nd obtained by subtracting from $\chi(T)$ of $NdPt_4Ge_{12}$ the susceptibility of non-magnetic $LaPt_4Ge_{12}$ and taking the inverse. From the straight line section above 100 K, the effective magnetic moment $\mu_{eff} = 3.72$ μ_B/f.u. and $\theta_{CW} = -31.1$ K followed. Again, the effective magnetic moment was quite close to the Hund's rule free-ion Nd^{3+} value of 3.62 μ_B. The close agreement between the experimental and theoretical effective moments indicates that the $4f$ electrons are rather strongly localized in this skutterudite. Because the Nd^{3+} ion is ten-fold degenerate, having $J = 9/2$, its multiplet gets split by CEF into a doublet Γ_5 and two quartets $\Gamma_{67}^{(1)}$ and $\Gamma_{67}^{(2)}$, assuming the local T_h symmetry. Simplifying the analysis of $NdPt_4Ge_{12}$ by assuming the cubic O_h symmetry, i.e., neglecting the last term in Equation 3.84, the CEF calculations returned two possible scenarios for the ground state: either a $\Gamma_{67}^{(2)}$ quartet or a Γ_5 doublet. A fit of $\rho_{4f}(T)$ and $1/\chi_{4f}(T)$ with the ground state quartet $\Gamma_{67}^{(2)}$ is shown in Figure 3.116(a) and Figure 3.119(a), respectively. A clear differentiation between the schemes with the ground state quartet and the ground state doublet came from magnetization measurements in Fig, 3.119(b), where the $\Gamma_{67}^{(2)}$ quartet provides a distinctly better alternative. This was reaffirmed by subsequent measurements of the magnetic susceptibility and the specific heat by Nicklas et al. (2011), which specifically focused on the low-temperature behavior and extended down to below 0.5 K. Instead of an anticipated single anomaly in the specific heat at the ordering temperature, two closely-lying sharp peaks were observed at $T_{N1} = 0.67$ K and $T_{N2} = 0.58$ K, indicating a more complex ordering than expected. The higher temperature peak was in excellent agreement with the originally reported onset of a rapidly decreasing resistivity at 0.67 K by Gumeniuk et al. (2008). Magnetic entropy extracted from the specific heat data reached $R\ln 4$ at 4 K, providing solid evidence for the ground state being the quartet $\Gamma_{67}^{(2)}$. The rapidly rising magnetic susceptibility concurred with the data in Figure 3.119(a). However, being able to attain much lower temperatures, the authors were able to observe a large peak near 0.7 K, followed by rapidly decreasing $\chi(T)$ with a notable change of slope near 0.6 K. As is usual in the case of AFM ordering, the transition temperature is taken as the onset of the decreasing susceptibility, and it coincided with

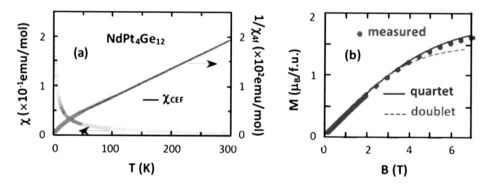

FIGURE 3.119 (a) Magnetic susceptibility $\chi(T)$ (left scale) and inverse magnetic susceptibility of the $4f$ electron contribution $\chi_{4f}^{-1}(T)$ (right scale) of $NdPt_4Ge_{12}$. (b) Field dependence of magnetization. Experimental data are shown by red solid circles, and fitting using CEF schemes with the ground state quartet is shown by a solid blue curve, and with the ground state doublet is shown by a green dashed curve. Adapted and redrawn from M. Toda et al., *Journal of the Physical Society of Japan* **77**, 124702 (2008). With permission from the Physical Society of Japan.

T_{N1} obtained from specific heat measurements. The slope change in the susceptibility agreed with T_{N2}. The magnetization was measured at 0.5 K and was essentially linear up to near 4 T, followed by a tendency to saturation. Already at 7 T it reached 1.88 μ_B/f.u., a value significantly larger than the saturation moment expected for the ground state doublet Γ_5 (1.33 μ_B/f.u.), providing further support for the $\Gamma_{67}^{(2)}$ quartet being the ground state.

3.3.5.3   Sm-Filled [Pt$_4$Ge$_{12}$] Framework

The smallest "trivalent" rare earth with which the [Pt$_4$Ge$_{12}$] framework was successfully filled is Sm. The task was accomplished by Gumeniuk et al. (2010), who had to use a high pressure of 5 GPa generated by an octahedral multi-anvil press and a temperature of 1070 K. The resulting SmPt$_4$Ge$_{12}$ had the lattice constant a = 8.6069 Å. Powder X-ray analysis indicated a large ADP of Sm of 1.80 Å^2, in comparison to that of Pt (0.685 Å^2) and Ge (1.01 Å^2). Apparently, a small Sm ion had enough space to "rattle" even in the smaller icosahedral cage of Ge. Energy dispersive spectroscopy indicated a nearly single-phase skutterudite with a small amount of impurities (2% of PtGe$_2$ and 2% of Ge). It should be noted that the window of useful pressure is apparently quite limited, as the pressure of 4 GPa resulted in a much larger content of impurities, while the synthesis at pressures above 5 GPa was not successful at all.

As noted in section 3.3.4.5.3, the valence of Sm might deviate from being strictly trivalent and might even be temperature-dependent, as in SmOs$_4$Sb$_{12}$. Therefore, it was essential to ascertain what happens to it in the environment of [Pt$_4$Ge$_{12}$]. Careful X-ray absorption spectroscopy, monitoring the Sm L_3 edge over a broad range of temperatures down to 5 K, indicated a small admixture of the $4f^6$ state, i.e., a contribution of divalent Sm^{2+} ions, for an average valence of about 2.75 in SmPt$_4$Ge$_{12}$.

Electrical resistivity of SmPt$_4$Ge$_{12}$ is metallic throughout, but a sublinear temperature dependence above about 10 K extends up to 300 K. With $\rho_{300\ K} \approx 100\ \mu\Omega$cm and the residual resistivity $\rho_0 \approx 27\ \mu\Omega$cm, the resistivity ratio RRR = 3.7 is quite small. Below 10 K, a broad shoulder develops and the resistivity decreases faster. At liquid helium temperatures, the resistivity attains a quadratic temperature dependence, $\rho(T) = \rho_0 + AT^2$, with the coefficient A = 0.138 $\mu\Omega$cmK^{-2}. SmPt$_4$Ge$_{12}$ thus displays a Fermi-liquid behavior.

The rising magnetic susceptibility as the temperature decreases shows a typical van Vleck characteristics given its ^{6}H$_{5/2}$ ground state multiplet with the participation of thermal excitations in the first excited ^{6}H$_{7/2}$ multiplet of Sm^{3+} corresponding to the total angular momentum J = 7/2. Consequently, there is no interval on the $\chi^{-1}(T)$ vs. T curve where the Curie–Weiss behavior holds. As discussed in section 3.3.4.5, the ground state multiplet is split by the CEF into a doublet Γ_5 (Γ_7) having the purely magnetic degree of freedom and a quartet Γ_{67} (Γ_8) having both magnetic and electronic degrees of freedom. The states are specified using the T_h point group symmetry with the O_h symmetry designation given in the brackets. There are no anomalies discerned on the magnetic susceptibility down to 0.5 K that might indicate some form of magnetic ordering taking place, and the magnetization measured at 1.8 K in the field of 7 T yielded only a small moment of 0.052 μ_B/f.u. with no signs of saturation. The value of the magnetic moment is much smaller than the saturated moment of 0.238 μ_B expected for the ground state doublet.

More insight into the relevant CEF scheme was provided by specific heat measurements. As we have seen in the analysis of most of the non-magnetic filled skutterudites, the specific heat consists of the Debye contribution from phonons of the framework, $C_{[Pt4Ge12]}$, the Einstein contribution from weakly bonded filler ions, C_{filler}, and the electronic contribution characterized by the Sommerfeld coefficient γ. If the structure is magnetic, the specific heat of the isostructural but non-magnetic skutterudite is subtracted to reveal the $4f$ contribution, C_{4f}. A plot of the combined $4f$ and electronic contributions to the specific heat below 10 K,

$$C_{4f} + C_{el} = C_{exp} - C_{\left[Pt4Ge12 \right]} - C_{filler}, \tag{3.112}$$

in the form $(C_{4f} + C_{el})/T$ vs. T is shown in Figure 3.120. It indicates a broad peak at 2.9 K and a hump near 4.4 K. The fit, employing the term γT and the Schottky-like two-level contribution representing

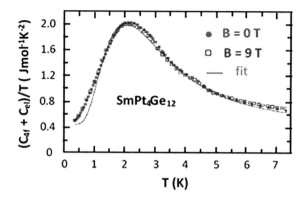

FIGURE 3.120 Plot of $(C_{4f} + C_{el})/T$ *vs. T* for $SmPt_4Ge_{12}$ below 7 K in zero magnetic field (solid red circles) and in the field of 9 T (open blue squares). The solid blue line indicates a fit using a linear electronic term and a two-level Schottky term to simulate a two-level CEF scheme. Note an insensitivity of the specific heat to the effect of the applied magnetic field. Adapted and redrawn from R. Gumeniuk et al., *New Journal of Physics* **12**, 103035 (2010). Deutsche Physicalische Gesellschaft. With permission from IOP Publishing.

the CEF level scheme, is shown by a solid blue curve. The fit returned a large Sommerfeld coefficient $\gamma = 450$ mJmol^{-1}K^{-2} and the two levels separated by $\Delta E/k_B = 6.6$ K. Although not reaching the twice as large value of γ observed in $SmOs_4Sb_{12}$, such a large electronic specific heat places $SmPt_4Ge_{12}$ in a category of heavy fermion systems. Considering the evolution of the magnetic entropy and its value below 6 K swayed the authors to conclude that the ground state in $SmPt_4Ge_{12}$ is the doublet Γ_5. Regarding the heavy fermion nature of both $SmOs_4Sb_{12}$ and $SmPt_4Ge_{12}$ and its remarkable insensitivity to the magnetic field, there is considerable debate with respect to the origin of the heavy fermion state. Among the possibilities are the intermediate valence of Sb, hybridization between $4f$ electrons and conduction electrons, and the proposed enhanced hybridization induced by a rattling filler (Sm ion) with a very low Einstein temperature of $\theta_E = 40$ K, as described by Hotta (2008).

3.3.5.4 Eu-Filled [Pt₄Ge₁₂] Framework

$EuPt_4Ge_{12}$ with the lattice constant $a = 8.6363$ Å was briefly mentioned by Gumeniuk et al. (2008) as undergoing AFM ordering at 1.7 K. A more detailed account of the magnetic properties of this skutterudite came from studies by Grytsiv et al. (2008), which included both experimental work and DFT-based electronic structure calculations. Assuming a divalent state of europium (indirectly affirmed by a very similar volume of $EuPt_4Ge_{12}$ with that of $SrPt_4Ge_{12}$, where Sr is unquestionably divalent), the magnetic Eu^{2+} ion has a total angular momentum $j = s = 7/2$, i.e., no orbital moment contributes to its total angular momentum and, consequently, there is no participation of CEF effects to magnetic properties of $EuPt_4Ge_{12}$. An extensive Curie–Weiss range of $\chi^{-1}(T)$ yielded a large effective moment $\mu_{eff} = 7.35$ μ_B/Eu^{2+} and a negative $\theta_{CW} = -11.5$ K. The effective magnetic moment is somewhat smaller than the theoretical Eu^{2+} free-ion value of 7.9 μ_B, possibly due to a small fraction of non-magnetic Eu^{3+} ions present or perhaps a less than full occupancy of the voids. Magnetization is essentially a linear function of the applied field (consistent with the absence of CEF effects) with no hint of saturation up to 6 T. A sign that the structure undergoes magnetic ordering came originally from the low-temperature electrical resistivity shown in Figure 3.121(a), and the trend in the magnetoresistance is shown in Figure 3.121(b). The missing evidence of magnetic ordering in $EuPt_4Ge_{12}$ was provided by subsequent measurements of the specific heat and magnetic susceptibility by Nicklas et al. (2011) that extended to below 0.5 K. The authors also directly confirmed the divalent state of europium by using XAS. Specific heat and magnetic susceptibility are depicted in Figures 3.122(a) and 3.122(b). At least three anomalies are identified in the specific heat below 2 K. The transition at $T_{N1} = 1.78$ K closely matches the previously reported onset of a rapidly

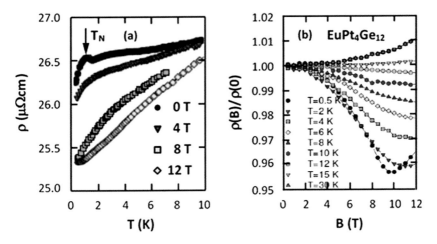

FIGURE 3.121 (a) Low-temperature electrical resistivity of $EuPt_4Ge_{12}$ in zero field and fields indicated. The Néel temperature is indicated by an arrow. (b) Normalized isothermal magnetoresistance of $EuPt_4Ge_{12}$ at various temperatures. Note a minimum on the two curves at the lowest temperatures (0.5 K and 2 K) believed to indicate a reorientation of the ordered AFM spin structure. At high temperatures, above 12 K, the magnetoresistance attains the classical positive values. Adapted and redrawn from A. Grytsiv et al., *Journal of the Physical Society of Japan* **77**, Suppl. A, 121 (2008). With permission from the Physical Society of Japan.

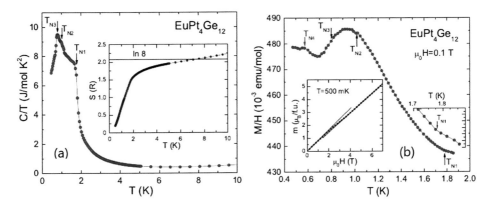

FIGURE 3.122 (a) Specific heat of $EuPt_4Ge_{12}$ plotted as C/T vs. T al low temperatures. Note three peaks revealing three ordering temperatures, $T_{N1} = 1.78$ K, $T_{N2} = 1.03$ K, and $T_{N3} = 0.8$ K. The inset shows the magnetic entropy associated with the $j = 7/2$ ground state multiplet expected to reach $Rln8$. (b) Temperature dependence of magnetic susceptibility of $EuPt_4Ge_{12}$ at low temperatures. Ordering temperatures obtained from the specific heat measurements are marked on the curve of $\chi(T)$. The reader might note that a distinct anomaly marked as T_{N1} on the specific heat leaves hardly any mark on the magnetic susceptibility. Moreover, the transition at T_{N2} on the specific heat shows up as a barely notable anomaly on $\chi(T)$ and requires magnification, such as provided in the right-hand-side inset where it appears as a change of slope. The main inset shows a substantially linear magnetization with a barely visible step near 2.5 T, perhaps indicating some reorientation of the magnetic moments. Reproduced from M. Nicklas et al., *Journal of Physics: Conference Series* **272**, 012118 (2011). With permission from the Institute of Physics and the first author.

decreasing resistivity, but leaves no mark on the magnetic susceptibility shown in Figure 3.122(b). The transition at $T_{N2} = 1.03$ K is a barely notable feature on the C/T vs. T plot and also on the $\chi(T)$ vs. T dependence and requires a closer inspection, such as provided by an expanded form of the plot in the inset of Figure 3.122(b), where it shows as a change in slope. A well-developed sharp peak at $T_{N3} = 0.8$ K maps on the susceptibility curve a bit below its maximum. There might also be yet another transition, showing as a small peak on $\chi(T)$ below 0.6 K and marked as T_{N4}, which, on the specific heat curve, appears as a change in the slope. Clearly, the richness of the phases is far greater than one might have assumed, and their detailed assessment requires further investigations, especially focusing on measurements of a single crystal form of the structure.

In an interesting study by Jeon et al. (2017), the authors explored a crossover from the unconventional, TRS-breaking $PrPt_4Ge_{12}$ superconductor to an AFM $EuPt_4Ge_{12}$ skutterudite when Eu substitutes at the site of Pr and a full range of $Pr_{1-x}Eu_xPt_4Ge_{12}$ solid solutions with $0 \leq x \leq 1$ is formed. SC aspects of this work, with the SC state extending to at least $x = 0.5$ and perhaps even coexisting with the AFM-ordered state over a limited range of x values around 0.5, are described in section 2.6.3, where an overall phase diagram is also included as Figure 2.59. Here, I focus on the magnetically ordered state of the high Eu content solid solutions.

As the concentration x of Eu increases, the temperature-dependent electrical resistivity undergoes a change of slope at low temperatures from the initially Bloch-Grüneisen T^5 power law behavior for $x \approx 0$ to a weaker quadratic temperature dependence near $x = 0.5$, consistent with the Fermi liquid behavior, and then to a linear, i.e., NFL dependence for values of $x \approx 1$. The evolution of the power exponent n, the coefficient of the temperature term A_n, and the residual resistivity ρ_0 is depicted in Figs 3.123(a)–(c).

Magnetic susceptibility data were collected in a field of 0.1 T, and the extracted effective magnetic moment μ_{eff} and the Curie–Weiss temperature θ_{CW} as a function of the content x of Eu revealed a smooth variation, shown in Figures 3.124(a) and (b). The red line in Figure 3.124(a) is calculated based on Equation 3.113

$$\mu_{eff}(x) = \left[\left(\mu_{Pr^{3+}} \right)^2 (1-x) + \left(\mu_{Eu^{2+}} \right)^2 x \right]^{\frac{1}{2}}, \tag{3.113}$$

and agrees well with the experimental data.

In discussions of magnetic properties of isovalent $EuFe_4Sb_{12}$ in section 3.3.4.6.3, it was noted that this skutterudite ordered ferrimagnetically at quite a high temperature of 84 K. The difference between $EuFe_4Sb_{12}$ and $EuPt_4Ge_{12}$ is the framework structure of the former skutterudite, which is nearly FM and couples antiferromagnetically to the moments of Eu^{2+}, while the $[Pt_4Ge_{12}]$ framework is non-magnetic. This suggests that in $EuPt_4Ge_{12}$ the spin-only moments of Eu^{2+} are subject to strong fluctuations that are not stabilized because the polyanion has no magnetic moment and, consequently, the magnetic order is easily wiped out by thermal fluctuations.

3.3.5.5 Actinide-Filled [Pt$_4$Ge$_{12}$] Framework

The first actinide-filled $[Pt_4Ge_{12}]$ framework, $ThPt_4Ge_{12}$ with the lattice constant $a = 8.5924$ Å, was synthesized by Kaczorowski and Tran (2008). Unlike ThT_4X_{12} skutterudites in either phosphide or

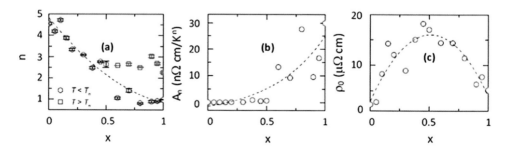

FIGURE 3.123 Parameters of the temperature-dependent electrical resistivity $\rho(T) = \rho_0 + A_n T^n$ of solid solutions of $Pr_{1-x}Eu_xPt_4Ge_{12}$. (a) Exponent n, (b) coefficient A_n, and (c) the residual resistivity ρ_0. Adapted from I. Jeon et al., *Physical Review B* **95**, 134517 (2017). With permission from the American Physical Society.

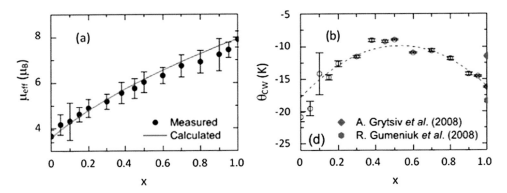

FIGURE 3.124   (a) Effective magnetic moment of $Pr_{1-x}Eu_xPt_4Ge_{12}$ solid solutions as a function of the content x of Eu. The red curve was calculated based on Equation 3.113. (b) The Curie–Weiss temperature θ_{CW} as a function of the Eu content. The dashed curve is a guide for the eyes. Adapted from I. Jeon et al., *Physical Review B* **95**, 134517 (2017). With permission from the American Physical Society.

arsenide forms that remained paramagnetic down to the lowest temperatures, $ThPt_4Ge_{12}$ became a superconductor upon cooling below 4.62 K. Its SC properties are discussed in section 2.6.2.

Single crystalline UPt_4Ge_{12} with the lattice constant $a = 8.5887$ Å was reported on by Bauer et al. (2008). In contrast to $ThPt_4Ge_{12}$, the uranium-filled cousin displayed neither magnetic ordering nor superconductivity to the lowest temperature of the experiment. Even though the icosahedral cage formed by Ge atoms is considerably smaller, a rather small U ion apparently had plenty of space to develop a significantly larger ADP in comparison to that of both Pt and Ge. The distinction between the SC ground state of $ThPt_4Ge_{12}$ and the paramagnetic state of UPt_4Ge_{12} left a clear imprint on the temperature dependence of resistivity, Figure 3.125(a), which showed a considerably more pronounced "bowing" and perhaps even a tendency to saturation above 300 K. The actual room temperature resistivities were $\approx 190\ \mu\Omega$cm for UPt_4Ge_{12} and $\approx 145\ \mu\Omega$cm for $ThPt_4Ge_{12}$. At low

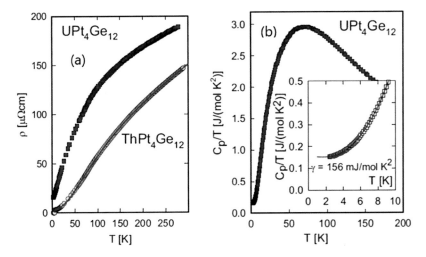

FIGURE 3.125   (a) Temperature dependence of the electrical resistivity of UPt_4Ge_{12} shown in comparison to that of $ThPt_4Ge_{12}$. (b) Specific heat of UPt_4Ge_{12} plotted as C_p/T *vs.* T. The inset shows the behavior at low temperatures together with a least-squares fit based on a spin fluctuation model, Equation 3.114, which returned the Sommerfeld coefficient of 156 mJmol^{-1}K^{-2}. Reproduced from E. Bauer et al., *Physical Review B* **78**, 064516 (2008). With permission from the American Physical Society.

temperatures, the resistivity of UPt_4Ge_{12} attained the form $\rho(T) = \rho_0 + AT^n$, with $\rho_0 = 14.5\ \mu\Omega cm$, $A = 0.42\ \mu\Omega cmK^{-1.5}$, and $n = 1.5$, i.e., a distinctly NFL behavior. Specific heat measurements, plotted in Figure 3.125(b) as C/T vs. T, depicted a smooth variation with no indication of any phase transition, in accordance with the resistivity data. Treating UPt_4Ge_{12} as a simple Debye solid, a least-squares fit of the form

$$C\left(T\right) = \gamma\,T + \beta\,T^3 + \delta\,T^3 \ln\!\left(T\!\Big/_{T^*}\right),\qquad\qquad (3.114)$$

shown in the inset of Figure 3.125(b), with the last term standing for spin fluctuations expected to play a role in this NFL structure, returned a rather large Sommerfeld coefficient of the electronic specific heat $\gamma = 156\ mJmol^{-1}K^{-2}$ and the temperature $T^* = 2.7$ K. Although a reliable value of β, i.e., of the Debye temperature θ_D, could not be obtained from the fit, the observed temperature dependence of the ADP parameter of uranium allowed an estimate of the Einstein temperature $\theta_E = 59$ K. DFT calculations performed by the authors indicated the DOS at the Fermi level of UPt_4Ge_{12} of 16.21 states $eV^{-1}f.u.^{-1}$, much larger than 9.63 states $eV^{-1}f.u.^{-1}$ in the case of $ThPt_4Ge_{12}$. A much smaller Sommerfeld coefficient $\gamma_{calc} = 38.2\ mJmol^{-1}K^{-2}$ calculated from such DOS in comparison to its experimental value $\gamma_{exp} = 156\ mJmol^{-1}K^{-2}$ supported the argument that spin fluctuations are the dominant feature in UPt_4Ge_{12} at low temperatures.

The discovery of $[Pt_4Ge_{12}]$-based skutterudites was a surprising and welcome broadening of structures with the $Im\bar{3}$ skutterudite lattice. The purely Ge-based icosahedral cages are smaller than the cages formed by Sb atoms, and the filler species, particularly the larger Sr and Ba ions that stabilize the structure electronically, are more tightly bonded to the framework, resulting in ADPs that are more-or-less similar for all atoms of the structure. Apart from this notable difference *vis-à-vis* large ADPs of fillers in the pnicogen-based framework, there is also a significant difference in the electronic structure of $[Pt_4Ge_{12}]$-based and $[T_4X_{12}]$-based skutterudites. Electronic states near the Fermi energy in the former skutterudite structure are governed by the dominant Ge-$4p$ states with only a small participation of Pt-$5d$ states and no influence of the filler's states. While a rather weak hybridization of Ge-$4p$ and Pt-$5d$ states brings into play relativistic spin-orbit coupling, which may tune the peak in the DOS to a near coincidence with the Fermi level, the hybridization is nowhere near the strength typical in pnicogen-based skutterudites. Moreover, the distinctly directional Ge-Ge bonding contrasting with a spherical (metallic) electron distribution around Pt atoms invalidates the Zintl concept, which was essential and useful when considering the stability of pnicogen-based filled skutterudites. Because the $[Pt_4Ge_{12}]$ framework is clearly non-magnetic, as opposed to some Fe-containing pnicogen frameworks, there is a stronger tendency for developing a SC ground state in the former rather than the latter structure, as documented by the case of Th-filled skutterudites. For the same reason, magnetic properties of APt_4Ge_{12} skutterudites depend solely on the magnetic moment of the filler species with no assistance from the polyanion lattice. This is the reason why one observes considerably weaker magnetism in $EuPt_4Ge_{12}$ skutterudites compared to $EuFe_4Sb_{12}$ structures, where the ordering temperatures of the pnicogen-based skutterudites are an order of magnitude higher than skutterudites having the $[Pt_4Ge_{12}]$ framework.

3.4 CONCLUDING REMARKS

As the readers have seen, magnetic properties of skutterudites cover a broad spectrum of magnetic orderings, reflecting interactions of the filler species with the framework atoms of the structure. While binary skutterudites, except for NiP_3, are diamagnetic substances, the presence of fillers in the crystalline lattice gives rise to paramagnetism and often a variety of ordered magnetic structures. In principle, magnetism in filled skutterudites arises due to moments associated with the framework and with the magnetic filler ions. Most of the skutterudite frameworks are nonmagnetic, and magnetism in skutterudites is then due to the magnetic moment associated with the filler ions.

Except for La, all rare-earth elements carry a magnetic moment and overwhelmingly dominate the magnetic response of the skutterudite. However, an interesting situation arises when Fe is part of the frameworks. There is strong experimental evidence for Fe in most of the phosphide and antimonide skutterudites to have a zero or insignificant magnetic moment. In contrast, experimental studies indicate that in arsenides the polyanion $[Fe_4As_{12}]$ does carry a magnetic moment and contributes to the overall magnetism of such filled skutterudites.

I have presented a rather detailed account of magnetism in skutterudites, and I trust the readers will find it useful in their studies of skutterudites.

NOTE

1   This is strictly true within the first-order perturbation theory. When the second-order perturbation theory is applied, all states above the ground state mix in with the ground state, and this leads to a small paramagnetic susceptibility given by $\chi_{vV} = \frac{2N\mu_B^2}{V} \sum_n \frac{\left|\langle 0|(L_z + gS_z)|n\rangle\right|^2}{E_n - E_0}$ and known as the van Vleck paramagnetism.

REFERENCES

Abe, K., H. Sato, T. D. Matsuda, T. Namiki, H. Sugawara, and Y. Aoki, *J. Phys.: Condens. Matter* **14**, 11757 (2002).

Ackermann, J. and A. Wold, *J. Phys. Chem. Solids* **38**, 1013 (1977).

Adroja, D. T., J.-G. Park, K. A. McEwen, N. Takeda, M. Ishikawa, and J.-Y. So, *Phys. Rev. B* **68**, 094425 (2003).

Adroja, D. T., J.-G. Park, E. Goremychkin, N. Takeda, M. Ishikawa, K. McEwen, R. Osborn, A. Hillier, and B. Rainford, *Physica B* **359–361**, 983 (2005).

Adroja, D. T., J.-G. Park, E. A. Goremychkin, K. A. McEwen, N. Takeda, B. D. Rainford, K. S. Knight, J. W. Taylor, J. Park, H. C. Walker, R. Osborn, and P. S. Riseborough, *Phys. Rev. B* **75**, 014418 (2007).

Alleno, E., D. Bérardan, C. Godart, and P. Bonville, *Physica B* **378–380**, 237 (2006).

Anno, H., K. Hatada, H. Shimizu, K. Matsubara, Y. Notohara, T. Sakakibara, H. Tashiro, and K. Motoya, *J. Appl. Phys.* **83**, 5270 (1998).

Anno, H., H. Tashiro, and K. Matsubara, *Proc. 18th Int. Conf. on Thermoelectrics*, IEEE Catalog Number 99TH8407, Piscataway, NJ, p. 169 (1999).

Aoki, D., Y. Haga, Y. Homma, H. Sakai, S. Ikeda, Y. Shiokawa, E. Yamamoto, A. Nakamura, and Y. Onuki, *J. Phys. Soc. Jpn.* **75**, 073703 (2006a).

Aoki, Y., T. Namiki, T. D. Matsuda, K. Abe, H. Sugawara, and H. Sato, *Phys. Rev. B* **65**, 064446 (2002a).

Aoki, Y., T. Namiki, S. Ohsaki, S. R. Saha, H. Sugawara, and H. Sato, *J. Phys. Soc. Jpn.* **71**, 2098 (2002b).

Aoki, Y., T. Namiki, S. Ohsaki, S. R. Saha, H. Sugawara, and H. Sato, *Physica C* **388–389**, 557 (2003).

Aoki, Y., W. Higemoto, S. Sanada, K. Ohishi, S. R. Saha, A. Koda, K. Nishiyama, R. Kadono, H. Sugawara, and H. Sato, *Physica B* **359–361**, 895 (2005a).

Aoki, Y., H. Sugawara, H. Hisatomo, and H. Sato, *J. Phys. Soc. Jpn.* **74**, 209 (2005b).

Aoki, Y., S. Sanada, H. Aoki, D. Kikuchi, H. Sugawara, and H. Sato, *Physica B* **378–380**, 54 (2006b).

Aoki, Y., S. Sanada, D. Kikuchi, H. Sugawara, and H. Sato, *J. Phys. Soc. Jpn.* **76**, 113703 (2007).

Aoki, Y., W. Higemoto, Y. Tsunashima, Y. Yonezawa, K. H. Satoh, A. Koda, T. U. Ito, K. Ohishi, R. H. Heffner, D. Kikuchi, and H. Sato, *Physica B* **404**, 757 (2009).

Aoki, Y., T. Namiki, S. R. Saha, T. Tayama, T. Sakakibara, R. Shiina, H. Shiba, H. Sugawara, and H. Sato, *J. Phys. Soc. Jpn.* **80**, 054704 (2011a).

Aoki, Y., S. Sanada, D. Kikuchi, H. Sugawara, and H. Sato, *J. Phys. Soc. Jpn.* **80**, SA013 (2011b).

Baenitz, M., R. Sarkar, R. Gumeniuk, A. Leithe-Jasper, W. Schnelle, H. Rosner, U. Burkhardt, M. Schmidt, U. Schwarz, D. Kaczorowski, Y. Grin, and F. Steglich, *Phys. Stat. Solidi (b)* **247**, 740 (2010).

Bauer, E., A. Galatanu, H. Michor, G. Hilscher, P. Rogl, P. Boulet, and H. Noel, *Eur. Phys. J. B* **14**, 483 (2000).

Bauer, E., S. Berger, A. Galatanu, H. Michor, C. Paul, G. Hilscher, V. H. Tran, A. Grytsiv, and P. Rogl, *J. Magn. Magn. Mater.* **226–230**, 674 (2001a).

Bauer, E., S. Berger, A. Galatanu, M. Galli, H. Michor, G. Hilscher, C. Paul, B. Ni, M. M. Abd-Elmeguid, V. H. Tran, A. Grytsiv, and P. Rogl, *Phys. Rev. B* **63**, 224414 (2001b).

Bauer, E. D., A. Slebarski, R. P. Dickey, E. J. Freeman, C. Sirvent, V. S. Zapf, N. R. Dilley, and M. B. Maple, *J. Phys.: Condens. Matter* **13**, 5183 (2001c).

Bauer, E. D., A. Slebarski, E. J. Freeman, C. Sirvent, and M. B. Maple, *J. Phys.: Condens. Matter* **13**, 4495 (2001d).

Bauer, E., S. Berger, C. Paul, M. D. Mea, G. Hilscher, H. Michor, M. Reissner, W. Steiner, A. Grytsiv, P. Rogl, and E. W. Scheidt, *Phys. Rev. B* **66**, 214421 (2002a).

Bauer, E., S. Berger, A. Galatanu, C. Paul, M. D. Mea, H. Michor, G. Hilscher, A. Grytsiv, P. Rogl, D. Kaczorowski, L. Keller, T. Herrmannsdörfer, and P. Fisher, *Physica B* **312–313**, 840 (2002b).

Bauer, E. D., N. A. Frederick, P.-C. Ho, V. S. Zapf, and M. B. Maple, *Phys. Rev. B* **65**, 100506(R) (2002c).

Bauer, E. D., A. Slebarski, N. A. Frederick, W. M. Yuhasz, M. B. Maple, D. Cao, F. Bridges, G. Giester, and P. Rogl, *J. Phys.: Condens. Matter* **16**, 5095 (2004).

Bauer, E., A. Grytsiv, X.-Q. Chen, N. Melnychenko-Koblyuk, G. Hilscher, H. Kaldarar, H. Michor, E. Royanian, G. Giester, M. Rotter, R. Podloucky, and P. Rogl, *Phys. Rev. Lett.* **99**, 217001 (2007).

Bauer, E., X.-Q. Chen, P. Rogl, G. Hilscher, H. Michor, E. Royanian, R. Podloucky, G. Giester, O. Sologub, and A. P. Gonçalves, *Phys. Rev. B* **78**, 064516 (2008).

Baumbach, R. E., P.-C. Ho, T. A. Sayles, M. B. Maple, R. Wawryk, T. Cichorek, A. Pietraszko, and Z. Henkie, *J. Phys.: Condens. Matter* **20**, 075110 (2008).

Becke, A. D. and E. R. Johnson, *J. Chem. Phys.* **124**, 221101 (2006).

Beeman, D. and P. Pincus, *Phys. Rev.* **166**, 359 (1968).

Bérardan, D., C. Godart, E. Alleno, S. Berger, and E. Bauer, *J. Alloys Compds.* **351**, 18 (2003).

Bethe, H., *Ann. Phys. Lpz.* **3**, 133 (1929).

Bhat, T. M. and D. C. Gupta, *J. Solid State Chem.* **266**, 274 (2018).

Blaha, P., K. Schwarz, G. K. H. Madesen, D. Kvasnicka, and J. Luitz, WIEN2K, Vienna University of Technology (2001), http://www.wien2k.at.

Bickers, N. E., D. L. Cox, and J. W. Wilkins, *Phys. Rev. Lett.* **54**, 230 (1985).

Bloch, F., *Zeit. Phys.* **57**, 545 (1929).

Blundell, S., in *Magnetism in Condensed Matter*, Oxford Master Series in Condensed Matter Physics, Oxford Press (2001).

Braun, D. J. and W. Jeitschko, *Acta Crystallogr.* **33**, 3401 (1977).

Braun, D. J. and W. Jeitschko, *J. Solid State Chem.* **32**, 357 (1980a).

Braun, D. J. and W. Jeitschko, *J. Less-Common Met.* **72**, 147 (1980b).

Butch, N. P., W. M. Yuhasz, P.-C. Ho, J. R. Jeffries, N. A. Frederick, T. A. Sayles, X. G. Zheng, M. B. Maple, J. B. Betts, A. H. Lacerda, F. M. Woodward, J. W. Lynn, P. Rogl, and G. Giester, *Phys. Rev. B* **71**, 214417 (2005).

Carra, P., B. T. Thole, M. Altarelli, and X. D. Wang, *Phys. Rev. Lett.* **70**, 694 (1993).

Chen, B., J. H. Xu, C. Uher, D. T. Morelli, G. P. Meisner, J.-P. Fleurial, T. Caillat and A. Borshchevsky, *Phys. Rev B.* **55**, 1476 (1997).

Chen, Z., J. Yang, R. H. Liu, L. L. Xi, W. Q. Zhang, and Jihui Yang, *J. Electron. Mater.* **42**, 2492 (2013).

Cheng, J.-G., J.-S. Zhou, K. Matsubayashi, P. P. Kong, Y. Kubo, Y. Kawamura, C. Sekine, C. Q. Jin, J. B. Goodenough, and Y. Uwatoko, *Phys. Rev. B* **88**, 024514 (2013).

Cheng, Z.-Z., B. Xu, and Z. Cheng, *Commun. Theor. Phys.* **49**, 1049 (2008).

Chi, S., P. Dai, T. Barnes, H. J. Kang, J. W. Lynn, R. Bewley, F. Ye, M. B. Maple, Z. Henkie, and A. Petraszko, *Phys. Rev. B* **77**, 094428 (2008).

Chikazumi, S., *Physics of Ferromagnetism* Oxford University Press, New York (1997).

Cichorek, T., A. Rudenko, P. Wisniewski, R. Wawryk, L. Kepinski, A. Pietraszko, and Z. Henkie, *Phys. Rev. B* **90**, 195123 (2014).

Curnoe, S. H., H. Harima, K. Takegahara, and K. Ueda, *Phys. Rev. B* **70**, 245112 (2004).

Danebrock, M. E., C. B. H. Evers, and W. Jeitschko, *J. Phys. Chem. Solids* **57**, 381 (1996).

Dedkov, Y. S., S. L. Molodtsov, H. Rosner, A. Leithe-Jasper, W. Schnelle, M. Schmidt, and Y. Grin, *Physica B* **460–462**, 698 (2007).

DeLong, L. E. and G. P. Meisner, *Solid State Commun.* **53**, 119 (1985).

Deminami, S., Y. Kawamura, Y. C. Chen, M. Kanazawa, J. Hayashi, T. Kuzuya, K. Takeda, M. Matsuda, and C. Sekine, *J. Phys.: Conf. Ser.* **950**, 042032 (2017).

Dilley, N. R., E. J. Freeman, E. D. Bauer, and M. B. Maple, *Phys. Rev. B* **58**, 6287 (1998).

Dilley, N. R., E. D. Bauer, M. B. Maple, S. V. Dordevic, D. N. Basov, F. Freibert, T. W. Darling, A. Migliori, B. C. Chakoumakos, and B. C. Sales, *Phys. Rev. B* **61**, 4608 (2000).

Ding, Q.-P., K. Rana, K. Nishine, Y. Kawamura, J. Hayashi, C. Sekine, and Y. Furukawa, *Phys. Rev. B* **98**, 155149 (2018).

Dordevic, S. V., N. R. Dilley, E. D. Bauer, D. N. Basov, M. B. Maple, and L. Degiorgi, *Phys. Rev. B* **60**, 11321 (1999).

Dordevic, S. V., D. N. Basov, N. R. Dilley, E. D. Bauer, and M. B. Maple, *Phys. Rev. Lett.* **86**, 684 (2001).

Dyck, J. S., W. Chen, J. Yang, G. P. Meisner, and C. Uher, *Phys. Rev. B* **65**, 115204 (2002).

Effantin, J. M., J. Rossat-Mignod, P. Burlet, H. Bartholin, S. Kunii, and T. Kasuya, *J. Mag. Magn. Mater.* **47 & 48**, 145 (1985).

Evers, C. B. H., L. Boonk, and W. Jeitschko, *Z. Anorg. Allg. Chemie* **620**, 1028 (1994).

Evers, C. B. H., W. Jeitschko, L. Boonk, D. J. Brown, T. Ebel, and U. D. Scholz, *J. Alloys Compd.* **224**, 184 (1995).

Fujiwara, K., K. Ishihawa, K. Miyoshi, J. Takeuchi, C. Sekine, and I. Siro, *Physica B* **281–282**, 296 (2000).

Fujino, T., Y. Nakanishi, K. Ito, T. Yaegashi, M. Nakamura, K. Matsuhira, S. Takagi, C. Sekine, I. Shirotani, and M. Yoshizawa, *J. Phys. Soc. Jpn.* **77**, 306 (2008).

Fukazawa, H., K. Hachitani, M. Shimizu, R. Kobayashi, Y. Kohori, I. Watanabe, K. Akahira, C. Sekine, and I. Shirotani, *J. Phys.: Conf. Ser.* **150**, 042035 (2009).

Fulde, P. and I. Peschel, *Adv. Phys.* **21**, 1 (1972).

Fultz, B., Mössbauer Spectroscopy, in *Characterization of Materials*, ed. E. Kaufmann, John Wiley, NY (2011).

Gajewski, D. A., N. R. Dilley, E. D. Bauer, E. J. Freemany, R. Chau, M. B. Maple, D. Mandrus, B. C. Sales, and A. H. Lacerda, *J. Phys.: Condens. Matter* **10**, 6973 (1998).

Galéra, R. M., C. Opagiste, M. Amara, M. Zbiri, and S. Rols, *J. Phys.: Conf. Ser.* **592**, 012011 (2015).

Galván, D. H., N. R. Dilley, M. B. Maple, A. Posada-Amarillas, A. Reyes-Serrato, and J. C. Samaniego-Reyna, *Phys. Rev. B* **68**, 115110 (2003).

Galván, D. H., *J. Supercond. Nov. Magn.* **24**, 1957 (2011).

Gérard, A., F. Grandjean, J. A. Hodges, D. J. Braun, and W. Jeitschko, *J. Phys. C: Solid State Phys.* 16, 2797 (1983).

Giri, R., C. Sekine, Y. Shimaya, I. Shirotani, K. Matsuhira, Y. Doi, Y. Hinatsu, M. Yokoyama, and H. Amitsuka, *Physica B* **329–333**, 458 (2003).

Goremychkin, E. A., R. Osborn, E. D. Bauer, M. B. Maple, N. A. Frederick, W. M. Yuhasz, F. M. Woodward, and J. W. Lynn, *Phys. Rev. Lett.* **93**, 157003 (2004).

Goto, T., Y. Nemoto, K. Onuki, K. Sakai, T. Yamagichi, M. Akatsu, T. Yanagisawa, H. Sugihara, and H. Sato, *J. Phys. Soc. Jpn.* **74**, 263 (2005).

Grandjean, A. G., A. Gérard, J. Hodges, D. J. Braun, and W. Jeitschko, *Hyperfine Interact.* **15–16**, 765 (1983).

Grandjean, A. G., A. Gérard, D. J. Braun, and W. Jeitschko, *J. Phys. Chem. Solids* **45**, 877 (1984).

Grytsiv, A., P. Rogl, S. Berger, C. Paul, E. Bauer, C. Godart, B. Ni, M. M. Abd-Elmeguid, A. Saccone, R. Ferro, and D. Kaczorowski, *Phys. Rev. B* **66**, 094411 (2002).

Grytsiv, A., X.-Q. Chen, N. Melnychenko-Koblyuk, P. Rogl, E. Bauer, G. Hilscher, H. Kaldarar, H. Michor, E. Royanian, R. Podloucky, M. Rotter, and G. Giester, *J. Phys. Soc. Jpn.* **77**, Suppl. A, 121 (2008).

Guertin, R. P., C. Rossel, M. S. Torikachvili, M. W. McElfresh, M. B. Maple, S. H. Bloom, Y. S. Yao, M. V. Kuric, and G. P. Meisner, *Phys. Rev. B* **36**, 8665 (1987).

Gumeniuk, R., W. Schnelle, H. Rosner, M. Nicklas, A. Leithe-Jasper, and Y. Grin, *Phys. Rev. Lett.* **100**, 017002 (2008).

Gumeniuk, R., M. Schöneich, A. Leithe-Jasper, W. Schnelle, M. Nicklas, H. Rosner, A. Ormeci, U. Burkhardt, M. Schmidt, U. Schwarz, M. Ruck, and Y. Grin, *New J. Phys.* **12**, 103035 (2010).

Gumeniuk, R., K. O. Kvashnina, W. Schnelle, M. Nicklas, H. Borrmann, H. Rosner, Y. Skourski, A. A. Tsirlin, A. Leithe-Jasper, and Y. Grin, *J. Phys.: Condens. Matter* **23**, 465601 (2011).

Hachitani, K., H. Fukazawa, Y. Kohori, I. Watanabe, Y. Yoshimitsu, K. Kumagai, R. Giri, C. Sekine, and I. Shirotani, *J. Phys. Soc. Jpn.* **75**, 124717 (2006a).

Hachitani, K., H. Fukazawa, Y. Kohori, I. Watanabe, C. Sekine, and I. Shirotani, *Phys. Rev. B* **73**, 052408 (2006b).

Hachitani, K., H. Amanuma, H. Fukazawa, Y. Kohori, K. Koyama, K. Kumagai, C. Sekine, and I. Shirotani, *J. Phys. Soc. Jpn.* **75**, 124712 (2006c).

Hachitani, K., H. Fukazawa, Y. Kohori, I. Watanabe, K. Kumagai, C. Sekine, and I. Shirotani, *Physica B* **378–380**, 230 (2006d).

Hachitani, K., H. Amanuma, H. Fukazawa, Y. Kohori, I. Watanabe, K. Koyama, K. Kumagai, C. Sekine, and I. Shirotani, *J. Phys. Chem. Solids* **68**, 2080 (2007).

Hao, L., K. Iwasa, M. Nakajima, D. Kawana, K. Kuwahara, M. Kohgi, H. Sugawara, T. D. Matsuda, Y. Aoki, and H. Sato, *Acta Phys. Pol. B* **34**, 1113 (2003).

Hao, L., K. Iwasa, K. Kuwahara, M. Kohgi, H. Sugawara, Y. Aoki, and H. Sato, T. D. Matsuda, J.-M. Mignot, A. Gukasov, and M. Nishi, *Physica B* **359–361**, 871 (2005).

Harima, H. and K. Takegahara, *Physica B* **312**, 843 (2002).

Harima, H. and K. Takegahara, *Physica B* **328**, 26 (2003).

Harima, H., K. Takegahara, K. Ueda, and S. H. Curnoe, *Acta Phys. Pol. B* **34**, 1189 (2003).

Hattori, K., Y. Hirayama, and K. Miyake, *J. Phys. Soc. Jpn.* **74**, 3306 (2005).

Hayashi, J., K. Akahira, K. Matsui, H. Ando, Y. Sugiuchi, K. Takeda, C. Sekine, I. Shirotani, and T. Yagi, *J. Phys.: Conf. Ser.* **215**, 012142 (2010).

Heisenberg, W., *Zeit. Phys.* **38**, 441 (1926).

Henkie, Z., M. B. Maple, A. Pietraszko, R. Wawryk, T. Cichorek, R. E. Baumbach, W. M. Yuhasz, and P.-C. Ho, *J. Phys. Soc. Jpn.* **77**, Suppl. A, 128 (2008).

Hermann, R. P., R. Jin, W. Schweika, F. Grandjean, D. Mandrus, B. C. Sales, and G. J. Long, *Phys. Rev. Lett.* **90**, 135505 (2003).

Hidaka, H., I. Ando, H. Kotegawa, T. C. Kobayashi, H. Harima, M. Kobayashi, H. Sugawara, and H. Sato, *Phys. Rev. B* **71**, 073102 (2005).

Hidaka, H., H. Kotegawa, S. Fukushima, N. Wada, T. C. Kobayashi, H. Harima, K. Fujiwara, D. Kikuchi, H. Sato, and H. Sugawara, *J. Phys. Soc. Jpn.* **75**, 094709 (2006).

Higashinaka, R., K. Takeda, T. Namiki, Y. Aoki, and H. Sato, *J. Phys. Soc. Jpn.* **82**, 114710 (2013).

Ho, P.-C., W. M. Yuhasz, N. P. Butch, N. A. Frederick, T. A. Sayles, J. R. Jeffries, M. B. Maple, J. B. Betts, A. H. Lacerda, P. Rogl, and G. Giester, *Phys. Rev. B* **72**, 094410 (2005).

Ho, P.-C., T. Yanagisawa, W. M. Yuhasz, A. A. Dooraghi, C. C. Robinson, N. P. Butch, R. E. Baumbach, and M. B. Maple, *Phys. Rev. B* **83**, 024511 (2011).

Ho, P.-C., J. Singleton, M. B. Maple, H. Harima, P. A. Goddard, Z. Henkie, and A. Pietraszko, *New J. Phys.* **9**, 269 (2007).

Ho, P.-C., J. Singleton, P. A. Goddard, F. F. Balakirev, S. Chikara, T. Yanagisawa, M. B. Maple, D. B. Shrekenhamer, X. Lee, and A. T. Thomas, *Phys. Rev. B* **94**, 205140 (2016).

Hoshino, S., J. Otsuki, and Y. Kuramoto, *J. Phys. Soc. Jpn.* **80**, 033703 (2011).

Hotta, T., *J. Phys. Soc. Jpn.* **74**, 1275 (2005).

Hotta, T., *Phys. Rev. Lett.* **96**, 197201 (2006).

Hotta, T., *J. Phys. Soc. Jpn.* **76**, 034713 (2007).

Hotta, T., *J. Phys. Soc. Jpn.* **77**, 103711 (2008).

Huang, K., L. Shu, I. K. Lum, B. D. White, M. Janoschek, D. Yazici, J. J. Hamlin, D. A. Zocco, P.-C. Ho, R. E. Baumbach, and M. B. Maple, *Phys. Rev. B* **89**, 035145 (2014).

Hulliger, F., *Helv. Phys. Acta* **34**, 782 (1961).

Ikeno, T., A. Mitsuda, T. Mizushima, T. Kuwai, Y. Isikawa, and I. Tamura, *J. Phys. Soc. Jpn.* **76**, 024708 (2007).

Ikeno, T., K. Tanaka, D. Kikuchi, K. Kuwahara, Y. Aoki, H. Sato, and Y. Isikawa, *J. Phys. Soc. Jpn.*, **77**, Suppl. A, 309 (2008).

Imada, S., H. Higashimichi, A. Yamasaki, M. Yano, T. Muro, A. Sekiyama, S. Suga, H. Sugawara, D. Kikuchi, and H. Sato, *Phys. Rev. B* **76**, 153106 (2007).

Indoh, K., H. Onodera, C. Sekine, I. Shirotani, and Y. Yamaguchi, *J. Phys. Soc. Jpn.* **71**, 243 (2002).

Itoh, T. U., W. Higemoto, K. Ohishi, R. H. Heffner, N. Nishida, K. Satoh, H. Sugawara, Y. Aoki, D. Kikuchi, and H. Sato, *J. Phys. Chem. Solids* **68**, 2072 (2007).

Iwasa, K., Y. Watanabe, K. Kuwahara, M. Koghi, H. Sugawara, T. D. Matsuda, Y. Aoki, and H. Sato, *Physica B* **312–313**, 834 (2002).

Iwasa, K., L. Hao, M. Nakajima, M. Koghi, H. Sugawara, Y. Aoki, H. Sato, and T. D. Matsuda, *Acta Phys. Pol. B* **34**, 1117 (2003).

Iwasa, K., L. Hao, M. Nakajima, M. Koghi, H. Sugawara, T. Takagi, K. Horiuchi, Y. Mori, Y. Murakami, K. Kuwahara, S. R. Saha, Y. Aoki, and H. Sato, *J. Phys. Soc. Jpn.* **74**, 1930 (2005a).

Iwasa, K., L. Hao, K. Kuwahara, M. Koghi, S. R. Saha, H. Sugawara, Y. Aoki, H. Sato, T. Tayama, and T. Sakakibara, *Phys. Rev. B* **72**, 024414 (2005b).

Iwasa, K., S. Itobe, T. C. P. Yang, Y. Murakami, M. Koghi, K. Kuwahara, H. Sugawara, H. Sato, N. Aso, T. Tayama, and T. Sakakibara, *J. Phys. Soc. Jpn.*, **77**, Suppl. A, 318 (2008).

Iwasa, K., A. Yonemoto, S. Takagi, S. Itoh, T. Yokoo, S. Ibuka, C. Sekine, and H. Sugawara, *Phys. Procedia* **75**, 179 (2015).

Jeitschko, W. and D. J. Braun, *Acta Cryst. B* **33**, 3401 (1977).

Jeitschko, W. and D. J. Braun, *J. Less-Common Mater.* **76**, 33 (1980).

Jeitschko, W., A. J. Foeker, D. Paschke, M. V. Dewalsky, C. B. H. Evers, B. Kunnen, A. Lang, G. Kotzyba, U. C. Rodewald, and M. H. Moller, *Z. Anorg. Allg. Chemie* **626**, 1112 (2000).

Jeon, I., K. Huang, D. Yazici, N. Kanchanavatee, B. D. White, P.-C. Ho, S. Jang, N. Pouse, and M. B. Maple, *Phys. Rev. B* **93**, 104507 (2016).

Jeon, I., S. Ran, A. J. Breindel, P.-C. Ho, R. B. Adhikari, C. C. Almasan, B. Luong, and M. B. Maple, *Phys. Rev. B* **95**, 134517 (2017).

Kaczorowski, D. and V. H. Tran, *Phys. Rev. B* **77**, 180504(R) (2008).

Kadowaki, K. and S. B. Woods, *Solid State Commun.* **58**, 307 (1986).

Kaiser, J. W. and W. Jeitschko, *J. Alloys Compd.* **291**, 66 (1999).

Kawaguchi, M., H. Tou, M. Sera, K. Kojima, T. Ikeno, and Y. Isikawa, *J. Phys. Soc. Jpn.* **75**, 093702 (2006).

Kanai, K., N. Takeda, S. Nozawa, T. Yokoya, M. Ishikawa, and S. Shi, *Phys. Rev. B* **65**, 041105(R) (2002).

Kawamura, N., S. Tsutsui, M. Mizumaki, N. Ishimatsu, H. Maruyama, H. Sugawara, and H. Sato, *J. Phys.: Conf. Ser.* **190**, 012020 (2009).

Kawamura, Y., Y. Kiyota, C. Sekine, M. Wakeshima, and K. Matsuhira, *J. Phys. Soc. Jpn.* **81**, SB047 (2012).

Kawamura, Y., Y. Q. Chen, T. Nakayama, R. Shirakawa, J. Hayashi, K. Takeda, and C. Sekine, *J. Phys.: Conf. Ser.* **592**, 012033 (2015).

Kawamura, Y., S. Deminami, L. Salamakha, A. Sidorenko, P. Heinrich, H. Michor, E. Bauer, and C. Sekine, *Phys. Rev. B* **98**, 024513 (2018).

Kawana, D., K. Kuwahara, M. Sato, M. Takagi, Y. Aoki, M. Kohgi, H. Sato, H. Sagayama, T. Osakabe, K. Iwasa, and H. Sugawara, *J. Phys. Soc. Jpn.* **75**, 113602 (2006).

Keiber, T., F. Bridges, R. E. Baumbach, and M. B. Maple, *Phys. Rev. B* **86**, 174106 (2012).

Keller, L., P. Fischer, T. Herrmannsdörfer, A. Dönni, H. Sugawara, T. D. Matsuda, K. Abe, Y. Aoki, and H. Sato, *J. Alloys Compd.* **323–324**, 516 (2001).

Keppens, V., D. Mandrus, B. C. Sales, B. C. Chakoumakos, P. Dai, R. Coldea, M. B. Maple, D. A. Gajewski, E. J. Freeman, and S. Bennington, *Nature* **395**, 876 (1998).

Khenata, R., A. Bouhemadou, A. H. Reshak, R. Ahmed, B. Bouhafs, D. Rached, Y. Al-Douri, and M. Rérat, *Phys. Rev. B* **75**, 195131 (2007).

Kihou, K., I. Shiratoni, Y. Shimaya, C. Sekine, and T. Yagi, *Mater. Res. Bull.* **39**, 317 (2004).

Kihou, K., C. Sekine, I. Shirotani, C.-H. Lee, K. Hijiri, and K. Takeda, *Physica B* **359–361**, 859 (2005).

Kikuchi, D., M. Kobayashi, H. Sugawara, Y. Aoki, H. Sato, H. Shishido, R. Settai, Y. Onuki, *Physica B* **359–361**, 874 (2005).

Kikuchi, D., H. Sugawara, H. Aoki, K. Tanaka, S. Sanada, Y. Aoki, and H. Sato, *Physica B* **378–380**, 226 (2006).

Kikuchi, D., K. Tanaka, H. Aoki, K. Kuwahara, Y. Aoki, H. Sugawara, and H. Sato, *J. Magn. Magn. Mater.* **310**, e225 (2007a).

Kikuchi, D., M. Tagikawa, H. Sugawara, and H. Sato, *J. Phys. Soc. Jpn.* **76**, 043705 (2007b).

Kikuchi, D., H. Sugawara, K. Tanaka, H. Aoki, M. Kobayashi, S. Sanada, K. Kuwahara, Y. Aoki, H. Shishido, R. Settai, Y. Onuki, H. Harima, and H. Sato, *J. Phys. Soc. Jpn.* **77**, 114705 (2008a).

Kikuchi, D., S. Tatsuoka, K. Tanaka, Y. Kuwahito, M. Ueda, A. Shinozawa, H. Aoki, K. Kuwahara, Y. Aoki, and H. Sato, *Physica B* **403**, 884 (2008b).

Kimura, S., H. J. Im, Y. Sakurai, T. Mizuno, K. Takegahara, H. Harima, K. Hayashi, E. Matsuoka, and T. Takabatake, *Physica B* **383**, 137 (2006).

Kimura, S., H. J. Im, T. Mizuno, S. Narazu, E. Matsuoka, and T. Takabatake, *Phys. Rev. B* **75**, 245106 (2007).

Kiss, A. and Y. Kuramoto, *J. Phys. Soc. Jpn.* **75**, 103704 (2006).

Kiss, A. and Y. Kuramoto, *J. Phys. Soc. Jpn.* **78**, 124702 (2009).

Kitagawa, H., S. Kondo, K. Oda, M. Hasaka, and T. Morimura, *Proc. 17th Int. Conf. on Thermoelectrics*, IEEE Catalog Number 98TH8365, Piscataway, NJ, p. 319 (1998).

Kjekshus, A. and G. Pedersen, *Acta Cryst.* **14**, 1065 (1961).

Kjekshus, A., D. G. Nicholson, and T. Rakke, *Acta Chem. Scand.* **27**, 1307 (1973).

Kjekshus, A. and T. Rakke, *Acta Chem. Scand. A* **28**, 99 (1974).

Kondo, J., *Physica B+C* **84**, 40 (1976), ibid. **84**, 207 (1976).

Konno, S., A. Suzuki, K. Nihei, K. Kuwahara, D. Kawana, T. Yokoo, and S. Itoh, *J. Phys.: Conf. Ser.* **592**, 012029 (2015).

Kontani, H., *J. Phys. Soc. Jpn.* **73**, 515 (2004).

Kotegawa, H., H. Hidaka, Y. Shimaoka, T. Miki, T. C. Kobayashi, D. Kikuchi, H. Sugawara, and H. Sato, *J. Phys. Soc. Jpn.* **74**, 2173 (2005).

Kotegawa, H., H. Hidaka, T. C. Kobayashi, D. Kikuchi, H. Sugawara, and H. Sato, *Phys. Rev. Lett.* **99**, 156408 (2007).

Kotegawa, H., K. Tabira, Y. Irie, H. Hidaka, T. C. Kobayashi, D. Kikuchi, K. Tanaka, S. Tatsuoka, H. Sugawara, and H. Sato, *J. Phys. Soc. Jpn.* **77**, Suppl. A, 90 (2008).

Krishnamurthy, V. V., J. C. Lang, D. Haskel, D. J. Keavney, G. Srajer, J. L. Robertson, B. C. Sales, D. G. Mandrus, D. J. Singh, and D. I. Bilc, *Phys. Rev. Lett.* **98**, 126403 (2007).

Krishnamurthy, V. V., D. J. Keavney, D. Haskel, J. C. Lang, G. Srajer, B. C. Sales, D. G. Mandrus, and J. L. Robertson, *Phys. Rev. B* **79**, 014426 (2009).

Kumagai, T., Y. Nakanishi, H. Sugawara, H. Sato, and M. Yoshizawa, *Physica B* **329–333**, 471 (2003).

Kumigashira, H., T. Takahashi, S. Yoshii, and M. Kasaya, *Phys. Rev. Lett.* **87**, 067206 (2001).

Kunitoshi, H., T. D. Matsuda, R. Midorikawa, R. Higashinaka, K. Kuwahara, Y. Aoki, and H. Sato, *J. Phys. Soc. Jpn.* **85**, 114708 (2016).

Kuramoto, Y. and A. Kiss, *J. Phys. Soc. Jpn.* **77**, Suppl. A, 187 (2008).

Kuwahara, K., K. Iwasa, M. Kohgi, K. Kaneko, S. Araki, N. Metoki, H. Sugawara, Y. Aoki, and H. Sato, *J. Phys. Soc. Jpn.* **73**, 1438 (2004).

Kuwahara, K., K. Iwasa, M. Kohgi, K. Kaneko, N. Metoki, S. Raymond, M.-A. Méasson, J. Flouquet, H. Sugawara, Y. Aoki, and H. Sato, *Phys. Rev. Lett.* **95**, 107003 (2005).

Kuwahara, K., M. Takagi, K. Iwasa, S. Itobe, D. Kikuchi, Y. Aoki, M. Kohgi, H. Sato, and H. Sugawara, *Physica B* **403**, 903 (2008).

Laulhé, C., K. Saito, K. Iwasa, H. Nakao, and Y. Murakami, *J. Phys.: Conf. Ser.* **200**, 012102 (2010).

Lea, K. R., M. J. M. Leask, and W. P. Wolf, *J. Phys. Chem. Solids* **23**, 1381 (1962).

Lee, C. H., H. Oyanagi, C. Sekine, I. Shirotani, and M. Ishii, *Phys. Rev. B* **60**, 13253 (1999).

Lee, C. H., H. Matsuhata, A. Yamamoto, T. Ohta, H. Takazawa, K. Ueno, C. Sekine, I. Shirotani, and T. Hirayama, *J. Phys.: Condens. Matter* **13**, L45 (2001).

Lee, C. H., H. Matsuhata, H. Yamaguchi, C. Sekine, K. Kihou, T. Suzuki, T. Noro, and I. Shirotani, *Phys. Rev. B* **70**, 153105 (2004).

Lee, C. H., S. Tsutsui, K. Kihou, H. Sugawara, and H. Yoshizawa, *J. Phys. Soc. Jpn.* **81**, 063702 (2012).

Leithe-Jasper, A., D. Kaczorowski, P. Rogl, J. Bogner, M. Reissner, W. Steiner, G. Wiesinger, and C. Godart, *Solid State Commun.* **109**, 395 (1999).

Leithe-Jasper, A., W. Schnelle, H. Rosner, N. Senthilkumaran, A. Rabis, M. Baenitz, A. Gippius, E. Morozova, J. A. Mydosh, and Y. Grin, *Phys. Rev. Lett.* **91**, 037208 (2003).

Leithe-Jasper, A., W. Schnelle, H. Rosner, M. Baenitz, A. Rabis, A. A. Gippius, E. N. Morozova, H. Borrmann, U. Burkhardt, R. Ramlau, U. M. Schwarz, J. A. Mydosh, Y. Grin, V. Ksenofontov, and S. Reiman, *Phys. Rev. B* **70**, 214418 (2004).

Leithe-Jasper, A., W. Schnelle, H. Rosner, R. Cardoso-Gil, M. Baenitz, J. A. Mydosh, Y. Grin, M. Reissner, and W. Steiner, *Phys. Rev. B* **77**, 064412 (2008).

Leithe-Jasper, A., W. Schnelle, H. Rosner, W. Schweika, and O. Isnard, *Phys. Rev. B* **90**, 144416 (2014).

Long, G. J., D. Hautot, F. Grandjean, D. T. Morelli, and G. P. Meisner, *Phys. Rev. B* **60**, 7410 (1999).

Long, G. J., B. Mahieu, B. C. Sales, R. P. Hermann, and F. Grandjean, *J. Appl. Phys.* **92**, 7236 (2002).

Magishi, K., H. Sugawara, T. Saito, K. Koyama, and H. Sato, *Physica B* **378–380**, 175 (2006).

Magishi, K., Y. Iwahashi, T. Horimoto, H. Sugawara, T. Saito, and K. Koyama, *J. Mag. Magn. Mater.* **310**, 951 (2007a).

Magishi, K., H. Sugawara, I. Mori, T. Saito, and K. Koyama, *J. Phys. Chem. Solids* **68**, 2076 (2007b).

Magishi, K., H. Sugawara, T. Saito, K. Koyama, C. Sekine, K. Takeda, and I. Shirotani, *J. Phys. Soc. Jpn.* **77**, 300 (2008).

Magishi, K., K. Nagata, Y. Iwahashi, H. Sugawara, T. Saito, and K. Koyama, *J. Phys.: Conf. Ser.* **200**, 012110 (2010).

Magishi, K., H. Sugawara, M. Takahashi, Takahito Saito, K. Koyama, Takashi Saito, S. Tatsuoka, K. Tanaka, and H. Sato, *J. Phys. Soc. Jpn.* **81**, 124706 (2012).

Magishi, K., R. Watanabe, A. Hisada, Takahito Saito, K. Koyama, Takashi Saito, R. Higashinaka, Y. Aoki, and H. Sato, *J. Phys. Soc. Jpn.* **83**, 084712 (2014).

Maisuradze, A., M. Nicklas, R. Gumeniuk, C. Baines, W. Schnelle, H. Rosner, A. Leithe-Jasper, Y. Grin, and R. Khasanov, *Phys. Rev. Lett.* **103**, 147002 (2009).

Maisuradze, A., W. Schnelle, R. Khasanov, R. Gumeniuk, M. Nicklas, H. Rosner, A. Leithe-Jasper, Y. Grin, A. Amato, and P. Thalmeier, *Phys. Rev. B* **82**, 024524 (2010).

Mandrus, D., A. Migliori, T. W. Darling, M. F. Hundley, E. J. Peterson, and J. D. Thompson, *Phys. Rev. B* **52**, 4926 (1995).

Maple, M. B., P.-C. Ho, N. A. Frederick, V. S. Zapf, W. M. Yuhasz, and E. D. Bauer, *Acta Phys. Pol. B* **34**, 919 (2003a)

Maple, M. B., P.-C. Ho, N. A. Frederick, V. S. Zapf, W. M. Yuhasz, and E. D. Bauer, A. D. Christianson, and A. H. Lacerda, *J. Phys.: Condens. Matter* **15**, S2080 (2003b).

Maple, M. B., P.-C. Ho, V. S. Zapf, N. A. Frederick, E. D. Bauer, W. M. Yuhasz, F. M. Woodward, and J. W. Lynn, *J. Phys. Soc. Jpn.* **71** Suppl., 23 (2002).

Maple, M. B., E. D. Bauer, V. S. Zapf, and J. Wosnitza, in *Superconductivity in Nanostructures, High T_C and Novel Superconductors, Organic Superconductors*, edited by K. H. Bennemann and J. B. Ketterson, *The Physics of Superconductors*, Vol. II, Springer-Verlag, Berlin (2004), Ch. 8, p. 555.

Maple, M. B., N. P. Butch, N. A. Frederick, P.-C. Ho, J. R. Jeffries, T. A. Sayles, T. Yanagisawa, W. M. Yuhasz, S. Chi, H. J. Kang, J. W. Lynn, P. Dai, S. K. McCall, M. W. McElfresh, M. J. Fluss, Z. Henkie, and A. Pietraszko, *Proc. Natl. Acad. Sci. USA* **103**, 6783 (2006).

Maple, M. B., Z. Henkie, R. E. Baumbach, T. A. Sayles, N. P. Butch, P.-C. Ho, T. Yanagisawa, W. M. Yuhasz, R. Wawryk, T. Cichorek, and A. Pietraszko, *J. Phys. Soc. Jpn.* **77**, Suppl. A, 7 (2008).

Martins, G. B., M. A. Pires, G. E. Barberis, C. Rettori, and M. S. Torikachvili, *Phys. Rev. B* **50**, 14822 (1994).

Masaki, S., T. Mito, M. Takemura, S. Wada, H. Harima, D. Kikuchi, H. Sato, H. Sugawara, N. Takeda, and G. Q. Zheng, *J. Phys. Soc. Jpn.* **76**, 043714 (2007).

Masaki, S., T. Mito, M. Takemura, S. Wada, H. Harima, D. Kikuchi, H. Sato, H. Sugawara, N. Takeda, and G. Q. Zheng, *Physica B* **403**, 1630 (2008a).

Masaki, S., T. Mito, S. Wada, H. Sugawara, G. Kikuchi, and H. Sato, *Phys. Rev. B* **78**, 094414 (2008b).

Matsuda, T. D., A. Galatanu, Y. Haga, S. Ikeda, E. Yamamoto, M. Hedo, Y. Uwatoko, T. Takeuchi, K. Sugiyama, K. Kindo, R. Settai, and Y. Onuki, *J. Phys. Soc. Jpn.*, **73**, 2533 (2004).

Matsuda, T. D., K. Abe, F. Watanuki, H. Sugawara, Y. Aoki, H. Sato, Y. Inada, R. Settai, and Y. Onuki, *Physica B* **312–313**, 832 (2002).

Matsuhira, K., T. Takikawa, T. Sakakibara, C. Sekine, and I. Shirotani, *Physica B* **281–282**, 298 (2000).

Matsuhira, K., Y. Hinatsu, C. Sekine, and I. Shirotani, *Physica B* **312–313**, 829 (2002a).

Matsuhira, K., Y. Hinatsu, C. Sekine, T. Togashi, H. Maki, I. Shirotani, H. Kitazawa, T. Takamatsu, and G. Kido, *J. Phys. Soc. Jpn.* **71**, 237 (2002b).

Matsuhira, K., Y. Doi, M. Wakeshima, Y. Hinatsu, H. Amitsuka, Y. Shimaya, R. Giri, C. Sekine, and I. Shirotani, *J. Phys. Soc. Jpn.* **74**, 1030 (2005a)

Matsuhira, K., Y. Doi, M. Wakeshima, Y. Hinatsu, K. Kihou, C. Sekine, and I. Shirotani, *Physica B* **359–361**, 977 (2005b).

Matsuhira, K., C. Sekine, K. Kihou, I. Shirotani, M. Wakeshima, Y. Hinatsu, Y. Sei, A. Nakamura, and S. Takagi, *Physica B* **378–380**, 235 (2006).

Matsuhira, K., M. Wakeshima, Y. Hinatsu, C. Sekine, I. Shirotani, D. Kikuchi, H. Sugawara, and H. Sato, *J. Mag. Magn. Mater.* **310**, 226 (2007).

Matsuhira, K., C. Sekine, M. Wakeshima, Y. Hinatsu, T. Namiki, K. Takeda, I. Shirotani, H. Sugawara, D. Kikuchi, and H. Sato, *J. Phys. Soc. Jpn.* **78**, 124601 (2009).

Matsumura, M., G. Hyoudou, H. Kato, T. Nishioka, E. Matsuoka, H. You, T. Takabatake, and M. Sera, *J. Phys. Soc. Jpn.* **74**, 2205 (2005).

Matsumura, T., S. Michimura, T. Inami, Y. Hayashi, K. Fushiya, T. D. Matsuda, R. Higashinaka, Y. Aoki, and H. Sugawara, *Phys. Rev. B* **89**, 161116(R) (2014).

Matsumura, T., S. Michimura, T. Inami, K. Fushiya, T. D. Matsuda, R. Higashinaka, Y. Aoki, and H. Sugawara, *Phys. Rev. B* **94**, 184425 (2016).

Matsunami, M., H. Okamura, T. Nanba, H. Sugawara, and H. Sato, *J. Phys. Soc. Jpn.* **72**, 2722 (2003).

Matsunami, M., M. Takimoto, H. Okamura, T. Nanba, C. Sekine, and I. Shirotani, *Physica B* **359–361**, 844 (2005).

Matsunami, M., K. Horiba, M. Taguchi, K. Yamamoto, A. Chainani, Y. Takata, Y. Senba, H. Ohashi, M. Yabashi, K. Tamasaku, Y. Nishino, D. Miwa, T. Ishikawa, E. Ikenaga, K. Kobayashi, H. Sugawara, H. Sato, H. Harima, and S. Shin, *Phys. Rev. B* **77**, 165126 (2008a).

Matsunami, M., H. Okamura, K. Senoo, S. Nagano, C. Sekine, I. Shirotani, H. Sugawara, H. Sato, and T. Nanba, *J. Phys. Soc. Jpn.* **77**, Suppl. A, 315 (2008b).

Matsunami, M., R. Eguchi, T. Kiss, K. Horiba, A. Chainani, M. Taguchi, K. Yamamoto, T. Togashi, S. Watanabe, X.-Y. Wang, C.-T. Chen, Y. Senba, H. Ohashi, H. Sugawara, H. Sato, H. Harima, and S. Shin, *Phys. Rev. Lett.* **102**, 036403 (2009).

Matsuoka, E., K. Hayashi, A. Ikeda, K. Tanaka, T. Takabatake, and M. Matsumura, *J. Phys. Soc. Jpn.* **74**, 1382 (2005).

Matsuoka, E., S. Narazu, K. Hayashi, K. Umeo, and T. Takabatake, *J. Phys. Soc. Jpn.* **75**, 014602 (2006).

Meisner, G. P., *Physica B* **108**, 763 (1981).

Meisner, G. P., M. S. Torikachvili, K. N. Yang, M. B. Maple, and R. P. Guertin, *J. Appl. Phys.* **57**, 3073 (1985).

Meissner, G. P., D. T. Morelli, S. Hu, J. Yang, and C. Uher, *Phys. Rev. Lett.* **80**, 3551 (1998).

Mito, T., S. Masaki, N. Oki, S. Noguchi, S. Wada, N. Takeda, D. Kikuchi, H. Sato, H. Sugawara, and G. Q. Zheng, *Physica B* **378–380**, 224 (2006).

Mitsumoto, K. and Y. Ono, *Physica C* **426–431**, 330 (2005).

Miyake, A., K. Shimizu, C. Sekine, K. Kihou, and I. Shirotani, *J. Phys. Soc. Jpn.* **73**, 2370 (2004).

Miyake, A., I. Ando, T. Kagayama, K. Shimizu, C. Sekine, K. Kihou, and I. Shirotani, *J. Alloys Compd.* **408–412**, 238 (2006).

Mizumaki, M., S. Tsitsui, H. Tanida, T. Uruga, D. Kikuchi, H. Sugawara, and H. Sato, *J. Phys. Soc. Jpn.* **76**, 053706 (2007).

Mombetsu, S., T. Yanagisawa, H. Hidaka, H. Amitsuka, S. Yasin, S. Zherlitsyn, J. Wosnitza, P.-C. Ho, and M. B. Maple, *J. Phys. Soc. Jpn.* **85**, 043704 (2016).

Mounssef, B. Jr., M. R. Cantarino, E. M. Bittar, T. M. Germano, A. Leithe-Jasper, and F. A. Garcia, *Phys. Rev. B* **99**, 035152 (2019).

Möchel, A., I. Sergueev, H. C. Wille, J. Voigt, M. Prager, M. B. Stone, B. C. Sales, Z. Guguchia, A. Shengelaya, V. Keppens, and R. P. Hermann, *Phys. Rev. B* **84**, 184306 (2011).

Morelli, D. T., T. Caillat, J.-P. Fleurial, A. Borshchevsky, J. Vandersande, B. Chen, and C. Uher, *Phys. Rev. B* **51**, 9622 (1995).

Morelli, D. T. and G. P. Meisner, *J. Appl. Phys.* **77**, 3777 (1995).

Morelli, D. T., G. P. Meisner, B. Chen, S. Hu, and C. Uher, *Phys. Rev. B* **56**, 7376 (1997).

Mori, I., H. Sugawara, K. Magishi, T. Saito, K. Koyama, D. Kikuchi, K. Tanaka, and H. Sato, *J. Magn. Magn. Mater.* **310**, 277 (2007).

Moriya, T. and T. Takimoto, *J. Phys. Soc. Jpn.* **64**, 960 (1995).

Mössbauer, R., *Zeit. Phys. A* **151**, 124 (1958).

Mössbauer Effect Data Index, ed. J. G. Stevens and V. E. Stevens, Plenum, New York (1975).

Mutou, T. and T. Saso, *J. Phys. Soc. Jpn.* **73**, 2900 (2004).

Nakai, Y., K. Ishida, H. Sugawara, D. Kikuchi, and H. Sato, *Phys. Rev. B* **77**, 041101(R) (2008).

Nakamura, S., T. Goto, S. Kunii, K. Iwashita, and A. Tamaki, *J. Phys. Soc. Jpn.* **63**, 623 (1994).

Nakanishi, Y., T. Simizu, M. Yoshizawa, T. Matsuda, H. Sugawara, and H. Sato, *Phys. Rev. B* **63**, 184429 (2001).

Nakanishi, Y., T. Kumagai, M. Yoshizawa, H. Sugawara, and H. Sato, *Phys. Rev. B* **69**, 064409 (2004).

Nakanishi, Y., T. Tanizawa, T. Fujino, P. Sun, M. Nakamura, H. Sugawara, D. Kikuchi, H. Sato, and M. Yoshizawa, *J. Phys.: Conf. Ser.* **51**, 251 (2006).

Nakanishi, Y., T. Kumagai, M. Oikawa, T. Tanizawa, M. Yoshizawa, H. Sugawara, and H. Sato, *Phys. Rev. B* **75**, 134411 (2007).

Nakanishi, Y., K. Ito, T. Kamiyama, M. Nakamura, M. Yoshizawa, M. Ohashi, G. Oomi, M. Kosaka, C. Sekine, and I. Shirotani, *Physica B* **404**, 3271 (2009).

Nakanishi, Y., G. Koseki, D. Tamura, K. Kurita, T. Saito, M. Koseki, M. Nakamura, M. Yoshizawa, Y. Koyota, C. Sekine, and T. Yagi, *J. Korean Phys. Soc.* **62**, 1855 (2013).

Nakotte, H., N. R. Dilley, M. S. Torikachvili, H. N. Bordallo, M. B. Maple, S. Chang, A. Christianson, A. J. Schultz, C. F. Majkrzak, and G. Shirane, *Physica B* **259–261**, 280 (1999).

Namiki, T., Y. Aoki, T. D. Matsuda, H. Sugawara, and H. Sato, *Physica B* **329–333**, 462 (2003a).

Namiki, T., Y. Aoki, H. Sugawara, and H. Sato, *Acta Phys. Pol. B* **34**, 1161 (2003b).

Namiki, T., Y. Aoki, H. Sato, C. Sekine, I. Shirotani, T. D. Matsuda, Y. Haga, and T. Yagi, *J. Phys. Soc. Jpn.* **76**, 093704 (2007).

Namiki, T., C. Sekine, K. Matsuhira, M. Wakeshima, and I. Shiratoni, *J. Phys. Soc. Jpn.* **79**, 074714 (2010).

Nanba, T., M. Hayashi, I. Shirotani, and C. Sekine, *Physica B* **259–261**, 853 (1999).

Nanba, Y., M. Mizumaki, and K. Okada, *J. Phys. Soc. Jpn.* **82**, 104712 (2013).

Narazu, S., Y. Hadano, T. Takabatake, and H. Sugawara, *J. Phys. Soc. Jpn.* **77**, 238 (2008).

Néel, L., *Ann. Phys.* (Paris) **18**, 5 (1932).

Néel, L., *J. Phys.* **4**, 225 (1954).

Nicklas, M., R. Gumeniuk, W. Schnelle, H. Rosner, A. Leithe-Jasper, F. Steglich, and Y. Grin, *J. Phys.: Conf. Ser.* **273**, 012118 (2011).

Nicklas, M., S. Kirchner, R. Borth, R. Gumeniuk, W. Schnelle, H. Rosner, H. Borrmann, A. Leithe-Jasper, Y. Grin, and F. Steglich, *Phys. Rev. Lett.* **109**, 236405 (2012).

Nickel, E. H., *Chem. Geol.* **5**, 233 (1969).

Niikura, F. and T. Hotta, *J. Phys. Soc. Jpn.* **81**, 114720 (2012).

Nishine, K., Y. Kawamura, J. Hayashi, and C. Sekine, *Jap. J. Appl. Phys.* **56**, 05FB01 (2017).

Nitta, K., Y. Omori, D. Kikuchi, T. Miyanaga, K. Takegahara, H. Sugawara, and H. Sato, *J. Phys. Soc. Jpn.* **77**, 063601 (2008).

Nordström, L. and D. J. Singh, *Phys. Rev. B* **53**, 1103 (1996).

Nowak, B., O. Žogal, A. Pietraszko, R. E. Baumbach, M. B. Maple, and Z. Henkie, *Phys. Rev. B* **79**, 214411 (2009).

Ogawa, Y., H. Sato, M. Watanabe, T. Namiki, S. Tatsuoka, R. Higashinaka, Y. Aoki, K. Kuwahara, J.-I. Yamaura, and Z. Hiroi, *J. Phys. Soc. Jpn.* **83**, 034710 (2014).

Ogita, N., T. Kondo, T. Hasegawa, Y. Takasu, M. Udagawa, N. Takeda, K. Ishikawa, H. Sugawara, D. Kikuchi, H. Sato, C. Sekine, and I. Shirotani, *Physica B* **383**, 128 (2006).

Öğüt, S. and K. M. Rabe, *Phys. Rev. B* **54**, R8297 (1996).

Okamura, H., I. Matsutori, A. Takigawa, K. Shoji, K. Miyata, M. Matsunami, H. Sugawara, H. Sato, C. Sekine, I. Shirotani, T. Moriwaki, Y. Ikemoto, and T. Nanba, *J. Phys. Soc. Jpn.* **80**, SA092 (2011).

Okane, T., S. Fujimori, K. Mamiya, J. Okamoto, Y. Muramatsu, A. Fujimori, Y. Nagamoto, and T. Koyanagi, *J. Phys.: Condens. Matter* **15**, S2197 (2003).

Osakabe, T., K. Kuwahara, D. Kawana, K. Iwasa, D. Kikuchi, Y. Aoki, M. Kohgi, and H. Sato, *J. Phys. Soc. Jpn.* **79**, 034711 (2010).

Pauli, W., *Zeit. Phys.* **31**, 765 (1925).

Pauli, W., *Zeit. Phys.* **43**, 601 (1927).

Perdew, J. P. and Y. Wang, *Phys. Rev. B* **45**, 13244 (1992).

Pfau, H., M. Nicklas, U. Stockert, R. Gumeniuk, W. Schnelle, A. Leithe-Jasper, Y. Grin, and F. Steglich, *Phys. Rev. B* **94**, 054523 (2016).

Pleass, C. M. and R. D. Heyding, *Canad. J. Chem.* **40**, 590 (1962).

Rajan, V., *Phys. Rev. Lett.* **51**, 308 (1983).

Ravot, D., U. Lafont, L. Chapon, J. C. Tedenac, and A. Mauger, *J. Alloys Compds.* **323–324**, 389 (2001).

Rayjada, P. A., A. Chainani, M. Matsunami, M. Taguchi, S. Tsuda, T. Yokoya, S. Shin, H. Sugawara, and H. Sato, *J. Phys.: Condens. Matter* **22**, 095502 (2010).

Reissner, M., E. Bauer, W. Steiner, and P. Rogl, *J. Magn. Magn. Mater.* **272–276**, 813 (2004).

Reissner, M., E. Bauer, W. Steiner, G. Hilscher, A. Grytsiv, and P. Rogl, *Physica B* **378–380**, 232 (2006).

Reissner, M., E. Bauer, W. Steiner, and P. Rogl, A. Leithe-Jasper, and Y. Grin, *Hyperfine Interact.* **182**, 15 (2008).

Rhodes, P. and E. P. Wohlfarth, *Proc. Roy. Soc. London, Ser. A* **272**, 247 (1963).

Rudenko, A., Z. Henkie, and T. Cichorek, *Solid State Commun.* **242**, 21 (2016).

Saha, S. R., W. Higemoto, A. Koda, K. Ohishi, R. Kadono, Y. Aoki, H. Sugawara, and H. Sato, *Physica B* **359–361**, 850 (2005).

Saito, K., C. Laulhé, T. Sato, L. Hao, J.-M. Mignon, and K. Iwasa, *Phys. Rev. B* **89**, 075131 (2014).

Saito, T., H. Sato, K. Tanaka, S. Tatsuoka, M. Ueda, R. Higashinaka, T. Namiki, Y. Aoki, Y. Utsumi, K. Kuwahara, and T. Hosoya, *J. Phys. Soc. Jpn.* **80**, 063708 (2011).

Sakakibara, T., T. Tayama, J. Custers, Hidekazu Sato, T. Onimaru, H. Sugawara, Y. Aoki, Hideyuki Sato, and S. R. Saha, *Physica B* **359–361**, 836 (2005).

Sakurai, A., M. Matsumura, H. Kato, T. Nishioka, E. Matsuoka, K. Hayashi, and T. Takabatake, *J. Phys. Soc. Jpn.* **77**, 063701 (2008).

Sales, B. C., D. Mandrus, and R. K. Williams, *Science* **272**, 1325 (1996).

Sales, B. C., B. C. Chakoumakos, and D. Mandrus, *Phys. Rev. B* **61**, 2475 (2000a).

Sales, B. C., B. C. Chakoumakos, and D. Mandrus, *Mater. Res. Soc. Symp. Proc.* **626**, Z7.1.1 (2000b).

Sales, B. C., in *Handbook on the Physics and Chemistry of Rare Earths*, ed. by K. A. Gschneidner, Jr., J.-C. G. Bünzli, and V. K. Pecharsky, Elsevier, Amsterdam, Vol. 33, Ch. 211, pp. 1–34 (2003).

Sanada, S., Y. Aoki, H. Aoki, A. Tsuchiya, D. Kikuchi, H. Sugawara, and H. Sato, *J. Phys. Soc. Jpn.* **74**, 246 (2005).

Sato, H., Y. Abe, H. Okada, T. D. Matsuda, K. Abe, H. Sugawara, and Y. Aoki, *Phys. Rev. B* **62**, 15125 (2000).

Sato, H., Y. Abe, T. D. Matsuda, K. Abe, T. Namiki, H. Sugawara, and Y. Aoki, *J. Mag. Magn. Mater.* **258–259**, 67 (2003a).

Sato, H., H. Sugawara, T. Namiki, S. R. Saha, S. Osaki, T. D. Matsuda, Y. Aoki, Y. Inada, H. Shishido, R. Settai, and Y. Onuki, *J. Phys.: Condens. Matter* **15**, S2063 (2003b).

Sato, H., H. Sugawara, Y. Aoki, and H. Harima, in *Handbook of Magnetic Materials*, Vol. 18, Elsevier, pp. 1–110 (2009).

Sayles, T. A., W. M. Yuhasz, J. Paglione, T. Yanagisawa, J. R. Jeffries, M. B. Maple, Z. Henkie, A. Pietraszko, T. Cichorek, R. Wawryk, Y. Nemoto, and T. Goto, *Phys. Rev. B* **77**, 144432 (2008).

Sayles, T. A., R. E. Baumbach, W. M. Yuhasz, M. B. Maple, L. Bochenek, R. Wawryk, T. Cichorek, A. Pietraszko, Z. Henkie, and P.-C. Ho, *Phys. Rev. B* **82**, 104513 (2010).

Schnelle, W., A. Leithe-Jasper, M. Schmidt, H. Rosner, H. Borrmann, U. Burkhardt, J. A. Mydosh, and Y. Grin, *Phys. Rev. B* **72**, 020402(R) (2005).

Schnelle, W., A. Leithe-Jasper, H. Rosner, R. Cardoso-Gil, R. Gumeniuk, D. Trots, J. A. Mydosh, and Y. Grin, *Phys. Rev. B* **77**, 094421 (2008).

Sekine, C., T. Uchiumi, I. Shirotani, and T. Yagi, *Phys. Rev. Lett.* **79**, 3218 (1997).

Sekine, C., H. Saito, T. Uchiumi, A. Sakai, and I. Shirotani, *Solid State Commun.* **106**, 441 (1998).

Sekine, C., H. Saito, A. Sakai, and I. Shirotani, *Solid State Commun.* **109**, 449 (1999).

Sekine, C., T. Inaba, I. Shirotani, M. Yokoyama, H. Amitsuka, and T. Sakaka, *Physica B* **281–282**, 303 (2000a).

Sekine, C., M. Inoue, T. Inaba, and I. Shirotani, *Physica B* **281–282**, 308 (2000b).

Sekine, C., T. Uchiumi, I. Shirotani, K. Matsuhira, T. Sakakibara, T. Goto, and T. Yagi, *Phys. Rev. B* **62**, 11581 (2000c).

Sekine, C., T. Inaba, K. Kihou, and I. Shirotani, *Physica B* **281–282**, 300 (2000d).

Sekine, C., T. Uchiumi, I. Shirotani, and T. Yagi, *Science and Technology of High Pressure*, Universities Press, Hyderabad, India, p. 826 (2000e).

Sekine, C., I. Shirotani, K. Matsuhira, P. Haen, S. de Brion, G. Chouteu, H. Suzuki, and H. Kitazawa, *Acta Phys. Pol. B* **34**, 983 (2003).

Sekine, C., K. Kiho, I. Shirotani, K. Matsuhira, Y. Doi, M. Wakeshima, and Y. Hinatsu, *Physica B* **359–361**, 856 (2005).

Sekine, C., N. Hoshi, I. Shirotani, K. Matsuhira, M. Wakeshima, and Y. Hinatsu, *Physica B* **378–380**, 211 (2006).

Sekine, C., N. Hoshi, K. Takeda, T. Yoshida, I. Shirotani, K. Matsuhira, M. Wakeshima, and Y. Hinatsu, *J. Magn. Magn. Mater.* **310**, 260 (2007a).

Sekine, C., T. Yoshida, T. Kimura, T. Namiki, I. Shirotani, K. Matsuhira, M. Wakeshima, and Y. Hinatsu, *J. Mag. Magn. Mater.* **310**, 229 (2007b).

Sekine, C., H. Ando, Y. Sugiuchi, I. Shirotani, K. Matsuhira, and M. Wakeshima, *J. Phys. Soc. Jpn.* **77**, Suppl. A, 135 (2008a).

Sekine, C., R. Abe, K. Takeda, K. Matsuhira, and M. Wakeshima, *Physica B* **403**, 856 (2008b).

Sekine, C., K. Akahira, K. Takeda, Y. Ohishi, and P. Haen, *J. Phys.: Conf. Ser.* **150**, 042179 (2009a).

Sekine, C., K. Akahira, K. Ito, and T. Yagi, *J. Phys. Soc. Jpn.* **78**, 093707 (2009b).

Sekine, C., T. Kachii, T. Yoshida, R. Abe, K. Akahira, K. Matsui, and K. Ito, *Phonon Factory Activity Report #26, Part B* (2009c).

Sekine, C., K. Ito, K. Matsui, and T. Yagi, *J. Phys.: Conf. Ser.* **273**, 012120 (2011a).

Sekine, C., M. Takusari, and T. Yagi, *J. Phys. Soc. Jpn.* **80**, SA024 (2011b).

Sekine, C., Y. Kiyota, Y. Kawamura, and T. Yagi, *J. Phys.: Conf. Ser.* **391**, 012061 (2012).

Sekine, C., T. Ishizaka, K. Nishine, Y. Kawamura, J. Hayashi, K. Takeda, H. Gotou, and Z. Hiroi, *Phys. Procedia* **75**, 383 (2015).

Shankar, A. and R. K. Thapa, *Physica B* **427**, 31 (2013).

Shankar, A., D. P. Rai, Sandeep, and R. K. Thapa *J. Alloys Compnd.* **578**, 559 (2013).

Shankar, A., D. P. Rai, Sandeep, and R. K. Thapa, *Phys. Procedia* **54**, 127 (2014).

Shankar, A., D. P. Rai, Sandeep, R. Khenata, and R. K. Thapa, *Phase Trans.* **88**, 1062 (2015).

Shankar, A., D. P. Rai, Sandeep R. K. Thapa, and P. K. Mandal, *J. Appl. Phys.* **121**, 055103 (2017).

Sharath Chandra, L. S., M. K. Chattopadhyay, S. B. Roy, and S. K. Pandey, *Phil. Mag.* **96**, 2161 (2016).

Sheet, G., H. Rosner, S. Wirth, A. Leithe-Jasper, W. Schnelle, U. Burkhardt, J. A. Mydosh, P. Raychaudhuri, and Y. Grin, *Phys. Rev. B* **72**, 180407(R) (2005).

Shenoy, G. K., D. P. Noakes, and G. P. Meisner, *J. Appl. Phys.* **53**, 2628 (1982).

Shiina, R. and H. Shiba, *J. Phys. Soc. Jpn.* **79**, 044704 (2010).

Shiina, R., *J. Phys.: Conf. Ser.* **273**, 012141 (2011).

Shiina, R., *J. Phys. Soc. Jpn.* **81**, 105001 (2012).

Shiina, R., *J. Phys. Soc. Jpn.* **82**, 083713 (2013).

Shiina, R., *J. Phys. Soc. Jpn.* **83**, 094706 (2014).

Shiina, R., *J. Phys. Soc. Jpn.* **85**, 124706 (2016).

Shiina, R., *AIP Adv.* 8, 101301 (2018).

Shimizu, M., H. Amanuma, K. Hichitani, H. Fukazawa, Y. Kohori, C. Sekine, and I. Shirotani, *J. Phys. Soc. Jpn.* **76**, 104705 (2007).

Shimizu, M., H. Amanuma, K. Hichitani, H. Fukazawa, Y. Kohori, T. Namiki, C. Sekine, and I. Shirotani, *J. Phys. Soc. Jpn.* **77**, Suppl. A, 229 (2008).

Shirotani, I., K. Adachi, K. Tachi, S. Todo, K. Nozawa, T. Yagi, and M. Kinoshita, *J. Phys. Chem. Solids* **57**, 211 (1996).

Shirotani, I., T. Uchiumi, K. Ono, C. Sekine, Y. Nakazawa, K. Kanoda, S. Todo, and T. Yagi, *Phys. Rev. B* **56**, 7866 (1997).

Shirotani, I., T. Uchiumi, C. Sekine, M. Hiro, and S. Kimura, *J. Solid State Chem.* **142**, 146 (1999).

Shirotani, I., K. Ono, C. Sekine, T. Yagi, T. Kawakami, T. Nakanishi, H. Takahashi, J. Tang, A. Matsushita, and T. Matsumoto, *Physica B* **282–283**, 1021 (2000).

Shirotani, I., J. Hayashi, T. Adachi, C. Sekine, T. Kawakami, T. Nakanishi, H. Takahashi, J. Tang, A. Matsushita, and T. Matsumoto, *Physica B* **322**, 408 (2002).

Shirotani, I., Y. Shimaya, K. Kihou, C. Sekine, N. Takeda, M. Ishikawa, and T. Yagi, *J. Phys.: Condens. Matter* **15**, S2201 (2003a).

Shirotani, I., Y. Shimaya, K. Kihou, C. Sekine, and T. Yagi, *J. Solid State Chem.* **174**, 32 (2003b).

Shirotani, I., N. Araseki, Y. Shimaya, R. Nakata, K. Kihou, C. Sekine, and T. Yagi, *J. Phys.: Condens. Matter* **17**, 4383 (2005).

Shirotani, I., K. Takeda, C. Sekine, J. Hayashi, R. Nakada, K. Kihou, Y. Ohishi, and T. Yagi, *Z. Naturforsch. B* **61**, 1471 (2006).

Shull, C. G. and J. S. Smart, *Phys. Rev.* **76**, 1256 (1949).

Sichelschmidt, J., V. Voevodin, H. J. Im, S. Kimura, H. Rosner, A. Leithe-Jasper, W. Schnelle, U. Burkhardt, J. A. Mydosh, Y. Grin, and F. Steglich, *Phys. Rev. Lett.* **96**, 037406 (2006).

Singh, D. J. and W. E. Pickett, *Phys. Rev. B* **50**, 11235 (1994).

Singleton, J., P.-C. Ho, M. B. Maple, M. Harima, P. A. Goddard, and Z. Henkie, *Physica B* **403**, 758 (2008).

Slack, G. A., in *CRC Handbook of Thermoelectrics*, ed. D. M. Rowe, CRC Press, Boca Raton, FL, pp. 407–440 (1995).

Steinheimer, R. M., *Phys. Rev.* **159**, 266 (1967).

Stetson, N. T., S. M. Kauzlarich, and H. Hope, *J. Solid State Chem.* **91**, 140 (1991).

Stevens, K. W. H., *Proc. Phys. Soc. London* **A65**, 209 (1952).

Sugawara, H., Y. Abe, Y. Aoki, H. Sato, M. Hedo, R. Settai, Y. Onuki, and H. Harima, *J. Phys. Soc. Jpn.* **69**, 2938 (2000).

Sugawara, H., Y. Abe, T. D. Matsuda, Y. Aoki, H. Sato, R. Settai, and Y. Onuki, *Physica B* **312–313**, 264 (2002a).

Sugawara, H., T. D. Matsuda, Y. Abe, Y. Aoki, H. Sato, S. Nojiri, Y. Inada, R. Settai, and Y. Onuki, *Phys. Rev. B* **66**, 134411 (2002b).

Sugawara, H., S. Osaki, S. R. Saha, Y. Aoki, H. Sato, Y. Inada, H. Shishido, R. Settai, Y. Onuki, H. Harima, and K. Oikawa, *Phys. Rev. B* **66**, 220504(R) (2002c).

Sugawara, H., M. Kobayashi, S. Osaki, S. R. Saha, T. Namiki, Y. Aoki, and H. Sato, *Phys. Rev. B* **72**, 014519 (2005a).

Sugawara, H., S. Osaki, M. Kobayashi, T. Namiki, S. R. Saha, Y. Aoki, and H. Sato, *Phys. Rev. B* **71**, 125127 (2005b).

Sugawara, H., S. Yuasa, A. Tsuchiya, Y. Aoki, H. Sato, T. Sasakawa, and T. Takabatake, *Physica B* **378–380**, 173 (2006).

Sugawara, H., Y. Iwahashi, K. Magishi, T. Saito, K. Koyama, H. Harima, D. Kikuchi, H. Sato, T. Endo, R. Settai, Y. Onuki, N. Wada, H. Kotegawa, and T. C. Kobayashi, *J. Phys. Soc. Jpn.* **77**, Suppl. A, 108 (2008a).

Sugawara, H., R. Settai, and Y. Onuki, *J. Phys. Soc. Jpn.* **77**, 297 (2008b).

Sugawara, H., Y. Iwahashi, K. Magishi, T. Saito, K. Koyama, H. Harima, D. Kikuchi, H. Sato, T. Endo, R. Settai, and Y. Onuki, *Phys. Rev. B* **79**, 035104 (2009).

Sugiyama, K., N. Nakamura, T. Yamamoto, D. Honda, Y. Aoki, H. Sugawara, H. Sato, T. Takeuchi, R. Settai, K. Kindo, and Y. Onuki, *J. Phys. Soc. Jpn.* **74**, 1557 (2005).

Sun, P., Y. Nakanishi, M. Nakamura, M. Ohashi, G. Oomi, C. Sekine, I. Shirotani, and M. Yoshizawa, *J. Mag. Magn. Mater.* **310**, e169 (2007).

Suzuki, M., H. Ikeda, and P. M. Oppeneer, *J. Phys. Soc. Jpn.* **87**, 041008 (2018).

Takabatake, T., E. Matsuoka, S. Narazu, K. Hayashi, S. Morimoto, T. Sasakawa, K. Umeo, and M. Sera, *Physica B* **383**, 93 (2006).

Takahashi, Y. and T. Moriya, *J. Phys. Soc. Jpn.* **44**, 850 (1978).

Takeda, N. and M. Ishikawa, *J. Phys. Soc. Jpn.* **69**, 868 (2000a).

Takeda, N. and M. Ishikawa, *Physica B* **281–282**, 388 (2000b).

Takeda, N. and M. Ishikawa, *J. Phys.: Condens. Matter* **13**, 5971 (2001).

Takeda, N. and M. Ishikawa, *J. Phys.: Condens. Matter* **15**, L229 (2003).

Takeda, N., H. Mitamura, and K. Kindo, *J. Phys. Soc. Jpn.* **77**, Suppl. A, 209 (2008).

Takeda, K., N. Hoshi, J. Hayashi, C. Sekine, S. Kagami, I. Shirotani, and T. Yagi, *J. Phys.: Conf. Ser.* **215**, 012130 (2010).

Takeda, K., N. Hoshi, J. Hayashi, C. Sekine, and T. Yagi, *J. Phys. Soc. Jpn.* **80**, SA029 (2011).

Takegahara, K, H. Harima, and A. Yanase, *J. Phys. Soc. Jpn.* **70**, 1190 (2001).

Takegahara, K. and H. Harima, *Physica B* **329–333**, 464 (2003).

Takegahara, K., M. Kudoh, and H. Harima, *J. Phys. Soc. Jpn.* **77**, 294 (2008).

Takimoto, T., *J. Phys. Soc. Jpn.* **75**, 034714 (2006).

Tamura, I., T. Ikeno, T. Mizushima, and Y. Isikawa, *J. Phys. Soc. Jpn.* **75**, 014707 (2006).

Tamura, I., T. Ikeno, T. Mizushima, and Y. Isikawa, *J. Phys. Soc. Jpn.* **76**, 065004 (2007).

Tamura, I., T. Ikeno, T. Mizushima, and Y. Isikawa, *J. Phys. Soc. Jpn.* **81**, 074703 (2012).

Tanaka, K., Y. Kawahito, Y. Yonezawa, D. Kikuchi, H. Aoki, K. Kuwahara, M. Ichihara, H. Sugawara, Y. Aoki, and H. Sato, *J. Phys. Soc. Jpn.* **76**, 103704 (2007).

Tanikawa, S., H. Matsuura, and K. Miyake, *J. Phys. Soc. Jpn.* **78**, 034707 (2009).

Tatsuoka, S., H. Sato, D. Kikuchi, K. Tanaka, M. Ueda, H. Aoki, T. Ikeno, K. Kuwahara, Y. Aoki, H. Sugawara, and H. Harima, *J. Phys. Soc. Jpn.* **77**, 033701 (2008).

Tatsuoka, S., K. Tanaka, T. Saito, T. Namiki, K. Kuwahara, Y. Aoki, and H. Sato, *Physica B* **404**, 2912 (2009).

Tayama, T., T. Sakakibara, H. Sugawara, Y. Aoki, and H. Sato, *J. Phys. Soc. Jpn.* **72**, 1516 (2003).

Tayama, T., J. Custers, H. Sato, T. Sakakibara, H. Sugawara, and H. Sato, *J. Phys. Soc. Jpn.* **73**, 3258 (2004).

Tayama, T., T. Sakakibara, H. Sugawara, and H. Sato, *J. Mag. Magn. Mater.* **310**, 274 (2007).

Tayama, T., Y. Isobe, T. Sakakibara, H. Sugawara, and H. Sato, *J. Phys. Soc. Jpn.* **78**, 044708 (2009).

Tayama, T., W. Ohmachi, M. Wansawa, D. Yutani, T. Sakakibara, H. Sugawara, and H. Sato, *J. Phys. Soc. Jpn.* **84**, 104701 (2015).

Thole, B. T., P. Carra, F. Sette, and G. van der Laan, *Phys. Rev. Lett.* **68**, 1943 (1992).

Toda, M., H. Sugawara, K. Magishi, T. Saito, K. Koyama, Y. Aoki, and H. Sato, *J. Phys. Soc. Jpn.* **77**, 124702 (2008).

Tokunaga, Y., T. D. Matsuda, H. Sakai, H. Kato, S. Kambe, R. E. Walstedt, Y. Haga, Y. Onuki, and H. Yasuoka, *Phys. Rev. B* **71**, 045124 (2005).

Tokunaga, Y., S. Kambe, H. Sakai, H. Chudo, T. D. Matsuda, Y. Haga, H. Yasuoka, D. Aoki, Y. Homma, Y. Shiokawa, and Y. Onuki, *Phys. Rev. B* **79**, 054420 (2009).

Torikachvili, M. S., C. Rossel, M. W. McElfresh, M. B. Maple, R. P. Guertin, and G. P. Meisner, *J. Magn. Magn. Mater.* **54–57**, 365 (1986).

Torikachvili, M. S., J. W. Chen, Y. Dalichaouch, R. P. Guertin, M. W. McElfresh, C. Rossel, and M. B. Maple, *Phys. Rev. B* 36, 8660 (1987).

Tsubota, M., S. Tsutsui, D. Kikuchi, H. Sugawara, H. Sato, and Y. Murakami, *J. Phys. Soc. Jpn.* **77**, 073601 (2008).

Tsuda, S., T. Yokoya, T. Kiss, T. Shimojima, S. Shin, T. Togasi, S. Watanabe, C. Q. Zhang, C. T. Chen, H. Sugawara, H. Sato, and H. Harima, *J. Phys. Soc. Jpn.* **75**, 064711 (2006).

Tsujii, N., H. Kontani, and K. Yoshimura, *Phys. Rev. Lett.* **94**, 057201 (2005).

Tsutsui, S., H. Kobayashi, T. Okada, H. Haba, H. Onodera, Y. Yoda, M. Mizumaki, H. Tanida, T. Uruga, C. Sekine, I. Shirotani, D. Kikuchi, H. Sugawara, and H. Sato, *J. Phys. Soc. Jpn.* **75**, 093703 (2006).

Tsutsui, S., M. Mizumaki, M. Tsubota, H. Tanida, T. Uruga, Y. Murakami, D. Kikuchi, H. Sugawara, and H. Sato, *J. Phys.: Conf. Ser.* **150**, 042220 (2009).

Tsutsui, S., H. Kobayashi, Y. Yoda, H. Sugawara, C. Sekine, T. Namiki, I. Shirotani, and H. Sato, *Hyperfine Interact.* **206**, 67 (2012).

Tsutsui, S., N. Kawamura, M. Mizumaki, N. Ishimatsu, H. Maruyama, H. Sugawara, and H. Sato, *J. Phys. Soc. Jpn.* **82**, 023707 (2013).

Tsutsui, S., J. Nakamura, Y. Kobayashi, Y. Yoda, M. Mizumaki, A. Yamada, R. Higashinaka, T. D. Matsuda, and Y. Aoki, *J. Phys. Soc. Jpn.* **88**, 023701 (2019).

Uchiumi, T., I. Shirotani, C. Sekine, S. Todo, T. Yagi, Y. Nakazawa, and K. Kanoda, *J. Phys. Chem. Solids* **60**, 689 (1999).

Ueda, M., Y. Kawahito, K. Tanaka, D. Kikuchi, H. Aoki, H. Sugawara, K. Kuwahara, Y. Aoki, and H. Sato, *Physica B* **403**, 881 (2008).

Uher, C., B. Chen, S. Hu, D. T. Morelli, and G. P. Meisner, *Mat. Res. Soc. Symp. Proc.* **478**, 315 (1997).

Uher, C., in *Semiconductors and Semimetals*, Vol. **69**, Academic Press, p. 139 (2001).

Uher, C., in *Thermoelectric Skutterudites*, CRC Press, Taylor & Francis, Boca Raton, FL (2021).

Verma, H. C. and G. N. Rao, *Hyperfine Interactions* **15**, 207 (1983).

van Vleck, J. H., in *The Theory of Electric and Magnetic Susceptibilities*, Oxford University Press, London (1932).

Viennois, R., F. Terki, A. Errabbahi, S. Charar, M. Averous, D. Ravot, J. C. Tedenac, P. Haen, and C. Sekine, *Acta Phys. Pol. B* **34**, 1221 (2003).

Viennois, R., D. Ravot, F. Terki, C. Hernandez, S. Charar, P. Haen, S. Paschen, and F. Steglich, *J. Magn. Magn. Mater.* **272–276**, e113 (2004).

Viennois, R., S. Charar, D. Ravot, P. Haen, A. Mauger, A. Bentien, S. Paschen, and F. Steglich, *Eur. Phys. J. B* **46**, 257 (2005).

Viennois, R., L. Girard, L. C. Chapon, D. T. Adroja, R. I. Bewley, D. Ravot, P. S. Riseborough, and S. Paschen, *Phys. Rev. B* **76**, 174438 (2007).

Vladar, K. and A. Zawadowski, *Phys. Rev. B* **28**, 1564 (1983), ibid. **28**, 1582 (1983).

Vollmer, R., A. Faist, C. Pfleiderer, H. V. Löhneysen, E. D. Bauer, P.-C. Ho, V. Zapf, and M. B. Maple, *Phys. Rev. Lett.* **90**, 057001 (2003).

Wakeshima, M., Y. Hinatsu, K. Matsuhira, C. Sekine, and I. Shirotani, *Proc. Joint Workshop NQP-Skutterudites and NPM in Multi-approach*, Hachioji, Tokyo, PB28 (2005).

Watanabe, M., K. Tanaka, S. Tatsuoka, T. Saito, R. Miyazaki, K. Takeda, T. Namiki, K. Kuwahara, R. Higashinaka, Y. Aoki, and H. Sato, *J. Phys.: Conf. Ser.* **200**, 012222 (2010).

Wawryk, R., O. Zogal, A. Pietraszko, S. Paluch, T. Cichorek, W. M. Yuhasz, T. A. Sayles, P.-C. Ho, T. Yanagisawa, N. P. Butch, M. B. Maple, and Z. Henkie, *J. Alloys Compd.* **451**, 454 (2008).

Wawryk, R., Z. Henkie, A. Pietraszko, T. Cichorek, L. Kepinski, A. Jezierski, J. Kaczkowski, R. E. Baumbach, and M. B. Maple, *Phys. Rev. B* **84**, 165109 (2011).

Wawryk, R., O. Zogal, A. Rudenko, T. Cichorek, Z. Henkie, and M. B. Maple, *J. Alloys Compnd.* **688**, 478 (2016).

Weiss, P., *J. de Phys.* (Paris) **6**, 667 (1907).

White, B. D., K. Huang, and M. B. Maple, *Phys. Rev. B* **90**, 235104 (2014).

Wigner, E. P., *Trans. Farad. Soc.* **34**, 678 (1938).

Wisniewski, P., A. Gukasov, Z. Henkie, and M. B. Maple, *J. Phys. Soc. Jpn.* **80**, SA012 (2011).

Xue, J. S., M. R. Antonio, W. T. White, L. Soderholm, and S. M. Kauzlarich, *J. Alloys. Compd.* **207–208**, 161 (1994).

Yamada, T., H. Nakashima, K. Sugiyama, M. Hagiwara, K. Kindo, K. Tanaka, D. Kikuchi, Y. Aoki, H. Sugawara, H. Sato, R. Settai, Y. Onuki, and H. Harima, *J. Magn. Magn. Mater.* **310**, 252 (2007).

Yamamoto, A., S. Wada, I. Shirotani, and C. Sekine, *J. Phys. Soc. Jpn.* **75**, 063703 (2006).

Yamamoto, A., S. Wada, I. Shirotani, and C. Sekine, *J. Mag. Magn. Mater.* **310**, 835 (2007).

Yamamoto, A., S. Iemura, S. Wada, K. Ishida, I. Shirotani, and C. Sekine, *J. Phys.: Condens. Matter* **20**, 195214 (2008).

Yamaoka, H., I. Jarrige, N. Tsujii, J. Lin, T. Ikeno, Y. Isikawa, K. Nishimura, R. Higashinaka, H. Sato, N. Hiraoka, H. Ishii, and K. Tsuei, *Phys. Rev. Lett.* **107**, 177203 (2011).

Yamasaki, A., S. Imada, H. Higashimichi, H. Fujiwara, T. Saita, T. Miyamachi, A. Sekiyama, H. Sugawara, D. Kikuchi, H. Sato, A. Higashiya, M. Yabashi, K. Tamasaku, D. Miwa, T. Ishikawa, and S. Suga, *Phys. Rev. Lett.* **98**, 156402 (2007).

Yanagisawa, T., Y. Yasumoto, Y. Nemoto, T. Goto, W. M. Yuhasz, P.-C. Ho, M. B. Maple, Z. Henkie, and A. Pietraszko, *J. Phys. Soc. Jpn.* **77**, Suppl. A, 225 (2008a).

Yanagisawa, T., W. M. Yuhasz, T. A. Sayles, P.-C. Ho, M. B. Maple, H. Watanabe, Y. Yasumoto, Y. Nemoto, T. Goto, Z. Henkie, and A. Pietraszko, *Phys. Rev. B* **77**, 094435 (2008b).

Yanagisawa, T., P.-C. Ho, W. M. Yuhasz, M. B. Maple, Y. Yasumoto, H. Watanabe, Y. Nemoto, T. Goto, *J. Phys. Soc. Jpn.* **77**, 074607 (2008c).

Yanagisawa, T., T. Mayama, H. Hidaka, H. Amitsuka, A. Yamaguchi, K. Araki, Y. Nemoto, T. Goto, N. Takeda, P.-C. Ho, and M. B. Maple, *Physica B* **404**, 3235 (2009).

Yang, C., M. Koghi, K. Iwasa, H. Sugawara, and H. Sato, *J. Phys. Soc. Jpn.* **74**, 2862 (2005).

Yang, J., G. P. Meisner, D. T. Morelli, and C. Uher, *Phys. Rev. B* **63**, 014410 (2000).

Yasumoto, Y., A. Yamaguchi, T. Yanagisawa, Y. Nemoto, T. Goto, and A. Ochiai, *J. Phys. Soc. Jpn.* **77**, 242 (2008).

Yogi, M., H. Kotegawa, Y. Imamura, G.-Q. Zheng, Y. Kitaoka, H. Sugawara, and H. Sato, *Phys. Rev. B* **67**, 180501(R) (2003).

Yogi, M., H. Kotegawa, G.-Q. Zheng, Y. Kitaoka, S. Ohsaki, H. Sugawara, and H. Sato, *J. Phys. Soc. Jpn.* **74**, 1950 (2005).

Yogi, M., H. Niki, H. Mukuda, Y. Kitaoka, H. Sugawara, H. Sato, and N. Takeda, *J. Phys. Soc. Jpn.* **77**, Suppl. A, 321 (2008).

Yogi, M., H. Niki, M. Yashima, H. Mukuda, Y. Kitaoka, H. Sugawara, and H. Sato, J. Phys. Soc. Jpn. 78, 053703 (2009).

Yoshii, S., E. Matsuoka, K. Hayashi, T. Takabatake, M. Hagiwara, and K. Kindo, *Physica B* **378–380**, 241 (2006).

Yoshizawa, M., Y. Nakanishi, T. Kumagai, M. Oikawa, C. Sekine, and I. Shirotani, *J. Phys. Soc. Jpn.* **73**, 315 (2004).

Yoshizawa, M., Y. Nakanishi, M. Oikawa, C. Sekine, I. Shirotani, S. R. Saha, H. Sugawara, and, H. Sato, *J. Phys. Soc. Jpn.* **74**, 2141 (2005).

Yoshizawa, M., Y. Nakanishi, T. Tanizawa, A. Sugihara, M. Oikawa, P. Sun, H. Sugawara, S. R. Saha, D. Kikuchi, and H. Sato, *Physica B* **378–380**, 222 (2006).

Yoshizawa, M., Y. Nakanishi, T. Fujino, P. Sun, C. Sekine, and I. Shirotani, *J. Mag. Magn. Mater.* **310**, 1786 (2007).

Yoshizawa, M., P. Sun, M. Nakamura, Y. Nakanishi, C. Sekine, I. Shirotani, D. Kikuchi, H. Sugawara, and H. Sato, *J. Phys. Soc. Jpn.* **77**, Suppl. A, 84 (2008).

Yoshizawa, M., H. Mitamura, F. Shichinomiya, S. Fukuda, Y. Nakanishi, H. Sugawara, T. Sakakibara, and K. Kindo, *J. Phys. Soc. Jpn.* **82**, 033602 (2013).

Yotsuhashi, S., M. Kojima, H. Kusunose, and K. Miyake, *J. Phys. Soc. Jpn.* **74**, 44 (2005).

Yuhasz, M. W., N. A. Frederick, P.-C. Ho, N. P. Butch, B. J. Taylor, T. A. Sayles, M. B. Maple, J. B. Betts, A. H. Lacerda, P. Rogl, and G. Giester, *Phys. Rev. B* **71**, 104402 (2005).

Yuhasz, M. W., N. P. Butch, T. A. Sayles, P.-C. Ho, J. R. Jeffries, T. Yanagisawa, N. A. Frederick, M. B. Maple, Z. Henkie, A. Pietraszko, S. K. McCall, M. W. McElfresh, and M. J. Fluss, *Phys. Rev. B* **73**, 144409 (2006).

Yuhasz, M. W., P.-C. Ho, T. A. Sayles, T. Yanagisawa, N. A. Frederick, M. B. Maple, P. Rogl, and G. Giester, *J. Phys.: Condens. Matter* **19**, 076212 (2007).

Zhang, J. L., Y. Chen, R. Gumeniuk, M. Nicklas, Y. H. Chen, L. Yang, B. H. Fu, W. Schnelle, H. Rosner, A. Leithe-Jasper, Y. Grin, F. Steglich, and H. Q. Yuan, *Phys. Rev. B* **87**, 064502 (2013).

Index